TECHNOLOGY
AND THE MAKING OF THE
NETHERLANDS

THE AGE OF CONTESTED MODERNIZATION, 1890–1970

EDITED BY JOHAN SCHOT • ARIE RIP • HARRY LINTSEN

WALBURG PERS

MIT Press

The research presented in this book was supported by the Netherlands Organisation for Scientific Research (NWO), the Foundation for the History of Technology, and Eindhoven University of Technology. NWO also supported the translation of the Dutch edition in English.

This book was set in Garamond by Adobe and TheSans by LucasFonts with design by ViaMare, Marijke Maarleveld, Zutphen.

Printed and bound in The Netherlands.

Library of Congress Cataloging-in-Publication Data

Technology and the making of The Netherlands : the age of contested modernization, 1890–1970 / edited by Johan Schot, Harry Lintsen, and Arie Rip.
      p. cm.
   Includes bibliographical references and index.
   ISBN 978-0-262-01362-8 (MIT Press : hbk. : alk. paper) — ISBN 978-90-5730-630-3 (Walburg Pers : hbk. : alk. paper) 1. Technology—Social aspects—Netherlands—History—20th century. 2. Industrialization—Netherlands—History—20th century. 3. Netherlands—History—20th century. I. Schot, J. W. II. Lintsen, Harry, 1949–. III. Rip, Arie, 1941–.
T26.N4.T457 2010
338.9492'07—dc22

                                                                                              2009008601

ISBN 978-90-5730-630-3 (Walburg Pers)
ISBN 978-0-262-01362-8 (The MIT Press)
NUR 950

10 9 8 7 6 5 4 3 2 1

5

Contents

7

Preface

The aim of this study is twofold: to provide a twentieth-century history of technology in the Netherlands, and a contemporary history of the Netherlands through the lens of technology. The narrative deals as much with infrastructures, machines, office and household technologies, chemical plants, research and development, cars, and radios as with the development of national income and living standards, industrialization, physical capital, consumption patterns, urban development, the knowledge society, and colonial relationships—all of which are implicated in technological change. Throughout the text and illustrations, this book communicates an array of specifically Dutch aspects of twentieth-century historical experience: the omnipresence of water, the pervasive influence of urbanization coupled with a high-tech agricultural sector, and the specific role of Dutch governments and their reliance on intermediary actors that helped to manage the relationships between public and private spheres and between products and consumers. Earlier history has shaped such a political culture. This includes the Netherlands being a global power during its Golden Age, in the seventeenth century, and the legacy of the country's colonial relationships.

Thus, the book offers a new interpretation of Dutch history in the twentieth century. The notion of contested modernization is introduced as an overarching concept for examining the various themes and developments. This concept is also used to provoke discussion of alternative paths to modernization. The book's chapters reveal the various struggles involved in embracing modernization, examining specific attempts to create an industrial nation, materially connected through infrastructures; analyzing the contestation that came with the development of mass production and consumer society; and providing insights into the discourse on planning for the disposition and design of new and old land in a low-lying country. The overall argument includes attention to design and innovation as well as to the social implications and user aspects of twentieth-century modernization efforts.

This book is a translation of the seventh volume of *Techniek in Nederland in de Twintigste Eeuw*, a series published in between 1998 and 2003 by the Stichting Historie der Techniek (Foundation for the History of Technology) and the publisher Walburg Pers. The first six volumes contain fourteen separate "books." Each book develops its own set of historical interpretations and covers a separate area like water management, office technologies, energy, chemical industry, mining, food, agriculture, transport, communication, household technology, medical technology, industrial production, construction, and city technologies. The books, and the series as a whole, represent the cumulative wisdom of a community of almost eighty scholars working together for over a decade. Within the series, the seventh volume presents further analyses of these themes as developed in the fourteen books. It stands firmly on their shoulders—as the footnotes to this volume's chapters make abundantly clear. Without them the last volume could not have been written. For easy bibliographic handling, a shortened reference form, "Schot et al., *Techniek in Nederland*," is used to streamline the notes so as to avoid constant repetition of "Johan Schot et al., eds., *Techniek in Nederland in de Twintigste Eeuw* [Technology in the Netherlands in the twentieth century], 7 vols. (Zutphen: Walburg Pers, 1998–2003)." "Et al." refers to the members of the editorial team responsible for the publication of the entire series. We want to acknowledge them and their outstanding work individually here: Adri Albert de la Bruhèze, Liesbeth Bervoets, Jan Bieleman, Aad Bogers, Mila Davids, Nil Disco, Ben Gales, Ernst Homburg, Eddy Houwaart, Jan Korsten, Ruth Oldenziel, Anneke van Otterloo, Geert Verbong, and Onno de Wit. They have been responsible for much of the research and writing in the series.

The seventh volume was presented to Queen Beatrix on November 12, 2003, at the Royal Tropical Institute in Amsterdam, and on this occasion five leading Dutch corporations—Philips, Shell, Unilever, Akzo, and DSM—announced that they would support the distribution of the entire series to all of the high school libraries in the Netherlands. The initial print run was sold out by the end of 2008, but the complete series will be made available digitally in Dutch through its inclusion in the national Dutch digital library (see www.dbnl.org).

The work on the *Techniek in Nederland* book series began with several brainstorming meetings, but really got off the ground when Adri Albert de la Bruhèze, Nil Disco, Ruth Oldenziel and Johan Schot, at the University of Twente, drafted an initial plan. Ernst Homburg, Harry Lintsen, Arie Rip, and Geert Verbong participated in the planning. The discussions within this group of four were wide-ranging, passionate, and exciting, generating many of the themes that are very visible in this translated seventh volume, such as the attention to big as well as small technologies, attention to high culture and daily life, and the care taken to examine issues relating to production as well as to consumption. We would like to mention especially Ruth Olden-ziel, who had just arrived from the United States and introduced the group to some of the latest issues discussed among the historians of technology community on the other side of the Atlantic. The series plan was adopted by the Foundation for the History of Technology (Stichting Historie der Techniek) and led to the formation of a team of editors who were asked to develop their own research programs, which also became the basis for fundraising efforts. Eventually the Foundation was in a position to support a large group of researchers, ranging from Ph.D. students and post-docs and full professors. The unfolding research program evolved into a unique collaboration not only among the eighty scholars from nine universities who conducted research but also between the researchers and the many project sponsors, including companies, governmental agencies, and the Netherlands Organization for Scientific Research NWO. Thus, the overall project can also be seen as an example of the power of collaborative research. Ongoing financial support made it possible to give each volume in the series a firm foundation of several years of in-depth primary and archival research. Results of this research were discussed at conferences and in workshops and meetings where draft chapters were presented. All the participants benefited greatly from picture research by Giel van Hooff, and the organizational skills of Adri Albert de la Bruhèze, the project's editorial secretary, but also active researcher, author, and co-editor. Both could occasionally be heard to grumble about all the work but also persistently showed their dedication to ensure the accomplishment of the projects. The program not only resulted in major publications but also helped to create a larger Dutch network of historians of technology who have produced a wide range of articles and books (see the list of further reading at the end of this book). Finally, the Dutch network was also instrumental in creating a new transnational network of scholars and a new project, *Tensions of Europe*, also under the auspices of the Foundation for the History of Technology. This transnational group is currently preparing a multivolume series on European history (for more information see www.tensionsofeurope.eu).

The research project on twentieth-century technology in the Netherlands and the resulting publications (nicknamed TIN-20) were a continuation of the pathbreaking multivolume work, *History of Technology in the Netherlands* (*Geschiedenis van de Techniek in Nederland*, or TIN-19), edited by Harry Lintsen, Martijn Bakker, Ernst Homburg, Dick van Lente, Johan Schot, and Geert Verbong, which covered the nineteenth century. The current volume is thus a further, and major, milestone in a long history of research collaboration, a modus operandi for which we see a bright future.

Work on the seventh volume of TIN-20, which here appears in translation, began just before the close of the twentieth century. The research was made possible principally by a large program grant from the Netherlands Organization for Scientific Research, which allowed us to hire ten researchers who for several years worked collectively on this volume, with Johan Schot serving as program director. Many of these researchers became authors of chapters in this book (two of them, Henk van den Belt and Milena Veenis, who contributed to our many exciting and rewarding discussions, published their work elsewhere). This group of researchers also enjoyed many rich and intellectually stimulating conversations with the editors of the entire TIN-20 series, to whom we are grateful for their contributions.

Translating volume 7 brought many challenges but also opportunities to update and fine-tune some of the chapters. We have been most fortunate in having Ton Brouwers translate the entire book, making it available to an English-reading audience, and in having Katherine Scott as our copy editor, who substantially improved the text with her eye for detail and relentless queries. We also express our gratitude to our sponsors; to all the TIN-20 researchers; to the Board of the Foundation for the History of Technology; and to Pieter Schriks, publisher of the Dutch volumes. At the MIT Press we thank our editor, Marguerite Avery, without whom this volume would not have been published internationally, and Mel Goldsipe, who coordinated communications between the press and the authors in the Netherlands. Finally, we owe a large debt to the staff of the Foundation for the History of Technology, which has supported this project consistently throughout the years. We mention in particular Jan Korsten for his dedication and efforts to see all the work done that is necessary before a volume such as this one can be released. It is deeply satisfying to see this book in print and our work made accessible to a broad international audience. Many of the researchers involved in TIN-20 still talk about the enriching experiences of participating in the program. The reader can now experience the fruits of their enthusiasm, dedication, and passion.

Johan Schot
Harry Lintsen
Arie Rip
Eindhoven/Enschede, June 2009

This photo, taken around 1965, shows a supermarket in a new neighborhood in the town of Schiedam. By the 1960s, in the wake of preliminary experiments in the interwar period, the new consumer society was definitely in place in the Netherlands. Housewives were seen as major consumers but also as passive consumers. The supermarket was one of the symbols of the forward thrust of this new kind of society.

1 Inventing the Power of Modernization

Johan Schot and Arie Rip

Technology figures prominently in people's accounts of the twentieth century. The following recollection is typical when people look back, at the end of the century, and describe what happened:

> When I was growing up in my village we were still living in the era of paraffin light, pit toilets, and water from the pump, while gaslight was just being introduced. Upon moving to Zeist, the miracle of electricity awaited us there. Later historians will undoubtedly refer to the twentieth century as the age of big inventions, for the list of things I have seen introduced and changed is endless...aircraft; ice cream wafers; vacuum cleaners, washstands, and central heating; electric trains, neon light, and the Kodak box camera ("You press the button; we do the rest"); the gramophone; radio (homemade at first); television; color photography on paper; X-rays; the defeat of tuberculosis as public enemy number 1; the use of insecticides, vitamins, hormones, and antibiotics; the movies; the atomic bomb; and nuclear plants.[1]

This book aims to provide a better understanding of the role of technology in Dutch modern history. Although existing accounts written from economic, political, or cultural perspectives frequently mention technological developments, their authors treat technology mainly as a context for other processes such as industrialization, state formation, colonization, or the rise of mass consumption. By contrast, our study shows that to understand those historical processes, historians need to analyze the constitutive role of technology. Consequently we analyze the nature and impacts of technological choices, and we make visible the work of industrialists, engineers, consumers, governments, and the social groups that actively pursued the mutual shaping of technology and Dutch politics, economy, and culture.

This ambition to integrate technology into a general history is not new. Particularly in the case of the United States, several historians have tried to do this. Those attempts either aim at a broad overview or try to advance a

HOLLANDSCH FABRIKAAT.

STRIJKT ELECTRISCH

met een

„INVENTA"-STRIJKIJZER.

TWEE JAAR GARANTIE.

Electric clothes irons were introduced on a large scale in the 1920s, before most homes had sufficient power outlets. In this 1919 ad for an electric iron (*strijkijzer*) the power comes from an outlet on the light fixture called a "thief." The Amsterdam Municipal Power Company provided these power "thieves" free of charge to make it possible to piggyback on the power supplied to a single electric light. During the war the government encouraged the use of electrical lighting instead of gas lighting, to conserve coal. This advertisement, stemming from the period immediately after World War I, appeals to then prevalent nationalist sentiments with its commendation "Hollandsch Fabrikaat"—"Dutch Product."

specific interpretation based on the analysis of a limited set of technologies.[2] We have opted for a strong interpretation as well as a broad scope. Our ambition is to render visible the co-construction of technology and society for a wide range of different technologies and several major twentieth-century developments such as the unification of the Netherlands, industrialization and mass production, and the rise of a consumer society.

Historians and sociologists have used the concept of modernization as a way to interrelate these developments. At the same time this concept has met with sharp criticism, notably when it relies on the presumption that all countries go through the same modernization process.[3] We share this criticism; therefore, when we speak of modernization we refer to the modernization process characterized as such by historical actors themselves.[4] Using the concept of modernization as an actor category has led us to a specific periodization of the twentieth century. In this introduction we argue that the decades between 1890 and 1970 constitute a period in which a variety of actors sought to modernize Dutch society by promoting the development and use of specific technologies. This strategy was reflected in the adoption of the very notion of technology.

Instead of referring to specific technologies, "technology" became an abstract and generic issue used in the singular, not only in the Netherlands but in the entire Western world.[5] We depart from the common characterization and periodization of twentieth-century European history. For example, in *The Age of Extremes* the British historian Eric Hobsbawm takes the life cycle of the Soviet Union as the century's determining factor, implying that the twentieth century begins in 1917 and ends in 1991. In his account, World Wars I and II, the economic depression in the interwar years, the economic growth and the cold war after 1945 serve as its main events.[6] In fact, many histories of twentieth-century Europe have focused on the conflicts between fascism, communism, and democracy, and the experience of catastrophic wars and the decades of unanticipated economic expansion following World War II.[7] Although both world wars certainly play an important role in our narrative, their role as clear markers of beginnings or endpoints for a history of the twentieth century is less clear cut. Rather, we interpret the wars as events that influenced an ongoing struggle about how to modernize.

The modernization of the Netherlands was embodied and explicitly propagated by a wide array of actors, including engineers, accountants, management consultants, industrial information officials, household professionals, architects, and artists. These modernizers performed an important role in adapting their country to what they defined as the demands of the modern era.[8] The painter Theo van Doesburg, an influential member of the De Stijl modernist movement, expressed the mood of the modernizers very well:

> You long for wildernesses and fairytales? I show you the order of the engine rooms and the fairytale of modern production. Each product is a real miracle. You long for heaven? I show you climbing airplanes with their quiet pilots. You long for nature? Its corpse is lying at your feet. You have defeated it on your own. Your tall mountains have changed into skyscrapers. Where once the wings of your windmill were turning, there is now a chimney. Today's humming automobiles have taken the place of yesterday's stagecoaches.[9]

The leading modernizers did not always agree on how to modernize. Moreover when they tried to implement their plans the unruliness of specific practices came to the fore. Accordingly the actual path taken toward modernization was shot through with tensions and often turned out differently than

An apartment in the Coevordenstraat in The Hague in 1957. The Modernist doctrine of air and light had great influence on the construction of residences in this neighborhood. The apartments already were supplied with a series of outlets for electrical devices, such as floor lamps and radio sets; units with just one electrical connection for the light above the dining table no longer existed. What did not change, however, was the focus on housing construction geared to the nuclear family, with housewives in charge of their children's moral education.

planned. Modernization unfolded in a social dynamic. For example, Dutch consumers embraced the notion of a modern home, but they interpreted it according to their own views. Not everyone liked living in a home with a "rational layout" or having a modern, efficient electric stove. Some consumers preferred instead what they considered "cozy" technology. Listening to the radio with the whole family, for instance, could be more important than doing household chores efficiently; and many people used the telephone for chatting with friends rather than for household "managing."[10] In other words, consumers formulated and acted on their own modernizing priorities, and did not merely march to the drumbeat of the modernizing elites.

For this reason we label the period from 1890 to 1970 the Age of Contested Modernization. Beginning around 1890, the ambition to modernize using the technological avenue was shared by a large majority of the Dutch population, even though the actual path toward modernization reflected ongoing struggles and technological choice. Time and again the question was: How do we modernize? Which technological choices should be made? From the 1970s onward, the fusion of modernization with progress and technology began to fall apart, and encountered heavy criticism. Whether this will lead to a radical new definition of modernization remains to be seen, but it is highly likely that more reflexive approaches to the major effects of technological development will be incorporated in the modernization project.[11]

In this introduction we first provide a short history of the Dutch linking of modernization and technology and what it meant for Dutch society. Then we introduce the various chapters and the window they provide on the Age of Contested Modernization.

Inventing the Modern Project, 1890–1914

In 1913 the Dutch historian C. te Lintum framed the recent history of the Netherlands as "a century of progress," *een eeuw van vooruitgang*.[12] He argued that in the nineteenth century the development of Dutch society was determined by ongoing improvements in all aspects of social life. The core of this progress was technological development, particularly in traffic and communication. In the chapter "Steam" he wrote:

> As to the locomotive's accomplishments, even the influence of a Napoleon pales by comparison. Napoleon was active a few years and only in Europe, but the locomotive has continued to progress and conquer terrain on all continents. ...Opposition against the locomo-

tive will gradually disappear altogether because [the locomotive] brings people closer together, spreading a sense of unity throughout the human world. Who follows its triumphal advance can no longer feel any hostility vis-à-vis the modern era's machines.[13]

The steam technology that so much thrilled Lintum was a nineteenth-century technology. In the twentieth century steam would be supplemented with a whole series of new key technologies such as electricity, the internal combustion engine, the telephone, and synthetic materials. Steam engines also came relatively late to the Netherlands and they never played such a dominant role as in England or Belgium.[14] The Netherlands pursued its own path toward industrialization, one that in the nineteenth century relied less on steam power, and more on wind and water power and the use of peat as a fuel. Dutch industrialization had a broad scope—that is, it occurred in many sectors simultaneously, exploited the availability of a large trading network stemming from its colonial past, and relied on the appropriation of technical knowledge from elsewhere.[15]

The first wave of industrialization, around 1860, involved metals, food, and foodstuffs, sectors that profited from the growing domestic market resulting from the population increase, the rise of incomes, the expansion of the transport system (railroads, canals), the drop of the price of coal, and the disappearance of all sorts of institutional barriers that had impeded the introduction of steam engines and other innovations. After 1890, Dutch industrialization received an extra boost from the discovery and exploitation of coal in the province of Limburg and the country's overseas colonies, notably the Dutch East Indies. This second wave of industrialization can be related to the development of several key transportation, communication, and energy technologies and resulted in much new commercial activity, including a whole series of utility companies that exploited these technologies.[16] Both early and late waves helped to create a new economic structure in which industry flourished alongside agricultural and service sectors, yet only after 1890 we can speak of the Netherlands as an industrial nation.

With the second round of industrialization came new modes of acquiring and diffusing technology, and this led to the establishment of research departments in companies, governmental testing stations, and new technical educational facilities at home and overseas. The first industrial laboratory was set up in 1885 by J.C. van Marken of NGSF (Nederlandse Gist- en Spiritusfabrieken), a yeast and spirits producer, and Philips and Royal Dutch/Shell set up laboratories in 1905 and 1906.[17] At the end of the nineteenth century, the Dutch government became very active in organizing the process of knowledge development, initiating knowledge-related projects at various levels. These initiatives came at the end of a fierce debate about the role of government in society. A main issue was whether it was proper for the government to engage in business-related activities, such as maintaining testing stations for agriculture as a response to the agricultural crisis of the 1880s, but also whether the government should own companies, which became known as a public utilities. By the end of the nineteenth century it had become widely accepted that municipalities would provide water, gas, transport, and communication technology (telegraph and telephone). Finally, after a longer period of trying to encourage the private sector to expand coal mining, the Dutch Parliament decided to establish a large-scale facility itself. Ironically, by the time the Dutch began to mine coal the century of coal was almost over.

Many historians attribute the government's proactive stance to the rise of new ideologies. In the Netherlands a new kind of radical liberalism called on the state to intervene in society in order to create the proper conditions for individual development.[18] A coalition of the confessional parties and the left-liberals created a new framework for government to play a dynamic role in the establishment of public utilities and laboratories. This coalition also passed landmark legislation such as the 1889 Labor Act and the 1901 Housing Act. In this book we point to another relevant factor, often overlooked in political history: the push to modernize on the part of engineers, household professionals, and other experts who were often the ones to put specific problems related to industrialization on the political agenda. They asked the state to intervene and modernize Dutch society. They broadened the industrialization agenda and, by including energy, transport, and communication infrastructure, urban development, public housing, and healthcare, turned it into a modernization program. They also were made responsible for the implementation of this program. Public utilities run by the municipalities were staffed by engineers and other experts who took the lead in developing plans for sewer systems, ports, roads, housing, transport, and drinking water systems. The experts managed to get a license to modernize while they continued their power struggles. For example, urban developers came into conflict with physicians who earlier as part of the hygienic movement had made an effort to improve houses and cities.[19] But such conflicts did not harm the experts' credibility with the public.

In addition to the newly emerging public tasks, experts also put the private sphere of households on the political agenda. Socialist feminists, domestic science teachers, and female housing inspectors presented themselves as the new household professionals and organized public discussions on the dif-

At the start of the twentieth century the port of Rotterdam changed radically through the introduction of mechanized transshipment technology, such as the grain elevator, visible here on a photo from 1937. In new ports such as the Maashaven, where many ships could tie up side by side in calm water, goods were loaded from one ship directly into another rather than first being brought ashore. The elevator symbolizes the new transit port.

ferent possibilities for organizing household tasks more hygienically and efficiently with the help of gas, electricity and the telephone. These debates
focused on the advantages and disadvantages of adopting such new technologies and also related issues such as the outsourcing of household work,
the role of servants, and the possible collectivization of households. The
expertise of the new household professionals rested in their training and their
particular knowledge about and affinity for the domestic domain. Like most
experts, household professionals had high expectations of government intervention, but also emphasized self-help. Accordingly, they set up new schools,
wrote for professional journals, and published household handbooks, and
they founded new professional organizations such as the Dutch Association
of Housewives, which organized many public exhibits and demonstrations
on the modern household.

Many such developments had their origins in the 1860s, but we argue
that they really took off after 1890, when a thoroughly positive image of the
potential of technology began to be widely shared by the political, economic,
and cultural elites, giving rise to the prevailing optimism of Dutch culture
around 1900.[20] In the Netherlands, not the fin-de-siècle atmosphere of doom
and decline but the vitality and patriotism of la Belle Époque became the
dominant trend. Things were going to be better and new developments in
science and technology would show the way. At the end of the century, many
Dutch people believed the newly available technologies of electricity (tram,
electric motor, and lighting), communication (movies and the telephone),
and transport (tram and automobile) embodied new and exciting opportunities. Technological development had to be embraced, even if its short-term
effects would prove disastrous. For example, when traders in Rotterdam
pushed for a fully mechanized transshipment process in the port of Rotterdam, causing many boatmen to lose their jobs, the ensuing national debate
in the Dutch press revealed that Dutch elites across the political spectrum
agreed that technological advances should be given free rein. Stopping them
would hamper the overall project of the country's modernization.[21] In the
twentieth century, the Dutch elite no longer predominantly looked to the
past to try to recover and revitalize the trade capitalism that had defined the
country's seventeenth-century Golden Age. From the 1890s onward, Dutch
political leaders and new professionals collectively began to look "ahead" and
to try to find ways to modernize the country, reserving a central role for technology in this endeavor. But only during the Great War did this project of
modernization through technology became a genuine national agenda.

The Netherlands and World War I

In retrospect World War I can be seen as being a major catalyst in the modernization of the Netherlands. The war conditions created more room for government intervention, allowing for the adoption of plans that had been in the making for some time already. The plan to close the Zuiderzee to the North Sea, under discussion for thirty years, was ratified by Parliament in 1918. Cornelis Lely, the minister of water management, wrote to his son: "We are ready to do something big and because the war has changed our notion of money so radically, people are no longer as afraid of taking up large works."[22] The government also intervened extensively in the economy, agriculture, and the food supply. Under war conditions, the government ordered farmers to cultivate more grain, and they were forced to turn over to the government grain and potatoes they harvested along with their seeds. Bread, dairy products, and fuel were rationed. The government established two entirely new companies, the heavy-industrial combine Hoogovens and the Netherlands Salt Industry. Although the aim was to reduce dependency on imports, the effect was to give a boost to industrialization. Regulation of international trade was left up to the Netherlands Overseas Trust Company (NOT), which was managed by large banks and trading companies. Because government was not involved, the "neutral status" of the Dutch state could be maintained. The NOT successfully negotiated commodity imports and exports with both the German and English governments.[23]

Large segments of the Dutch economy profited from the country's neutral status. Agriculture and industry both produced sizable wartime profits. Profits between 1915 and 1919 were 40 percent higher than before the war, a result of strongly rising prices and stagnating wages.[24] The only industry sectors that suffered where those that depended on imported raw materials. During the war years Dutch economic growth, in terms of the gross domestic product per man-hour, was markedly higher than elsewhere in Europe, including other neutral countries such as the Scandinavian countries and Switzerland.[25] Jan Pieter Smits in chapter 8, "Technology, Productivity, and Welfare," shows that in 1916 there was a break in the growth pattern of the Dutch economy, notably in the growth of labor productivity. His argument is that the wartime profits were mainly invested in machinery, and this investment spurred continuing economic growth long after the war.

A new energetic spirit among government and business led to new forms of cooperation as well as to a realization that intervention in the economy was possible, desirable, and even necessary for meeting the new national ambitions. Drastic economic regulation led to close deliberations among government, unions, and business, including the banks. These parties came together often in various commissions and state agencies. Their deliberations gave rise to new networks and new relationships, and had a lasting effect on Dutch economy and society. Even though most bodies and commissions set up during the war did not formally survive it, the informal networks persisted and the call for order, modernization, and more potent economic intervention remained popular, and an awareness of a shared national interest grew stronger.

The new networks made room for various kinds of professionals who gladly embraced the craving for order.[26] The war situation provided new professionals the opportunity to advance plans that might help the country in its time of need. Some of the plans may not have been new, yet the vigorous national ambition emanating from the plans was really quite novel. For example, a new national water management agenda was defined, which included not only the Zuiderzee Works but also the Twente channels, the canalization of the Maas River, and the construction of the Noordersluis in IJmuiden.[27] Leaders in the chemical industry were dreaming of a nationally integrated chemical-industrial complex on the German model, which would allow the Netherlands to operate more independently of other countries.[28] Mass production also came to the housing sector, especially in public housing. In 1916 a central standardization commission was set up to stimulate mass production.[29] The Royal Dutch Academy of Science established a Scientific Commission for Advice and Research in the interest of public prosperity and resilience, to mobilize "all strength and experience in the Netherlands to draw maximal use from the few available natural resources and means of production."[30] Such wide-ranging plans were well received. The war established the project of turning the Netherlands into a modern nation. This did not mean, however, that people agreed on how to modernize. There were major differences of opinion, notably on the desirability and dangers of mass production and mass consumption.

Controlled Modernization, 1918–1939

Following World War I, the Netherlands made a very successful switch from a war economy to a peace economy. The postwar demand for products and capital goods was enormous, and Dutch industry and agriculture were both ready to supply them. The economy's growth rate (gross domestic product) reached higher levels, and Dutch companies began to do much better in international markets. From 1913 to 1929 the share of Dutch exports in world

The Dutch population grew increasingly enthralled with modern technology, which was reflected, among other things, in the huge popularity of flying and airplanes. This is a photo of an aviation event at Welschap Airport near Eindhoven in 1937.

trade increased more than 30 percent.[31] Consumer spending increased strongly as well, almost 3 percent annually between 1923 and 1929, and employees got more leisure as a result of the state-mandated eight-hour working-day.[32] The growth of consumer buying power persisted even after 1929, although more slowly. Those with a job in the 1920s and 1930s could spend part of their income on new luxuries, such as vacations, the movie theater, an automobile, a radio, a telephone, or new electric household appliances such as vacuum cleaners. The accessibility of these consumer goods gave rise to a new dynamic of mass consumption. The new products enabled a new set of experiences as well as consumer expertise in dealing with new technology. New ideas about comfort, hygiene, beauty, adventure, personal development, and quality of life changed everyday life, especially that of the middle classes. These experiences became defined and experienced as participation in modern life. At the same time, the elites were somewhat apprehensive about the drawbacks of too much consumption, the demoralizing and undermining of what they saw as the strength of the nation (*volkskracht*). Educating the young was seen as a remedy to overindulgence; they were encouraged to join clubs and societies that would teach them the art of "controlled pleasure."[33] Modernization was increasingly conceptualized as *controlled* modernization, and the family played a key role in this control. Notably the housewife served as a source of both economic and moral revival. In well-run households, housewives would have enough time to support their husbands in their productive tasks and raise their children to become "community individuals."[34] This overriding view on the role of the family, and the housewife in particular, implied the demise of experiments involving collective use of new consumer goods. Mass consumption became defined as consumption within the family context.

The emergence of mass consumption ran parallel to the rise of mass production in industry and the service sector. Leading Dutch companies such as Shell, Unilever, Philips, Dutch State Mines (DSM), Algemene Kunstzijde Unie (AKU, later part of Akzo Nobel), and Hoogovens gained a central economic position, their share in industrial employment tripling in the interwar period. But many medium-sized companies also made significantly contributions to production, exports, and employment. In this era the Netherlands developed its characteristic "dual structure," with half a dozen multinationals—those just mentioned above—and a large number of much smaller companies. The six leading companies attached much value to their own research and by the 1950s they were paying for almost half of the total industrial R&D in the Netherlands. Yet the smaller companies also engaged in research, frequently collaborating with the government in networks. Business expansion led to expansion of internal and external administrative support. Bank offices experienced explosive growth, which gave rise to new forms of administrative mass production that would later evolve into the information economy.[35]

The trend toward mass production and consumption was emphatically stimulated and propagated by the new experts, who had been active before and during World War I, and who now received a broader mandate to develop their ideas. The ideal they embodied was characterized by a striving for planning, efficiency, and rationality, often summarized as modernity.[36] These "scientific principles" would lead to higher productivity, and also to more harmony between labor and capital. Between 1914 and 1918 Taylorism was much discussed in the Netherlands, and after the war the new principles inspired a whole series of experiments in mines, factories, offices, cities, and households. The experiments involved mass production in housing construction and water management projects, the reorganization and mechanization of offices and factories, urban expansion, and experiments with new home designs and furniture. They prompted the application and diffusion of a series of new technologies, from concrete to electricity, punch-card machines, archival systems, household appliances, kitchens, and blueprints. The experiments were not always successful, as illustrated by the fiasco of the reorganization and mechanization based on punch-card machines by the Postcheque- en Girodienst, a process supervised by the Mechanical-Administrative Business Organization (MABO), a major promoter of the efficiency

In the course of the twentieth century, Dutch State Mines developed into a major chemical company that continued to search for ways of making new products from coal. In the foreground the polychemistry plant and in the background the Maurits Mine in Beek, in 1963.

movement.[37] Yet there were also success stories, such as the application of scientific management in the State-owned coal mines, which transformed them into an international model of efficient management.[38]

The diffusion of the principles of efficiency and mass production made it clear that practice could be unruly and that the ideals articulated by the experts could prove difficult to realize. Frequently their views reflected major ideological differences and the experts fought each other as to the proper interpretation and implementation of the ideal of modern society. In addition, those who used the experts' ideas—companies, consumers, and government institutions—were not always sensitive to the argument that mass production was the one and only ideal. Manufacturing companies advocated flexibility and in fact pointed to the risks of economies of scale. Many small and medium-sized businesses continued to operate on a small-scale basis thanks to the electric motor, which explains the wide implementation of the electromotor in the Netherlands. At least 50 percent of the positive achievements of the Dutch economy can be attributed to this technological diffusion, demonstrating that efficiency and small-scale business could go together well.[39] Dutch agriculture also explicitly opted for efficient small-scale farming operations. Economies of scale were achieved by cooperative purchasing, processing, and matching of products, while the government funded research, public information, and education. Networks among businesses were more important than integration and made it possible to absorb the disadvantages of working on a small scale. Similarly, some household experts advocated collectivizing household tasks, which can be regarded as a form of economies of scale, but this hardly proved successful.

During the 1920s and 1930s the need for modernization as such was hardly debated, but decisions on the actual path of modernization triggered heated discussion. The idea of modernization seen exclusively as mass production and mass consumption was sharply criticized. The alternative, modernity as expressed in small-scale production and the need to control the modernization process, gained approval and turned out to be feasible and successful in practice.

The Heyday of Modernization

After World War II the idea of modernization through scale increase, mass production, and mass consumption grew dominant. This was partly an effect of the war, and partly a matter of elaborating interwar achievements. The Dutch economy did well in particular in the first three years of the war, partly because the demand in Germany for Dutch goods was high and German contracts enabled large-scale investments, in the metal industry and shipbuilding. The capital goods stock had not declined by 40 percent, as was the estimate in 1945, but by only 7 percent.[40] It turned out that the damage to the infrastructure (roads, railways, harbors, airports, and dikes) resulting from the war could be quickly repaired. Between 1945 and the late 1950s the economy grew nearly without interruption. Industrial production bounced back to its prewar level by 1947, and agriculture did so by 1950. During the postwar period the Netherlands once again managed to profit from the chaotic international situation; for instance, the country received more Marshall Aid than most other European countries. Economic historians have called the following period, from 1950 until the late 1960s, an economic miracle, a long period of sustained and substantial growth. This miracle meant the definitive democratization of prosperity for the Dutch. Having a telephone, radio, an automobile, a refrigerator, a washer, and central heating turned from a privilege into a right of all citizens. It became a natural thing to live a modern life in a modern society. As a result, memories of the world of crisis and war faded rapidly.[41]

During the war there has been much talk among the cabinet in exile in London, as well as among the resistance movement, about the possibility of a radical break in the Dutch "pillarized" political relationships.[42] Yet after the war the prewar political parties simply resumed control. There was, however, a radical change in the country's economic policy, as all parties now agreed on the need to focus on economies of scale, planning, mass production, and mass consumption. A national industrialization policy was formulated that was geared toward this need, accompanied by a controlled wage policy to enable such an overall shift. Tellingly, economies of scale was now also deemed desirable in agriculture. Engineers and other experts were afforded ample room to implement their plans for public housing, water management, and large infrastructural projects, for an expanded port of Rotterdam, a larger national airport, a denser highway network, and the Delta Works project. This particular trajectory of modernization was basically taken for granted because of the many economic successes that had already been achieved. Yet in the course of the 1960s new criticism was formulated regarding the effects of economic growth on the environment, the quality of working conditions, and the implications of this mode of modernization for democracy. The process of modernization, which had seemed completed, once again became subject to debate.

This book further develops the history of the contested modernization in the Netherlands. The chapters that follow this introduction offer different vistas on the quite productive covenant between technology and modernization, including examinations of the major roles of the various groups of modernizers.

The Age of Contested Modernization: Nine Perspectives

Erik van der Vleuten explores the physical integration of the Netherlands in chapter 2. He shows how transport, energy, and communication networks became major vectors in unifying the Netherlands as a spatial ensemble. Although at the end of the nineteenth century many Dutch infrastructure systems were poorly integrated nationally, in the twentieth century entirely new national infrastructures came into being. Among many potential system builders, the national government increasingly dominated infrastructural development, defining infrastructure development as a national task. This

In the twentieth century a number of new infrastructures were developed. The Netherlands became increasingly serviced by new communication, transportation, and energy infrastructures, both below- and aboveground. This was visible in particular in cities, where infrastructures converged and required coordination. In this photo, taken in December 1928, a train, a tram, bicyclists, cars, pedestrians, and a horse and carriage jostle for space at the Catharijnebrug in Utrecht.

view was tied to concerns related to the economy, safety, and accessibility, but also to the ideal of realizing the creation of a national Dutch space. This ideal became a reality by the 1950s, when many infrastructures took on a national dimension. Yet Van der Vleuten shows that the use of infrastructure—despite all the national rhetoric—was often not national in scope. In 1950 it was possible to reach all corners of the Netherlands by road or railroad, but the average travel distance continued to be just 50 kilometers. This meant that the mobility of the Dutch was mainly confined to the regional level. Similarly, a national electricity grid was created, but it actually served only as backup for the province-based electricity supply systems. The contested character of infrastructure development was expressed in the competition among various parties, including the various levels of government (national, regional, local), concerning the authority over infrastructures and also the role of certain infrastructures in the entire ensemble. In some cases this led to the demise of certain infrastructures. For example the regional

tramway network was built and later removed as a result of the rise of automobiles. After 1970 the mission to realize a national set of infrastructures
increasingly became subject to debate, as did the role of the national government as central systems builder. On the one hand, international (European) and regional levels gained even more prominence. On the other hand,
the lack of citizen involvement and the environmental effects of infrastructure development were put on the agenda. In response, a new mode of infrastructure development emerged, one with more participation from social
interest groups and a more facilitating role of government. Under this new
regime the national mission reappeared in the 1990s. The Netherlands was
again turned into a construction site, as dikes were reinforced, new railroads
were put in (the Betuweroute between the Port of Rotterdam and Germany),
and a new national cell phone network was developed. In this phase, nature,
too, was eventually transformed into an infrastructure in a new plan geared
toward integrating all of the country's nature reserves into a single national
ecological system. Yet national ambitions were more muted and tempered
by ambitions to construct regional and international spaces.

Infrastructures not only played a major role in realizing a national space,
but also influenced the design of cities, households, offices and airports, as
Adrienne van den Bogaard shows in chapter 3, where she discusses site-specific innovations such as kitchen, airports, offices and cities. In many such
locations space is scarce (especially in the densely populated Netherlands), and
different energy, communication and transport infrastructures are used side
by side. The simultaneous development of several infrastructures called for
site-specific innovations to realize a workable exploitation of space. Ad hoc
solutions for integrating different infrastructures led to efforts aimed at finding integrated solutions for a location as a whole, such as at Schiphol Airport
or in inner cities. After the 1920s it became increasingly common to work
with detailed blueprints and careful plans—a sustained planning approach
that was an innovation in its own right. New professional groups, such as
household experts and urban developers, claimed the domain of planning
for the modernization of specific locations. The planning ideology that further developed after World War II as part of the country's modernization was
thus partly based on prewar problem-solving experience in specific locations.
After World War II, top-down and integral planning approaches became the
standard, even though in actual sites practices continued to be unpredictable,
so that plans frequently had to be recast. In the 1970s planners abandoned
top-down planning, as well as the implied goal of planning everything as
much as possible in advance. Instead, they adopted more incremental styles
of planning.

In this photo, from 1956, dozens of women produce undergarments at the Tricotage-fabriek (knitting factory) Frans Beeren & Zonen in Weert. The participation of women in the industrial process in the twentieth century has always been relatively low in the Netherlands. Many women worked in the household, especially after marriage. Only younger, unmarried women worked outside the house, many of them in the garment industry.

Chapter 4, by Rienk Vermij, "Scale Increase and Its Dynamic," presents a history of mass production and economies of scale in the Netherlands. Mass production has been an obvious trend in the process industry, but not in agriculture or manufacturing. Although assembly-line production was used in manufacturing, it was mainly in smaller companies such as in the clothing sector and served primarily as a way to discipline workers. A major argument made in chapter 4 is that Dutch businesses opted for alternatives to scaling up, for example, through mutual coordination and cooperation, for example, in the creation of agricultural cooperatives and in standardization of products, a goal pursued since World War I. In the interwar period, the desirability of scale increase was fiercely debated; many felt it would be at the expense of operational flexibility or they rejected its implied speculative element, based as it was on not-yet-realized sales. Cooperation and cartels were therefore seen as better business strategies, bringing fewer risks and allowing more flexibility. At the same time, modernizers were pushing scale increase for the manufacturing sector. Their general focus was on rationalization of production, business management, and society at large. This was already evident at the start of the century, when professionals who embraced modernist ideals aimed to prevent waste and social misbehavior, increase labor safety and efficiency, and also to resolve contradictions between labor and capital. To them, scale increase was part of this rationalization. After World War II, business expansion was mainly implemented for economic reasons, and its wider ideals receded to the background. Dutch companies mainly embraced scale increase in order to compete with the United States, where much higher productivity levels were realized; the opening up of markets in Europe was another driving force. After two decades of ongoing expansion in the size of facilities, technological and economic ceilings were reached by the early 1970s. The oil crisis in particular contributed to a reversal. Other forms of expansion, such as the pursuit of cooperation and networks among businesses, became significant again. Next to standardization, certification gained increasing importance to the development of collaborative efforts, because thus it became possible to guarantee the quality of sales and purchasing more effectively, even if these activities did not occur within the same companies.

In chapter 5, "The Rise of a Knowledge Society," Peter Baggen, Jasper Faber, and Ernst Homburg focus on the development of a new knowledge infrastructure in the Netherlands. Specifically the authors address the growth of technical education, the development of research laboratories, and the formulation of a national technology policy. After 1875 the demand for skilled labor increased, causing rapid expansion of the basic, intermediary, and higher technical education systems. Initially this education was geared toward practice, but around 1900 frictions emerged between those who where practice-oriented and those who favored more theoretical and general mathematical and scientific training. The debate about the modernization of education was most intense in the sugar industry and the construction sector. The theoretical approach in education eventually prevailed, giving technical education a dynamic of its own, moving it away from the direct demands for labor and allowing professional groups more room for the development of their own expertise and prestige. A similar trend is visible in the development of research laboratories. At Royal Dutch/Shell and Philips, research efforts were initially geared toward solving specific production problems and developing patents, the basis was laid for a Dutch knowledge infrastructure in the 1920s and 1930s, when company labs were integrated into company strategies. Contacts between universities and industry flourished, while government more actively intervened in knowledge development. This resulted in the establishment of both the Dutch Organization for Applied Scientific Research (TNO) in 1932 and the provincial economic-technological institutes. The development of this knowledge infrastructure received a strong boost after World War II because science and technology were seen as decisive factors in the allied victory and hence they needed to be acknowledged as national strategic factors. This led to an embracing of science and technology as the endless frontier. In the Netherlands, like other western countries, an array of activities unfolded, including major support of fundamental research, as reflected in the establishment, in 1952, of the Netherlands Organization for Pure Scientific Research (Zuiver Wetenschappelijk Onderzoek, ZWO). The debate on the priority of pure scientific research or stimulation of scientific-technological research at the service of social objectives continued for the entire period, leading to clashes in the 1970s. The discussion was not so much about whether it was desirable to stimulate science and technology but about who—the market or society—should do so.

The Dutch government and entrepreneurs also tried to build a knowledge infrastructure in the Dutch colonies, notably in the East Indies. In chapter 6, "Technology and the Colonial Past," Harro Maat shows that at first this effort mainly took place in laboratories for botanical and chemical research. Around 1900, the modernization ambition was transformed into a "civilizing mission," and state intervention became more prominent. Centering his analysis on the case study of sugar cane cultivation and production, Maat argues that although the civilizing mission and implicated colonial technology development was pushed by the colonizers, it can only be understood against the backdrop of colonial society. Dutch engineers contributed to

In the Dutch East Indies, Dutch companies pursued new transportation modes, to avoid the necessity of relying on local facilities. The construction of a bridge for the steam tram across the Kali Pakis is one example. Members of the local population were brought in to service the trams, which allowed them to gain experience with Western technology. After decolonization this familiarity enabled these systems, which had meanwhile become integrated into Indonesian society, to continue in operation.

technology development in the East Indies, as did local workers and technicians active in the fields and factories. It was the interactions among these various cultural groups that produced colonial technological development.

Local Chinese technicians and colonial experts collaborated to improve the sugar-cooking process. This local input in innovation processes, which caused local workers to develop new skills and technical knowledge, contributed to the survival of local industry and research in the postcolonial era, after Indonesian independence and the forced departure of the Dutch experts.

The chapters so far have brought into view several professional groups of modernizers, active in various locations, including the colonies, and with divergent objectives. The project of modernization through technology was pushed ahead not only on the basis of specific economic concerns, problem-solving efforts, and the striving for more output or efficiency, but also by more specifically modernist ideals—ideals about expansion, national integration, the role of knowledge, social development, and the particular role of engineers and other professional groups. In chapter 7, "Technology as Politics: Engineers and the Design of Dutch Society," Dick van Lente and Johan Schot further spotlight the relationship between modernizing engineers and the Dutch state. They analyze how the rise and recognition of the engineering profession and the importance of technology at the end of the nineteenth century contributed to the emergence of technocratic ambitions among engi-

Increasingly, technological ingenuity became a feature of Dutch identity. The country was proud of its accomplishments in hydraulic engineering such as the Zuiderzee Works and, later, the Delta Works, but also of the many innovations in social housing. The visit of Queen Juliana to an exhibition in the new city hall of The Hague, "Fifty Years of Housing," in which Modernist housing construction figured prominently, symbolized this newly acquired status of technology in Dutch society. The exhibition was held to mark the fifty-year anniversary of the passage of the Housing Act.

neers. Tellingly, Cornelis Lely wrote in a 1904 letter to his son that "engineers should not merely act as the technical servants" of those in charge of management, nor should they just receive technical training; they ought to be "social engineers" as well. This social ideal among the new generation of engineers was embodied in the Social-Technical Association of Democratic Engineers and Architects organized in 1904. The authors argue that technocracy existed when, because of their expertise, a broad mandate was given to these engineers to carry out government tasks autonomously. The growing mandate for engineers emerged in several sectors, including water management, housing construction, and urban development. After World War I engineers gradually managed to gain autonomy vis-à-vis politics, an autonomy that increased further after World War II. Engineers and other experts attained a broad mandate in the process of postwar reconstruction of the

Netherlands, although concerns about what was called "instrumentalist" and "materialist" approaches persisted. In the 1960s, however, their unquestioned power declined. Engineers were forced to communicate much more with the broader public in more open and interactive planning procedures. Van Lente and Schot conclude that during the entire twentieth century the rise of technocracy never implied the end of political discussions or the straightforward pursuit of the best technical solution; on the contrary, the debate simply continued, albeit mainly among the experts involved.

In chapters 8, 9, and 10 the analytical scope is broadened to economic growth, industrialization, and consumption. In chapter 8, "Technology, Productivity, and Welfare," Jan Pieter Smits makes use of statistical analysis to examine the relationship among technological development, labor productivity, economic growth, and the competitiveness of the Dutch economy in the world market. He shows that the Depression of the 1930s did not reflect a discontinuity in Dutch economic development. In these years Dutch businesses, for example, managed to benefit from electro motors, precisely because steam technology had never been diffused widely in the Netherlands. Moreover, electromotors could be profitably deployed in sectors such as the food industry that were well developed in the Netherlands, as well as in the country's many small-scale businesses. If other Western countries emphasized capital-intensive industries, the Dutch held on to labor-intensive business sectors such as textiles.

After World War II the Dutch government developed an energy policy in which rates for the industry were kept artificially low, supported by the discovery of huge natural gas deposits near Slochteren. Dutch industries, more than those in other countries, adopted energy-intensive modes of production, from which the chemical industries in particular profited. In the late 1960s, labor-intensive industries gradually began to lose their competitiveness and in the oil crisis of the 1970s the Dutch energy-intensive industries met with serious challenges. This slowdown can also be explained in part by the exhaustion of existing technological opportunities; industry proved unable to cope with rising prices through innovation within the existing paradigm. It is striking that in these years Dutch industry failed to take sufficient advantage of the development and application of new information and communication technology.

To complete the picture, Smits analyzes welfare effects and the significance of household work in addition to measuring conventional economic growth indicators. He shows that income inequality decreased in the Netherlands between 1890 and the 1990s, which he interprets as a strong indicator of increased well-being. He also shows that households and women in par-

Infrastructures often have prolonged effects on society that are hard to undo. For example, in nineteenth-century Rotterdam an above-ground railroad had been built that cut across the city, but by the end of the twentieth century an underground railroad was preferred. This photo was taken during construction of the underground railroad where it crosses the Maas River in Rotterdam.

ticular significantly contributed to economic development, precisely because
of the introduction of new technologies in households. Finally, he assesses
the influence of environmental damage on Dutch economic development.
His calculation indicates that since the 1980s it has begun to contribute negatively to the country's economic growth. This chapter's conclusion is that
technological development has led to economic growth and a higher standard
of living, even though in the closing decades of the twentieth century this
process ran up against its environmental limits.

In chapter 9, "Technology, Industrialization, and Contested Modernization of the Netherlands," Johan Schot and Dick van Lente take up the
broader process of modernization. They explore the rise of twentieth-century Dutch industry, the modernization of the agrarian sector, and the rise
of consumer society. Largely historiographical in nature, this chapter shows
explicitly how the history of technology can contribute to a political and
economy history of the Netherlands.

The main contribution is defined by the struggle between different modes
of modernization. They argue that technological development in agriculture
before 1950 was geared toward making smaller businesses more productive by
raising soil, crop, and livestock outputs, through, for instance, the introduction of improved feed and fertilizer and through drainage and improvement of cultivation and breeding techniques. This mode of modernizing
agriculture was embedded in a cooperative structure in which feed, fertilizer,
and other products could be purchased and farm products such as milk could
be processed and sold. Technical development also occurred within networks
of businesses, governments, and laboratories, in which the farmers themselves played a crucial role. The new techniques they developed could be diffused quickly through farmers' own networks. Moreover, the role of the
Dutch national government was crucial because it funded much of the
research efforts involved. After World War II the modernizing emphasis
shifted more toward mechanization, specialization, and increasing the size
of farming businesses. Research efforts were no longer keyed toward the success of small mixed farming operations and focused instead on the new ideal
of producing as cheaply as possible for mass markets.

This struggle between the ideal of mass production and the particular
advantages of small-scale production is also visible in Dutch industry. Industrial activity grew from 1860, and after 1890 big industry won terrain. This
process not only involved very large companies with thousands of employees such as Philips and Shell but also especially smaller companies, with a
size of fifty to one hundred employees. These companies managed to do well
by cooperating and by exploiting specific niche markets. Frequently their

management was not looking to maximize outputs, but to find their company's optimal scale. Flexibility was crucial, and in this respect the electromotor provided a crucial advantage.

After World War II, the emphasis shifted toward maximizing scale size after all, just as it had in agriculture—partly as a result of the Marshall Plan. The share of Dutch workers who were employed in industry kept increasing until the mid-1960s, when this trend was reversed again and precisely the companies that had put all their cards on scale increase encountered problems.

Rival views on the best approach to modernize and industrialize competed not only in agriculture and industry but also in the household sector. Schot and Van Lente argue that Dutch consumers had already been experimenting with forms of mass consumption during the interwar years. In this period two conflicting modes of shaping mass consumption were discussed and tried out: consumption through collective services to households or consumption in an individual household context. This last model became dominant because fostering family life was regarded as the best strategy for softening the negative effects of burgeoning modern mass society generated by mass production and consumption.

In the final chapter, "Between Sensation and Restriction: An Emerging Technological Consumer Culture," Irene Cieraad offers an analysis of the ongoing cultural critique of consumption. Critics have interpreted the twentieth century as the age of the masses; with their naïve materialism and pleasure seeking they have been held responsible for social evils such as moral decay and lack of community sense. Time and again critics stressed the dangers of the entertainment provided by movies, radio, jazz, and television. Stigmatizing the working class as the mechanized masses "driven by cogwheels" or women who go shopping as merely "driven by desire," critics interpreted culture and technology by and large as incompatible categories.

Although "the masses" began to fade as a concept in the 1970s, intellectuals continued to point to the risks of commercialization and unbridled consumption, their commentary creating a basis for a new environmental criticism. As a result of this critique of consumer behavior, the need to legitimize consumption remained in place. "Modern" had already become an ambivalent notion by the 1920s: it stood for what is new cool, and youthful, but also for comfort, luxury, and passiveness. The dominant mode of Dutch society was a form of "controlled hedonism." In addition to analyzing the critique of mass production and consumption, Cieraad provides insight into new, technology-induced cultural sensitivities for space, sound, speed, and image. A new cultural infrastructure emerged around technological appli-

The introduction of new media such as film, radio, and television time and again led to moral panic among the country's pillarized elites. In their view, these new technologies embodied hazards of modern society that ought to be controlled. One such control site was the family, which was expected to integrate new products in proper ways. This 1958 photo of family life issued by KEMA (Keuring Electrotechnisch Materieel Arnhem, founded in 1927 as the Dutch electricity industry's test institute for electrical equipment) highlights the role of the various new electrical devices.

At the start of the twentieth century it seemed questionable whether all the new projects on the water management agenda were going to be implemented. Soon, however, it became clear that they would. The Netherlands was and continued to be a *work in progress*, driven by an eagerness to reclaim land and put in roads and waterways everywhere. This photo was taken during the construction of the Linge pumping station along the Amsterdam-Rijnkanaal in 1938.

ances that made it possible for people to enjoy consumption, even though in the Netherlands the ethic of moderation would prevail.

Conclusion

The trajectory of modernization through technology, an idea that developed in the decades around 1900, could have evolved in a variety of ways. In the interwar period the ideal of modernization became formulated and predominantly practiced in the Netherlands as mass production, mass consumption, and economies of scale. At the same time this ideal was heavily contested. Although after World War II the efforts toward mass production and mass consumption gained firm ground, criticism would never dissipate entirely and grew stronger again in the 1970s. During the heyday of modernization, in the 1950s and 1960s, the risks of its expansive and technology-driven principles were hardly considered, partly because the overriding view was that any potentially negative effects would be removed again through new and more sophisticated technology. This assumption proved misguided. From the 1970s onward, new experts began to translate environmental and social problems into an agenda marked by the ideal of what became known as sustainable development rather than outright modernization. Arguably, then, the spirit of Dutch society during the closing decades of the twentieth century is best characterized as a mode of "reflexive modernization". Today, the modernization ideal is still very much alive, while in many areas its perceived ties with technological progress seem as strong as before. Actors, however, are a great deal more aware of the detrimental effects of technological development. No longer do we automatically associate technological development with progress and modernization.

Acknowledgements

This introductory chapter is the result of a lengthy process involving many discussions to which the editors and authors of this volume as well as Ed Taverne have made inspiring contributions. The final draft also received valuable commentary from Adri Albert de la Bruhèze, Harry Lintsen, and Geert Verbong.

Notes

1 G. A. Klinkenberg, *Wetenschap als natuurverschijnsel* (Utrecht: Prisma, 1983), 106–107.

2 See, for example, textbooks by Pauline Maier, Merrit Roe Smith, Alexander Keyssar, and Daniel J. Kevles, *A History of the United States: Inventing America* (New York: Norton, 2003); Ruth Schwartz Cowan, *A Social History of American Technology* (New York: Oxford University Press, 1997); Carroll Pursell, *The Machine in America: A Social History of Technology* (Baltimore: Johns Hopkins University Press, 1995). Some books intended for academic and general readers are Thomas P. Hughes, *American Genesis: A Century of Invention and Technological Enthusiasm* (New York: Viking, 1989); David Nye, *Electrifying America* (Cambridge, Mass.: MIT Press, 1999). There are virtually no national histories of European countries that attempt to integrate the role of technology; there are a few that focus on a specific theme and shorter time period. See, for example, Gabrielle Hecht, *The Radiance of France: Nuclear Power and National Identity After world War II* (Cambridge, Mass.: MIT Press, 1998).

3 For an overview of the main literature on and objections to the concept of modernization, see Thomas J. Misa, Philip Brey, and Andrew Feenberg, eds., *Modernity and Technology* (Cambridge, Mass.: MIT Press, 2003), especially chapter 1; see also H. van der Loo and W. van Reijen, *Paradoxen van modernisering* (Bussum: Coutinho 1997); D. van Lente et al., "Techniek en modernisering," in H. W. Lintsen et al., *Geschiedenis van de techniek in Nederland: De wording van een moderne samenleving, 1800–1890*, vols. 1–6 (Zutphen: Walburg Pers, 1992–1995), vol. 1, 19–38.

4 For other examples of public responses to new technologies in terms of modernization see Bernhard Reiger, *Technology and the Culture of Modernity in Britain and Germany 1890–1945* (New York: Cambridge University Press, 2005); Mikael Hard and Andrew Jami-

son, *The Intellectual Appropriation of Technology: Discourses on Modernity, 1900–1939* (Cambridge, Mass.: MIT Press, 1998).

5 L. Marx, "The Idea of 'Technology' and Postmodern Pessimism," in Y. Ezrahi, E. Mendelsohn, and H. Segal, *Technology, Pessimism, and Postmodernism* (Amherst: University of Massachusetts Press, 1994), 11–28; see also "Technology," special issue, *Technology and Culture* 47, no. 3 (2006): 477–535.

6 Eric Hobsbawm, *The Age of Extremes: A History of the World, 1914–1991* (New York: Knopf, 1996).

7 See, for example, Harold James, *Europe Reborn: A History, 1914–2000* (Harlow, UK: Pearson-Longman, 2003); Mark Mazower, *Dark Continent: Europe's Twentieth Century* (Harmondsworth, U.K.: Penguin, 1998); Clive Ponting, *The Pimlico History of the Twentieth Century* (New York and London: Random House/Pimlico, 1999).

8 C. Hartveld, *Moderne zakelijkheid: Efficiency in wonen en werken in Nederland 1918–1940* (Amsterdam: Het Spinhuis, 1994), chapter 2.

9 Quoted in T. Anbeek, *Geschiedenis van de Nederlandse literatuur 1885–1985*, 2nd ed. (Amsterdam: Arbeiderspers, 1990), 123.

10 On the history of the telephone in the Netherlands, see O. de Wit, "Telefonie in Nederland 1877–1940: Opkomst en ontwikkeling van een grootschalig systeem," Ph.D. diss., Technical University Delft, 1998. On the role of Dutch consumers in technical change see Ruth Oldenziel, Adri Albert de la Bruhèze, and Onno de Wit, eds., *Manufacturing Technology/Manufacturing Consumers: The Making of Dutch Consumer Society* (Amsterdam: Aksant, 2009). Similar analyses on the role of consumers have been made for other countries. Compare, for example, S. Bowden and A. Offer, "The Technological Revolution That Never Was: Gender, Class, and the Diffusion of Household Appliances in Interwar England," in V. de Grazia and E. Furlough, eds., *The Sex of Things: Gender and Consumption in Historical Perspective* (Berkeley: University of California Press, 1996), 244–276; Claude S. Fischer, *America Calling: A Social History of the Telephone to 1940* (Berkeley: University of California Press, 1992).

11 The notion of reflexivity refers to the awareness by historical actors of the causes and impacts of technical change and modernization. Ulrich Beck introduced the notion of reflexive modernization for the post-1970 period. See Ulrich Beck, *Risk Society: Towards a New Modernity* (London: Sage, 1992). For a historical analysis of the rise of reflexive approaches, see Johan Schot, "The Contested Rise of a Modernist Technology Politics," in Thomas J. Misa, Philip Brey, and Arie Rip, eds., *Technology and Modernity* (Cambridge, Mass.: MIT Press, 2003), 257–278.

12 C. te Lintum, "Een eeuw van vooruitgang, 1813–1913," *Utrechts Provinciaal: En Stedelijk Dagblad*, n.p., n.d. [1913]).

13 Te Lintum, *Eeuw van vooruitgang*, 111.

14 The importance of steam technologies to British industrialization is also debated. It is argued that England enjoyed a wide spectrum of change and many incremental innovations took place. See, for example, Kristine Bruland, "Industrialisation and Technological Change," in R. Floud and P. Johnson, *The Cambridge Economic History of Modern Britain*, vol. 1: *Industrialisation, 1700–1860* (Cambridge: Cambridge University Press, 2004), 117–146.

15 On the specific characteristics of Dutch industrialization see Lintsen et al., *Geschiedenis van de techniek in Nederland*; On the basis of this series J. Schot published "The Usefulness of Evolutionary Models for Explaining Innovation: The Case of the Netherlands in the Nineteenth Century," *History and Technology* 14 (1998): 173–200. This is a slightly adapted version of one of the concluding chapters in the series' sixth volume and contains a lot of the case material and references concerning the specific Dutch path to industrialization. See also J. L. van Zanden and A. van Riel, *The Structures of Inheritance: The Dutch*

Economy in the Nineteenth Century (Princeton: Princeton University Press, 2004). For a recent similar argument on French industrialization see Jeff Horn, *The Path Not Taken: French Industrialization in the Age of Revolution, 1750–1830* (Cambridge, Mass.: MIT Press, 2006).

16 See J. A. de Jonge, *De industrialisatie in Nederland tussen 1850 en 1914* (Nijmegen: Sun, 1996), especially chapter 17.

17 See chapter 5, this volume.

18 For the history of Dutch liberalism, see Siep Stuurman, *Wacht op onze daden: Het liberalisme en de vernieuwing van de Nederlandse staat* (Amsterdam: Bert Bakker 1992).

19 See H. Buiter, "Werken aan sanitaire en bereikbare steden, 1880–1914," in Schot et al., *Techniek in Nederland*, vol. 6, 24–49. (Throughout this volume a shortened reference, "Schot et al., *Techniek in Nederland*," refers to the the seven-volume Johan Schot et al., eds, *Techniek in Nederland in de twintigste eeuw* [Technology in the Netherlands in the twentieth century]. Please see preface and bibliography for complete bibliographical information.) For the struggle between urban developers and physicians, see E. M. L. Bervoets, "Betwiste deskundigheid: De volkswoning 1870–1930," in Schot, *Techniek in Nederland*, vol. 6, 128–130.

20 Jan Bank and Maarten van Buuren, *1900: The Age of Bourgeois Culture* (Assen-Basingstoke: Van Gorcum-Palgrave Macmillan, 2005), introduction. We follow G. Barraclough, *An Introduction to Contemporary History*, 2nd ed. (Harmondsworth, U.K.: Penguin, 1967), who argues that a new period begins when most of the contemporary people living at that time recognize that this is the case. We therefore consider a new period of Dutch history to have begun around 1890 because by that time almost the entire Dutch population had embraced the view that Dutch society had to modernize.

21 See Dick van Lente, "Dutch Conflicts: The Intellectual and Practical Appropriation of a Foreign Technology," in M. Hård and A. Jamison, eds., *The Intellectual Appropriation of Technology: Discourses on Modernity, 1900–1939* (Cambridge, Mass.: MIT Press, 1998), 189–224; see also Hugo van Driel and Johan Schot, "Radical Innovation as a Multilevel Process: Introducing Floating Grain Elevators in the Port of Rotterdam," *Technology and Culture* 46 (2005): 51–76.

22 G. L. Cleintuar, *Wisselend getij. Geschiedenis van de Zuiderzeevereniging 1886-1949* (Zutphen: Walburg Pers, 1982), 283.

23 On the NOT and the wider effects of World War I, see E. Homburg, "De Eerste Wereldoorlog: Samenwerking en concentratie binnen de Nederlandse chemische industrie," in Schot et al., *Techniek in Nederland*, vol. 2, 316–331. See also Paul Moeyes, *Buiten schot: Nederland tijdens de Eerste Wereldoorlog 1914–1918* (Amsterdam: Arbeiderspers, 2001).

24 R. J. van der Bie, "'Een doorlopende groote roes': De economische ontwikkeling van Nederland, 1913–1921," Ph.D. diss., University of Amsterdam, 1995, 85.

25 See Jan Luiten van Zanden, *The Economic History of the Netherlands in the 20th Century* (London: Routledge, 1997), 129.

26 See Oldenziel, "Ontstaan van het moderne huishouden," 34. On the new spirit among engineers, see Homburg, "Eerste Wereldoorlog." For example, in 1916 the Dutch Ministry of Agriculture, Industry and Trade hired sixty domestic science teachers to give lectures and demonstrations on economical cooking.

27 See C. Disco, "Een volk dat leeft, bouwt aan zijn toekomst," in Schot, *Techniek in Nederland*, vol. 1, 199–208.

28 Homburg, "Eerste Wereldoorlog."

29 See Bervoets, "Betwiste deskundigheid," especially 130–133.

30 *Een kwarteeuw TNO, 1932-1957. Gedenkboek bij de voltooiing van de eerste 25 jaar werkzaamheid van de organisatie TNO op 1 mei 1957* (Den Haag: TNO, 1957) 9.

31 See Jan Pieter Smits, "Economische ontwikkeling, 1800–1995," in Ronald de Bie and Pit

Dehing, eds., *Nationaal goed: Feiten en cijfers over onze samenleving (ca.), 1800–1999* (Heerlen: Centraal Bureau Statistiek, 1999), 22.

32 Van Zanden, *Economic History of the Netherlands*, 144.

33 See P. de Rooy, *Republiek van rivaliteiten: Nederland sinds 1813* (Amsterdam: Mets and Schilt Uitgevers, 2002), 172–177.

34 See Hartveld, *Moderne zakelijkheid*, particularly chapter 6.

35 On this development see O. de Wit and J. van den Ende, "Gemechaniseerde kantoor 1914–1940," in Schot et al., *Techniek in Nederland*, vol. 1, 236–629.

36 See De Wit and Van den Ende, "Gemechaniseerde kantoor 1914–1940," 238–247, and Hartveld, *Moderne zakelijkheid*.

37 De Wit and Van den Ende, "Gemechaniseerde kantoor 1914–1940," 260–263.

38 B. P. A. Gales, J. P. Smits, and R. Bisscheroux, "Steenkolen," in Schot et al., *Techniek in Nederland*, vol. 2, 53–54.

39 M. Davids, "Van stoom naar stroom: De veranderingen in aandrijfkracht in de industrie," in Schot et al., *Techniek in Nederland*, vol. 6, 271–284. On calculating the effect of electrification, see H. J. de Jong, *De Nederlandse industrie 1913–1965: Een vergelijkende analyse op basis van productiestatistieken* (Amsterdam: Aksant 1999), 194.

40 B. van Ark and H. J. de Jong, "Accounting for Economic Growth in the Netherlands Since 1913," *Economic and Social History in the Netherlands* 7 (1996): 199–242. See also Hein A. M. Klemann, *Nederland 1938–1948: Economie en samenleving in jaren van oorlog en bezetting* (Amsterdam: Boom, 2002).

41 Kees Schuyt and Ed Taverne, *1950- Prosperity and Welfare* (Assen-Basingstoke: Van Gorcum-Palgrave Macmillan, 2005).

42 Pillarization refers to the vertical division of Dutch society in several "columns," or pillars, according to different religions and ideologies. Each pillar had its own social institutions, newspapers, political parties, schools, hospitals, universities, and sports clubs.

The Netherlands as a networked society around 2000. The Dutch spatial configuration is held together and structured by an array of infrastructures. The more or less coordinated development of these networks has ensured the integration of communities and entire regions at the national level.

2 Networked Nation: Infrastructure Integration of the Netherlands

Erik van der Vleuten

"The word *landscape* literally refers to creating land and space. Without humans there would be no landscape. This is the essence of the Netherlands."[1] Thus the Dutch presentation of their country at the World Fair 2000 in Hanover, Germany. Several government documents further elaborate this image: this shaping of Dutch space, and also of the Dutch economy and society, particularly relies on human-built material infrastructure, or "networks."[2] The *Fifth National Policy Document on Spatial Planning 2000/2020*, published in 2001, and the *National Spatial Strategy* (in 2006) explicitly portray the Netherlands as an evolving "network society" and a "network economy."[3] Unlike modern network society theorists focusing exclusively on information and communication technology, Dutch policymakers emphasize that this process is carried by the "entirety of roads, railways, waterways, pipelines and sewers, digital networks, seaports, airports and transfer points."[4] These networks are conceived of as a spatial layer mediating between the natural condition and social life, shaping "where people live, work, and spend their leisure time."[5] For one, "the activities of [Dutch] citizens and businesses occur in increasingly larger spaces, in both a physical and a virtual sense."[6] Successive governments of different political stripes have long agreed on this pivotal role of networks in Dutch space and society and the need to sustain and expand them, despite recurrent disagreements about the role of the central government in this undertaking.

A survey of the current infrastructure landscape confirms this political assessment of the Netherlands as a country built on networks (see map on page 46). A variety of infrastructures for transport, communication, and energy supply, several of which rank among the densest in the world, create several features that make the country remarkable when viewed from an international perspective. These include a nearly complete cultivation of its territory (just a tiny percentage of its territory still counts as "natural"); remarkably high levels of population density, urbanization, intensive agriculture, and large-scale industries; a key position in European and global trade flows; and associated problems such as pollution, congestion, and vulnerability to technical failure.

The Netherlands is a networked society par excellence, with many different networks in a fairly small area, so that these networks are constantly intersecting. As part of the work on the A12 national highway, shown under construction in 1981, this aquaduct was built to carry the small Gouwe River over the automobile tunnel.

This Dutch infrastructure landscape, cherished by politicians and taken for granted by many Dutch citizens today, has a long and complex history. Some infrastructures are rooted in the Middle Ages and the early modern period. In the nineteenth and twentieth centuries infrastructures multiplied and proliferated, producing a veritable "networked nation."[7]

Often such infrastructure developments were pushed by politicians, industrialists, and engineers in search of ways to shape the Dutch polity, economy, or society. In other cases, infrastructure change was a byproduct of uncoordinated efforts targeting at different objectives. Either way, material infrastructures have become carriers of what the historical geographers Hans Knippenberg and Ben de Pater have labeled "the unification of the Netherlands": a social-spatial integration of its regions and communities, which accelerated in the nineteenth century and was by and large accomplished in the twentieth.[8]

This chapter aims to bring into focus the historical shaping of the Netherlands as a networked nation. It aims to interpret and synthesize the individual histories of a wide range of infrastructures and infrastructure-related societal changes into one coherent narrative. This effort is informed by, and is designed as a contribution to, the international literature on large technical systems, the main literature on infrastructure in historical and sociological technology studies.[9] Associated research aims include replacing a traditional history-of-technology focus on artifacts (automobiles, locomotives, telephones, dynamos) with a focus on spatially extended transport, communication, or energy supply *systems,* of which such artifacts are integrated parts; acknowledging and examining the intimate intertwining of system development and major societal changes; and analyzing system development by following key individuals and organizations that negotiated the shaping and uses of these systems. This literature has hitherto predominantly centered on the development of individual infrastructures.[10] This chapter, by contrast, aims to address the shaping of the Dutch infrastructure landscape in its entirety. Since the aim is to bring into view this infrastructure landscape without a priori limitations following from ownership (public or private), form (hierarchical versus weblike), or management structure (centralized or distributed), the terms "infrastructure," "network," and "system" are used interchangeably in their broad meaning of materially integrated, geographically extended structures.[11]

Infrastructure development is here intentionally interpreted in the historical context of nation-state building. This does not mean that the nation-state is taken for granted as the implicit, unproblematic, almost naturalized container for historic inquiry—which much infrastructure history unfortunately does.[12] Instead this chapter is an inquiry into how a networked nation emerged as an important historical category. It situates this development—if only briefly—against simultaneous and intertwining processes of international and local infrastructure building, and spotlights the limits and contested nature of national infrastructure integration.[13] To bring out contrast alongside connectedness, it repeatedly places the Dutch networked nation developments in comparative international perspective.

The Contested Integration of the Dutch Geography

Although this chapter focuses on nineteenth and particularly twentieth-century developments, interactions among infrastructures, landscape development, and societal change on what is currently Dutch territory have a much longer history. During the Middle Ages, settlers massively colonized and cultivated the extensive peat bogs in the coastal zones that covered perhaps half of the present country—the so-called Low Netherlands. They turned swamps into fertile agricultural lands by means of dense networks of drainage canals and dikes. Most of these still exist today unchanged.[14] By the late sixteenth and seventeenth centuries, the Low Netherlands had become the political and economic center of the Dutch Republic and the locus of a striking degree of urbanization and economic growth. As Jan de Vries, Ad van der Woude, and others have documented, the world's "first modern economy" was shaped along waterways that facilitated interurban transport systems as well as drainage: notable innovations include scheduled services along a dense inland navigation network for sail-powered freight ships, and horse-pulled passenger, parcel, and mail barges using dedicated canal systems. These boosted the republic's internationally competitive position by linking an exceptionally large hinterland to international trade flows at several large harbors. Peat provided power to the republic's industries, distributed via an extensive peat-shipping network.[15]

The political centralization and unification of the Netherlands during the French occupation (1795–1814) marked a decisive break in the country's history. A centralized, hierarchical state replaced the decentralized republic,

For the most part, Dutch natural gas and electricity facilities evolved from local and regional systems to a nationally integrated system. This process occurred through expansion and scale increase. The gas and electricity systems developed a layered structure consisting of a main network and extensive secondary and tertiary networks of local connections.

MAP 2-1: DEVELOPMENT OF NATURAL GAS AND ELECTRICITY NETWORKS

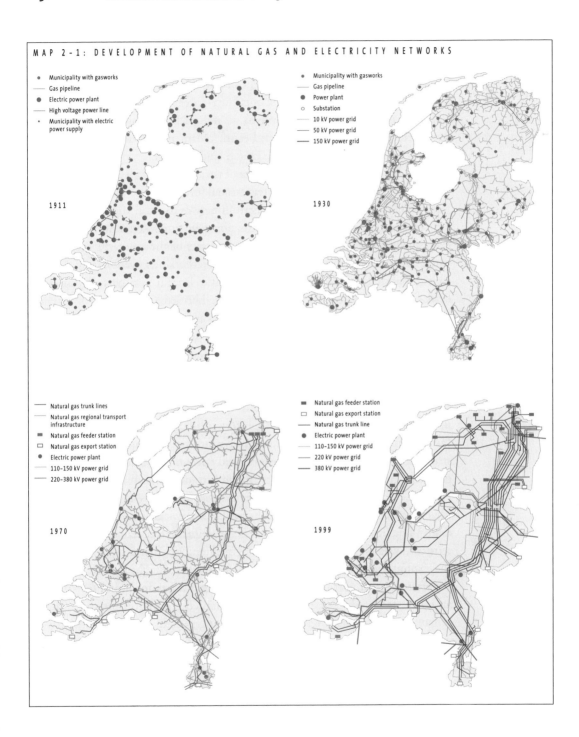

which had been characterized by urban and provincial autonomy; national citizenship replaced urban citizenship, and a national budget and debts replaced provincial ones. However, the country's new political unity did not automatically imply the economic or cultural integration of its people. Infrastructure integration would serve as a "material precondition" for such processes, but by 1800 it hard barely begun.[16] Many infrastructures that today seem an almost natural part of the Dutch landscape were still absent. Moreover, the infrastructures built in the era of the old republic were chiefly integrated on a intraprovincial and international level; interprovincial networks were poorly developed. Inland navigation networks—still the most important connecting infrastructure at this point in time—were found nearly exclusively in the Low Netherlands. Large parts of the country's more elevated eastern and southern areas, the High Netherlands, were hardly accessible, inhabited, or cultivated; its communities lived in relative isolation, its landscape covered with heaths and marshes. This led a prominent historian to characterize the Netherlands around 1800 as "empty lands."[17]

Not until the nineteenth and especially the twentieth century would infrastructure developments open up the country's inland regions. Wet infrastructures were expanded and interconnected on a national level, and also a variety of entirely new, nationally integrated networks was built. Eventually each and every part of the Dutch territory, dry- or wetland, became linked up in a nationally integrated and internationally connected geography. Households, factories, and farms were bound together by a host of infrastructures for energy supply, transport, and telecommunications. Even areas generally thought of as "nature" were integrated into this human-made space: like those of the Low Netherlands, the extensive marshes and heaths of the High Netherlands were transformed into cultivated woods, fields, and meadowland.[18] The flows of major Dutch rivers were equalized or canalized, and the territorial waters were divided into areas for shipping, fisheries, recreation, and the exploitation of sand, shell, oil and natural gas.[19] The Dutch sky was penetrated by air corridors and electromagnetic waves for radio, television, and data traffic. On the ground, the fragmented plots of "real nature" that remained were increasingly manipulated, engineered, and physically integrated by newly constructed "green corridors" into a so-called National Ecological Network, thus completing the human-built networked nation.[20]

The history of the infrastructure unification of the Netherlands is not merely a story of the unstoppable expansion of infrastructures in the service of nation building, however. The *Fifth National Policy Document on Spatial Planning 2000/2020* was subtitled *Making Space, Sharing Space*, implying that artificially created Dutch space is shared by many social groups that often have contradictory interests. These produce an ongoing dynamic of competing options and negotiated decisions in an ever-changing political, economic, and technical context, both in the past and present. Consider, for instance, the Dutch railroad network. In the last decades of the nineteenth century it became the national transport infrastructure par excellence, and its future development path was defined accordingly. In the 1920s and 1930s, however, automobility became a major competitor, requiring its own infrastructure. In a period of fierce competition, about half of the rail network was torn up, including most local lines and all inter-urban tram lines, and was replaced by roads.[21] Today, in the context of road congestion and environmental concerns, inter-urban light-rail networks seem to be making a comeback, as is the development of Trans-European Rail Networks. Such developments involved struggles between competing interests, the outcomes of which could not be anticipated. This chapter explores such struggles, examining a variety of actors involved in choices between alternative development path. It reminds the reader that the country's infrastructure integration was neither a strictly linear process nor an inevitable one.

I first map the development of the major infrastructure systems in the Netherlands. The following sections examine in detail a variety of social groups, negotiations, and struggles that shaped infrastructures and some of its uses. The chapter concludes with a historical characterization of the major regimes of infrastructure change.

Building the Networked Nation

Energy Infrastructure

To give the reader a sense of the vast changes in the infrastructure landscape in the twentieth century, I start this mapping exercise with energy infrastructures. In the course of the twentieth century nearly all buildings in the Netherlands were hooked up to two new energy infrastructures—electricity and gas supply—which were planned as nationally integrated networks.[22] These networks radically altered the Dutch energy geography: simply by pushing a button or turning a knob, everyone gained instant access to light and heat, regardless of social class or location. Other energy infrastructures, such as compressed air systems or district centralized heating, hardly made inroads in the Netherlands. This is clearly a matter of choice, not of naturally unfolding technological logic: in Denmark, for instance, more than half of the heating for homes and buildings is supplied by warm water or steam networks.[23]

The maps in map 2.1 outline the major transformations of the Dutch energy geography for selected years in the twentieth century. On the eve of World War I public electricity supply (here meaning supply *to* the public, not necessary public ownership), which had first been introduced only a few decades earlier, mainly consisted of a number of local systems in more densely populated areas. Some eighty power stations supplied electricity to consumers in their immediate surroundings via a local low-voltage cable network. Electricity supply delivered over longer distances by means of high-voltage transmission (higher transport voltages decrease relative transmission costs and thus increase the economically feasible supply range) had only made a modest beginning. The electrical map of the Netherlands showed mainly "blank" spots: over 900 of the 1,121 Dutch municipalities lacked public electricity supply. The much older system of gas supply had a local character as well. Since the first half of the nineteenth century, private and municipal companies had set up about 200 local systems that consisted of gasworks and pipe systems transporting so-called "city gas" to mainly local users. Some 330 municipalities were connected to such networks.[24]

Two decades later, the electricity map of the Netherlands had changed radically. By 1930 the electric power landscape was dominated by electricity supply systems that used high-voltage transmission—typically 10 kilovolts (kV) and sometimes 50 kV—to supply areas as extensive as entire provinces. Although these systems remained largely unconnected,[25] provincial electricity companies and a few municipal and private ones as key players had now electrified the country: 94 percent of all Dutch municipalities had access to an electricity supply network. There was almost one connection for every five residents.[26] By contrast, gas supply had hardly changed, even though the principle of long-distance supply had been introduced: the coke factories of Royal DSM (Dutch State Mines) and Royal Hoogovens (a leading steel producer in Europe) produced gas as a byproduct, which they supplied as "distance gas" to municipalities.

The energy map of 1970 shows that by that time, both electricity and gas infrastructures had been integrated nationally. The first electric interconnection of two provincial power plants dates from 1931, but especially during and right after World War II previously isolated provincial systems were interconnected. The first national power grid, coordinated by Cooperating Electricity Production Companies (Samenwerkende Electriciteitsproductiebedrijven, or SEP), established in 1949, and operating at 150 or 110 kV, was completed in 1953. During the late 1960s a second national grid with a still higher transport capacity of 380/220 kV was developed.[27] A similar development occurred in gas distribution. Soon after World War II the construction of a national gas transport network, powered by cokes factories and several large regional gas works, was being discussed.[28] By the mid-1950s three long-distance gas networks in the northern Netherlands were interconnected, but they remained isolated from two southern networks, run by Royal DSM and the new State Gas Company (Staatsgasbedrijf). The discovery of huge natural gas reserves in Slochteren near Groningen in the northern Netherlands in 1959, then considered the second largest gas field in the world and still the largest in Europe, triggered the establishment of a new national high-pressure transport grid to distribute Slochteren gas throughout the country, managed by the Nederlandse Gasunie. Local gas works gradually discontinued local production.[29]

Several patterns characterize the spatial development process of these two infrastructures. First, both experienced an evolution from local systems to "long-distance" systems to national and international systems. Infrastructure integration thus partly took place through expansion and scale increases.

Second, an important nuance to this expansion logic is the observation that national integration and international connection were mutually constitutive processes. The first attempt to connect Dutch electric power grids to foreign grids preceded national integration: during the German occupation (1940–1945) the occupying forces aimed to use Dutch power plants to help run the German war economy. Several delays meant that a large interconnection from The Hague in the west to the Southeastern mining district and its connection with German and Belgian grids was completed only after the war. Since then, the Dutch have completed a national grid while at the same time remaining on the forefront of European electrical integration as they have been with economic and political integration. The second national grid was originally, in the late 1950s, conceived of as part of a Western European grid; later, national concerns for security of supply took precedence. Still, today the Netherlands is one of the best-integrated countries in the European Union, with an import-export capacity of more than 20 percent of its national production capacity.[30] The gas grid was linked up with foreign systems in an even more ambitious way. The construction of a national grid went hand in hand with international connection, for the plan was to export about half the production of the newly discovered Slochteren gas field. In the 1970s exports began to Germany, Belgium, Luxemburg, France, Switzerland, and Italy, and the Netherlands rapidly emerged as a leading gas exporter.[31]

Third, network development was marked by an increasing degree of local branching. The density of the electricity and gas networks continued to increase during the expansion process; eventually almost every building in the Netherlands was connected. Because energy generation took place in a limited number of power plants or natural gas wells, the transmission networks connecting the sites of production and consumption were both extensive and dense. By the early 1990s the Netherlands counted 150,000 kilometers of gas lines, which is one and a half times the total length of all paved roads and is the highest gas-line density in Europe.[32] By 1992, electricity supply required no less than 230,000 kilometers of cable.[33] Both networks are multilayered and consist of "primary," "secondary," and "tertiary" networks, a terminology adopted in the 1930s to indicate the layered organization of infrastructure.[34]

Finally, the case of electricity serves as a reminder that nationally integrated infrastructures do not automatically produce national energy flows. For a number of decades the national electric power grid was used on a limited scale. Provincial power companies primarily planned and operated their systems to balance supply and demand in their own supply areas, using the grid chiefly for backup and additional power supply. By and large, electricity circulation continued to operate on a provincial scale. This changed with the introduction of national economic optimization (*landelijke economische optimalisatie*) in 1982, after which point all Dutch power plants should be deployed as part of a single system. Similarly, the Western European grid was initially used on a limited scale, such as for importing night-rate electricity and backup. In the 1960s Dutch electricity imports came to a near standstill. Only during the neoliberal era, when provincial power monopolies were deliberately broken up, did such exchanges become permanently integrated, involving, for instance, large-scale imports of French and Belgian nuclear power.[35]

Transport Infrastructure

The geography of Dutch transport also changed drastically in the nineteenth and twentieth centuries.[36] Dutch transportation infrastructures have a long history. Much of the northern and western Netherlands were blessed with natural waterways; the digging of interconnecting canals started in the medieval period and was followed by improvements to natural waterways and digging new canal networks for horse-pulled barges during the era of the Dutch Republic (1581–1795). By that time Dutch harbors were linked internationally through maritime shipping: trade with the Baltic region was gradually expanded to include Atlantic ports, the Mediterranean and, finally, Africa, Asia, Australia, and the Americas. Amsterdam became a global marketplace. An unpaved road system existed, including a few long-distance highways, in addition to the water infrastructure.

The country's large rivers presented major obstacles for networks of roads and railroads. Bridges serve as important elements in the national integration of these transportation infrastructures. The capacity of the network of bridges is regularly expanded. In 1976 the first bridge component of the old bridge across the Hollands Diep in Moerdijk was replaced by a new bridge section that was more than twice as wide as the old bridge.

By approximately 1800, transport infrastructures were better integrated at regional and international levels than at the national level. There were several obstacles to national integration. Inland navigation was complicated by wind, waves, currents, ice, shallows, sandbanks, narrow sluices, and bridges, while ditches and puddles made it hard to use unpaved roads.[37] In addition, few roads accessed the southern and eastern regions of the Netherlands. These circumstances translated into longer travel times: depending on weather conditions, freight shipped from the western Netherlands to the eastern town of Zwolle could take anywhere from two to fourteen days.[38]

By the late nineteenth century this situation had changed drastically (see map 2.2). The national government and some private companies had dug several new waterways and "improved" major rivers for navigation by standardizing their width and depth and eliminating curves. Such ways initially served the major trade routes between the North Sea, the main ports of

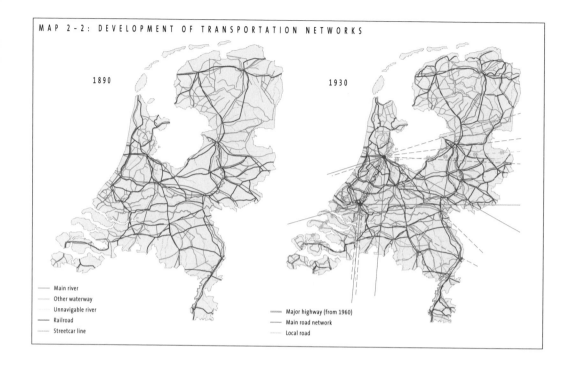

MAP 2-2: DEVELOPMENT OF TRANSPORTATION NETWORKS

1890

1930

— Main river
— Other waterway
— Unnavigable river
— Railroad
— Streetcar line

═ Major highway (from 1960)
— Main road network
— Local road

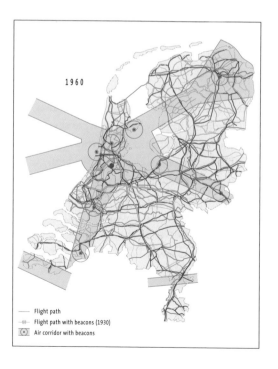

1960

— Flight path
-•- Flight path with beacons (1930)
⊙ Air corridor with beacons

On the eve of the twentieth century, railroads and waterways constituted the country's main transportation infrastructures. The development of the Dutch transportation landscape in the twentieth century can largely be seen as a process of expansion, scale increase, and concentration of existing and new transport infrastructures. Conversely, after World War II there was a decrease in the density of specific kinds of infrastructure, such as tramways and railroads, as tracks were taken up in several cities and regions.

Amsterdam and Rotterdam, and the German hinterland. In the last decades of the nineteenth century, however, a nationally integrated waterway network was established.[39] The road system, too, had been greatly expanded. King William I (on the throne from 1815 to 1840) had continued Napoleon's policy of building imperial roads on Dutch occupied territory, and a network of paved "national" roads was in place by 1850. In the century's second half, the network rapidly became more dense as provinces and municipalities paved smaller roads.[40] Finally, an entirely new transport modality was added to the transport landscape. In the first half of the nineteenth century, private companies built railroads between selected western cities with an eye to lucrative passenger transport, while the national government, as it had with waterways, financed rail connections between the ports of Amsterdam and Rotterdam and with Germany. The state subsequently pushed the development of a nationally integrated railroad network, which was in place by around 1880.

By the 1930s, all these networks were significantly more elaborated. After 1880 many "secondary" local railroads and interurban tramways, slower and built to lower construction standards, were established, as well as "tertiary" urban tramway systems. The total length of rail track peaked at some 6,500 kilometers. The total length of navigable waterways peaked in the late 1940s at 7,000 kilometers, categorized according to the size of the ships they could accommodate. For example, 1,300 kilometers of waterways were navigable for ships of over 1,500 tons (large Rhine barges) and 2,500 kilometers for ships with a cargo capacity of less than 150 tons (clipper, tjalk [canal boat or barge], motor ship, or flatboat). The principal international waterway was the Rhine, Europe's main water transport route, and to a lesser degree the Meuse River and several canals.[41] Ultimately the density of the paved-road network outstripped that of all other transport modes: already by 1920 at 20,000 kilometers it was the country's densest transport net.[42] Finally, aviation got off the ground in this period, starting with international connections. In 1920 the newly established Royal Dutch Airlines (KLM) started service to London, Hamburg, and Copenhagen.

In the second half of the twentieth century, the prewar picture of multiple transport networks of approximately equal density was replaced by a situation in which the road traffic network clearly prevailed.[43] Many local railroads, all inter-urban tramlines, most urban tram lines, and many smaller canals were turned into roads and highways. Road construction continued, resulting in a network that today has a total length of some 115,000 kilometers, producing a density surpassed only by those of Belgium and Japan.[44] At the same time a completely new network of freeways was built. The notion of automobile-only roads dates back to the 1920s, but only in the 1960s did their construction gain momentum; by 1990 the highway network would comprise over 2,000 kilometers. It was linked into the European E-road network, launched by the Economic Commission for Europe of the United Nations (1947) and subsequently developed by Europe's transportation ministers.[45] It is notable that during the twentieth century a national network of bicycle paths was created whose total length is comparable to the that of the motorway system's.[46] Finally, air traffic grew considerably and air corridors were defined in greater detail. In the 1970s, for instance, air lanes were commonly found at altitudes of 900 to 5,800 meters and had a width of ten nautical miles.[47]

   This brief survey allows several interpretations. First, the development of transport infrastructures, like energy infrastructures, can be represented in terms of processes of scale increase as well as enlarged density. Strikingly, nationally integrated, layered road and rail networks emerged via an expansion process with national and international dimensions, followed by a process of growing density and infilling. Air transport, however, followed a different development path. The first airlines were international in scope. Interwar visionaries anticipated a subsequent branching process, in which airplanes would ultimately succeed cars as leading means of transportation.[48] This did not happen, but postwar government policies of stimulating inland air traffic between secondary airports only had some branching effects. Still air traffic remains overwhelmingly internationally oriented.

   Second, in the case of railroads and waterways a period of steadily increasing density was followed by a process of "thinning," whereby secondary and tertiary lines were discontinued.[49] In the course of the twentieth century the total length of the Dutch network of railroad tracks was halved, though the system has recently begun to expand again. As briefly noted, road congestion and environmental concerns prompted new projects for interurban tramways, inspired by well-publicized German successes in Karlsruhe (1993) and Kassel (1996). Also, two entirely new primary rail lines are being developed, which compare to the freeway concept for roads. In 1997 construction was begun on the Betuwe railroad line, the so-called Betuweroute, a 160-kilometer "freight freeway" from the port of Rotterdam to the German border. It is accessible only to freight trains and is built for double-stacked container transport (two containers vertically stacked), and avoids city centers. This line provoked widespread protest but became operational in 2007. In addition, a high-speed line is planned to connect Amsterdam and Rotterdam to Brussels, Paris, and London, currently scheduled for commercial operation in 2009. These lines were planned as part of the Trans-European

Rail Network, meaning that they received limited EU funding and were built in accordance with anticipated EU traction and safety standards.

Third, in the wake of network expansion, individual mobility has increased to the degree that transport historians speak of a "mobility explosion."[50] Dutch passenger transport increased from 1 billion passenger-kilometers in 1900 to 17 billion at the eve of World War II to 190 billion in 2000. Still, as in the case of electricity supply, the national integration of transport networks did not necessarily imply national transportation flows. In the second half of the century, the average daily trip distance remained stable at some 50 kilometers. It seems that use of the new transport networks is increasing, but for local and regional rather than national travel. By contrast, freight flows have steadily mirrored network expansion, opening up formerly remote or isolated regions and stimulating their integration into the (inter)national economy.[51]

Communication Infrastructure

The Netherlands also became "covered with visible and invisible modes of communication."[52] Communication services had traditionally relied on transport infrastructure in the form of messengers and mail.

Yet, like the energy supply, communication gained infrastructure of its own in the nineteenth and twentieth centuries. The first newcomer was an electric telegraphy network (see map 2.3).[53] Prior to that, systems for optical telegraphy had been used in warfare, but these had had a temporary character. For example, during the Belgian Revolt of 1830 a chain of eleven optical telegraphs located in church steeples made it possible to send a message from the seat of government in The Hague in the west to 's-Hertogenbosch in the south in a matter of minutes, provided the weather was favorable. Electric telegraphy made transmission of letters even faster and more independent of weather conditions and daylight. Its pattern of development is similar to that of railroads: private initiative tried to cash in on several lucrative short routes, but the national government decided to set up a nationwide system. By 1855 major Dutch towns were connected and also linked to the networks of Belgium, Prussia, and Hanover. This was followed by a phase of increasing density. By 1900 over 600 telegraphy offices were interconnected through some 20,000 kilometers of telegraph wire. After the turn of the century a new international "primary grid" was added in the form of radiotelegraphic connections with ships and with other countries.[54] Where the telegraphy network had become nationally integrated and transnationally connected, the telephony network still had a local character. In 1895 there were just thirty-two local phone networks in the Netherlands and just eighteen of these were connected in a long-distance network.[55]

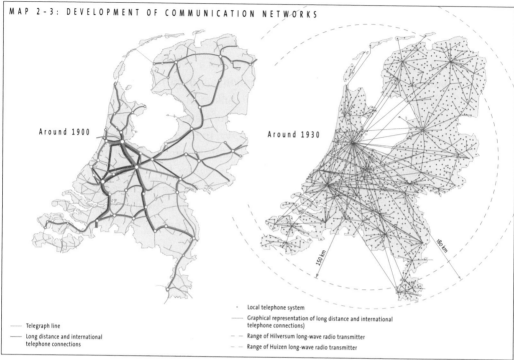

MAP 2-3: DEVELOPMENT OF COMMUNICATION NETWORKS

Around 1900

Around 1930

—— Telegraph line

—— Long distance and international
telephone connections

· Local telephone system

—— Graphical representation of long distance and international
telephone connections)

– – Range of Hilversum long-wave radio transmitter

– – Range of Huizen long-wave radio transmitter

Around 1960

Automated local telephone system

······ Telephone districts

—— Planned radio telephone network

—— Automated telephone network

– – Range of radio transmission (Medium Frequency)

—— Range of television transmitters

Pace of development of the Dutch communication infrastructure in the twentieth century. Communication infrastructures develop as networks almost by definition. The twentieth century saw an enormous increase in the density of communication infrastructures, and the number of means of communication also grew markedly.

By 1930 the telecommunications landscape had become much more crowded. First, the telephony system had been nationally integrated after the national government took over responsibility for long-distance connections and most local networks. About 1,500 local telephone networks were connected, via twenty district exchanges, into a national system. In 1937 five district switching stations were interconnected into a new national primary network. Many international connections were available via telephone lines; radiotelephony had been in operation since the late 1920s, and this huge network was now used for telegraphy as well.[56] In addition, a completely new system was added to the communications landscape: radio amateurs and manufacturers of radio appliances developed radiotelephony into broadcasting networks and transmitted the first programs, from The Hague in 1919 and from Hilversum in 1923. In 1925, using two radio towers donated by the large electro-technical manufacturer Philips, Hilversum attained a national range and became the basis for a national broadcasting system. A second national radio station in Huizen doubled the available broadcast time. By 1940, 65 percent of Dutch households owned a radio and were able to listen to national programming, either through their own radio set or via a local radio distribution system connecting a local receiver via a local wire network to loudspeakers in individual homes.[57]

By 1970 the telephone network had grown even denser. Between 1950 and 1980 the number of fixed connections increased from five to thirty-four per hundred inhabitants, practically connecting every Dutch household. In addition, a modest start had been made with mobile telephony. From 1949, the Dutch PTT (Post Telegraphy Telephony) set up its first mobile telephony system consisting of dozens of base stations that fixed phones could use to call a car or boat with radiotelephone. By the 1970s this network had 2,000 subscribers. Also, separate telegraph lines were put in for the successful telex system, originally set up in the 1930s using the telephone network but now in desperate need of additional capacity.[58] The ether had become busier as well, as a new system for "television" was introduced. Test broadcasts had been made by Philips starting in 1948, with a range of 40 to 50 kilometers around its home city of Eindhoven. In 1951 the Nederlandse Televisie Stichting (Dutch Television Foundation) began national television broadcasts from the radio broadcasting station in Lopik. A system of auxiliary transmitters guaranteed near national coverage by 1958. The rapid diffusion of TV sets (75 percent of all Dutch households owned a TV set by 1970) tied the Dutch population into a very influential new network. As earlier with radio, programs were received either via one's own TV set or via antenna stations with local cable infrastructure.

By 2000 it was clear that the rapid expansion of mobile phone networks had increased density of the telephony infrastructure still further. After the adoption of the European GSM-standard in 1994, a number of private companies in the Netherlands set up national networks consisting of base stations interconnected by underground cable networks. At the start of the twenty-first century, the Netherlands, with a population of 15 million, was home to 11 million cell phones. The fixed and mobile telephone networks also supported data transfer by acting as connecting links between computers or computer networks. Finally, many new radio and television stations became available, including local stations transmitted via local cable networks and international stations that could be received via satellite. At the end of the twentieth century, the telecom landscape consisted of a "colorful palette of services, infrastructures and actors."[59] Fiber optic networks, first announced in the mid-1980s, now carry different services between main nodes. However, on the so-called "last mile" to individual homes, copper (originally for telephony) and coaxial (originally for television) cables still dominate, increasingly carrying telephone, television, and data transport signals.

The physical integration of communications infrastructures can also be described in patterns of expansion and density. Two qualifications are relevant regarding these networks' flows. First, information can be transported on various types of infrastructure. Telegraphy pulses can be transmitted via the telephone lines; the telex service, set up in 1932, initially relied on the telephone network, until the success of this service justified putting in a separate system of telex lines. Also, some radio distribution exchanges distributed radio programs locally with the help of telephone networks. Thus, even before the era of digitalization, there was exchangeability between different communication networks.

Second, in Dutch broadcasting a distinction was made between physical infrastructure and the circulation of information. Various religious and political groups set up their own broadcasting organizations to cater to specific segments of the national population, but they used the same physical network. Households received programs specifically aimed at Catholics, Protestants, workers, or apolitical entertainment. Thus, broadcasting was nationally integrated yet socially fragmented—although it must be emphasized that individual households obviously could not be restricted to looking only at the programs that were intended for their specific social group.[60]

Nature as Infrastructure

The huge proliferation of infrastructure on, above, and under the ground has transformed the Netherlands into an artificial, human-made spatial configuration.[61] By opening up previously peripheral areas, transport systems promoted its nationwide integration. Telecom systems seemed to shrink time and space even further by facilitating simultaneous communication between connected points. Energy infrastructures had a similar effect. Where people used to be dependent on a local wind or water mill, a steam engine, peat or petroleum, now energy networks made light and power instantly available everywhere.

Remarkably, during this process areas that are commonly considered "natural" and antithetical to human-made space were infrastructurally integrated as well. The first examples are the creation and cultivation of the peat bogs in the coastal Low Netherlands since the medieval period and the opening to cultivation of the pristine wastes in the High Netherlands in the nineteenth and twentieth centuries. I also mentioned the manipulation of wet nature above. Natural watercourses were "improved" (deepened, standardized, canalized) and artificial ones were added. Starting in the Middle Ages this was undertaken at a local and regional level, and from the late eighteenth century this effort was extended to the country's major rivers. The first achievement with national ramifications involved the stable distribution of Rhine River water to its downstream branches the Waal, Lower Rhine, and IJssel rivers in a proportion of 6:2:1. This was followed in the nineteenth century by dredging of river beds and the construction of three new artificial river mouths, including the New Waterway (1863–1872) that became crucial to the competitive position of the Rotterdam harbor. The capstone of Dutch wet network building was the ambitious construction of a national system for manipulating water flows (1940–1971). A system of dams, weirs, and sluices distributed the incoming water of the Rhine and the Meuse rivers into the country's various smaller rivers, canals, and reservoirs. This system will be further discussed later; here it suffices to say that the natural wet ecosystem was transformed into a human-controlled infrastructure.[62]

In the second half of the twentieth century, the North Sea, too, was brought under human management. Waterways were dredged to keep open the ports of Rotterdam and Amsterdam, and a network of shipping routes was defined and marked with buoys and light markers. In the mid-1970s oil and natural gas exploitation started; by 1997 there were 121 permanent natural gas and oil platforms, connected to the mainland by 1,880 kilometers of pipelines. In addition, specific zones were reserved for mining seashell deposits and sand; sand was deposited where needed for coastal protection.[63]

The Netherlands' wet network cannot clearly be classified as belonging to one of the previously mapped types of infrastructures. It is essentially multifunctional: a single physical infrastructure is used for water discharge, water

supply, inland navigation, national defense, and fishing.[64] Still, the layered structure that characterizes other infrastructures can be seen here as well. The layered waterway network has already been mentioned, and the water discharge system is also layered. The major rivers and estuaries were connected to many local or regional subsystems were for water discharge and drainage, such as urban sewer systems put in since the nineteenth century, and the waste water systems of industry. These large rivers also were main arteries in the system for freshwater supply; many subsystems were hooked up to them, such as urban water services, agricultural irrigation systems, and water-intensive industries.[65]

By 1970, over a millennium of reclamation, cultivation, and networking had left only fragmented pieces of "untrodden nature," which together made up about 6 percent of the Dutch territory (in addition to human-made forests, which account for 8 percent).[66] Currently, these fragmented plots find their place in the Dutch networked nation as they are being interconnected and transformed into a green infrastructure. This development dates back to the 1970s. Up to then, Dutch nature conservation had developed a rich tradition predicated on extensive human interference: nature management techniques included fishing and hunting, grazing by sheep, tree cultivation, mowing and peat cutting, even maintaining water mills—in short, preserving preindustrial landscapes. Yet the 1970s witnessed a paradigm shift in nature management. This new paradigm was about creating new "real nature," defined by ecological standards as nature "as it would have looked without human interference."[67] This idea was inspired by the new disciplines of systems ecology and ecological engineering, developed by Eugene and Howard Ogdum in the United States. The Ogdums looked at nature in terms of ecosystems that run on solar energy, which after being fixed in green plants flows throughout the entire system via the various food chains, resulting, through several feedback mechanisms, in a natural balance. Nature conservationists should create the proper initial conditions, after which nature itself should evolve naturally, that is, without human interference.

This poster of the AZEM (N.V. Algemeene Zeeuwsche Electriciteits-Maatschappij, or General Zeeland Power Company), the provincial electricity company of Zeeland, informed the people that they could enjoy the advantaged of electricity everywhere in Zeeland. In the Netherlands, provincial electricity companies have been the main builders of the electricity system. For rural provinces such as Zeeland, North Brabant, and Groningen, the principal motivation for building an electricity network was not profit but land use: delivery of electricity to residents in all corners of the province was seen a way to prevent massive urbanization of rural parts of the country.

In this new style of "nature building," the size of natural areas was considered crucial to the abilities of species to survive on their own in "the wild." Creating larger nature development areas was thus the top priority. There were two ways to achieve this. First, existing nature zones could be expanded and improved. The Oostvaardersplassen—a sixty-square-kilometer (big by Dutch standards!) wetland on the large reclaimed South Flevoland polder northeast of Amsterdam—served as a major showpiece. Previous planned as an industrial zone, it was claimed for nature development, and the existing nature was "improved" to recreate the type of landscape that biologists thought typical to the prehistoric Netherlands—a mosaic of open and wooded patches home to a rich variety of species. To keep the landscape open biologists introduced large herbivores such as the Heck oxen (the result of retrobreeding aurochs from existing species in the 1920s in Germany), konics (European wild horses), and red deer.

A second method to increase the size of natural territory was to link up fragmented nature areas, including human-made forests, by means of so-called ecological corridors. The first initiative was the so-called Ooievaar Plan (1985), which aimed at setting aside lands in the Lower Rhine, Waal, and Meuse river regions for conservation. Two strategic junctions served as "generator sites" of biodiversity. From there, plant and animal species could migrate to other stretches along the rivers via smaller nodes, called stepping stones. Another example of such nature networking on a subnational scale is the defragmentation of the wooded Veluwe region by means of "gray-green crossings" such as badger tunnels and so-called ecoducts, viaducts for animal migration, to facilitate animals' access to a larger territory.

Ecological system building became a matter of national policy in the Nature Policy Plan (Natuurbeleidsplan) of 1990.[68] The plan proposed to establish a physical infrastructure for the circulation of plants and animals on a national scale. This so-called National Ecological Network should comprise core areas and nature development areas connected by ecological corridors, which are green or wet "robust corridors" facilitating the migration of species from one area to another (see map on page 70). One example of such a corridor is the wet corridor connecting northeastern wet zones to the southwestern Zeeland delta. Another example is the dry connection between the Oostvaardersplassen and Veluwe, which should facilitate the migration of red deer between these zones. Less profoundly, ecological corridors may take the form of, say, twenty-five-meter-broad strips of wild nature cutting through agricultural fields. The ecological backbone of the Netherlands is scheduled for completion in 2018 and it should be connected to a Pan-European ecological network, presented by the Council of Europe and originally endorsed by forty-nine countries, in 1995, with the Dutch example as a model.[69] Progress on these networks is slow and complicated, though, an issue to which I shall return. What is important here is that the Dutch seek to integrate even their country's last remaining areas into a single comprehensive, human-made and -controlled network geography. The National Ecological Network will complete the Dutch networked nation.

The Netherlands as Problematic Achievement

In the twentieth century, one historian recently wrote, the Netherlands has been "a colossal work in progress" in terms of its urban space, agricultural land, infrastructure, and nature.[70] Its geography became covered with networks of steel, stone, wiring, pipes, electro-magnetic waves, water, and air corridors—and a beginning was made on establishing green corridors. Often these national infrastructures were further developed and refined at provincial or local levels and connected to networks in other countries.

This achievement is striking, but not necessarily hailed by all. Some praise the ultimate victory of human might over the cruel order of nature. Thanks to its human-made landscape, a densely populated country such as the Netherlands can still remain habitable, livable, and rich. This was the message conveyed by the Dutch pavilion at the Hanover Fair 2002 and several governmental spatial planning documents. Others argue, however, that the extreme density of cars and the country's exploding energy use precisely and poignantly reveal the critical boundaries of a technological society for public health and the environment.[71] They point to a new kind of "emptiness" that has surfaced in a "full country."

As early as the 1930s the prominent Social Democrat Henri Polak described how major forces had deprived entire regions "of all that made them appealing, in ways that make them unrecognizable... crude banalities, stripped even of each memory of what once was." At mid-century, the author Nescio (the pen name of Jan Hendrik Frederik Grönloh) felt the reclaimed Wieringermeerpolder to be "barren throughout, a country of 'tractor devotees'"; only in villages in the southern part of the country did he still find a world without "steel and concrete" and without the familiar stores and commercial interests prevailing elsewhere."[72] In 1990 the novelist Willem van Toorn touched on these changes in the landscape and the loss of historical identity in his *Een leeg landschap* (*An Empty Landscape*).[73] Notably, supposedly benign ecological system building does not escape criticism as it potentially facilitates not only the migration of red dear and badgers, but also of ticks, foot-and-mouth disease, rabies, tuberculosis, and exogenous species that threaten indigenous species.

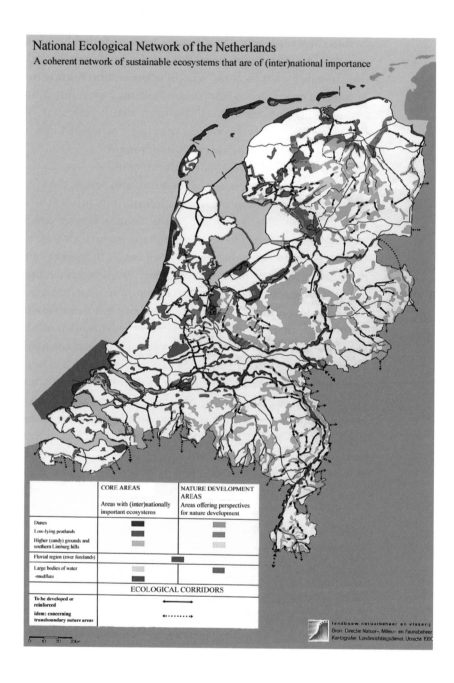

National Ecological Network of the Netherlands
A coherent network of sustainable ecosystems that are of (inter)national importance

	CORE AREAS	NATURE DEVELOPMENT AREAS
	Areas with (inter)nationally important ecosystems	Areas offering perspectives for nature development
Dunes	▆	▆
Low-lying peatlands	▆	▆
Higher (sandy) grounds and southern Limburg hills	▨	▨
Fluvial region (river forelands)	▆	
Large bodies of water	▨	▆
-mudflats	▆	
	ECOLOGICAL CORRIDORS	
To be developed or reinforced	←——→	
idem: concerning transboundary nature areas	◄·······►	

0 10 20 30km

landbouw, natuurbeheer en visserij
Bron: Directie Natuur-, Milieu- en Faunabeheer
Kartografie: Landinrichtingsdienst, Utrecht 1990

The growth of infrastructures has turned the Netherlands into an increasingly artificial country. The mounting pressure on the environment has also been integrated into the concept of networks. In 1990 the national government presented a first indication of the future basic ecological structure according to the Nature Policy Plan.

Infrastructures were not just contested a posteriori by outside observers, however. A critical analysis of the process of infrastructure change itself reveals multiple conflicts and choices that gave the Dutch networked nation its particular shape.

Social Latitude for Infrastructure Development

Several examples have already suggested that the re-creation of Dutch space and society by building infrastructures did not involve an automatic or unstoppable process, driven by an unambiguous logic of technological and societal progress. Instead, specific patterns of infrastructure development followed a host of choices involving a variety of social groups with divergent, if not opposing, objectives, in ever-changing political, economic, and technological contexts. Closer scrutiny of several cases of infrastructure change may convey a sense of the complex social dynamic that shaped the infrastructure integration of the Netherlands.

Analysis of the case of electricity supply and several other cases spotlights the various design choices made by "system builders," that is, private or public actors to whom infrastructure building was or became a core task.[74] Such choices were informed by specific goals, but these might change over time. Moreover, system builders often competed with each other on preferred modes of infrastructure development. In many cases it was not clear in advance which view, option, system builder, and system design would become dominant, in part because of continuous lobbying efforts by diverse players aimed at influencing system builder preferences and decision-making processes.

I also discuss the role of several sectoral or "institutional users" of infrastructures, such as the chemical industry, the food sector, and the Dutch military, as co-constructors of infrastructure and its uses. These actors contributed to the infrastructure integration of the Netherlands in at least two ways: they organized flows of people, things, energy, and information in existing infrastructure, and they did not refrain from building new infrastructure themselves if public infrastructure seemed insufficient.[75] They, too, were key players in the shaping of the networked nation.

Power games: A Second Look at Electricity Supply
The spatial scale increase of the electricity supply system in the Netherlands up to 1970 has often been described as a more or less linear process. Potential conflicts or alternative development trajectories remained hidden from view. Indeed, at first sight Dutch electrification appears to follow develop-

ments in leading countries such as the United States and Germany, confirming the standard historical narrative of electrification: electricity supply supposedly went from self-generation in individual factories to local public supply systems serving inner cities or villages, to systems covering entire urban or rural districts to national and ultimately international power pools.[76] This development seems quite universal and therefore subject to a quasi-autonomous technical-economic dynamic or, given the existence of exceptions, a 'soft-determinism.'[77] Several generations of Dutch historians of technology have followed a similar historiographical format, stressing or assuming the economic superiority of large-scale systems vis-à-vis their small-scale predecessors.[78] It was argued that "the introduction of larger and more efficient generation units and the beginnings of regional electricity supply" would have caused "many smaller power plants to close down"; "technological and economic developments eventually led to scale increase and the demise of a variety of systems."[79]

A number of studies, however, have called into question such quasi-linear development models, suggesting that they may result from a specific historiographical interest or bias. These studies instead advocate keeping an open eye for alternative development trajectories and not taking for granted technical or economic superiority of one of them in historical explanation, since the process might work the other way around: successful technological systems in time produce technological and economic advantages that they did not yet possess in their early development stages.[80] Indeed, an international comparison based on more recent literature on electrification history confirms that specific natural, social, and political circumstances inspired different electrification trajectories in various countries. This applies not only to development pace or spatial design, but also to the successive phases in the process of scale increase. In some countries scaling up met limited success. For example, after World War II Danish engineers deemed it nonsensical to deploy the standard ideology of scale increase and its associated technologies in Greenland. Its electrification took on the shape of many local and sometimes district systems that were not interconnected.[81] In other cases, supposedly outdated "development phases" remained important or even dominant. In Norway, electrically the most developed country in the world with a per capita electricity use twice that of the United States, decentralized systems remained dominant even after the national government set up a national system based on large-scale hydropower generation in the 1980s.[82] The argument also pertains to the first phase of industrial self-generation. Even in 1970, when large-scale thinking about electricity supply climaxed, industrial generation was responsible for no less than a third of the annual electricity production in advanced economies such as Germany and Belgium.[83]

International comparison also reveals how in various countries different actors managed to gain control of the building of electricity systems, with implications for national patterns of electrification. According to existing scholarship, in the United States small systems were brushed aside by a powerful alliance of expansive private electricity companies operating on a grow-and-build ideology; a strong electro-technical industry, which put all its cards on developing ever larger-scale equipment; and state governments, which granted monopolies to electric utilities in exchange for extending electrification of the entire state instead of merely its most densely populated—and lucrative—areas.[84] Systems for electricity supply developed into statewide systems and later interconnected individual state systems, but further nationwide integration was not deemed interesting. According to the U.S. Department of Energy, a U.S. national grid still does not exist, since the eastern, western, and Texas interconnected systems are poorly interlinked.[85] By contrast, in the early 1920s the French and British national governments set up state companies to enforce the construction of national power grids. These governments used their financial and legal power to build a national grid and then determined which power plants could be hooked up to it. A similar thing occurred in Sweden. As a result, these countries were quick to create national power pools.[86]

Denmark opted for a third alternative, and stands as the prime example of the temporally stable coexistence of large-scale and small-scale systems.[87] In the Danish electricity playing field, higher levels of government and multinational companies hardly gained a foothold. Instead, electricity supply was predominantly claimed and organized by municipal utilities and consumer-owned rural cooperative societies. Already in the 1920s, some of these collaborated in the construction of state-of-the-art large-scale power pools whose high-voltage networks soon covered large parts of the country. Until the 1960s, however, decentralized systems continued to thrive, even within the supply areas of these large-scale systems. Alongside the urban systems of most municipalities exist several hundred very small, consumer-owned local village systems. With sizes varying from a thousand down to a dozen connections, these proved technologically, economically, and socially feasible: time and again, decentralized expansion of production capacity was preferred to connecting to an external power grid. Although larger systems enjoyed economies of scale and a superior load factor, decentralized systems had advantages of their own: small direct-current systems might turn off their machinery at night and use batteries instead. They could also opt for power generated by wind, diesel, peat or steam, depending on fuel prices. Decentralized municipal systems often sold waste heat as district heating, allowing an energy efficiency and income that would not be possible when purchasing power from an external grid. Most important, decentralized systems did not require investment in costly power grids, the costs of which often made up well over half of the KWh price.[88]

With these examples in mind, a reexamination of Dutch electrical history indicates that here, too, social and political dynamics produced the particular Dutch electrification trajectory. In retrospect the 1910s and 1920s were decisive in producing a playing field that proved remarkable stable until the 1980s. At the beginning of this period, a number of players tried to lay claim to the future electrification of the Netherlands, including existing private and municipal electric power companies and industrial power producers, as well as potential new entrants—the state and provincial governments and even international organizations, which discussed the option of supranationally financed and owned power systems around 1930, as they would again in the late 1940s.[89] Yet in the 1920s it became clear that provincial electricity companies had become the dominant system builders.[90] It was they who engineered the move from local to district systems in the Netherlands and had electrified the country by 1930. It was they who would organize, establish and control the national power grid after World War II.

These provincial utility companies were set up by provincial governments, which unlike municipalities had little history of entrepreneurship. For them, profit was not the top priority, as it had been for many municipal councils. The provincial government of North Brabant, one of first to take on the electricity supply issue, explicitly rejected the profit argument. Instead, its rationale for engagement in electrification was the political objective of countering the countryside's depopulation. From this angle, rural development by means of electrifying the *entire* province became a preeminent task of provincial government. The province of Groningen took the same step to raise overall prosperity in its countryside, thereby ignoring an alternative scheme that its advisory commission had presented as economically superior. In the early 1920s most other Dutch provinces copied the model by setting up electricity companies and electrifying the entire province, typically using 10 kV cable distribution networks.

The emergence of provincial power companies as dominant system builders encountered resistance from the other candidates for electrification, resulting in fierce political and legal struggles. For example, the national government repeatedly tried to interfere in electricity supply to bring about faster electrification. When the first provincial companies were established, it feared—in retrospect correctly—that provincial boundaries would function as "electric barriers."[91] After World War I, top-down national electrification—as in France, England, and Sweden—was almost achieved.

Criticizing provincial utilities for insufficient cooperation, the Lely Commission, established in 1919, proposed to establish a national power company to build and run a national power pool in the national economy's interest (see figure on page 76). The government-proposed legislation to achieve this was, however, rejected by Parliament. Following intensive lobbying by municipalities and their organizations, Parliament decided that the bill's weak financial underpinnings did not justify undermining the autonomy of lower governments and their power systems. Several subsequent attempts to pass national legislation were rejected or were withdrawn in the last moment, and the Dutch government never became an electricity system builder. Accordingly, the proposed national power pool did not materialize.

At the same time, the provincial governments were successful in discouraging municipal and private system builders from exploiting smaller-scale systems. This was rather a matter of exercising political authority than fair competition based on technological or economic performance. An important political instrument was a kind of provincial concession system. The province of North Brabant, for example, obliged potential newcomers to obtain a concession to establish and exploit a new electricity company in 1912. Municipal and private players did not plainly accept this move and accused the provincial government of trying to monopolize the power business and of violating municipal autonomy and freedom of enterprise. The provincial government plainly responded that its taking over tasks from a lower level of government was not in violation of the constitution.[92] The municipalities of Breda and 's-Hertogenbosch, which both wanted to set up their own electricity supply systems, then formed a coalition and gained further support of local industrialists, the chambers of commerce of both towns, the Association of Dutch Municipalities, and twenty other municipalities nationwide. Jointly they lobbied the national government to reject the provincial legislation involved. The provincial government mobilized support as well, including 158 rural municipalities. The Minister of Interior Affairs and the Senate were inclined to support the municipalities, but the responsible minister of public works wished to support the provincial practice. As a compromise the provincial concession system was allowed, but municipalities were given the possibility to request the Crown for release. When the 's-Hertogenbosch municipality tried to do so, however, its request was rejected. A new appeal gained the support of the Council of State, the highest body for appeals against the state—but then, unlike now, it had only an advisory status. In the end, the appeal was again rejected.[93]

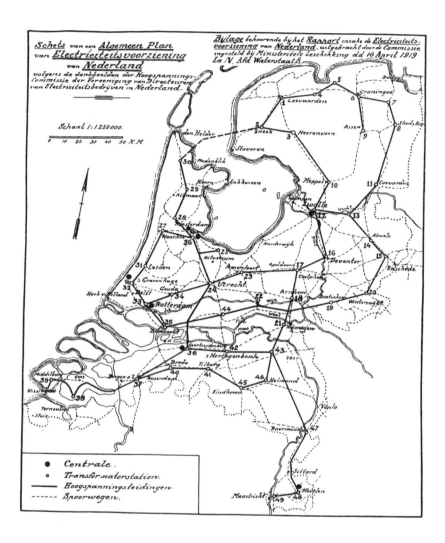

Right after the First World War the government proposed to make the building of a national electricity supply system a state policy: a state power company would put in a nationwide grid. The proposal was specified in a report that also contained the map shown here. It shows the proposed electricity supply system with power stations, a high-tension network, electricity substations and railways. But the plan was voted down by Parliament. Provincial governments turned out to be main players in the country's electrification.

Private companies also failed to make a dent in the emerging power monopoly of the provinces. In North Brabant, for instance, the N.V. Peel-centrale was forbidden by the province to cross provincial roads. Uncertain about its future prospects when unable to expand, this power company under-invested in its system. Technological and economic inferiority became a self-fulfilling prophecy. It lost customers and eventually went bankrupt. Similar circumstances doomed small companies in other parts of the country.[94]

One further group of potential system builders, the industries that generated their own power, advocated the possibility of very cheap decentralized electricity production in factory plants. Many factories already had power generators or produced a surplus of steam, but electricity generation would only be profitable if there was a market for the electricity generated. In 1925 A. W. Hellemans, the spokesman for the power-generating industries, proposed a national electrification scheme powered by industrial companies, even if it would mean "decentralization of power production in accordance with the power sources of Dutch industry."[95] In Hellemans's view, a national power pool was in fact a huge waste of national capital. Government support was needed to break the provincial and local hegemony, but for the time being the main government policy was to support the provinces, which represented at least some form of scale increase.

These struggles had profound effects on the development trajectory of Dutch electrification. They facilitated the rapid move from local to provincial district systems in the interwar period. By the 1930s, provincial utilities had produced a geographical density of power networks as well as a degree of centralization of power production that were exceptional in international perspective, even when compared to Germany and the United States.[96] At the same time, however, the dominant position and preferences of the provincial utilities thwarted further scale increase to a national system. While interconnected power pools were being set up in England, France, Germany, the United States, Denmark, and Sweden, the Dutch electrification process lingered at the stage of isolated provincial systems. National government plans for a national grid had been warded off, and provincial utilities themselves did not agree on the necessity of a common power grid. Some argued the economic advantages of interconnecting Dutch power plants in one power pool, but most utilities had doubts. For example, the influential J. C. van Staveren, whose office at the Dutch Association of Electricity Company Managers (Vereniging van Directeuren van Elektriciteitsbedrijven in Nederland) did calculations on this issue, believed that further scale increase to a national system would not be profitable because electricity production had already been centralized. Other economic benefits of a national system, such as higher load factors or savings in backup machinery, did not outweigh the huge investment in a national grid.[97]

World War II and increasing government pressure, including implicit threats of nationalization, finally triggered a utility consensus about the desirability of a national grid. Still this involved a redefinition of the purpose of such a grid, from economic benefits to increased reliability. In the postwar context, the possibility of drawing power from the grid in case of breakdown justified "the economic drawbacks" of the scheme.[98] The provincial utilities now decided to collectively build a national grid, the design of which reflected the lead motive of reliability. Outside the Netherlands, many grids were initially built in a star-shaped form, so as to cover the country with a minimum investment in power lines. In the Netherlands, by contrast, the national grid took the form of two connected rings allowing each major transformer station to draw in power from two sides (see map 2.1). The uses of the national grid reflected its purpose as well; as noted, for decades power exchanges remained limited to emergency and other occasional deliveries. The grid was not used as a true power pool until the neoliberal era.[99]

Finally, the mutual dependency of infrastructure development and social dynamics reemerged in the closing decades of the twentieth century. Influenced by the 1972 Club of Rome report *Limits to Growth*, the energy crises, and rising electricity costs, and an emerging neoliberal ideology, the coalition between provincial power companies and the Dutch government ended. In its 1974 and 1979 energy white papers, the government argued for electricity conservation and for strengthening its control over the electricity sector through covenants and legislation. The purpose of the electricity acts of 1989 and 1998 was increased efficiency through market liberalization and increased competition, which should be achieved by splitting up power companies by enforcing a radical separation of electricity generation and distribution. The government also now supported decentralized industrial power generation. Industrial proponents of this effort, formerly "voices in the wilderness," now actively contributed to policy making. Decentralized systems were allowed to supply electricity to the public system, much against the electricity generation companies' wishes: between 1989 and 1994, the share of current produced in decentralized systems increased from 14 percent to 22 percent.[100]

We see this change reflected in changing views of the infrastructure future. The notion of scale increase surfaced in Germany around World War I, in the context of the option to integrate remote hydropower and brown coal plants in the supply structure, and reached a high in the 1960s with the generation potential promised by nuclear plants and fusion reactors. But alongside this line of thinking a view of the future radically opposed to scale increase has emerged. It foresees an era of *micro-power:* by 2050 the electricity supply would be based on decentralized generation in millions of micro-turbines, solar cell installations, and fuel cells that function cleaner and more efficiently, cheaply, and reliably than large-scale systems.[101] Companies such as Siemens Westinghouse, Asea Brown Boveri, Shell, British Petroleum, and even Bayer and Akzo today invest in research into decentralized generation technologies with an eye to tapping future power markets. Whether this vision will become a reality is quite another matter; as in the first half of the twentieth century, the route to this outcome is highly contested.

Patterns and Choices in Infrastructure Development

The case of electricity supply networks illustrates several general features of infrastructure change. First, patterns of infrastructure development are not fixed in advance. The seemingly linear development trajectory of the Dutch electricity supply resulted not from an autonomous technical-economical logic, but from social and political dynamics and dominant system builder preferences and choices. The open character of infrastructure change is all the more visible when expected infrastructure developments do not materialize. In electricity supply, isolated power systems were carried far into an era where interconnected power pools were thought to represent the state of the art. Television presents a similar case on a European scale. Dutch television had a nationwide infrastructure from the start, and in the early 1950s a European system was envisioned. The European Broadcasting Union, founded by twenty-three national public broadcasters, set up Eurovision to develop an interconnected broadcasting system and prepare for the anticipated shift toward European programming.

Yet this initiative almost completely lost out to national programming—with the exception of an occasional coronation, soccer match, or song festival (see figure on page 80).[102] In aviation, the expected increase in density hardly materialized, despite Interwar visions that airplanes would soon be more important than automobiles.[103] Some trends were even reversed. In electricity supply, an era of expansion, scale increase, and closing down of smaller power stations was followed by a stagnation in the growth of power plant sizes and an upsurge in decentralized generation units. Railroads went from booming business to stagnation and ultimately declined by 50 percent in the period from 1930 to 1990. Later the railroads recovered some of their earlier dynamics.

Three aspects of the social dynamics affecting the infrastructure development trajectory deserve closer examination. First, frequently infrastructure development followed concerns and goals of the players involved, rather

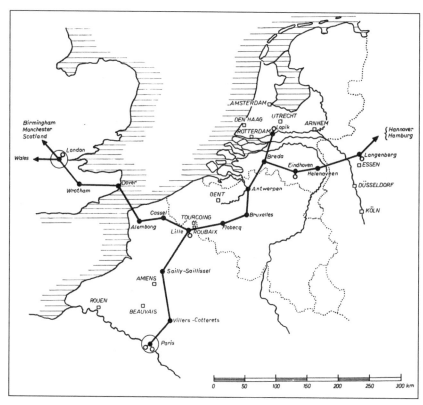

The coronation festivities of Queen Elizabeth II in London in 1953 were an opportunity for the first practical experiments with international television broadcasts. To transmit images, an improvised network of television stations (the circles on the map) was linked to transmitters and receivers (black dots on the map); telephone lines were used for sound. In the 1950s a European network of television connections was put in. In the early years there was virtually no European programming. Television would continue to be nationally and regionally oriented for some years to come—except for soccer ("football") matches and the Eurovision Song Contest.

than technological developmental laws that circulated widely in international engineering journals. Second, if the players involved held opposing goals, such goals had to be prioritized in processes of negotiation and sometimes open conflict. Here, strategy and power games became important. Finally, the design process itself might be adapted so as to align and accommodate multiple concerns and goals.

As for the concerns and goals of system builders, in the case of electricity supply they displayed a considerable variety. The goals of private, municipal, and provincial utilities and industrial generators included profit making, countering the urbanization process, improving the reliability of the supply, and fending off government interference. Goals could also vary over time. For instance, the national government pursued national economic development in the 1910s, reduced dependency on foreign energy markets in the 1970s, and market liberalization and CO_2 reduction in the 1990s.

Transport and telecommunication infrastructure developments reflected a similarly large variety of goals. The national government, for example, engaged in massive canal and railroad construction in the first half of the nineteenth century for reasons of economic policy. Infrastructure was seen as a "miracle drug" to restore the lost economic structure of the Golden Age. And the government's efforts to link the country's large ports to the German hinterland proved successful: for some decades Amsterdam became again a major European marketplace, notably for colonial products such as sugar, coffee, and indigo.[104] In the final decades of the twentieth century, the government pursued a comparable strategy using infrastructure connections to boost the position of the so-called "mainports" of Amsterdam Schiphol Airport and the port of Rotterdam. Recently this concept was expanded with the definition of "greenports" and "brainports," economic centers of agricultural and knowledge production for export, which are internationally connected to transport and communication infrastructures.[105]

By contrast, the same government invested in nationally integrated highway, railroad, telegraph, and telephone systems in part to create a national Dutch space. In the early nineteenth century, the national system of paved roads was meant to demonstrate that the provinces were interconnected.[106] In 1895 a Dutch member of Parliament, M. Tydeman, was partly successful in convincing the national government why the state should back the expansion of telephony, arguing that it was "such a preeminent means for reducing and removing the drawbacks of remoteness because it does away with distances; because it brings all corners of our country in direct contact with the centers of traffic; because, in combination with local railroads and streetcar lines, it will encourage those who are diligent and well-to-do to go living in the countryside and it will facilitate the establishment of factories and businesses, also in faraway places."[107] During much of the twentieth century the improvement of major roads was legitimized by the needs of automobile traffic and people's desire for increased mobility. After 1970, the building of freeways was linked to the policy of opening up the country nationally and regionally.[108]

There was a host of relevant motives in the field of water management: protecting the water supply, protection against flood disasters, and the exploitation of pristine areas. Frequently individual waterworks served several goals simultaneously, as exemplified by the Zuiderzee Works (1922–1975), which aimed at the closure and partial reclamation of the large inland sea, the Zuiderzee. In 1932 this inner sea was separated from the North Sea by a thirty kilometers long dam, which would remain the pride of Dutch civil engineering for decades to come. This dam served multiple goals: coastal protection, land reclamation and cultivation, opening up the northern Netherlands by means of a highway atop the dam, creating a fresh water buffer against seeping saltwater, ensuring a freshwater supply to cities and agriculture, boosting engineering pride, and employment relief in the context of the Great Depression.[109] Finally, the 1990 plan for a National Ecological Network added supporting biodiversity to the national government's impressive list of goals. Ten years later, however, its ecological priorities had shifted again: breaking with a purely conservationist agenda, the wet and green corridors would serve not only the circulation of red deer and otters, but also the leisure of sporty urban residents enjoying canoeing or bicycle-riding. From then on nature was explicitly developed for people's pleasure as well.[110]

Power Struggles over System Building

Often, infrastructure development involved many potential system builders and regulatory bodies with different, and sometimes contradictory, goals. So it is important to establish whose concerns gained priority in decision making processes, why, and how these affected infrastructure development trajectories. Diverging goals often led to negotiations that could erupt in open, sometimes public, conflict.

As described above, national, provincial, and local governments; private electricity companies; and industrialists competed for the opportunity to electrify the Netherlands. The prioritization of goals was settled in a number of power struggles, the result of which could not be anticipated. A vote in Parliament prevented the establishment of a state company and the early implementation of a national power pool. Provinces blocked municipal and private system expansion across provincial roads. The Crown rejected several appeals and chose to let this happen. In short, the prolonged dominance of the provincial companies in electricity supply rested on legal power plays and struggles.

A similar struggle as to who was going to exploit the natural gas reserves took place largely behind closed doors. Key players in the negotiations included the national government; the state-owned steel works Royal Hoogovens (currently Corus IJmuiden); the chemical company Royal DSM, which owned large gas distribution networks; and the oil companies Royal Dutch/Shell and Esso, which cooperated in Nederlandse Aardolie

This photo, taken around 1970, shows the control room of Gasunie in Groningen, where the central distribution of Dutch natural gas was coordinated. The control panel shows the network of pipelines across the Netherlands.

Maatschapij (NAM, established 1947). Negotiations were quite arduous. Several political parties desired full nationalization to secure the national interests. Conversely, other parties realized that state involvement at home would not help the Royal Dutch/Shell when negotiating for concessions in OPEC countries. Ultimately a structure was negotiated in which the Dutch state participated heavily but quite invisibly in the Dutch gas system: the oil companies acted as gas producer, and would own half of the distribution company, Gasunie (established 1963). The state-owned Royal DSM owned the other half. A separate company was set up to distribute the revenues; the state share would increase with increasing gas sales. In the 1970s the state treasury received some 70 percent of the revenues.[111]

The field of broadcasting, by contrast, has been characterized by open and highly visible conflict as well as the use of political, public, technological, and police force, from its very beginnings to the early twenty-first century. Dutch radio broadcasting was born in illegal practices.[112] In 1916 radio amateurs organized in the Dutch Association for Radio Telegraphy (Nederlandsche Vereeniging voor Radiotelegrafie) to obtain legal access to the ether. A decade later another conflict followed. The first national broadcaster, the Hilversum Draadloze Omroep (later the General Radio Broadcasting Association, Algemene Vereniging Radio-Omroep, or AVRO), which wanted to broadcast entertainment with a broad secular appeal, was given less broadcast time after Catholic, Protestant, and labor broadcasters gained access to the ether. The AVRO gathered 400,000 signatures in an attempt to maintain its broadcast time, but the confessional government legally confirmed the compartmentalized organization along socio-political lines of broadcasting in 1930. This triggered one of the largest social protests of the interwar period: as many as 130,000 people traveled to the seat of government in The Hague to demonstrate. The government did not yield. For many decades, successive confessional cabinets used their political power to control Dutch radio programming and, later on, television, to ensure that it conformed to Christian norms and not "moral decay."[113]

In the postwar era, resistance to this control of the radio waves took on another form. The general public preferred entertainment to "high culture" and moral values, and massively tuned in to the so-called pirate stations. A Radio Control Service had been set up in 1927 to fight this decentralization of the broadcast system. In response, and inspired by the Danish *Radio Merkur*, some pirate stations started broadcasting from ships anchored in the North Sea. In 1960 Radio Veronica began broadcasting from a ship anchored just outside Dutch territorial waters. In a similar way, the Reclame Exploitatie Maatschappij (REM) started TV transmission from a former oil platform. By November 1964, 350,000 Dutch television sets had special REM antennas.[114] The Dutch government, citing the "war against onrushing popular culture," passed an anti-REM emergency act enabling the Royal Navy to stop this form of "illegal" broadcasting by force.[115] It also signed the Strasbourg Treaty of 1965, also known as "anti-piracy treaty," in which several European countries agreed to forbid broadcasting from international waters. The turmoil was complete when the Dutch cabinet was brought down by disagreement among coalition partners on allowing commercial broadcasting in the public system; the Strasbourg Treaty was not ratified until 1973. In the end, maritime pirate broadcasters did close down, but some were reborn as legal commercial broadcasters within the public broadcasting system.

The struggle between illegal and legal broadcasting continued, however. The Dutch government banned foreign commercial TV programming geared to the Dutch market by forbidding local cable companies from transmitting such programming. Not until the late 1980s did the European Union, the European Court, and the Dutch Council of State permit the legalization of commercial television outside the public system. In the next ten years, the market share of the Dutch "public stations" plunged from 88 percent to 35 percent.[116] In radio broadcasting, some sort of technological arms race occurred between detection networks and illegal transmitters using unmanned studios, mobile studios, and small relay stations tapping electricity from lampposts.[117] By the early 1980s a study by the Scientific Council for Government Policy estimated the number of Dutch ether pirates at between 3,000 and 4,000; the number of TV pirate stations had peaked at around sixty. The public had a generally positive view of radio piracy; some 40 percent regularly tuned in (some 17 percent to TV pirate stations).[118] One estimate from the beginning of the twenty-first century is that there are thousands of active pirates. In light of the government policy for redistribution of frequencies in an "ether auction" in 2003, these pirates were attacked with renewed urgency. "Operation ether-flash," which involved a nationwide detection network, supposedly reduced the number of pirates by two-thirds, but it also produced an "operation counterflash": ether pirates established their own Association for Free Radio (Vereniging Vrije Radio) to lobby for a change in government policy. Meanwhile, many pirates remain active, and the successor of the Radio Control Service still takes out pirates on a daily basis.[119]

If the national government acts as system builder, Parliament is an important arena of conflict and contestation, as when it rejected the Lely Commission plan for a national electric power pool. At times, such conflicts over infrastructure between different state bodies have served as negotiating sites for the structure of the Dutch political system itself. In the first half of the nineteenth century, King Willem I repeatedly and illegally allocated money for building canals, against the wishes of Parliament. His strategy, as one historian put it, stood midway between "fraud" and "creative bookkeeping." The canals were dug, but at the cost of violating the constitution and, after it was disclosed, triggering furious reactions from both Parliament and the public. This breach of confidence was cited as a major reason for the Belgian secession (1830–1839) and changes of the Dutch constitution in 1840.[120] Since then, ministers have cosigned royal decisions.

The 1838 decision to build an Amsterdam–Cologne railroad line (called the Dutch Iron Rhine) was no less controversial. Parliament rejected legislation whereby the state would build the line 46–2, but the King went ahead and carried out the plan by royal decree, which did not require parliamentary consent. A large government loan was made available. Later it turned out this money was partly used to cover up financial holes in the state budget, so as to hide them from parliamentary oversight.[121] In these incidents the organization and make-up of the Dutch state system was at stake; as a result, the political rules were changed and tightened, and the king was relieved of virtually all governing functions.

The functioning of Dutch parliamentary democracy was also at stake in one of the major infrastructure-related controversies of recent times, the decision process involving construction of the Betuwe freight-only railroad line.

This railway line connecting the Rotterdam harbor to Germany emerged on the political agenda in the late 1980s; it opened for business on June 16, 2007. By the mid-1990s it was meeting with widespread resistance from environmental and residential groups and local governments. The Betuweroute Commission, established by two ministries in 1994 to investigate the planning process of the line, mentioned a "a referendum-like atmosphere.... You either favor it or you don't."[122] Proponents actively tried to keep criticism under wraps, while opponents used methods both legal (such as filing over 130 actions against the state) and illegal (such as sit-ins and sabotage) to derail the plan.

As for the relationship between government and Parliament, several reports of the Netherlands Court of Audit criticized the quality of information that served as the basis of the decision-making process as well as the subsequent budget control. In 2003 Parliament acknowledged that it had completely lost control of the decisionmaking process and the budget. The same was true for the high-speed line, which was being built simultaneously. Parliament felt it had been misinformed and bypassed by the government in major decisions; indeed, its role was reduced to debating and facilitating the implementation phase, making at best minor changes. A parliamentary inquiry followed. The investigation commission identified several mechanisms that had kept Parliament on the sidelines in key decisions. Also, progress reports provided by the government contained too many details and obscured relevant cost management information. The commission proposed a number of changes, which were accepted by the government. In the wake of this critique, the government also postponed, and later canceled, the construction of the planned northbound "Zuiderzee" high-speed line.[123]

The case of the Betuweroute leads to a final observation on prioritizing goals in infrastructure development. System builders and government concerns were systematically influenced by various lobby groups. According to opponents of the Betuweroute, the project was placed on the political agenda by a powerful, politically well-connected Rotterdam harbor lobby. The lobbyists' clients were the sole beneficiaries of the new line, but they wanted the taxpayers to foot the bill for the investment. The group included major Rotterdam interests such as Europe's largest container transshipment company, Europe Combined Terminals (ECT); several large maritime shipping companies and the Municipal Port Authority Rotterdam (Gemeentelijk Havenbedrijf). These teamed up with the Dutch Railroads (Nederlandse Spoorwegen), the chambers of commerce in the eastern Betuwe region, the provinces of South Holland and Gelderland, and the transport organizations. A dedicated transport lobby, Holland International Distribution Council

(1987), was founded to influence politicians. By the mid-1990s a network analysis completed by opponents revealed an intimate relationship between this lobby and relevant national government bodies and commissions, including the transport minister: a mere ten persons performed forty-three functions in nine key organizations and government commissions involved in the Betuweroute project.[124] The report of the parliamentary investigation commission, drawing on interviews of several key players, reliably confirmed the role of informal lobbying by Rotterdam interests in placing these railway projects on the political agenda.[125]

Although such practices seemed outrageous to opponents, to proponents they seem a natural—even necessary—element in every major achievement. As one ECT director, Gerrit Wormmeester, told the parliamentary Commission of Inquiry, "You have probably never been an entrepreneur…. A country without infrastructure has no future…. Heading a company with significant interests in this matter, I do my very best to lobby." In fact, Transport Minister Nelie Kroes encouraged and subsidized the establishment of the Holland International Distribution Council, a lobbying organization, stating that "this makes it much easier to canalize the multiple transport interests and get them represented in Parliament and the ministries."[126] To her, effective lobbying served a political need. Moreover, in this case policy makers tried to align commercial interests and political forces with a vision of sustainable economic growth that should truly benefit the country; developed in the late 1980s, this vision aimed to combine attracting international trade flows to the Rotterdam harbor (as well as transshipment terminals planned at the Dutch-German border) with using clean railways, not polluting trucks, for transport to the hinterland. Lobbying, aligning political, commercial, and societal interests, stressing (but not critically examining) the financial benefits, and keeping Parliament at a distance were seen as important strategies in getting such large projects done.

Lobbying is hardly a new strategy of course. The history of infrastructure development is full of it. To mention just a few instances: When, during the French Occupation, the emperor Napoleon visited the city of 's-Hertogenbosch in 1810, the local elites seized the opportunity to argue for improving their trade position. They lobbied in particular for the construction of a canal to the south and main roads to Antwerp, Nijmegen, and Luik.[127] Around the turn of the twentieth century, municipalities did a great deal of lobbying to attract local railways or inter-urban tram lines, then perceived as major levers of economic development, to their towns or villages. And in the 1920s another lobby of user groups of car owners, including the touring associations the General Netherlands Bicyclists Association (Algemene Nederlandsche Wiel-

rijders Bond, ANWB) and the Royal Automobile Club (Koninklijke Nederlandse Automobiel Club, KNAC), and oil companies managed to put speeding up road construction on the national government's agenda. Lobbying,
then, seems a normal aspect of infrastructure development.

Aligning Interests in Infrastructure Design

Opposing interests in infrastructure development could be resolved in several
ways. In the cases of electricity supply and broadcasting, various options were
possible; after some struggle, one was ultimately chosen. In many other infrastructure decisionmaking processes specific options were either fully adopted
or fully rejected. Ideas that were fully rejected include building a second national
airport in the planned Markerwaardpolder (which local resistance prevented
from ever being reclaimed), reclamation of the Wadden Sea in the 1970s, and
construction of new nuclear plants in the 1980s. Especially when technologies
were exposed to public controversy the resulting debate seemed to culminate
in a standoff between advocates and opponents. Here, power struggles were a
key mechanism influencing the infrastructure development trajectory.

Often, however, choices between alternatives were hardly straightforward
because project aims and technological features might change in the course
of the decisionmaking process. The introduction of wind turbines is a relevant example here. In the 1970s wind energy was mainly seen as a small-scale
alternative to large-scale power generation. However, two decades later wind
turbines were deployed in arrays of wind turbines as large-scale generation
units in large-scale electricity supply; following the Danish example, the
Netherlands is currently building it own wind power plant in the North Sea.
It was not just a matter of a choice between large-scale and small-scale. Single wind turbines, too, could be connected to the existing system. This gave
rise to a hybrid system, in which large-scale and small-scale generation units
operated side by side.[128]

Moreover, it is possible to adapt aims and design criteria of infrastructures so as to accommodate the interests of many social groups. An exemplary case of aligning opposing interests by design—by "technical fix"—is
the construction of the national freshwater distribution system in the 1940s,
'50s and '60s.[129] As noted, by the early 1970s the national water-management
system enabled the manipulation of water flows in the main Dutch water
arteries. To understand how the system was planned it is necessary to understand Dutch river geography, and the Rhine delta in particular. Rhine water
that enters the Netherlands from Germany can follow three possible routes
to the sea: (1) southwest, via the Waal River to the Hollands Diep, which is
also fed by Meuse River waters, to Haringvliet and the North Sea; (2) west

via the Lower Rhine and Lek rivers to Rotterdam's artificial river mouth, the New Waterway; and (3) north, via the IJssel River to the dammed IJsselmeer, the large inland sea. From there, discharge sluices in the dam feed it into the Wadden Sea and the North Sea.

As noted, the IJsselmeer was separated from the sea in 1932; by 1936 its water was declared fresh. Soon an emerging struggle over use of the newly fresh water basin became visible. The large city of Amsterdam, using the lake as a sewage outlet, now also planned to take its drinking water from the lake. Meanwhile, the farmers farming the new polders surrounding the lake, which were former sea bottoms impregnated with salt, demanded huge freshwater intakes to flush out saltwater and irrigate fields. Eel fisheries also desired a large freshwater discharge in spring through dam sluices to attract elvers in the lake. In view of these multiple claims on a still limited supply of fresh water, in 1940 the director of Amsterdam's municipal water works concluded, "The future manager of the IJsselmeer will bear a heavy burden."[130]

The discussion reached a national level when navigation interests became involved. The navigability of the IJssel River feeding the IJsselmeer was a recurring problem during periods of drought. The river was of major economic importance for the northern and eastern regions of the country, and in 1940, before the German occupation, plans for its canalization were submitted to Parliament. Canalization would enable artificially increased and manageable water levels to benefit IJssel navigation. Moreover, an artificially increased IJssel water level would push more Rhine water westward into the Lower Rhine–Lek–New Waterway system to benefit greenhouse agriculture in the Westland, the area roughly between Rotterdam and The Hague, the home of Dutch intensive agriculture. Here, agriculture needed freshwater to setoff salt intrusion from the coast. Pushing Rhine water westwards, however, implies reduced northbound freshwater flows through the IJssel, which would endanger the freshwater supply of the northern agricultural regions, the Amsterdam municipal waterworks, and the eel fisheries. Navigation, drinking water, western and northern agriculture, and fishery interests were at loggerheads in an increasingly fierce debate on the national distribution of fresh water.

This dilemma did not, however, result in a power struggle in which some won and others lost. Instead, the national civil engineering agency, the Rijkswaterstaat, developed a plan that seemed to satisfy all interests. In a first innovative step, in late 1940, the agency's director-general, L. R. Wentholt, defined a new concept of a national water management plan and listed twenty criteria of national water circulation that had to be satisfied. Thinking in terms of solving a national problem, the Rijkswaterstaat's chief engineers proposed several measures. The IJssel canalization was replaced with a canalization of the Lower Rhine. Depending on need, a weir at the IJssel–Lower Rhine junction could divert Rhine water (all of it if necessary) to the north to ensure adequate water for shipping, drinking water, agriculture, and fishing interests. Canalization would also ensure navigability of the Lower Rhine itself. In addition, the freshwater needs of Westland agriculture were served from another source and by another set of works: closing off the estuaries of the southwestern delta, which had been studied by the agency since the 1930s in the context of flood control. The system was completed with the Haringvliet dam and sluices (1971), closing off the upper estuary in the delta, creating new freshwater basins fed by Meuse and Waal water, and diverting these to the Westland where they entered the sea through Rotterdam's New Waterway.

Such design solutions to deal with conflicting interests, often at considerable financial expense, sometimes led to the resolution of conflicts and successful infrastructure construction. In other cases, however, conflicts persisted. The Betuweroute is a case in point. Many attempts were made to accommodate conflicting interests by design change, both in the initial government plans and during their repeated treatment in Parliament. Much of the opposition came from NIMBY ("Not in my backyard") and environmentalist groups. In response to their concerns, the major part of the new railroad was located directly alongside existing freeways, thus minimizing the need to open new land to transport needs. The track is flanked by noise reduction walls. The stretches that cut through green areas were upgraded with five tunnels totaling fifteen kilometers in length, and 190 wildlife passages were built.

Despite such measures the controversy did not go away, and opponents still perceive the project as mainly a project of Rotterdam capitalists at the expense of Dutch society and nature. Moreover, vast cost overruns of 100 percent were increasingly mentioned as a main point of criticism. The 2004 parliamentary inquiry suggests that opponents rightly criticize the poor financial planning of the project, but fail to acknowledge that subsequent budget overruns were the price tag for design changes to accommodate NIMBY and environmental concerns, and thus were the result of a democratic design process. These measures accounted for approximately half of the total overruns; the other half stemmed mainly from inflation.[131]

Finally, the example of automobile traffic illustrates how major problems may persist despite rules and designs meant to address these problems, and despite broad societal acceptance of a technology. Opponents of car traffic have repeatedly referred to the "slaughter on Dutch roads" in the wake of the success of road transportation.[132] Automobiles became the dominant transportation mode as part of a complex embedding process. Automobile organizations and the car industry utilized the leeway granted to them for introducing their own solutions to safety issue, including safer car designs, infrastructure for the separation of traffic flows, driving tests, and far-reaching control of bicyclists' and pedestrians' use of public ways: streets gradually became off-limits for pedestrians and playing children.[133] Still, the annual number of traffic fatalities in the Netherlands continued to increase, to 3,000 annually by 1970—one of the major causes of unnatural death. This figure has since been reduced to under 1,000 deaths annually, plus 15,000 serious injuries and countless minor injuries. Although relative to population these figures are low in international perspective—in Europe only Malta has a better score—Dutch road traffic still accounts for a number of deaths and injured that would seem to be completely unacceptable for any other infrastructure. It still is the main cause of death after suicide for those fifteen to twenty-four years of age.[134]

Using and Co-constructing Infrastructure

System builders and lobby groups were not the only actors giving shape and meaning to the infrastructure integration of the Netherlands. Even after infrastructure was built, users often had considerable room to maneuver, regularly interpreting and using systems in ways unforeseen by system builders. Often users, not system builders, created flows through systems; road transport and telephony are familiar examples. Sometimes the role of users was not limited to mobilizing and using systems for their purposes; if infrastructures were deemed insufficient, users or their representatives might engage in the system building process on their own, either through lobbying or by taking construction into their own hands.[135] One example of a user organization interfering in the infrastructure design process is the Dutch bicyclists association, the ANWB. Representing both bicyclists and car users, in the 1920s the ANWB was a pivotal force in the lobby that put accelerated road construction on the national agenda and arranged for its financing, and it took matters into its own hands by purchasing a motor roller for building bike paths.[136] Another such example is housewives and their organizations who expanded electrical infrastructure in private homes. Before the era of electric sockets, each house got only one hookup for current. Housewives introduced adapters to hook up electric irons, the first widely used electric household appliance. Later, organizations such as the Dutch Women's Electricity Association (Nederlandse Vrouwen Elektriciteits Vereeniging), established 1932, and local women's advisory committees worked for the proper distribution of electricity and gas in homes.[137]

Most user studies are concerned with the use of cars, telephones, arc lights, electric or gas stoves, and electric streetcars by local actors, often individuals, in homes, factories, farms, and cities. By contrast, I shall here consider the role of what I call institutional users in the infrastructure integration of the Netherlands. The chemical industry, the military, and the food sector are examples of collective actors that mobilized, used, and possibly changed existing infrastructures to transform such social institutions as chemical production, food supply, and national defense.[138]

Pipelines and Petrochemistry

The use and construction of infrastructure by the chemical industry took place in the context of the development of "chemical complexes."[139]

The formation of complexes is known in chemical history from the sulfuric acid complex in England in the second half of the eighteenth century, and the synthetic dyestuffs complex in Germany in the second half of the nineteenth century. In such complexes, products and by-products from one factory were locally supplied as raw material for the next, creating complex local flows of chemicals.

In the Netherlands such intertwining of production processes first emerged around the nitrogen fixation plant of Royal DSM in the southeast in the 1930s, and around the Shell Chemicals oil refinery in the Rotterdam port area in the 1950s. In part thanks to the active port policy of the city of Rotterdam, many other international and domestic chemical companies built factories in the Rotterdam port area; by the 1970s it had become one of the world's largest chemical complexes, stretching from Europort in the Rotterdam harbor to Antwerp (Belgium), and well integrated in global, European, and Dutch flows of chemicals.

To set up this industry, chemical companies made great use of existing rail, road, and waterway infrastructure by means of specially built rail wagons, barges, or trucks for transporting chemicals.[140] They also built new networks, mainly local company railroads and extended pipeline networks. Pipelines are a fascinating case, for the construction and operation of which dedicated pipeline companies were created. Initially the Rotterdam port area served as a major oil port and refining location for the German hinterland. Crude oil and naphtha were transported to the German Ruhr region via, among others, the pipeline of the Rotterdam-Rhine Pipeline Society, set up jointly by Shell, Mobil, Gelsenberg, Texaco, and Chevron in 1960.[141] Later, with the development of diverse chemical flows in the rapidly growing Rot-

The petrochemical industry is set up as a network: oil terminals and plants are connected through railroads and pipelines. In the Botlek region this network regularly cuts across major waterways. In the mid-fifties, a depot in Vlaardingen was connected to the premises of Shell Pernis by means of a so-called "swing connection," a five-hundred-meter-long bundle of pipes.

terdam harbor complex, a multitude of local pipelines were added to transport crude oil, naphtha, ethylene, chlorine, and oxygen. Eventually, over seventy factories and complexes in Europort, Rotterdam, and Antwerp were interconnected through pipelines and railroad connections. Ethylene and naphtha pipelines tied this huge complex to Royal DSM's complexes in the southeast, a new complex near Delfzijl in the northeast, and factories and complexes in Belgium and Germany. By 1995 the Dutch chemical industry's main pipeline network was 1,900 kilometers for petroleum and petroleum products and 1,400 kilometers for other chemical products; interconnected factories formed one giant chemical machine producing and circulating chemicals throughout Dutch territory and beyond.[142]

The production, transport, and use of chlorine is an instructive example of how chemical production structures can be shaped by means of pipelines and of the contested nature of this process. Chlorine is generally considered to be a key substance to the chemical industry, used as a catalyst or in the end product of 60 percent of all chemical processes. In the 1950s, Dutch chlorine production was completely dominated by the Royal Dutch Salt Company (Koninklijke Nederlandse Zoutindustrie, or KNZ; now Akzo Nobel) factory in Hengelo, near the Dutch-German border, whence trains transported it across the country to the largest chlorine user, Shell Chemicals, in the Rotterdam harbor complex.

In 1961 KNZ opened a separate chlorine factory in the Rotterdam port area, linked directly by pipelines to the Shell factories, though chlorine deliveries by train continued for the time being. In the early 1970s KNZ built a larger factory in the Rotterdam harbor complex for production of chlorine in the form of vinyl chloride. The new factory received as inputs hydrochloric acid and ethylene from Shell Chemicals and ethylene from Gulf Chemicals, and supplied its product to factories of Shell Chemicals and Herbicide-Chemicals Botlek.[143] Finally, KNZ–Akzo Nobel set up a chlorine factory in the northern harbor town of Delfzijl, where a smaller chemical complex was emerging around natural gas production. In this chlorine geography, local pipeline networks carried most of the chlorine flow; additional chlorine transports by rail between the Akzo factories in Rotterdam in the west, Hengelo in the east, and Delfzijl in the north balanced supply and demand. By 2000 over 90 percent of all Dutch chlorine was transported by pipeline, 8 percent by rail, and 2 percent by road. Incidentally, approximately half of the chlorine pipelines run across publicly owned property.[144]

From the 1970s on, environmental groups increasingly targeted chlorine as an extremely dangerous substance that should be abolished. They launched protests such as demonstrations and blocking chlorine transports. The main targets of the public controversy were chlorine transports by rail, dubbed "chlorine trains" and "rolling bombs" by opponents. In response to an intensified Greenpeace International campaign called "Chlorine kills," Belgian and Dutch chlorine industry proponents set up a Belgium-based association of "chlorophiles" to counter "disinformation by environmental groups" on this highly useful chemical. Its first action was a counterdemonstration in front of Greenpeace's Brussels office in 1994. Actions in the Netherlands followed in 1995. Protests and lobbying by both sides initially resulted in Dutch legislation mandating higher safety standards for chlorine railway car designs, continuous monitoring of chlorine transports, lowering of maximum speeds, a ban on shunting, and a ban on transports during daytime.

After a near accident with a chlorine train in 2000, environmentalist groups allied with the small but active radical Socialist Party (not to be confused with the center-left Social Democrats) and developed a campaign to mobilize municipal councils and citizens in the fifty-five municipalities along the chlorine transport lines. Their message: "They come at night and roll slowly. Only a few people ever see them. Let alone know how dangerous they really are: the Akzo Nobel chlorine trains."[145] An accident in a city center, politicians and citizens were told, might cause 5,000 deaths and 18,000 injured. Some 2.5 million people lived in the danger zone. The campaign was successful and chlorine transport became a major issue on the national political agenda, which resulted in an arrangement that according to the responsible ministry is internationally unique. Parliament authorized the minister of environmental affairs to buy out Akzo's chlorine rail transports, which were to be replaced by pipeline distribution. A subsidy of 65 million euros subsidy persuaded Akzo to close its eastern Hengelo plant and expand the capacity of its Delfzijl and Rotterdam harbor plants by 2006. In that year structural chlorine transports by rail stopped, though incidental transports may still take place.[146] In early 2009 the government negotiated a similar buyout with Royal DSM to stop structural ammonia transport across the country.[147]

Military Infrastructure

Many other institutional users also constructed infrastructure. Starting in the 1890s, electro-technical manufacturer Philips built local transport, energy, and communication infrastructures to establish local production complexes. From the 1940s on, infrastructure considerations and active lobbying for infrastructure construction and routing were integral to its strategy to set up a nationally decentralized production system. The growth of globally dispersed production was mirrored by global transport networks and a globe-spanning private telex system for the company.[148] In the services sector, exchanges and stockbroker organizations set up mail, telephone, radio, and telex infrastructure along which the system for trade in stock and bonds was organized.[149]

Another intriguing example concerns national defense. In the early 1950s, the North Atlantic Treaty Organization (NATO) decided to build the Central European Pipeline System, a military European pipeline system for storage and transport of motor fuel.[150] It was a lesson from World War II, when the German general Erwin Rommel lost the battle for North Africa mainly because of inadequate supply lines. In 1957 the Pipeline Agency of the Ministry of Defense started building the Dutch section of this system. The Defense Oil Center foundation, a civilian organization led by officials from oil companies, was set up to manage and run it. In 1983, the Ministry of Defense took over this task as the Defense Pipeline Organization (DPO) under the Royal Air Force. By 1980 some 1,000 kilometers of pipelines con-

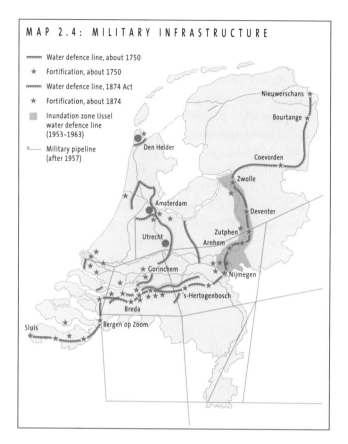

MAP 2.4: MILITARY INFRASTRUCTURE

〰〰 Water defence line, about 1750

★ Fortification, about 1750

〰〰 Water defence line, 1874 Act

★ Fortification, about 1874

▪ Inundation zone IJssel water defence line (1953–1963)

⚓— Military pipeline (after 1957)

Nieuwerschans ★

Bourtange ★

Den Helder

Coevorden

Zwolle

Amsterdam

Deventer

Zutphen

Utrecht

Arnhem

Gorinchem

Nijmegen

's-Hertogenbosch

Breda

Sluis

Bergen op Zoom

Military infrastructures developed early. From the time of the Dutch Republic until the 1960s, defense lines consisting of natural barriers, artificial water bodies, and fortifications played a prominent part in defending the Netherlands on the landside. A typical twentieth-century military infrastructure is the underground pipeline network for fuel supply set up in a NATO context in the 1950s.

nected strategically situated depots with refineries, tanker landing stages, and storage companies (see map 2.4).

Another telling example of military infrastructure creation but one with a quite different function is the construction of water defense lines (see map 2.4).[151] Water defense lines were designed to be activated only in times of crisis: by means of complex systems of inundation sluices and culverts, zones of low-lying areas could be flooded with a shallow layer of water, shallow enough that enemy troops could not deploy boats, but deep enough to hide submerged drainage canals from view and thus complicate the advance of enemy armies.

Water defense lines became a pillar of traditional Dutch land defense. It had enjoyed near-mythical status since 1672, when an improvised defense line stopped a massive invading army 200,000 strong of the Sun King, Louis XIV. In the following decades, a new national defense scheme was designed and implemented: natural barriers such as marshes and sand drifts were interconnected by water defense lines to shield national borders. Breaks in these defense lines such as roads and dikes were enclosed in fortresses or sconces. When the Dutch, in 1874, decided on a policy of strict neutrality in European military affairs, the defense line system was revised: a much cheaper system protected only the western part of the country, which was considered the political and economic "center." By then, Dutch military prowess had considerably slackened; the idea was that the new defense lines would slow down enemy troops until foreign armies could come to the rescue.[152]

Historians disagree about the military significance of water defense lines; perhaps they mainly had a delaying and deterrent effect.[153] Still, Dutch territory has been successfully invaded only twice since the proclamation of Dutch independence. The first time was when the French crossed the frozen defense lines in the winter of 1794–1795. The second was the German Invasion in 1940, which relied on massive airpower, which seemed to make water defense lines thoroughly obsolete. Surprisingly, and largely unknown, the Cold War inspired the greatest water defense line project ever. By the end of the 1940s, NATO had identified the Rhine River as its defense line against potential Russian aggression. The Dutch government pleaded to extend this line to the IJssel River and IJsselmeer (see map 2.4), so as to include the cherished Western Netherlands safely in the allied defense zone. Initially they met with skepticism from NATO; Britain's Field Marshall Bernard Montgomery was reported to say that "Your IJssel River is no obstacle at all. I can jump over it."[154] So the Dutch devised a grand scheme. Floating dams stood by to block the Rhine, diverting all water into the IJssel River. Here, a system of inundation sluices could create a wet barrier 110 kilometers long and 5 kilometers wide. This defense line was operational from 1953 to 1963. During the Cuba Missile Crisis of 1962, alarm phase 1 was activated, and over 200,000 residents, ignorant of the system's very existence, were close to being evacuated.[155] The line became redundant when NATO shifted its defense line eastward to the Wesel River and later to the Elbe River.

Food Chains and the Unification of the Dutch Meal

Infrastructure building and use, then, intertwined with Dutch economic, environmental, and military history. Institutional and individual uses of infrastructure in the food domain, moreover, demonstrate an important ele-

In the old days, daily transport of milk cans was done by so-called milk shippers, sometimes literally on ships, as shown on the top photo, of a Frisian cooperative dairy plant. The second photo shows the milk truck of the Menken firm, near Hekendorp, around 1965. Effective organization of milk transport from farm to factory provided an essential link in the operations of dairy factories. Because of the limited transportation options and the need to ensure freshness of milk, dairy processing plants were initially numerous and well distributed, close to their sources of milk. In the 1960s new cooling and transport options became available, coinciding with a major scale increase and leading to a wave of mergers. The new system gave rise to larger factories with larger service areas. Milk transport now took place twice a week in purpose-built refrigerated milk trucks.

ment of sociocultural integration with implications for such intimate elements of life as diet and eating habits.[156]

By the mid-nineteenth century, Dutch food chains reflected the rather fragmented nature of Dutch transport infrastructure. Areas with access to the waterway network were firmly tied into a global economy and thus had wide access to foodstuffs. By contrast, comparatively isolated village communities in the southern and eastern parts of the country were dependent on local economies; food flows were predominantly local, in the form of self-sufficiency or local exchanges of products via bartering or local markets. These structural constraints were reflected in a rich variety of local eating patterns: the number, timing, and contents of meals differed from place to place.

As previously discussed, transport infrastructure was subsequently integrated on a national scale. In due time, Dutch food distribution expanded accordingly. This was not due to an inherent "homogenizing effect" of national infrastructure that some authors have assumed. Instead, our understanding of infrastructure and societal change suggests that this national integration of food distribution networks resulted from agency and choices of institutional users of infrastructure, in this case stakeholders in the food sector.

From the 1880s on, new players such as industrial firms and distribution companies inserted themselves into food chains. At the infrastructure level, new food factories were literally tied into food flows deploying trucks, riverboats, and railway cars. For instance, the transition of Dutch dairy production from farms to local factories, triggered by the successful takeover by Danish factory butter of the English market, mobilized road and water networks to collect raw milk from farms and, after processing, return the skimmed milk. It is telling that butter factory inventories listed milk barges or milk trucks next to buildings and crucial equipment such as centrifugal separators. Lists of employees included "milk collectors" alongside other types of staff.[157] Also the subsequent merging of local dairy works into a few large companies with much bigger markets has an important infrastructure component. Centralization of dairy production required long-distance refrigerated tank trucks for transport between farm and factory. To keep transport costs down, these trucks came with a second innovation: farms were to be equipped with deep-cooling tanks for local storage, so that daily milk collection could be replaced with weekly collection. Farmers did not always adapt smoothly to this new regime; the 1970s even witnessed a so-called "tank war" that climaxed in 1978 when farmers in Hoogeveen, refusing to abandon their milk jug system for the deep-cooling tank system, took the board members of their dairy company hostage. Such actions failed to turn the tide, however.[158]

Food distribution companies likewise used infrastructure to organize themselves. Imitating the British "satellite system," the leading Dutch retailer, Albert Heijn (which as Royal Ahold became a global player), set up a branch system in which local branches were stocked from a central warehouse in Zaandam. By 1920 the company had developed a network of over fifty branches. To create food flows between this warehouse and the branches, the company purchased its own fleet of trucks, serviced in its own garages by company mechanics. Later on, having parking available proved an essential precondition for the success of supermarkets, which as major nodes were linked up with households through the road system.[159]

Studies of these food companies show a dominant tendency toward organization on a national scale for reasons of competitive advantages, developing national markets for raw materials and food supply. These strategies were rewarded when export markets fell prey to several global crises and the protectionism that followed. In particular the Great Depression triggered a renewed focus on the home market of Dutch export-minded food industries. In the first decades after World War II, renewed protectionism further strengthened the national circulation of foodstuffs despite several attempts to construct a European food system.[160] By the 1960s, national-level food flows had become quite normal. Throughout the country, an elaborate and standardized assortment of food had become available in all provinces, in both city and countryside, and to all social classes.

This development of food distribution networks coincided with a striking convergence of food and meal patterns by the 1960s, which Dutch food historians have dubbed the "unification of the Dutch meal": across the country and social class lines as well the Dutch ate regular meals three times a day, featuring a bread meal accompanied by milk or churned milk for breakfast and lunch, and a hot meal for dinner. The hot meal invariably consisted of soup, a main course of potatoes, vegetables, and a rather small piece of meat or fish, and a desert. Again, this event cannot be uncritically attributed to the homogenizing effects of either transport infrastructure or food chains. Nationally integrated transport networks and food chains were a necessary, but not sufficient preconditions for this convergence. Just as the emergence of predominantly nationally oriented food chains resulted from decisions made by food companies in a context of global protectionism, the unification of the Dutch meal followed choices of individual consumers—overwhelmingly represented by housewives doing the actual shopping —choosing foodstuffs from the available assortment in a remarkably homogeneous way. These choices were in turn shaped by decades of intense information campaigns explicitly aimed at Dutch housewives, involving home economics teachers, advertising campaigns, and so forth. Over the decades Dutch housewives were told quite consistently what and how to cook and eat in the service of national health and the home market of Dutch agriculture and food industries. Again the Great Depression was important, for the problems of Dutch food export industries triggered sectoral organizations as well as the national government to develop the domestic food market in terms of marketing, advertising and propaganda. After British example, the crisis office of the dairy industry (Crisis Zuivelbureau, established 1934) also managed to introduce schoolmilk in this period.[161] Such efforts had produced a remarkable convergence in consumer preferences by the 1960s. The nondeterministic character of this development was revealed decades later, when meal and food patterns again fragmented following new social and marketing dynamics. For instance, a new labor culture came with associated food habits such as "grazing": picking up snacks whenever possible throughout a busy day.

Periodization: Three Historical Regimes of System Building

So far our examination of the shaping of the Dutch networked nation has involved mapping its infrastructure integration; exploring the contested dynamics of this process, which did not follow an inevitable logic of development but instead resulted from the struggle between competing system builders' interests and priorities and the outcomes of conflicts and compromise; and examining the roles of institutional and individual users in shaping infrastructure as well as its societal implications. Now I shall combine these threads to attempt a historical periodization.

In a long-term perspective, we may identify three successive regimes for system building, each featuring a dominant social and regulatory structure that moderated the access of various groups to the system-building process, thus affecting specific choices and development tendencies in infrastructure change.[162] We can identify a fragmented system-building regime in the era of the Dutch Republic; a centralized system-building regime that marked the nineteenth and most of the twentieth century; and finally an "open" system-building regime that evolved in the era of participation and (neo)liberalization.

Most nationally integrated infrastructure was established during the centralized system-building regime. Its roots, however, go back to the times of the Dutch Republic (1581–1795). At that time the building of infrastructures was controlled by a number of parties who operated independently, generally on local and provincial levels and not on a national level. This is illustrated well by

The growing volume of long-distance telephone communications in the early years of the twentieth century required a large number of connection switches and lines. For instance, around 1925, the connection between Amsterdam and nearby Diemen required more than three hundred phone lines. Later, more conversations per "pair of cores" would become possible (120 around 1970) with the help of carrier wave telephony (alternating current in line); with beam connections the volume could increase even more (900 per beam in 1970).

the domain of water management.[163] From the Middle Ages, water-related interests pertaining to drainage, inland navigation, and fisheries had been assessed and coordinated by village councils of the newly established settler communities that cultivated the swamps and peat bogs. From the late medieval period, however, the institutional framework of wet system building became increasingly differentiated. First, from the 12th and 13th centuries drainage works were gradually taken over by specialized agencies, the provincial and, later, local water boards. These were dominated by large landowners, who gave priority to drainage, and regularly came into conflict with shipping or fisheries interests—for instance, concerning dams that blocked rivers or canals. The dams remained in place, and work-around solutions included carrying small boats over them.

Next, inland navigation became predominantly a concern of autonomous towns, which had become a key power factor in the late Middle Ages. Autonomous towns started to control (and levy tolls on) nearby waterways and from the sixteenth century on, some thirty towns in the western and northern Netherlands organized the regular barge service network for freight transport and the horse-pulled barge network for passenger and mail services. All connections were established as bilateral agreements between towns, which hired skippers and maintained the waterway infrastructure. The republic's wealthy merchants became a third group of system builders, increasingly setting up drainage projects as major investment projects, which yielded high returns in the form of land sales. The strong and autonomous provinces, who had created the republic as a loose confederation, constituted a fourth group, which, for example, worked to improve sections of larger rivers. The famous water defense line, the *Hollandse waterlinie,* was a provincial project. Finally, national entities such as the States-General (a kind of confederal government) had authority only to build the military system. National authorities created the eighteenth-century national belt of defensive water lines, natural barriers, and fortresses. Some defense projects, such as keeping the moors in the northeastern Dutch-German border artificially flooded, were regularly sabotaged by landowners wanting to cultivate the land. The decentralized and fragmented political structure of the Dutch Republic was thus reflected in a fragmented system-building regime, where different groups, with their own distinctive goals, built distinctive systems. Priorities were weighed and conflicts resolved on an ad hoc basis.

Toward a Centralized System-Building Regime

The French invasion and occupation (1795–1813) introduced a centralized state, and accordingly made the national government a more potent system builder. The new constitution of 1798 put the "condition of Dikes, Roads and Waters" under state jurisdiction. That same year a national agency for waterworks and management was established. After independence was regained, the Kingdom of the Netherlands retained this policy. The 1815 constitution formally placed centralized control of infrastructure works of "general importance" with the king. The national waterworks agency grew to a major system builder under its new name, the Rijkswaterstaat, the State Water Authority.[164]

The fragmented system-building regime gradually gave way to a centralized one, where national actors—the Crown, ministers, the Rijkswaterstaat, and others—assessed and prioritized various interests and made key investment and design choices. It was not immediately clear, however, in how centralized a fashion infrastructure building and management should be organized, and how the phrase "works of general importance" should be interpreted. In other words, which tasks were national government tasks, and which should be left to the private sector or lower bodies of government. Here a comparison is instructive. In his pathbreaking study of infrastructure regimes in Sweden, Arne Kaijser described the emergence of a Swedish "national regime" for infrastructure construction and governance in the nineteenth century. Railroads served as a paradigmatic case, which was subsequently copied to other infrastructure sectors: the state was responsible for main arteries and nodes (harbors, airfields), whereas other actors built and managed secondary lines and nodes.[165]

In the Netherlands, by contrast, a national model remained absent. Although national involvement clearly increased, its form was negotiated on a case-by-case basis. Some works were immediately labeled as state responsibilities, including the maintenance of major rivers and estuaries. A list of such works of general importance, or "state works," was drawn up in 1803.[166] Smaller works were handled by the municipalities or provincial or local water boards. For instance, the national government put in "state canals," but the old barge-canals were still owned and maintained by municipalities. The paving of "state roads" had been seen as a state task during the French occupation as well. This policy's main outline was simply taken over by King William I when he came to power. This task, too, was executed by the Rijkswaterstaat.

In the second half of the nineteenth century two new infrastructures were added to the list of state works: telegraphy and the railroads. Notably, this so-called liberal era in Dutch political history hardly meant a break in national infrastructure policy. Dutch liberal policy in this period is characterized less by the internationally known slogan *Laisser faire, laisser passer* than by "Mark our deeds" (*Wacht op onze daden*), the activist parole of the leading liberal politician, Johan Rudolf Thorbecke, who headed three liberal cabinets between the 1850s and the 1870s and is generally regarded as the founding father of Dutch parliamentary democracy.[167] His mantra also applies to infrastructure. Regardless of whether liberals or conservatives headed the government, infrastructure was conceived as a political instrument of progress. Of course, liberals preferred construction and operation by private companies as long as national interests would be secured. Still, Thorbecke's first liberal cabinet initiated major state-led river improvements and took the decision to establish a state telegraphy system. The first telegraph lines had been introduced by private companies, but, he stated, "the government will do whatever it can to engage the entire nation in the new means of communication."[168] The Telegraph Act (1852) was passed without significant opposition. A national telegraphy agency, the *Rijkstelegraaf*, was established, which took over all private systems to complete the national network by 1884. The state had become the sole telecommunications system builder.

In railroad construction the same concerns emerged, but negotiations and political processes produced a different outcome. In railways, too, the first lines were constructed by private enterprise. Prior to that, a state commission (1836) had argued that the state should take responsibility for railway lines of national interest, and the state indeed financed the Amsterdam-Cologne line, but in addition to the problems previously noted, it proved a financial disaster. In the end it was taken over and completed by a private company. By the mid-century, with this experience in mind, Dutch politicians studied the variety of alternative foreign models for rail development. In Belgium the state put in a railroad system. In Prussia the state was the main stockholder. In France the state was in charge of building main lines, and private companies constructed secondary lines. In England the construction of roads, canals, and railroads was entirely left up to the private sector. The Dutch liberals and conservatives could not reach consensus. A liberal–conservative cabinet settled on a plan to encourage private railroad construction (1860) through financing. Its predominantly conservative successor managed to get Parliament to adopt a Railway Act in 1860, which authorized the state to develop a national railroad system. A new liberal cabinet in 1863 decided that the exploitation of this network should exclusively be in private hands. Later acts (1873 and 1875) confirmed this distribution of tasks.[169] In the resulting institutional railway framework, private companies continued to build their own lines while the Dutch state built a national rail network to be operated by private railway companies.

In the first decades of the twentieth century, the state stepped up its involvement in infrastructure building. This was in part triggered by World War I, which created economic problems but also led to increased acceptance of state involvement in economic life. Increasing state involvement in infrastructure again took different forms.

Private participation in the railways was gradually phased out. During the war the largest private companies sought to counter economic distress by setting up a cooperation, the Dutch Railways (Nederlandse Spoorwegen, NS). After the war economic performance did not improve. Since railways were considered to be of national interest, but the weak financial condition of the state prevented a full takeover, the government opted to become the largest shareholder of the NS. In 1937 the state finally became the sole shareholder.[170]

In telephony, which also had been pioneered by private enterprise, the government nationalized long-distance services in 1897. From 1913 on it increasingly took over local private systems and, later on, municipal systems, in addition to building entirely new local systems.[171] By the late 1920s the state controlled the entire telephony network except for the large municipal networks of Amsterdam, Rotterdam, and The Hague, which successfully resisted a takeover. The state-owned network was managed by a state-owned company, Post & Telegraphy (P&T, 1913), in 1928 renamed Post, Telegraphy, and Telephony (PTT). In 1940 the remaining municipal systems were nationalized under the German occupation.

Regarding broadcasting infrastructure, the state gained a say when the PTT teamed up with private broadcasters to set up the company NOZEMA (Nederlandsche Omroep-Zendermaatschappij) to build and maintain the broadcasting infrastructure, a move that was orchestrated by the national government. Leading manufacturers of broadcasting equipment were kept out of the deal, despite their protests.

National players also emerged in civil aviation and electricity supply. Here, state influence was crucial but it took a more invisible form. In aviation, the Royal Dutch Airlines (Koninklijke Luchtvaart Maatschappij, KLM) established in 1919, counts internationally as one of the very few older airlines set up by private rather than state initiative. The national government supported the enterprise in the background, however, by covering a large part of the airline's operating losses; not until 1946 did the airline actually make a profit. After World War II the government took over 51 percent of KLM's capital stock.[172]

In electricity supply, immediately after World War I Parliament rejected government plans for a state-owned company that would establish a nationwide power grid. After several failed attempts, the central government managed to get an Electricity Act passed by Parliament on the eve of World War II. This legislation provided for the electricity companies to voluntarily develop a national network; if they did not do so, it gave the government the authority to force them to do so. The act was never formally enacted and served as a threat only, but after World War II, this threat certainly informed the electric utilities' decision to jointly establish a national grid, although—as we have seen—they did not actually use it fully until the 1980s.[173]

The regime of centralized system building peaked in the 1950s and 1960s. Interests were weighed—and investment and design decisions made—at the national level in state agencies such as the Rijkswaterstaat, state-owned companies (the PTT), private companies with more or less state participation (KLM, NS, NOZEMA, Gasunie), or nationally cooperating companies such as the electric utilities. Decision making had increasingly assumed a technocratic character—Rijkswaterstaat historians speak of a "technocratic-scientific period."[174] Rijkswaterstaat chief engineers internally developed their stance to the importance, necessity, priority and designs of specific works;

ministers and parliament had limited influence, being involved only at the later stage when major decisions had already been made. This applies to other centralized system builders as well. The decision processes were in the hands of national experts and often hidden from public view and democratic control, while in the background government was working to further scale increases.

Crisis

In his analysis of system building in the United States, the American historian Thomas P. Hughes argued that the late 1960s and 1970s were characterized by severe crisis. Not least as a result of the compromising of military systems in the Vietnam War and emerging counterculture values, large-scale technocratic system building was increasingly condemned by environmentalist and civil rights groups. As these gained public and political support, several system builders thought the days of large infrastructure projects were over. Out of this crisis, however, a new mode of system building emerged that was responsive to counterculture values. A new participative or "open" process gave interest groups access to the design process and became a key element of what Hughes termed "postmodern system building," comprising the various social and political complexities of the postindustrial world.[175]

In the Netherlands a similar crisis occurred. Here, too, centralized system building came under great pressure, causing serious delays in infrastructure development, large budget cuts for public works, and cancellation of some major projects. In the 1960s civil rights and environmental issues made their way onto the political agenda.[176] In contrast to the approaches of earlier political innovators, such as the mid-nineteenth-century liberals, the new social critique was directly aimed at infrastructure. Radical groups such as the Provo (from "provoke") movement declared war on the "asphalt terror." Mainstream organizations as the Dutch motorist and bicyclist association, the ANWB, also changed course. Its chairman stated in 1965: "Man discovers a large and gifted species of salamander, loves it, takes it in as a pet, raises it to be his equal, and soon the salamander ambitiously starts generating offspring. Its number increases along with its intelligence, and it threatens to start ruling mankind. Replace "salamander" with "automobile" and you get some idea on this era's problems."[177] The ANWB even left the car lobby. Meanwhile, protest groups rallied against concrete infrastructure projects on both the local and national level, gaining significant popular and political support.

This movement was further supported by new rules and procedures. Henceforth, plans for highway routes became subject to prior public scrutiny, the results of which were taken into account in the decisionmaking process. Also, environmental groups were invited to join in the planning process. In the context of openness, the center-left cabinet Den Uyl (1973–1977) legally enshrined the public's right to participation in system building. Projects of national significance were marked as "key planning decisions," which entailed the obligation to arrange hearings involving various councils and agencies, to follow public participation procedures, and to present the ultimate decision to Parliament. Soon, such projects had to be integrated into provincial and municipal zoning schemes, each having their own participation procedures.[178]

The effects of resistance, extensive participation procedures, and changed priorities are unmistakable. For example, the Rijkswaterstaat share of the state budget dropped from almost 8 percent in 1971 to 2.8 percent in 1981 and 1.6 percent in 1993.[179] In the 1970s and 1980s its construction activities did not come to a complete standstill, but the emphasis clearly shifted to maintaining and managing existing infrastructure rather than initiating new projects.[180] A number of approved projects were reconsidered. Plans for reclaiming the Wadden Sea and the large Markerwaard polder in the IJsselmeer, as well as for building a second national airport, gave rise to major protests and were cancelled.

Other system builders faced symptoms of crisis as well, but the implications for infrastructure development differed from one case to the next. Although the second national airport was canceled, the expansion of Amsterdam Schiphol Airport with a fifth runway remained on the table. This runway, under discussion since the late 1960s, met fierce resistance as well, and was not opened until 2003.[181] The electric power sector for the first time encountered real government interference in the wake of two energy crises and massive public protests against nuclear power. The Ministry of Economic Affairs initiated what it termed a "broad societal debate" on nuclear power from 1981 to 1983 to elicit the opinions of stakeholders and the public. The debate identified a clear majority against nuclear power. The national government overruled this majority, but after the highly profiled Chernobyl disaster in 1986, its plans for nuclear expansion were shelved.[182]

Occasionally, institutional users and their systems came under fire as well. The extension of the food distribution network, which previously had been supported by government, agriculture, industry, and households, was increasingly criticized for being unnatural and harmful because of the use of pesticides and additives and the general evils of large-scale production and capitalism. The call for "natural food" and a "life free of chemicals" was expressed in a fairly small counter movement favoring small-scale food flows or self-sufficiency and uncontaminated products.[183] Furthermore, Greenpeace and other environmentalist groups mounted campaigns against the chemical industry.

Remarkably, in this period some types of infrastructure expanded rapidly. Unaffected by the public debate, the size of the gas supply system doubled in the 1970s, as did the chemical industry's pipeline systems and, strikingly, the highway system: even though some newly planned highways were fiercely contested, thousands of kilometers of highways were in fact built, completing almost half of the 5,300-kilometer system announced in the mid-1960s.

The Contested Shaping of an "Open" System-Building Regime

As occurred in the United States, this crisis ultimately produced a new mode of system building that responded to the new demands. It remained centralized, but technocracy gave way to participation of a large number of new players. The national system builders lost their monopolies and the system-building process opened up, but the degree and form of access and openness remained highly contested—which is why we put "open" in quotation marks. Still, infrastructure construction itself received a major boost, particularly in the 1990s. Two processes played a major role in shaping the new regime: the (neo)liberalization of the economy, and learning to build systems within a context of public participation.

The neoliberal current began to reveal itself in the Netherlands in the 1980s.[184] The initial effort to reduce the national budget deficit evolved into the ideology that government should not engage in tasks that the market could perform as well, if not better. In complete contrast with the 1950s and '60s, government and corporate monopolies and cartels were now viewed as obstacles to lower prices and high-quality services, consumers' free choice, and client-centeredness, not guarantors of them. Now it was believed that government would serve the common interest best by ensuring open markets, not by acting as a system builder. The context within which these developments took place was that of the European Communities (later European Union), which was encouraging the development of European inner markets.

Again, the institutional framework was negotiated on a case-by-case basis. In theory, state-owned monopolies should be privatized and infrastructure building and management should be separated from services; different service providers, be it telephone, energy, or transport companies, should compete on the basis of equal access to the fixed infrastructure. In highway and waterworks construction, however, liberalizing had no effects on ownership;

In the 1960s, Gasunie put in a network of gas mains to distribute and sell natural gas throughout the country. This infrastructure was partly visible when it was installed, but soon completely disappeared from view forever.

here it merely entailed contracting with companies to design and construct works, which were still planned and financed by the responsible ministry and operated and maintained by the Rijkswaterstaat. Examples are the Amsterdam ring road, improvements of the Waal River, and the Maeslandt storm surge barrier in the New Waterway, which currently inspires storm surge barrier construction in post-Katrina New Orleans.[185]

In railroads, telecommunications, and electric power generation and distribution, privatization and the separation of infrastructure and services were implemented to varying degrees. In telecommunications, the state-owned PTT was gradually privatized after 1989. The state retained the final decisionmaking power during a transition period, but eventually all shares were sold. Negotiations on liberalization of the railway sector ultimately resulted in the state's becoming the direct owner of the infrastructure, while the NS operated rail service. The state remained its only shareholder, though. This was a case of marketization, not privatization: structural subsidies were withdrawn, and the company increasingly had to survive on market terms. Still, the NS gained a monopoly on operating service on the main network until 2015; on a few secondary lines service is operated by other companies. The electric utilities were forced to split production and transport functions. As with the railways, the state became the owner of the national power grid via the Dutch transmission system operator TenneT (established in 1999). For power generation, a series of mergers and purchases led to the existence of several competing suppliers.[186] These different outcomes reflect the fact that these processes were heavily contested and negotiated; indeed, they are still in flux.

A second pillar of the new regime of "open" system building is the participation of stakeholder groups. Three projects served as high-profile learning experiences in open system building: the extension of national motorway A27 from Breda-Vianen to Hilversum, the successful adaptation of the Oosterschelde Dam, and the Betuweroute railway.

The expansion of the A27 motorway from Breda to Hilversum, which was planned to cut through the Amelisweerd Woods, near Utrecht, initially seemed to be an exemplary case of open, ecotechnical system building.[187] In the early planning stages the wishes of both the National Forest Service and the agency for aesthetic design of national roads were incorporated in the design. When a local action group, Working Group Amelisweerd, in 1971 proposed to divert the road around the woods, the design was adapted accordingly. The ANWB even gave the working group and award in appreciation of its constructive contribution to road planning. In the course of the 1970s, however, resistance revived, allegedly because the Rijkswaterstaat did not keep its promise to spare as many trees as possible. New action groups emerged and tried to delay or block the project, taking advantages of the new participation regulations as well as using outright sabotage. Squatters built a village of tree huts in the contested woods; they were ultimately removed by antiriot police with support from the Dutch army. The motorway was finally completed in 1986. Despite its environmentally friendly design, both government and opponents looked back at the decision process with mixed feelings.[188]

The building of the Oosterschelde Dam, by contrast, became a showpiece of the new style of system building. The dam to close the Oosterschelde estuary was the largest project in the Delta Works, the ambitious scheme to close the southwestern delta area in Zeeland in the wake of the 1953 flood that had killed over 1.800. Whereas earlier works had been planned and executed relatively smoothly, this last project in the Delta Works was scheduled for the 1970s and ran into massive opposition. Environmentalists and biologists teamed up with representatives of the mussel and oyster industries, fishermen, and pleasure craft owners in a plea to keep the sea arm open, and prevent its turning into a freshwater basin behind a closed dam. The government was responsive to this pressure and set up an independent commission, which, tellingly, did not include a single Rijkswaterstaat representative. The commission proposed a semi-permeable storm surge barrier that would be closed only in case of a flooding threat. The government approved, and Rijkswaterstaat managed to design and build it. Completed in 1986, the barrier is an excellent example of integrating opposing concerns—flood protection and environmentalism—by design; the American Society of Civil Engineers counts the barrier as one of the seven wonders of the modern world. This marvelous achievement came with a price tag though; the cost was twice that of all earlier projects of the Delta Works combined.[189]

The contested decision-making process regarding the Betuweroute freight railway line, discussed earlier, involved experiments with new public participation procedures. These were in fact restricted; the system-building regime became less "open." The background was the government's fear of huge delays arising from extensive participation procedures. Even in the absence of serious opposition, national-level decisionmaking would take at least seven and a half years. But this decision would not be binding on provinces and municipalities, so negotiations and procedures would continue even after that. Following several studies, the government decided to experiment with new legislation to achieve "larger efficiency and time gains" in "large projects of national significance." Decision-making on such projects, including public participation, was to take place exclusively on a national level. Henceforward, the results of this national-level procedure would be binding on provincial and municipal zoning planning, so environmentalists, NIMBY groups, and municipalities could not delay and frustrate such projects' local procedures. National decisions could only be appealed to the Council of State.[190] The Betuweroute decision-making process was the new legislation's first test case. It did help the project get through, despite massive opposition, but the new legislation itself provoked opposition. Such opposition was allegedly triggered by blunt NS spokespersons at local information meetings and negotiations, who referred to the new legislation in warning provinces, municipalities, and citizens that resistance was futile: the project could not be altered, except perhaps with the addition of some "camouflage by a few trees or sound screens," as one provincial delegate testified to the parliamentary Commission of Inquiry.[191]

In the wake of these learning experiences, infrastructure development in the Netherlands revived in the 1990s and early 2000s, but the regime of "open" system building" remains more open to some than to others. Policymakers embraced the concept of the "network society," and new players in the liberalized telecom market installed several nationwide mobile telephony networks within a few years. The Dutch sections of new European freight and passenger railroad networks are being completed, and a National Ecological Network is under way. However, there is still friction between national planners, system builders, and other constituencies. The mobile telephony providers were heavily fined in 2002 for making illegal cartel agreements. The telephony and internet provider KPN, the private successor of the PTT, was repeatedly charged with and fined for abusing its ownership of last-mile telephone connections to frustrate access on equal terms by new market entrants. State-funded projects such as the Betuweroute, high-speed train lines, and the expansion of Amsterdam Schiphol Airport have benefited from the legislation preventing local participation. By contrast, the National Ecological Network is not prioritized and has to take the slow and difficult route of implementation in local zoning schemes, where it runs into small-scale village corruption and resistance from farmers. Even in the Netherlands, judged by international standards to be an extremely advanced networked nation, infrastructure change is as contested as ever.

Acknowledgments

Special thanks are due to Johan Schot, Geert Verbong, Mila Davids, Henk van den Belt, the other contributors to this volume, and the editorial team of the Techniek in Nederland research program for commenting on earlier drafts. Henk van den Belt first suggested studying the National Ecological Network as a Large Technical System, and I have drawn on his research on this topic. Furthermore, this chapter explicitly uses, interprets, and synthesizes parts of the Techniek in Nederland research program that deal with infrastructure. Intellectual and empirical debts are acknowledged in the footnotes. This chapter's research was sponsored by the Netherlands Organization for Scientific Research NWO and the Technische Universiteit Eindhoven.

Notes

1 Michiel Schwartz, "Holland schept ruimte," *Holland schept ruimte: Het Nederlandse paviljoen op de wereldtentoonstelling EXPO 2000 te Hannover* (Blaricum: V + K Publishing, 1999), 56.

2 Some of the conclusions of this chapter are also presented in Erik van der Vleuten and Geert Verbong, eds., "Networked Nation: Technology, Society and Nature in the Netherlands in the 20th Century," *History and Technology* 20 (special issue), no. 3 (2004).

3 English summaries of these policy documents have been published by Ministry of Housing, Spatial Planning, and the Environment, *Making Space, Sharing Space: Fifth National Policy Document of the Netherlands 2000/2020—Summary* (The Hague: 2001); and Ministry of Housing, Spatial Planning and the Environment et al., *National Spatial Strategy: Creating Space for Development—Summary* (The Hague: 2006).

4 Ministry of Housing, Spatial Planning, and the Environment, *Making Space, Sharing Space*, p. 11.

5 Ibid.

6 Ministerie van Volkshuisvesting, Ruimtelijke Ordening en Milieubeheer, *Ruimte maken, ruimte delen: Vijfde nota ruimtelijke ordening* (The Hague: 2001), part 1, chapter 1, 2 (author's translation).

7 Erik van der Vleuten, "In search of the Networked Nation: Transforming Technology, Society and Nature in the Netherlands in the 20th Century," *European Review of History* 10 (2003): 59–78.

8 Hans Knippenberg and Ben de Pater, *De eenwording van Nederland: Schaalvergroting en integratie sinds 1800*, 2nd ed. (Nijmegen: SUN 1990). Compare with Ben de Pater, ed., *Eenwording en verbrokkeling: Paradox van de regionale dynamiek* (Assen: Van Gorcum, 1995), which speaks of "incorporation" rather than "integration" to highlight the contested nature of this process.

9 For discussion and references see Erik van der Vleuten, "Understanding Network Societies: Two Decades of Large Technical Systems Studies," in Erik van der Vleuten and Arne Kaijser, eds., *Networking Europe: Transnational Infrastructures and the Shaping of Europe, 1850–2000* (Sagamore Beach, Mass.: Science History Publications, 2006), 279–314.

10 Thus complained Arne Kaijser, whose work is an inspiring exception. Arne Kaijser, *I fädrens spår:Den svenska infrastrukturens historiska utveckling och framtida utmaningar* (Stockholm: Carlssons, 1994); Arne Kaijser, "The Helping Hand: In Search of a Swedish institutional Regime for Infrastructure Systems," in Lena Andersson-Skog and Olle Kranz, eds., *Institutions in the Transport and Communications Industries: State and Private Actors in the Making of Institutional Patterns 1850–1990* (Canton, Mass.: Science History Publications, 1999), 223–244.

11 Narrow definition leads to controversy and sometimes to a priori exclusion of important aspects of the infrastructure landscape. For a discussion see Erik van der Vleuten and Arne Kaijser, "Prologue and Introduction," in Van der Vleuten and Kaijser, *Networking Europe*, 5–6.

12 Erik van der Vleuten and Arne Kaijser, "Networking Europe," *History and Technology* 21, no. 1 (2005): 21-48. Compare Erik van der Vleuten, "Toward a Transnational History of Technology: Meanings, Promises, Pitfalls," *Technology and Culture* 49, no. 4 (2008): 974-994.

13 For explicitly transnational infrastructure history see, for example, Van der Vleuten and Kaijser, *Networking Europe*; Johan Schot, ed., "Building Europe on Transnational Infrastructures," *Journal of Transport History* 28 (special issue), 2 (2007): 167-228; Alexander Badenoch and Andreas Fickers, eds., *Europe Materializing? Transnational Infrastructures and the Project of Europe* (Basingstoke: Palgrave Macmillan, in press). See also www.tie-project.nl.

14 G. P. van de Ven, ed., *Man-made Lowlands: History of Water Management and Land Reclamation in the Netherlands* (Utrecht: Matrijs, 1993); Erik van der Vleuten and Cornelis Disco, "Water Wizards: Reshaping Wet Nature and Society," *History and Technology* 20, no. 3 (2004): 291–309.

15 Clé Lesger, "Interregional Trade and the Port System in Holland, 1400–1700," *Economic and Social History in the Netherlands* 4 (1992): 186–218; Jan de Vries and Ad van der Woude, *The First Modern Economy: Success, Failure, and Perseverance of the Dutch Economy, 1500–1815* (Cambridge: Cambridge University Press, 1997).

16 Knippenberg and De Pater, *Eenwording van Nederland*, 13, 19, and 209.

17 Auke van der Woud, *Het lege land: De ruimtelijke orde van Nederland 1798–1848* (Amsterdam: Meulenhof, 1987).

18 Taken together these produced 5,000 square kilometers of additional cultivated land, one seventh of the surface area of the Low Netherlands and considerably more than the yields of large reclamation projects in the Low Netherlands during the same period. See Van de Ven, *Man-made Lowlands*, 224–225.

19 Ministry of Housing, Spatial Planning, and the Environment, *Making Space, Sharing Space*, chapter 6.

20 Van der Vleuten, "In Search of the Networked Nation." Compare Luuk Boelens, ed., *Nederland netwerkenland: Een inventarisatie van de nieuwe condities van planologie en stedebouw* (Rotterdam: Nai uitgevers, 2000).

21 J. W. Schot et al., "Concurrentie en afstemming: Water, rails, weg, en lucht," in Schot et al., *Techniek in Nederland*, vol. 5, 19–43.

22 The main reference work for Dutch energy history is Geert Verbong, ed., "Energie," in Schot et al., *Techniek in Nederland*, vol. 2, part 2.

23 In 2000 the Netherlands possessed thirty-nine district heating systems connecting 3,000 kilometers of pipeline and 220,000 consumers. This is less than the district heating system of the Danish capital, Copenhagen, alone. *Energie in Nederland 2000* (Arnhem: EnergieNed, 2000).

24 G. P. J. Verbong, "Grote technische systemen in de energievoorziening," in Schot et al., *Techniek in Nederland*, vol. 2, 115–123; A. N. Hesselmans and G. P. J. Verbong, "Schaalvergroting en kleinschaligheid: De elektriciteitsvoorziening tot 1914," in Schot et al., *Techniek in Nederland*, vol. 2, 124–139; *Eerste Nederlandse systematisch ingerichte encyclopaedie*, 1950, s.v. "Electriciteitsvoorziening"; Hans Schippers, "Statistiek van de gasvoorziening," unpublished manuscript, Eindhoven, table 3a. Copy in the author's collection.

25 "Unified systems" in the terminology of Thomas Hughes; see Hughes, *Networks of Power: Electrification of Western Society 1880–1930* (Baltimore: Johns Hopkins University Press, 1983).

26 A. N. Hesselmans et al., "Binnen provinciale grenzen: De elektriciteitsvoorziening tot 1940," in Schot et al., *Techniek in Nederland*, vol. 2, 157.

27 A. N. Hesselmans et al., "Electriciteitsvoorziening, overheid en industrie 1940–1970," in Schot et al., *Techniek in Nederland*, vol. 2, 222–232.

28 *Eerste Nederlandse systematisch ingerichte encyclopaedie*, 1950, s.v. "Nutsbedrijven."

29 A. Correljé and G. P. J. Verbong, "The Transition to Natural Gas," in B. Elzen, F. W. Geels, and K. Green, *System Innovation and the Transition to Sustainability: Theory, Evidence and Policy* (Cheltenham, U.K.: Edward Elgar, 2005), 114–135; J. L. Schippers and G. P. J. Verbong, "De revolutie van Slochteren," in Schot et al., *Techniek in Nederland*, vol. 2, 203–219; Hans Schippers, *De Nederlandse gasvoorziening in de twintigste eeuw tot 1975* (Eindhoven: Stichting Historie der Techniek, 1997).

30 Geert Verbong, "Dutch Power Connections: From German Occupation to the French Connection," in Van der Vleuten and Kaijser, *Networking Europe*, 217–244. Vincent Lagendijk, *Electrifying Europe: The Power of Europe in the Construction of Electricity Networks* (Amsterdam: Aksant, 2008).

31 Schippers and Verbong, "Revolutie van Slochteren"; Informatie- en Documentatiecentrum voor de Geografie van Nederland, *Kortfattet oversigt over Nederlands geografi* (The Hague and Utrecht: 1974), 28.

32 P. J. M. Groot, *Goederenvervoer per pijpleiding* (Amsterdam: Economisch Instituut voor de Bouwnijverheid, 1991), 22.

33 *Elektriciteit in Nederland 1996* (Arnhem: Samenwerkende Elektriciteitsproductiebedrijven, 1997), 5.

34 W. E. Boerman, "Inleiding," in W. E. Boerman et al., eds., *Het verkeer in Nederland in de XXe eeuw: Tijdschrift van het Koninklijk Nederlands Aardrijkskundig Genootschap* 50 (1933): 333.

35 Verbong, "Dutch Power Connections"; Geert Verbong and Frank Geels, "The Ongoing Energy Transition: Lessons from a Socio-Technical, Multi-Level Analysis of the Dutch Electricity System (1960–2004)," *Energy Policy* 35 (2007): 1025-1037.

36 J. W. Schot, ed., "Transport," in Schot et al., *Techniek in Nederland*, vol. 5, part 1, 13–149.

37 De Vries and Van der Woude, *First Modern Economy*, 13ff.; Van de Woud, *Lege land*, 144–147; R. Filarski, *Kanalen van de koning-koopman: Goederenvervoer, binnenscheepvaart en kanalenbouw in Nederland en België in de eerste helft van de negentiende eeuw* (Amsterdam: Nederlandsch Economisch-Historisch Archief, 1995).

38 H. W. Lintsen, ed., *Twee eeuwen Rijkswaterstaat 1798–1998* (Zaltbommel: Europese Bibliotheek, 1998), 13; Knippenberg and De Pater, *Eenwording van Nederland*, 55–57. Compare Filarski and Mom, *Transportrevolutie*, vol. 1, chapter 1.

39 Not all authors agree on this characterization of the inland navigation network as "national." Compare Rainer Fremdling, "The Dutch Transportation System in the Nineteenth Century," *The Economist* 148 (2000): 521–537; see especially 526ff.

40 Ewout Frankema and Peter Groote, "De modernisering van het Nederlandse wegenet: Nieuwe perspectieven op de ontwikkeling voor 1940," *NEHA Jaarboek* 65 (2002): 305–328; J. R. Luurs, "De aanleg van verharde wegen in Drenthe, Groningen en Friesland, 1825–1925," *NEHA jaarboek* 59 (1996): 211–237.

41 F. L. Schlingemann, "Het verkeer te water," in Boerman et al., *Het verkeer in Nederland in de XXe eeuw*, 334–419; *Eerste Nederlandse systematisch ingerichte encyclopaedie*, 1950, vol. 7, s.v. "Verkeer en vervoer."

42 R. Loman, "De wegen voor gewoon verkeer en het gebruik daarvan," in Boerman et al., *Het verkeer in Nederland in de XXe eeuw*, 480–481 and 489–490.

43 J. W. Schot, "De mobiliteitsexplosie in de twintigste eeuw," in Schot et al., *Techniek in Nederland*, vol. 5, 13–17; Filarski and Mom, *Van transport naar mobiliteit*, vol. 2.

44 Kees Schuyt and Ed Taverne, *1950- Prosperity and Welfare* (Assen-Basingstoke: Van Gorcum-Palgrave/Macmillan, 2004) 157-158; Ministerie van Economische Zaken, *Toets op het concurrentievermogen, Juni 1985*, chapter 4, 4. The railway and waterway networks had been reduced to 2,760 kilometers and 5,000 km, respectively, by 1980.

45 Frank Schipper, *Driving Europe: Building Europe on Roads in the 20TH Century* (Amsterdam: Aksant, 2008).

46 By 1999, 1,920 kilometers. From the early twentieth century on, bike path associations

constructed bicycle paths in nature areas. Starting in the 1920s, separate bike paths often were created when national and provincial roads were upgraded. After 1950 the needs of automobile traffic consumed nearly all the construction efforts, but starting in the 1980s the bike path network was improved, partly by creating connections by means of deserted country roads. Today there is a national long-distance bike path network. (Thanks to Frank Veraart for this information.) See Frank Veraart and Adri Albert de la Bruheze, "Fietsen in de Nederlandse bergen: Achterblijvend fietsgebruik in het zuiden van Limburg in historisch perspectief," in *Studies over de sociaal-economische geschiedenis van Limburg* 46 (Maastricht, 2001), 133–157.

47 *Summa encyclopaedie*, 1976, s.v. "Luchtvaart."

48 Schot, "Concurrentie en afstemming," 30–31.

49 Knippenberg and De Pater, *Eenwording van Nederland*, 49.

50 J. W. Schot, "De mobiliteitsexplosie," in Filarski and Mom, *Van Transport naar mobiliteit*, vol. 2: Gijs Mom and Ruud Filarski, eds., *De Mobiliteitsexplosie 1895-2005*.

51 Schot, "Mobiliteitsexplosie," 13.

52 W. O. de Wit, "Het communicatielandschap in de twintigste eeuw: De materiele basis," in Schot et al., *Techniek in Nederland*, vol. 5, 161). The authoritative reference work on Dutch communication history is W. O. de Wit, ed., "Communicatie," in Schot et al., *Techniek in Nederland*, vol. 5, part 2, 152–282.

53 W. O. De Wit, "Telegrafie en telefonie," in H. W. Lintsen et al., eds., *Geschiedenis van de techniek in Nederland: De wording van een moderne samenleving 1800–1890*, 6 vols. (Zutphen: Walburg, 1992-1995), vol. 4, 273–297.

54 R. De Boer, "De telegraaf," in W. E. Boerman et al., eds., *Het verkeer in Nederland in de XXe eeuw: Tijdschrift van het Koninklijk Nederlands Aardrijkskundig Genootschap* 50 (1933): 633.

55 De Wit, "Communicatielandschap in de twintigste eeuw," 162.

56 Fuchs, "Verkeer en vervoer," 321; *Eerste Nederlandse systematisch ingerichte encyclopaedie*, 1950, s.v. "Telegrafie en telefonie."

57 De Wit, "Communicatielandschap in de twintigste eeuw," 174.

58 *Summa encyclopaedie*, 1978, s.v. "Telegrafie."

59 De Wit, "De ICT-revolutie," in Schot et al., *Techniek in Nederland*, vol. 5, 262.

60 De Wit, "Radio tussen verzuiling en individualisering," in Schot et al., *Techniek in Nederland*, vol. 5, 202–229; see especially 211ff.

61 Others speak of a human-made geography or space-time. See Alain Gras, *Les macro-systèmes techniques* (Paris: PUF, 1997); Thomas Hughes, "Historical Overview," Todd La Porte, ed., *Social Responses to Large Technical Systems: Control or Anticipation* (Dordrecht: Kluwer, 1991), 185–186.

62 Erik van der Vleuten and Cornelis Disco, "Water Wizards: Reshaping Wet Nature and Society," *History and Technology* 20, no. 3 (2004): 291–309; C. Disco, "De verdeling van zoet water over heel Nederland 1940–1970," in Schot et al., *Techniek in Nederland*, vol. 1, 110–121.

63 Willem van der Ham, *Heersen en beheersen: Rijkswaterstaat in de twintigste eeuw* (Zaltbommel: Europese Bibliotheek, 1999), 342–344.

64 Cornelis Disco and Erik van der Vleuten, "The Politics of Wet System Building: Balancing Interests in Dutch Water Management from the Middle Ages to the Present," *Knowledge, Technology & Policy* 14, no. 4 (2002): 21–40.

65 The Netherlands contain over 90,000 kilometers of water supply pipelines and 50,000 kilometers of sewers. See De Groot, *Goederenvervoer per pijpleiding*, 22.

66 Statistics Netherlands, *Vijfenegentig jaren statistiek in tijdreeksen 1899–1994* (The Hague: Centraal Bureau Statistiek, 1994), 12.

67 Henk van den Belt, "Networking Nature, or, Serenghetti Behind the Dikes," *History and Technology* 20, no. 3 (2004): 311–333.

68 Ministry of Agriculture, Nature Management, and Fisheries, *Nature Policy Plan of the Netherlands* (The Hague: 1990); Ministry of Agriculture, Nature Management, and Fisheries, *Natuur voor mensen: Mensen voor natuur—Nota natuur, bos en landschap in de 21e eeuw* (The Hague: 2000).

69 See, for example, "The Pan-European Ecological Network," special issue, *European Nature* 1 (1998).

70 Auke van der Woud, "Stad en land: Werk in uitvoering," in Douwe Fokkema and Frans Grijzenhout, eds., *Rekenschap: Nederlandse cultuur in Europese context 1650–2000* (The Hague: Sdu, 2001), 180.

71 M. Bierman, "Infra Nederland is bijna klaar," www.pz.nl/m.bierman/infra.htm (accessed March 29, 1999).

72 Quotes from Lintsen, *Twee eeuwen Rijkswaterstaat*, 14–15 (author's translation).

73 Schuyt and Taverne, *Prosperity and Welfare*, 123–124.

74 The system-builder concept is developed in Hughes, *Networks of Power*. For a discussion and modification, see Van der Vleuten, "Understanding Network Societies"; Erik van der Vleuten, Irene Anastasiadou, Frank Schipper, and Vincent Lagendijk, "Europe's System Builders: The Contested Shaping of Road, Rail, and Electricity Networks," *Contemporary European History*, 16, 3 (2007): 321-347.

75 For more on the notion of institutional users of systems see Van der Vleuten, "In Search of the Networked Nation," and Van der Vleuten, "Infrastructures and Societal Change."

76 Hughes speaks of "local," "unified," and "regional" systems. See Hughes, *Networks of Power*. For a critical discussion of this historiographical format see Erik van der Vleuten, "Electrifying Denmark: A Symmetrical History of Central and Decentral Electricity Supply Until 1970," Ph.D. diss., University of Aarhus, 1998.

77 Hughes, *Networks of power*, 465.

78 A influential study that makes this claim (on the basis of just a few cases) is P. H. J. van den Boomen and A. N. Hesselmans, "Van kleinschalige naar grootschalige electriciteitsvoorziening: Een analyse van vier electriciteitscentrales 1880–1925," in *Jaarboek voor de geschiedenis van bedrijf en techniek* 3 (1986): 230–251. Their conclusions are partially accepted in Verbong, "Energie."

79 Hans Buiter, *Nederland kabelland: De historie van de energiekabel in Nederland* (The Hague: Stichting Historie der Techniek, 1994), 17 (author's translation).

81 See, for example, Renate Mayntz, "Zur Entwicklung technischer Infrastruktursysteme," *Differenzierung und Verselbständigung: Zur Entwicklung gesellschaftlicher Teilsysteme* (Frankfurt: Campus, 1988), 233-260; Wiebe Bijker, *Of Bicycles, Bakelites and Bulbs: Toward a Theory of Sociotechnical Change* (Cambridge, Mass: MIT Press, 1995); Van der Vleuten, "Electrifying Denmark."

81 Birgitte Wistoft et al., *Elektricitetens aarhundrede: Dansk elforsynings historie*, vol. 2, *1940–1991* (Copenhagen: DEF, 1992), 153–157.

82 Arne Kaijser, "Controlling the Grid: The Development of High-Tension Power Lines in the Nordic Countries," Arne Kaijser and Marika Hedkin, eds., *Nordic Energy Systems: Historical Perspectives and Current Issues* (Canton, Mass.: Science History Publications, 1995), 48.

83 Statistical Office of the European Community, *Energy Statistics Yearbook 1969–1973* (Luxembourg: 1974).

84 I rely here on Richard Hirsh, *Technology and Transformation in the American Utility Industry* (Cambridge: Cambridge University Press, 1989), and Richard Hirsh, *Power Loss: The Origins of Deregulation and Restructuring in the American Electric Utility Industry System* (Cambridge, Mass.: MIT Press, 1999).

85 U.S. Department of Energy, "U.S. Power Grids," www.eere.energy.gov/de/us_power_grids.html (accessed March 24, 2009).

86 Erik van der Vleuten, "Constructing Centralized Electricity Supply in Denmark and the

Netherlands: An Actor-Group Perspective," in *Centaurus* 41 (1999), 3–36; see especially 11–13.

87 Erik van der Vleuten and Rob Raven, "Lock-in and change: Distributed Generation in Denmark in a Long-Term perspective," *Energy Policy* 34 (2006): 3739-3748.

88 In the 1950s and particularly the 1960s, large power companies purchased smaller ones or subsidized their connection to the power grid. Most decentralized producers disappeared, only to reappear in the 1980s and 1990s. Denmark currently has the highest share of distributed generation in the European Union. See Van der Vleuten and Raven, "Lock-in and Change."

89 Lagendijk, *Electrifying Europe*.

90 I draw here primarily on A. N. Hesselmans and G. P. J. Verbong, *De electriciteitsvoorziening in Nederland in de 20e eeuw*, vol. 2: *De electriciteitsvoorziening georganiseerd naar provinciale grenzen, 1914–1945* (Eindhoven: 1997), 10–12 and 28ff. Cf. Geert Verbong and Erik van der Vleuten, "Under Construction: Material Integration of the Netherlands," *History and Technology* 20, no. 3 (2004): 205–226.

91 A. N. Hesselmans et al., *Wisselende Spanning: Een historische verkenning naar de relatie tussen Rijksoverheid en electriciteitssector* (Voorburg: N.V. Electriciteitsbedrijf Zuid-Holland, 1996).

92 Hesselmans, "Binnen provinciale grenzen," 142.

93 Hesselmans and Verbong, *Electriciteitsvoorziening in Nederland*, 37–39 and 41–42.

94 Ibid., 31–36.

95 Hellemans quoted in Hans Buiter and Ton Hesselmans, *Tegendruk: De geschiedenis van Vereniging Krachtwerktuigen`haar bemoeienis met de electriciteitsvoorziening 1915–1998* (Eindhoven: Stichting Historie der Techniek, 1999), 63.

96 In the 1940s Germany had 2,000 power stations supplying the public (1941), the United States 3,800 (1947), that is, 200 to 400 per 8 million inhabitants, or four times the rate in the Netherlands. See Brian Bowers, "Electricity," in Trevor Williams et al., eds., *A History of Technology*, vol. 6 (Oxford: Clarendon Press, 1978), 289–290.

97 J. C. van Staveren, *Electriciteitsvoorziening van Nederland: Overgedrukt uit de Handelingen van het XXVIe Nederlandsch Natuur- en Geneeskundig Congres, gehouden op 30, 31 Maart en 1 April 1937 te Utrecht* (1937).

98 This argument also dates from the 1930s. See D. T. Wiersum, "Koppeling van electrische centrales," *De Ingenieur* 51 (1936): E43.

99 Denmark is a good example. See Van der Vleuten, "Constructing Centralized Electricity Supply."

100 Buiter and Hesselmans, *Tegendruk*, 81ff.; Verbong, "Systemen in transitie," 262–264; Verbong and Geels, "The Ongoing Energy Transition."

101 "The Electric Revolution" and "The Dawn of Micropower," *The Economist*, August 5 2000, 17–18 and 75–77; Fred Pearce, "People Power," *New Scientist*, online news, November 11, 2000, www.terrawatts.com/powerplant.htm; Seth Dunn, *Micropower: The Next Electrical Era* (Washington, D.C.: Worldwatch Institute, 2000).

102 Henri Beunders, "Het volk verovert de media," in Fokkema and Grijzenhout, *Rekenschap*, 303ff.

103 In 1950 the domestic air network accounted for just 157 kilometers of the total of 88,852 kilometers of Dutch airline routes; see Fuchs, "Verkeer en vervoer," 310.

104 J. P.Smits, "Economische groei en structuurveranderingen in de Nederlandse dienstensector, 1850–1913: De bijdrage van handel en transport aan het proces van 'moderne economische groei,'" Ph.D. diss., Amsterdam University, 1995.

105 Ministry of Housing, Spatial Planning and the Environment et al., *National Spatial Strategy*.

106 Van der Woud, *Lege land*, 166.

107 Tydeman quoted in Onno de Wit, *Telefonie in Nederland 1877–1940: Opkomst en ontwikkeling van een grootschalig technisch systeem* (Rotterdam: Cramwinckel, 1998), 92.

108 Van der Woud, *Lege land*; Dirk Maarten Ligtermoet, *Beleid en planning in de wegenbouw* (The Hague: Rijkswaterstaat, 1990).

109 E. Berkers, "Kustlijnverkorting en afsluittechniek," in Schot et al., *Techniek in Nederland*, vol. 1, 71–87; C. Disco, "Een volk dat leeft, bouwt aan zijn toekomst," in Schot et al., *Techniek in Nederland*, vol. 1, 199–207.

110 Ministerie van Landbouw, Natuur en Voedselkwaliteit, *Natuur voor mensen: Mensen voor natuur—Nota natuur, bos en landschap in de 21e eeuw* (The Hague: 2000), 22.

111 Arne Kaijser, "Striking Bonanza: The Establishment of a Natural Gas Regime in the Netherlands," in Olivier Coutard, ed., *The Governance of Large Technical Systems* (London: Routledge, 1999), 38–57; Schippers and Verbong, "Revolutie van Slochteren."

112 De Wit, "Radio tussen verzuiling en individualisering," 208ff.

113 Beunders, "Volk verovert de media," 312.

114 De Wit, "Televisie en het initiatief van Philips," in Schot, *Techniek in Nederland*, vol. 5, 231–260; see especially 248–252.

115 Beunders, "Volk verovert de media," 314.

116 Ibid., 315.

117 Aart Veldman, *Etherpiraterij: De speurtocht en hun prooi* (Author: 2006).

118 Intomart/Scientific Council for Government Policy, *Etherpiraten in Nederland* (The Hague: Staatsuitgeverij, 1982).

119 Agentschap Telecom, *Eindrapport operatie etherflits* (The Hague: 2003); "Radiopiraten bestaan nog steeds," December 3, 2005, www.radio.nl/2003/home/medianieuws/010.archief/2005/12/101528.html (accessed March 26, 2009).

120 Filarski, *Kanalen van de Koning-Koopman*, 366.

121 Van der Woud, *Lege land*, 176.

122 Statement of the Commissie Betuweroute, cited in Robert Coops and Frank van Heijst, *Sporen naar een nationaal project: De Betuweroute—Een procesbeschrijving over de totstandkoming van de Planologische Kernbeslissing Betuweroute en het Tracebesluit Betuweroute* (The Hague: Adviesbureau Awareness, 1998), 30 (author's translation).

123 Tijdelijke Commissie Infrastructuur, *Hoofdrapport* (The Hague: Sdu, 2004). The Betuweroute case was examined in detail in Tijdelijke Commissie Infrastructuur, *Reconstructie Betuweroute: De besluitvorming uitvergroot* (The Hague: Sdu, 2004).

124 Frank Siddiqui, "Een duistere club: De lobby achter de betuwelijn," *Intermediair* 32, no. 51 (1996): 11.

125 Tijdelijke Commissie Infrastructuur, *Reconstructie Betuweroute*, 30–35.

126 Ibid., 32–33 (author's translation).

127 Maarten Duijvendak, *Rooms, rijk of regentesk: Elitevorming en machtsverhoudingen in oostelijk Noord-Brabant (circa 1810–1914)* ('s-Hertogenbosch: Het Noordbrabants Genootschap, 1990), 7–8.

128 Matthias Heymann, "A Fight of Systems? Wind Power and Electric Power Systems," *Centaurus* 41 (1999), 112–136.

129 Van der Vleuten and Disco, "Water Wizards."

130 Cited in ibid., 298.

131 Tijdelijke Commissie Infrastructuur, *Hoofdrapport*, 17–18. For a sample of the rhetoric used by project opponents on the European level, see Action for Solidarity, Equality, Environment and Development Europe, *Free Way for the Free Market? MATA Map of Activities on Transport in Europe*, pamphlet (Prague: A SEED Europe and Car Busters, 1999). Documents relating to Dutch opposition are available at http://geen.betuwelijn.nu.

132 S. A. Reitsma, former member of the board of directors of Dutch Railways, cited in Gijs Mom, Johan Schot, and Peter Staal, "Werken aan mobiliteit: De inburgering van de auto,"

in Schot et al., *Techniek in Nederland*, vol 5, 65.

133 This process started with urban tram lines, before the age of automobility. See Hans Buiter, *Riool, rails en asfalt: 80 jaar straatrumoer in 4 Nederlandse steden* (Zutphen: Walburg Pers, 2005).

134 Eurostat, *Energy, Transport and Environment Indicators: Data 1992–2002* (Luxembourg: European Communities, 2005), 128; Statistics Netherlands, *Vijfennegentig jaren statistiek in tijdreeksen 1899–1994* (Voorburg: Centraal Bureau Statistiek, 1994), 109; Statistics Netherlands, *Overledenen per belangrijke primaire doodsoorzaken* (Voorburg: Centraal Bureau Statistiek, 2006), available online at http://statline.cbs.nl (figures include direct casualties only).

135 The study of users as agents giving meaning to infrastructure was pioneered in David Nye, *Electrifying America: Social Meanings of a New Technology* (Cambridge: MIT Press, 1990), and Claude Fischer, *America Calling: A Social History of the Telephone to 1940* (Berkeley: University of California Press, 1992). The study of users as co-constructors of technical change is prominent in Nelly Oudshoorn and Trevor Pinch, eds., *How Users Matter: The Co-Construction of Users and Technology* (Cambridge, Mass.: MIT Press, 2003).

136 Mom, Schot, and Staal, "Werken aan mobiliteit," 61.

137 On women's advisory committees, see Wiebe E. Bijker and Karin Bijsterveld, "Women Walking Through Plans: Technology, Democracy, and Gender Identity," *Technology and Culture* 41, no. 3 (2000): 485–515; R. Oldenziel, ed., "Huishoudtechnologie," in Schot et al., *Techniek in Nederland*, vol. 2, part 1; E. M. L. Bervoets, ed., "Bouw," in Schot et al., *Techniek in Nederland*, vol. 6, part 2.

138 Van der Vleuten, "In Search of the Networked Nation," and Van der Vleuten, "Infrastructures and Societal Change." These infrastructure users are "institutional" in the sense of being formal organizational structures (as opposed to informal institutions) that structure social order, are identified by a social purpose and permanence, and transcend the individual level. The concept is inspired by the notion of second-order large technical systems; see Ingo Braun, "Geflügelte Saurier: Zur intersystemische Vernetzung grosser technischer Netze," in Ingo Braun and Bernward Joerges, eds., *Technik ohne Grenzen* (Frankfurt: Suhrkamp, 1994), 446–500.

139 I rely mainly on Ernst Homburg, ed., "Chemie," in Schot et al., *Techniek in Nederland*, vol. 2, part 2, and E. Wever, "Olieraffinaderij en petrochemische industrie: Ontstaan, samenstelling, en voorkomen van petrochemische complexen," Ph.D. diss., Groningen University, 1974.

140 Homburg, "Chemie," 390ff.

141 C. J. Klaverdijk, *Het vervoer van olie per pijpleiding* (Rotterdam: s.n., 1969), 17.

142 See map in G. P. J. Verbong et al., "Ter introductie," in Schot, *Techniek in Nederland*, vol. 2, 10.

143 Homburg, "Chemie," 389–390 and 397.

144 The chlorine distribution network as of 2000 is described in Adviesraad gevaarlijke stoffen, *Chloor, opslag en gebruik: Advies over PGS 11* (The Hague: 2006), appendix 1, available at www.adviesraadgevaarlijkestoffen.nl/getfile.asp?id=65 (accessed March 29, 2009).

145 Rob Janssen, "Stop de rijdende bommen," *Tribune*, February 16, 2001; see Ronald van Raak, *Rood sein voor de chloortrein: Akzo Nobel als maatschappelijk verantwoord ondernemer* (Rotterdam: Socialist Party Scientific Committee, 2000); Ronald van Raak, Harry Voss, and Vincent Mulder, *Rood sein voor de chloortrein: De onmogelijke verantwoordelijkheid van gemeenten* (Rotterdam: Comité rood sein voor de chloortrein, 2001).

146 Ministry of Housing, Spatial Planning and the Environment, "Het principeaccoord tussen AKZO en de overheid," press release EV2002.044867, July 5, 2002.

147 Kim van Keken, "Laatste rit voor de ammoniaktrein', *De Volkskrant*, March 19, 2009.

148 Mila Davids, "The Fabric of Production: The Philips Industrial Network," *History and Tech-*

nology 20, 3 (2004): 271–290.

149 Janneke Hermans and Onno de Wit, "Bourses and Brokers: Stock Exchanges as ICT Junctions," *History and Technology* 20, no. 3 (2004): 227–247.

150 M. Bolk and B. Straatman, *40 jaar DPO: Defensie Pijpleiding Organisatie* ([s.l.], [s.n.]. 1997); Groot, *Goederenvervoer per pijpleiding*, 43.

151 Van der Vleuten and Disco, "Water Wizards."

152 J. P. C. M. van Hoof, "Met een vijand als bondgenoot: De rol van het water bij de verdediging van het Nederlandse grondgebied tegen een aanval over land," *Bijdragen en mededelingen betreffende de geschiedenis der Nederlanden* 103 (1988): 622–651; G. B. Janssen, "De IJssellinie in historisch perspectief," in R. Beekmans and C. Schilt, eds., *Drijvende stuwen voor de landsverdediging: Een geschiedenis van de IJssellinie* (Zutphen: Stichting Menno van Coehoorn, 1997), 19–33.

153 Van Hoof, "Met een vijand als bondgenoot."

154 Quoted in J. C. H. Haex, "Inleiding," in Beekman and Schilt, *Drijvende stuwen*, 12.

155 J. R. Beekmans, "De gevolgen van de inundaties," in Beekmans and Schilt, *Drijvende stuwen*, 169–182.

156 Van der Vleuten, "In Search of the Networked Nation." The following section is based on Jozien Jobse-van Putten, *Eenvoudig maar voedzaam: Cultuurgeschiedenis van de dagelijkse maaltijd in Nederland* (Nijmegen: SUN, 1996), 499–506; A. H. van Otterloo, "Voeding" [Diet], in Schot et al., *Techniek in Nederland*, vol. 3, part 2; Jan Bieleman, ed., "Landbouw," in Schot et al., *Techniek in Nederland*, vol. 3, part 1. See also Adri Albert de la Bruhèze and Anneke van Otterloo, "The Milky Way: Infrastructures and the Shaping of Milk Chains," *History and Technology* 20, no. 3 (2004): 249–269.

157 For example, Jan Nieboer, "De Melkfabriek," Digitaal Dorp Zevenhuizen, available at "De geschiedenis van het dorp Zevenhuizen," www.7huizen.nl/geschied/Melkfabriek.htm (accessed February 19, 2009).

158 Priester, "Melkveehouderijbedrijf," 112–114.

159 Anneke van Otterloo, "Prelude op de consumptiemaatschappij in voor- en tegenspoed 1920-1960," in Schot, *Techniek in Nederland*, vol. 3, 264.

160 Erik van der Vleuten, "Feeding the Peoples of Europe: Transnational Food Transport Infrastructure in the Early Cold War, 1947-1960," in Alec Badenoch and Andreas Fickers, *Materializing Europe? Transnational Infrastructures and the Project of Europe* (London: Palgrave MacMillan, in press).

161 Ibid, 270–272.

162 Disco and Van der Vleuten, "Politics of Wet System Building." For a comparison of different regime concepts see Schot et al., "Methode en opzet van het onderzoek," in Schot et al., *Techniek in Nederland*, vol. 1, 38, and Kaijser, "Helping Hand."

163 Disco and Van der Vleuten, "Politics of Wet System Building."

164 Harry Lintsen, "Two Centuries of Central Water Management in the Netherlands," *Technology and Culture* 43, no. 3 (2002): 549–568; Lintsen, *Twee eeuwen Rijkswaterstaat*.

165 Kaijser, "Helping Hand."

166 Lintsen, *Twee Eeuwen Rijkswaterstaat*, 37.

167 Siep Stuurman, *Wacht op onze daden: Het liberalisme en de vernieuwing van de Nederlandse staat* (Amsterdam: Bakker, 1992), 13.

168 De Wit, "Telegrafie en telefonie," 282.

169 Lintsen, *Twee eeuwen Rijkswaterstaat*, 81; Augustus Veenendaal, *Railways in the Netherlands: A Brief History 1834–1994* (Palo Alto: Stanford University Press, 2001); A. J. Veenendaal Jr. *De ijzeren weg in een land vol water: Beknopte geschiedenis van de spoorwegen in Nederland 1834–1958* (Amsterdam: De Bataafsche Leeuw, 1998).

170 Fuchs, "Verkeer en vervoer," 295–296; Stieltjes, "De spoor- en tramwegen," 433–438.

171 De Wit, *Telefonie in Nederland*, 118–120.

172 Fuchs, "Verkeer en vervoer," 309.

173 Hesselmans and Verbong, *De electriciteitsvoorziening in Nederland*, 22–25.

174 Lintsen, *Twee eeuwen Rijkswaterstaat*, 285; Van der Ham, *Heersen en beheersen*, 316–317.

175 Thomas Hughes, *Rescuing Prometheus* (New York: Pantheon, 1998).

176 Schuyt and Taverne, *Prosperity and Welfare*, chapter 15.

177 Van der Linde van Sprankhuizen, cited in Hans Buiter and Kees Volkers, *Oudenrijn: De geschiedenis van een verkeersknooppunt* (Utrecht: Matrijs, 1996), 97.

178 Van der Ham, *Heersen en beheersen*, 321ff.

179 Ibid., 331.

180 Ibid., 315, 334.

181 M. L. J. Dierikx et al., "Van uithoek tot knooppunt: Schiphol," in Schot et al., *Techniek in Nederland*, vol. 5, 133.

182 G. Verbong et al., *Een kwestie van lange adem: De geschiedenis van duurzame energie in Nederland* (Boxtel: Aeneas, 2001), 105–109.

183 Van Otterloo, "Ingredienten"; de la Bruhèze and van Otterloo, "The Milky Way," 297–298.

184 Knippenberg and de Pater, *Eenwording van Nederland*, 210; Dorette Corbey, "Overheid moet grenzen verleggen," *De Staatscourant*, June 29, 2001; Minister of Economic Affairs Annemarie Jorritsma, "Privatisering: Overheid verleden tijd?," speech, Nieuwe Akademie, Groningen, April 11, 2002.

185 Van der Ham, *Heersen en beheersen*, 345–346.

186 Mila Davids, *De weg naar zelfstandigheid: De voorgeschiedenis van de verzelfstandiging van de PTT in 1989* (Hilversum: Verloren, 1989); Verbong and Geels, "Ongoing Energy Transition."

187 For the notion of ecotechnical system building see Thomas Hughes, *Human-Built World* (Chicago: University of Chicago Press, 2004).

188 Van der Ham, *Heersen en beheersen*, 328; Buiter and Volkers, *Oudenrijn*, 85 and 97.

189 See also Wiebe E. Bijker, "The Oosterschelde Storm Surge Barrier: A Test Case for Dutch Water Technology, Management, and Politics," *Technology and Culture* 43, no. 3 (2002): 569–584.

190 M. Wolsing, "De veronderstellingen achter het NIMBY beleid," *Beleid en Maatschappij* 1993, 143–151; C. Lambers et al., *Versnelling juridische procedures grote projecten: Voorstudies en achtergronden V85* (The Hague: Scientific Council for Government Policy, 1994); Scientific Council for Government Policy, *Besluiten over grote projecten*, report to the government no. 46 (The Hague: 1994); Adviesbureau Awareness et al., *Sporen naar een nationaal project*.

191 Tijdelijke Commissie Infrastructuur, *Reconstructie Betuweroute*, 39.

Washing day in Urk, in 1926. The limited public space on this island hardly allowed separate space for separate uses. It was common that the street was used for hanging out laundry, held up high with the help of boathooks. The shipper with his horse and carriage had to make his own way through the laundry on washing day.

3 Site-specific Innovation: The Design of Kitchens, Offices, Airports, and Cities

Adrienne van den Bogaard

To understand what the term "site-specific innovation" means, let's look at a simple and very Dutch example: a small urban space. Along the Oudezijds Kolk in Amsterdam, a narrow street squeezed between a canal on one side and large houses on the other, today's passerby may notice so-called "rail lamps" (see figure page 126). These lamps were a design solution to the problem of two artifacts competing for limited urban space. In the style of the city's seventeenth-century bridges the canal was lined with a rail to prevent people from falling in. When, in the 1910s, street lighting was also deemed important for safety reasons, engineers from the city's Department of Bridges and Sluices had to solve the issue of where to put the new lampposts. On the canal side of the rail there was clearly no space because of the water, and placing them on the other side of the rail would further narrow the street.

The solution was to place the lamps on the granite blocks that bordered the canal and to bend the existing rail around the lampposts, which allowed the rail to continue uninterrupted. This solution implied adaptation of the rail design but also of the lamppost because instead of being anchored in the ground, as would normally be the case, it had to be somehow fastened to the granite blocks. Thus specifically for this situation engineers from the city's Electricity Company designed a new and very slender lamppost base with screw bolt holes that could be screwed directly onto the edge of the granite blocks.

Entangled Technologies at Specific Locations

Our railing–lamp post solution to the problem of how to integrate lamps with the railing illustrates several relevant aspects of site-specific innovation. First, it involves combining two functions that were previously separated in a single artifact: integration of the lamp and the rail lining the canal. Site-specific innovation frequently results from the interference of technologies—in our example from two technologies' claim to scarce space. Their interference in a specific location posed a challenge to the city's departments, and this suggests a second aspect, namely, the need for coordination: the problem had

Limited space and new urban needs such as traffic safety and good street lighting occasioned the development of the "rail lamp" on a narrow street in Amsterdam, the Oudezijdskolk. Through collaboration between the Bridges and Sluices Department and the Municipal Power Company both the railing and the lamp post were adapted, resulting in a new technical device.

to be identified, discussed, and solved. In our example, coordination was needed between the city's Electricity Company and the Department of Bridges and Sluices, each having its own interests and responsibilities. In this case the combination of site-specific material demands and the interaction between different actors determined the problem's solution, and it may also have had an influence on subsequent problem solutions.

Of course, our rail-lamp example involves a rather minor innovation, a case of ad hoc adaptation of two technologies resulting in a new artifact with a very local application. However, the interlinking of technologies and the related challenges may also produce an altogether new artifact with more general applications. Consider, for instance, another innovation involving updating the city of Amsterdam's water-line system, which dates from the early twentieth century. The new water system somehow had to cross under bridges, but the city's low modern bridges did not always leave sufficient space for the water pipes without interfering with shipping.

In 1902 Amsterdam was improving the water supply system to guarantee safe drinking water for the growing population. In order to cross the city's existing low bridges the new main water pipe of thirty inches had to be replaced by four fifteen inch pipes as seen here on the Prins Hendrikkade. This so called *zeugstuk* was developed by the local water company and the Department of Bridges and Sluices.

To solve this problem, a single pipe with a diameter of thirty inches was replaced by four fifteen-inch-diameter pipes that together had the same capacity but would fit under the bridge (see photo).[1] Coordination between the local water company and the Department of Bridges and Sluices was needed to develop the new technology. Although the resulting artifact, the so-called *zeugstuk*, involved a very specific linking between the one big pipe and the four smaller pipes, it could also be deployed in other locations.

A characteristic feature of site-specific innovation is that aside from the functional requirements that already applied to the separate artifacts or systems, additional design requirements follow from the need to find a solution for the interference between heterogeneous systems and technologies. Although interference between technologies and systems involves more than just a struggle for scarce space, in cities, of course, space is always a crucial factor. Our third example, the Amsterdam Plan Zuid (Plan for Amsterdam South), further illustrates both this spatial aspect and the role of coordination and planning in site-specific innovation. This was a plan for an entirely new subdivision in the southern part of the city. When the project got started, in 1915, all the infrastructure still had to be put in: streets, utilities, public facilities, and so forth. The planners had set aside one square, the Hygiëaplein, to be the location of both a new school and an electricity substation, so several municipal departments had to collaborate to complete the plan, including the Department of Power, the Public Works Department, the Finance Department, and the Department of Education. Moreover, the city's Education Inspector had to become involved to ensure that the electricity substation didn't pose a hazard to the schoolchildren. Furthermore, the pedestrian routes providing access to the substation should be separate from those for the school. Accordingly, a short tunnel for access to the substation was added to the plan. This decision led to a new problem: How should the cost be divided between the city's Department of Power and the Department of Education?

Interference between the Hygiëaplein facilities and the new neighborhood's overall traffic system also led to new design considerations. The increasingly motorized traffic would pose a danger to the children, so the school board proposed to implement special traffic regulation fifteen minutes before and after school to keep the children separate from the motorized traffic. To guard the students' health and ensure good air quality, the school building was equipped with a new ventilation system that could filter out auto exhaust. Another safety precaution was the location of the entrance to the school in an inner courtyard—a solution that was subsequently used in many Dutch schools to effect a complete separation of the children's playground area from the space used by traffic. In this case the outcome of interfering technologies and systems at different sites show not just entanglement but also *disentanglement*, such as creating separate access routes for the school and the electricity substation and placing the playground in an inner courtyard. This disconnecting of systems also called for coordination, notably regarding designs' fine-tuning and decisions on who pays for the costs involved. The example underscores the large number and diversity of technologies in cities. By definition cities are unique sites with specific possibilities and problems, depending on the kind of technologies and systems in place or newly introduced. Various functions such as housing and traffic are simultaneously at issue, and the effort to integrate them leads to quite specific challenges and solutions. In the course of the twentieth century, as most Dutch cities continued to see steady growth, city administrators and other actors were increasingly forced to deal with this interplay of needs.

The city is not the only site that reveals this dynamic of interfering and interlinking technologies and the innovations developed in response. A modern home, too, serves as the end point of outputs of a great many large technical systems, including electricity, gas, telephone, water, and the sewer system. Spatiality, too, plays a key role in the home in a variety of ways. Where in 1890 a Dutch working-class home consisted of a single space with a number of box beds, in the twentieth century more separate spaces were added such as a kitchen, a living room, and bedrooms.

An Amsterdam laborer's home built prior to the Housing Act, in 1953, just before the dwelling was renovated. The outdated features are easily recognizable. Household functions were all carried out in one space. The water closet and the sink were installed in former cupboards. The box bed, prohibited by the Housing Act, had no ventilation; neither did the kitchen. Newly developed technologies were embedded in existing spaces. Apparently, blueprints for households only applied to new homes to be built, just as urban-planning designs mainly applied to the new suburbs. In existing homes, adaptations came about only gradually.

Around 1900 the sink was primarily conceived as a technology for hygienic discharging of wastewater—hence its name—but gradually this technology became part of efforts to design a modern, efficient kitchen. Thus the kitchen itself turned into a site that was meant to highlight the modern household. The kitchen was no longer merely a "physical construction with characteristic tools" but turned into an object for discussion of the "ideal kitchen."[2] In other words, technology development was not only geared toward providing ad hoc solutions to problems, but was also shaped by particular visions of the practices performed at the site at hand. The example of the kitchen thus reveals another aspect of entangling: it involves not only technologies and systems, space, and coordination, but also ideas about preferred practices and efforts at bringing about changes in behavior—what one could call social construction of behavior.[3]

There are numerous other sites that induce a pattern of linkage and innovation, such as offices, farms, seaports, and airports. Historians of technology have devoted much attention to the development of individual technologies used at these sites. Such a perspective, however, runs the risk of ignoring users who need to incorporate and integrate these new products and technologies into their daily practices. Ruth Schwartz Cowan has introduced the concept "consumption junction" to show that households serve as sites where production and consumption meet and decisions are made about the actual use of specific products; she looks specifically at household technologies.[4] Inspired by the work of Cowan, I introduce the notion "innovation junction" to refer to the phenomenon that producers, users, experts, and administrators simultaneously play a role in realizing innovations. Existing practices and new technologies interfere and get entangled in specific locations where challenges and ideals lead to innovations.[5]

For example, the factory was a type of location where innovations moved beyond ad hoc problem solving and where ideas and efforts to control unruly practices played a major role. Industrial engineers in particular developed general visions on how to optimze the production processes. The engineers proposed to achieve higher productivity not by mechanizing separate actions but by mechanizing the factory as a whole. They defined the factory in terms of flows that could be designed and mechanized in an integrated manner. The factory's physical space was eventually redesigned, leading to the typical late-nineteenth-century factory building. Factories convert input into output by means of human labor and technology, and this constitutes the core of a factory's processing, or its throughput, the material conversion of input-material into output-material. Industrial engineers offered the key to designing a factory as one "great efficient machine" through their focus on the coordination of the main links in the production process.[6] In other words, they moved from ad hoc mechanization to deliberate and integrated innovation on location. The views on scientific management by the American Frederick W. Taylor and his followers were geared toward measuring workers' contribution to the production process. Taylor started from the view that employers and workers unjustifiably considered each other as enemies. To succeed, he had to convince both parties that his science reflected their shared interests: "What constitutes a fair day's work will be a question for scientific investigation instead of a subject to be bargained and haggled over."[7] Taylor's basic management assumptions strongly influenced the thinking of engineers and managers, also in the Netherlands.[8] The combination of technical development and the new rational design of labor gave rise to a new type of factory.

The emergence of location-specific innovation patterns was a result of the recognition by administrators, professionals, and intermediary actors that their problems were not unique and that one could learn from others' experiences in other cities, households, factories, or airports. These actors visited one another's sites. In this way, new solutions could be further disseminated, for commercial as well as professional reasons. Starting in the 1910s and 1920s the role of professionals and administrators became more prominent in the creation of design and their adoption or implementation. In the case of cities, for example, the new technological discipline of urban planning emerged. Again, there was a tension between ongoing and slowly changing practices on the one hand and particular ideals and efforts toward "modernization" on the other.

In this chapter I investigate the innovation dynamic that moves beyond strictly local problem solving at particular sites: Are there any historical patterns in how innovations occurred in specific locations? I analyze this historical innovation dynamic for four types of location: the home, especially the kitchen; the office; the airport; and the city. For the household as a location, the period from 1900 to 1930 was especially important because of the transition during that time frame to so-called "modern" households. Being modern meant, among other things, following hygienic practices, listening together to the radio, establishing clearly separate roles for wives and husbands, and living in a house with separate rooms. In the discussion of the office the period from 1910 to 1950 is central because the integration of various functions, such as producing and reproducing information, resulted in the evolution of the modern office. Our analysis of Amsterdam Schiphol Airport's transformation shows the development from an airstrip in 1916 to a complex sociotechnical ensemble geared toward connecting multiple airside and landside functions in the 1970s. Finally, we turn to the city as an even more complex case of site-specific innovation, one that because it comprises numerous specific locations can be called a *junction of junctions*. Our discussion concentrates on the period from 1920 to 1940, when the ideas of a new group of urban developers who aimed at organizing this junction of junctions through the *art of synthesis* became active. Together, these cases offer a picture of long-term changes between 1910 and 1950. In the concluding section we reflect on the dynamic of site-specific innovation after 1950.

New Technologies and Rationalization in the Household

The household as a type of location has changed drastically in the course of the twentieth century, not because of changes in household functions—food preparation, clothing care, house cleaning, personal care, raising children, and leisure—but because of major changes in how these functions were performed. Take the house itself: around 1900 an average Dutch working-class dwelling consisted of a single space of some forty square meters with a number of box beds set in alcoves.[9] Water had to be obtained from a pump outside the home; the outhouse was outdoors, as its name implied; doing the laundry took a good part of the day; and the day's hot meal, which for poor families mostly consisted of potatoes, simmered on a stove that also heated the room. Hot water for the laundry or taking a bath was either purchased or heated on the same stove. A paraffin lamp or candles lighted the room. On average, eight people lived in such a space, where the housewife was in charge and carried out most household tasks.[10] Today a lower-income Dutch family has a house with a kitchen, a bathroom, a living room, and several bedrooms, totaling eighty to one hundred square meters. The average family size is 2.3 persons.[11] All homes have gas and electricity; kitchens have an electric or natural gas stove, and already in the early 1970s central heating was available in three quarters of all Dutch households. Doing the laundry still takes about as much time, but today's automatic washers have made it a significantly less burdensome task. The time spent on various tasks has slightly shifted: less time is devoted to household chores and slightly more time to taking care of children. Although this brief sketch ignores that functions were differentiated earlier and new technologies introduced sooner in wealthy households, and that clear differences existed between city and countryside, it is still possible to speak of general patterns. In this section we concentrate on the development of household practices such as cooking and ironing in the period from 1900 to 1930 to show how the interlinking of technologies at home occurred in interaction with efforts aimed at improving home-practices.

Technologies and Practices: Ad Hoc Entanglement, 1890–1918

Around 1900 an array of technologies was potentially applicable in households, but whether or not these were adopted was another matter. Various factors played a role, such as price, the embedding of new technologies in existing or new practices, and the images and ideals of being modern or being a model housewife. From the outside, the "mechanization of households" involved an obdurate process, because technological innovations nearly always challenged existing solutions.[12] For instance, at first the effort to promote gas as a new cooking technology was hardly successful because the coal or paraffin stoves already in use were less expensive, and these technologies could be further developed by building on already existing cooking practices.[13] Other factors besides housewives' habits and the costs of the technology came into play. The passage of the Dutch Housing Act in 1901 had drastic implications for the single-family home.[14] Among other things, this law banned the use of alcoves, which had previously been used as sleeping areas but were now seen to be unhygienic. Thus, the law stimulated the design of separate spaces in homes, such as kitchen, living room, and bedrooms. By introducing homes that had a separate kitchen, the function of meal preparation now got a separate space in the home, and heating and cooking became separate functions, each with its own artifacts. When living and preparing meals still occurred in one space, coal or paraffin stoves were used for both cooking and heating. But the introduction of separate kitchens turned the coal stove's advantage—it could build up a very high heat—into a disadvantage because it generated too much heat for a kitchen. As a result, energy companies eagerly began to develop natural gas and electric appliances specifically for cooking meals. Women simply had to learn how to handle these new stoves.[15] The kitchen stove's development, then, can be tied directly to spatial changes effected by the enactment of the Housing Act.

Producers of new technologies sometimes could profit from views advanced by household professionals. This new profession emerged after 1880, when the first schools were established for training professionals in domestic science. Schools already existed to teach girls how to become good housewives or housekeepers, but now a more advanced teacher training facility was set up for training household experts. Increasingly, household guides were published that women could check for advice on how to do everyday chores. This largely amounted to the translation of informal knowledge of household work into formal knowledge that was collected, developed, and guarded by household experts.[16] These experts, however, also wanted to modernize. Their handbooks diffused knowledge about all sorts of new technologies and artifacts and were quite detailed in providing technical explanations of modern developments. For example, one handbook indicated that "the new half-watt lamp [with a metal wire; developed around 1915] was an improvement on the tungsten light bulb because its wire does not burn in a vacuum, but in a gaseous atmosphere that chemically does not react with the wire."[17] Housewives were apparently expected to be interested in the technology behind such developments—as if the handbook's authors expected housewives to be like themselves.

These handbooks constitute a major source for establishing the extent of the entanglement of technologies and exploitation of scarce space. These technologies were generally treated separately. Some handbooks addressed the issue of the artifact's suitability for the room for which it was intended. In *Handleiding voor de huisvrouw* (*Manual for Housewives*) C. J. Wannée, a former instructor at the Amsterdam domestic science school, emphasized that housewives who purchased a heating device should take into account not only the technical aspects, but also the place, size, humidity conditions, and purpose of the room where the new device was to be placed. A heater was also expected to heat the room evenly, a requirement a hearth did not meet, which is why Wannée rejected the hearth.[18] The same manual also warned against interference of technologies, namely the unpleasant proximity of cessspool and drinking water:

> A rain reservoir is built into the ground with solid masonry so that no dirt from the soil can seep in. Especially dirt from the cesspool is reason for concern because the bacteria found in feces may easily end up in the water and contaminate it, which is why today, in building new outhouses, one has to make sure that the rain reservoir and the cesspool are ten meters apart.[19]

A specific design requirement was articulated, aimed at separating two sorts of flows: feces and drinking water. A similar recommendation was to ensure maximum distance between the rain reservoir and the sink with its drain.

It is obvious that manufacturers paid attention to the potential interference of technologies on location, even if they largely tried to provide ad hoc solutions, for example, regarding the separation of flows. Likewise, technologies were combined, but these basically involved special exploitations of possibilities, such as linking ironing and cooking. In the middle of the nineteenth century ironing clothes with a heated iron had become a common practice, using, for example, cast-iron boxes holding heated pieces of coal. (The mangle, used for mechanically pressing out water and flattening laundry, didn't require heating.) There were many disadvantages to irons heated with coal, and various producers concentrated on optimizing ways of heating the iron. Housewives who did their own laundry could heat their iron on their stoves. Etna, a Dutch stove manufacturer, developed a natural gas stove that integrated a gas-heated iron: a burner that served as holder and heater for the iron could be screwed onto a tiny gas burner in the middle, otherwise used for simmering food.[20] The development of the kitchen as a separate space for cleaning vegetables and preparing meals was also closely tied to the water supply: by around 1900 half of all Dutch households had a tap with clean water at their disposal.[21]

Planned Entanglement of Technologies in Kitchens 1918-1940
The development of the kitchen as a separate space, which partly resulted from the Housing Act, was accompanied by new technologies (heaters,

stoves, water faucets). This gave rise to a concept of the kitchen as a place where housewives performed specific activities and which therefore needed to be designed as optimally as possible. The kitchen space and the *flows* to be accommodated had to be brought in line. Various actors such as manufacturers, architects, household professionals, and housewives contributed to gradually transforming kitchens into *modern* kitchens.[22]

In the 1920s the kitchen was given explicit attention by Dutch architects and domestic science experts. One example is the architect L. Zwiers, who in 1924 published *Ons huis: Hygiëne en gerieflijkheid* (*Our Home: Sanitation and Comfort*), a book that addressed the building of single-family homes in the context of the Housing Act. The author discussed all spaces of a house, including the kitchen. Standards for kitchen design were indicated in detail. "Most kitchens have a rectangular shape. This is also the most efficient shape."[23]

And: "The water faucet should be as high as possible above the bottom of the sink, so that a bucket, jug or bottle will fit underneath. It is efficient to attach this faucet as close as possible toward one back corner of the sink and have the drain in the corner diagonally across from it. This arrangement will cause the sink to be rinsed automatically."[24] The use of the term "efficient" reveals the influence of efficiency thinking. The criteria were quite diverse, ranging from the way in which a tilting window had to be fastened to specific advice on avoiding baseboards so as to prevent dust traps. The glass of the windows should be light blue to keep flies out of the kitchen—a practice that had proved successful in abattoirs. The woodwork, too, had to be painted cobalt blue for the same reason.[25] Zwiers did not go as far as to define a kitchen prototype; he merely summed up a series of unrelated suggestions, which coincidentally were also provided by domestic science experts.

The first comprehensive kitchen designs came from Germany. Marja Berendsen has studied the development of the so-called *Hollandkeuken*, which was based on a German model, into the *Bruynzeelkeuken*.[26] There is a striking similarity between the transformation of factories and the transformation of kitchens into what was called efficient *machines*. The Institute for Domestic Science Advice, established in 1926 by the Dutch Association of Housewives (NVVH), advocated efficiency as one of the main criteria for new household technologies.[27] Tellingly, a 1929 commentary by the NVVH argued that the *Hollandkeuken* served the "household as factory" in a "rational" fashion.[28] New NVVH proposals on kitchen design reflected this same view. They sought to minimize distances between different kitchen areas, maximize hygiene, and make optimal use of modern technologies and materials. Accordingly, the NVVH redefined the household in terms of flows and practices that needed

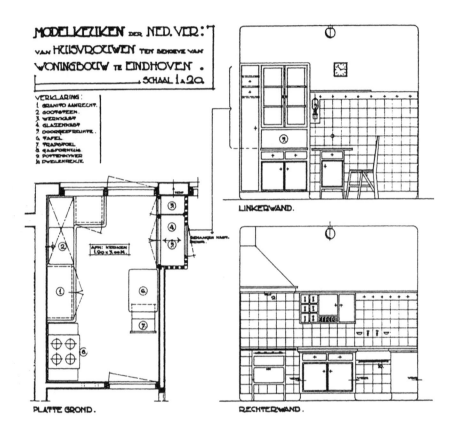

In the 1930s, the very idea of designing a kitchen was itself an innovation, with contributions from many actors such as architects and household professionals. In this design of a standard kitchen (*modelkeuken*), sponsored by the Dutch Association of Housewives (Nederlandse Vereniging van Huisvrouwen) for use in a 1934 housing project in Eindhoven, all household techniques and activities are brought into line spatially. The key: (1) granite counter, (2) sink, (3) utensil cabinet, (4) china cabinet, (5) serving hatch, (6) table, (7) chair, (8) stove, (9) light, (10) towel rack. The "serving hatch" optimized transport of meals to the dining area and of dirty dishes back to the kitchen. A clock made it possible to plan all the household activities carefully, and a ceiling fixture provided adequate light. The unit consisting of the sink with the countertop and kitchen cabinets is a typical example of a technical ensemble that integrated multiple functions. The plan of this model kitchen shows a bird's-eye view, and the right and left walls.

to be optimized, while also advocating further technological modernization.

The notion of the efficient kitchen led to small and large innovations. This is clear in the case of a kitchen co-designed by the architect K. Limperg (a member of "De 8," a group of architects), the engineer G. J. Meyers, and a household expert from Amsterdam, R. Lotgering-Hillebrand.[29] The first innovation involved the design for the comprehensive kitchen. A kitchen had to be viewed as a "modern workshop, a sort of laboratory." A small kitchen was a comfort, for one who did not have to go back and forth needlessly all the time. The team studied small kitchens in airplanes and railway dining cars—spaces where every inch counted, so that their layout (and vertical spatial use) would reflect others' efforts to achieve optimal spatial functionality. Existing kitchens were analyzed on the basis of visits and photos. This led to a distinction of three types of homes: the inexpensive flat, a middle-class flat, and a luxury flat, each with its own "ideal" kitchen design. This design might then be used to mass-produce standard kitchens for each of the three types. A design is a sketch of a planned order, and the design by Limperg, Meyers, and Lotgering-Hillebrand contains many examples of functional and spatial integration of technology. In some cases it involved a minor innovation that nicely captured the underlying thinking about the kitchen and the home:

> To keep out cooking smells and noisiness we recommend the application of double doors for the serving hatch in between kitchen and dining room. The door on the dining room side can be made foldable, so that in a folded-down position it may serve as table top.[30]

The kitchen was seen as a factory that converted input (groceries) into output (meals) by means of specific activities, technologies, and spatial distances. The process of this throughput had to be optimized. The meals that were prepared had to be brought into the adjoining living room or dining room. A direct door to the living room was not a viable option because of the spread of smells from the kitchen and because a door would mean a loss of space in the kitchen for the counter or cupboards.[31] The solution was a service hatch with double doors. This solved the two problems in one design: it provided a barrier to cooking odors and served as a direct connection between dining room and kitchen. In the case of scarce space yet another functional requirement could be added by having the little hatches function as a table top—which in turn came with various requirements involving the solidity and hardness of the wood used. Preferably the distance between the stove and the hatch was as short as possible. By integrating the hatch into a china cabinet, the drawer with cutlery could be placed right below the hatch, which allowed for easy handing over of clean forks, knives, and dishes to those in the dining room.[32] The serving hatch is an example of functional technology entanglement resulting from an integral vision on kitchens as specific locations that should fit most efficiently into their immediate environment. Another example involved a kitchen counter design requirement:

> The sink should be located in the middle of the counter.... Next to the sink, on the side where one puts the cups and plates after rinsing, the counter has to have "drip grooves" that empty in the sink.... Along the counter's outer edge there has to be a groove that empties into the sink as well, so that no water will drip down the fronts of the cupboards below the counter.[33]

Converting dirty dishes (input) into clean dishes (output) was another process that kitchens had to facilitate optimally. Thus the counter had to function as working space during cooking and as mechanism for discharging wastewater. Moreover, the counter top should stick out at least five centimeters, so that women would have room for their knees and not hurt them by hitting the cupboards underneath. A glass window above the counter at the side of the kitchen allowed a housewife to check on what was going on in the dining room.

Today many of these design standards that originated in the 1920s and 1930s come across as self-evident if not trivial. This suggests that efficiency as ideology has become materialized in the space and technology of the kitchen. In some other respects efficiency as ideology has become obsolete. For instance, household guides provided increasingly detailed work schedules for housewives to help them improve their efficiency.[34] Such schedules now seem dated. And who should enforce these schedules? Factory and office employees generally adhered to work schemes because these workers were part of a hierarchical organization and had to obey superiors. Housewives had no such superiors, but efforts to influence their behavior were made nevertheless. R. Lotgering-Hillebrand felt that girls had to be disciplined to learn how to be perfect housewives, by means of training in domestic science. Domestic science experts tried to professionalize household tasks, but also to discipline (future) housewives. Cooking demonstrations were held to inform and instruct women, while modern media technologies were deployed to reach housewives; from 1925 on, Lotgering-Hillebrand and other well-intentioned experts who wanted to educate them about modern life took to the radio waves.[35]

This kitchen from the mid-1950s shows efficient use of the wall. The groove for letting excess water flow from the work surface is clearly visible as well. Still, this kitchen did not yet meet all the modern standards: the kitchen towels behind the stove are definitely a fire hazard.

The first mass-produced kitchen was a standard kitchen designed, produced, and marketed by the Dutch timber company Bruynzeel in 1937. It could be ordered with variously sized cupboards which, depending on the kitchen's spatial features, could be optimally combined, and it was characterized as a "complex, small company" designed and produced according to the principles of "efficiency" and "laborsaving."[36] The Bruynzeel designers aimed to produce a kitchen that allowed enough room for doing many tasks and storing many artifacts in a small space (very likely inspired partly by Cornelis Bruynzeel's extensive experience as a successful sailor, bluewater racer, and boat designer). Flexible "stacking" of space offered one solution to space constraints. Housewives in turn designed their own practice and identity, for better or worse, based on the space afforded to them, the available technologies, the instructions they received, and the prevailing views on the perfect housewife.

In the 1920s and 1930s kitchens became a pervasive phenomenon in every home, rather than being found only in the homes of the well-to-do, where they served as a work space for servants. Housewives became responsible for what took place in their kitchens. The interplay of actors—producers, household experts, engineers, and architects who all increasingly spoke the same language—contributed to constructing both the "modern" kitchen and the "perfect housewife."

After World War II these concepts stabilized as the outcome of a combined technological and cultural standard, one that served as basis for further developments. The story of the growth of the kitchen in the 1920s and 1930s is exemplary of the unfolding of location-tied innovation that moves beyond strictly local problem solving. The performance of household tasks builds on traditions, yet new technological possibilities, combined with new ideologies, have to be integrated into daily practice as well. In the interwar period Dutch designers and housewives' organizations articulated specific views on kitchens, partly based on their efficiency thinking. New innovations, such as the counter, which integrates multiple functions, or the cupboard as a reflection of stacked space, had to be brought into line with each other, as well as with the various interrelated tasks. Although no central actor coordinated everything from a single viewpoint, the net result was the co-construction of new kitchen designs and new household practices, including their cultural claims. The particular innovations involved had stabilized in both in the Netherlands and in other countries by the fifties. Roles had crystallized and new technologies were developed for generally accepted household functions. The differences between middle-class and working-class households had grown smaller, notably in terms of their technological facilities and related patterns of expectation.

Technology Linkage and Functional Integration in Offices

As with the kitchen, new technological artifacts and systems have drastically changed the office as a specific site in the twentieth century. A major difference between the domestic sphere and the office was the increase in scale. At the beginning of the twentieth century the number of white-collar workers skyrocketed and the standards and requirements for various sorts of recording and bookkeeping were radically transformed. This has led to important innovation challenges and problems in offices. Office activities have a long history, but the office as a separate location grew more prominent only in the second half of the nineteenth century, notably in banking, clerical service organizations, government agencies, and manufacturing companies.[37] Demand for administrative activities increased, for example, because of new services and new legal requirements for bookkeeping. New opportunities for organizing the administrative work resulting from new technologies, and the division of labor made the demand for clerical activities grew even more. The size of offices grew accordingly. By the 1920s offices with one thousand employees were no longer a rarity in the Netherlands. This growth forced

This 1911 photo shows the typing room of a gauge factory in Dordrecht. The introduction of the typewriter early on gave rise to the new phenomenon of the typing room in offices. Young women, who were deemed preeminently suitable for typing work because of their refined motor skills, were supervised by a male manager, who in this photo is sitting behind a glass wall in a separate room. He did not have to be there in person: the young women would feel his presence behind their backs all the time anyway.

offices and clerical services to address the issue of organizing their data flow efficiently. This became even more important as the volume of paperwork and information exploded and standards as to the quality of information (for example, of bookkeeping and accounting) improved. The various administrative processes such as the production and reproduction of documents, bookkeeping (including calculation and data processing), filing, archiving, and communication (notably via mail and telephone) relied on a host of technologies that changed over time.[38] To perform these various functions, office workers and managers increasingly turned to technologies developed outside their own location. The market for office technologies was supported not only by journals in which many of these technologies were discussed but also by trade fairs where users could acquaint themselves with the latest inventions.

Ad Hoc Integration of Technologies, 1890–1914

The design and development of office technologies in the Netherlands was initially determined by office equipment salesmen, importers, manufacturers, and actual users who might actively pursue higher efficiency or try to integrate various functions in new machines. Most equipment was in fact imported from outside the office. Issues relating to scarce space and the interference of technologies did not play a significant role in the early years of office development. Before World War I, offices generally developed or adopted technologies that replaced a few labor-intensive tasks or that made specific activities less time-consuming. The main innovation was the introduction and further development of the typewriter. Increasingly, manufacturers began to look for ways to combine several technologies in new ways. Some functions became entangled on an ad hoc basis. For example, around the turn of the century the use of carbon paper in combination with typewriters grew common, integrating text production and reproduction. This new option was widely advertised by typewriter manufacturers at the time.[39]

Functional Integration

After 1914 the office as a specific location entered a new phase in which larger offices began to work with newly developed and introduced technologies, such as punch cards, and their further development and search for more efficiency led to further integration of individual functions. One indicator of this trend was the emergence of a broader "midfield" of office managers, office machine salesmen, business consultants, and efficiency engineers. These professionals began to work systematically on improving specific office functions and integrating office technologies. "Rationalization" and "efficiency" were keywords in optimizing the management of office work and office capital, so that by reducing labor and costs it was still possible to maximize outputs.[40]

Quite soon connections were drawn between scientific management as propagated by Taylor and operational management:

> One might point to the extent to which bookkeeping becomes easier and more accurate, and particularly how much its value for business accountancy control goes up, if the business organization itself improves, which in turn better guarantees the accuracy of the numbers ... as well as their stability.[41]

For the sake of good bookkeeping, this author, an accountant named J. G. Ch. Volmer, aspired to reorganize businesses to allow for comparing the accounts over a period of several years. To achieve this, the office itself had to be transformed. Volmer suggested it was important to know the advantages and disadvantages of using carbon paper, fountain pens, pencils and styluses, copiers, cyclo styles, and hectographs. Apparently, he considered these tools as separate technologies and he didn't write about them from a grand vision about the office as organization and practice. Volmer's contribution is an early example of how professionals were active in raising bookkeeping standards.

This photo was taken in 1946 at A. & N. Mutsaers, in Tilburg, a manufacturer of woolen textiles. In addition to a central or manager's office, in larger companies one or more offices would be situated right near where production took place. Visibility was important in the design, and walking distances had to be kept to a minimum. Although a direct connection with the work floor was important, physically the spaces were still separate units.

Meanwhile, more and more new technologies were developed and marketed; the market's growth was visible in the expansion of the number of brands of office machinery and supplies. The contents of a journal on office work, *Administratieve Arbeid* (*Administrative Work*), conveys this clearly. In 1923 this journal reviewed thirty-four kinds of counting machines and all sorts of calculators, card systems, dictating machines, sorting machines, postage meters, printing machines, payment machines, typewriters, and an internal telephone system. All sorts of combinations seemed possible: the journal published a review of Brunsviga's writing-counting machine, Monarch's counting typewriter, Moon Hopkins's type and counting machine, Smith Premier's counting typewriter, Underwood's typewriter with automatic paper feed, Urania Vega's counting typewriter, Yost's counting typewriter, and many others. When in 1922 the international exhibition "Het Kantoor" ("The Office") was held in Amsterdam, thirteen different companies displayed their latest typewriter models.

As the supply of technological improvements diversified, the striving for efficiency led to efforts at integrating technologies. From the 1920s the deployment of technologies was increasingly put at the service of optimal office management. Innovators viewed the office as a whole system of flows (data, documents) to be processed. The challenge was to optimize the throughput. This focus on efficiency characterized a broad group of professionals such as engineers, accountants, office managers, office machine traders, business consultants and psychologists.

Before long, "schematization"—meaning the creation of plans, models, or protocols (called schemas) of how information flows should be organized—became the standard approach to rationalizing office management. In the 1920s, for instance, the Amsterdam branch of Rotterdamsche Bankvereeniging (Rotterdam Bank Association), seeking to restructure its activities for organizational and financial reasons, developed work "schemas" (flow charts) and machine "schemas" to prescribe how tasks and processes should take be done. J. Goudriaan, cofounder of the Technical Economics section of the Dutch Royal Institute of Engineers, felt schemas should help create a "clear and simple picture." The article in which he introduced a typology was based on schemas for personnel, functions, production, administration, and processes that had been set up by all sorts of factories and organizations. Thus, his typology reflected a practice that had developed already.[42] A good administration schema (flow chart) had to contain the following elements:

1. The *stations* to be passed through
2. The *tasks* to be performed in these stations
3. The *material* (such as the paper forms to be used for the tasks)
4. Last but not least, an unambiguous indication of the *sequence* in which the stations are passed through, either as a single chain of consecutive links, one after the other, or as partially separate chains, side by side, that further down the line possibly merge again.[43]

The results of this analysis were represented as a map drawn to a specific scale so that the distances between the stations could be measured. The schemas would accurately indicate the machinery as well. Like blueprints for kitchens, such maps were innovations in their own right and show that the organization of space had become a focus of attention.

As management and bookkeeping were redefined in terms of flows to be processed, increasingly such processing, including the activities involved, was seen as a matter of "technical" tasks (technical in the sense of procedural tasks). From here it was only a small step to developing technologies to per-

Modeling workflows was a major tool for managing the increasing complexity of locations and organizations. Goods and information are described as parts of flows. Mapping flows that represented processes made interconnections visible. This diagram, called a traffic administration scheme (*schema vervoersadministratie*), represents the layout of the automation introduced in 1959 by KLM Royal Dutch Airlines.

form them. J. G. de Jongh, a professor at the Technical University in Delft, put this quite explicitly as early as 1923:

> Bookkeeping is a process that comes down to an alternation of collecting and sorting. Sorting and collecting are technical procedures. In technical procedures machines can play a major role. The art of organizing clerical procedures, then, must be tied to bringing the material to be processed in the most favorable condition to allow for maximally efficient performance of sorting and collecting processes.... Procedures can be done by a machine if their frequency warrants it.[44]

New practices such as bookkeeping, technological innovation, and organizational change were increasingly designed in an integrated fashion by defining the office in terms of core processes to be optimized. The mechanization of office tasks moved on toward functional integration of separate technologies, such as writing cash registers, writing counting machines, bookkeeping machines, and addressing machines. Increasingly, office managers decided to purchase these kinds of machines, what later would be called "systems machines." They became even more relevant during the economic crisis of the 1930s.

The growth of offices and the increased attention paid to work and its organization also led to more attention being paid to the interference of technologies and the office's spatial organization, which occasionally led to innovations. For instance, a sound-suppressing typewriter was developed to reduce the interference between typewriter and desk.

A kind of damper absorbed the typewriter's noise and prevented the table from functioning as a sounding box. Another example involved fastening the typewriter to the desk so it could be folded away, which made it possible at any moment "to change a normal desk quickly into a typewriter desk and vice versa." This innovation was a response to scarce space, when it was more logical to fold away a typewriter than to use a larger desk.[45] In addition to the "system machines" (one machine that combined two functions), the newly developed "machine systems" consisted of different specialized yet integrated machines that performed multiple clerical activities. The main example of such a system was the punch-card system.[46] For example, the introduction of the punch-card reader (a Hollerith machine) by the central bookkeeping unit of the Department of Public Works of the City of Amsterdam implied a transition from separate technologies to machine systems:

> Although the clerical department of Public Works already used mechanical aids for bookkeeping and calculating before 1930, it gradually became clear that it still lacked devices for grouping various countable data in all sorts of desirable ways. By employing Hollerith machines, which took off in early 1930, it has become possible for the clerical department to process even the minutest data more rapidly, and it has also proved possible to perform more control activities more easily and to obtain more statistical materials.[47]

Such functional integration was also guided by rising bookkeeping standards, formulated by outside parties (such as government) and also by the professionals involved. In the *Technisch Gemeenteblad,* a journal for municipal engineers, in 1932, the introduction of Hollerith machines was explained with the help of a schema to show how they contributed to the processing of data. In the 1920s and 1930s there was a question of how far the rationalization of offices and clerical departments could be pushed and, in particular, how employees would respond to new technologies and related changes.

Sustained critique of the instrumental way of dealing with employees would not gain prominence until the 1950s, when a "human relations" approach began to win terrain.

Until the early 1930s, office technologies and their further development were the concerns mainly of the various "midfield" actors. However, firms or organizations established specific staff sections whose job was to deal with such issues when their capacity mandated it—for example, the State Office Machine Center (Rijkskantoormachinecentrale, KMC), established in 1928 as part of the state Postal, Telegraph, and Telephone Service, and the Amsterdam

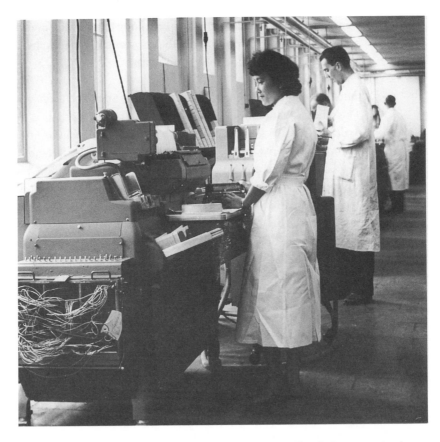

Hollerith machines, which use punch cards, were in use until 1960 for processing large quantities of data, as here, in the office of KLM Royal Dutch Airlines. The strong increase in the volume of air traffic put pressure on KLM to work ever more efficiently.

City Bureau for Organization and Efficiency. The creation of the State Office Machine Center was prompted by the desire to coordinate the deployment of all office technology within the government and have an in-house framework for consultancy.[48] The task of the Postal and Telephone Services was to deal with municipal issues related to management structure and training, work methods enhancement, function description, work classification, pricing, planning and routing, and administrative organization.[49] In 1953 the business economist Geertman claimed that in large organizations such staff sections had become a general phenomenon. This process of systematic analysis and innovation persisted after World War II. Scale increases and clerical activities were increasingly combined with efforts at centralization, representing a context for the gradual rise of computers and computerization in managing and processing data, which in turn stimulated further changes, culminating in the substantial yet not always successful automation efforts of the 1960s and 1970s.

Innovation in the Office: Integration from a Specific Vision

The schematization of flows, activities, and programs such as those required to deploy punch-card machines was a most interesting innovation. The interferences between various activities and technologies were no longer solved ad hoc but conceptualized and represented as one simplified schema. This schema could be used by managers to further rationalize work processes, and could be used to investigate further mechanization. Such schematizations also occurred in other locations, as in blueprints for the ideal modern kitchen. In offices their effect has been substantial, partly because the actors interested in these schematizations included office managers as well as office suppliers. Another intriguing aspect is that the schematizations paved the way for what today is called systems analysis and these schematizations—notably in their technical expression such as programs for punch-card systems—served as a preliminary stage of computer programming.[50] This is visible in factories at various levels, from scientific management to the modeling of processes in chemical plants.

Apart from site-specific innovation as a response to interlinking technologies and interference, then, innovation also results from reflection by actors themselves in a given location: these actors have specific visions of how it should ideally function. In other words, the schematizations and the related planning were not necessarily exogenous, as interventions in traditional practices, but could also be part of the intrinsic development of such practices.

Schiphol Airport: Connecting Airside and Landside

Schiphol Airport, situated in a polder near Amsterdam, began operations in 1916. At that time it was largely an empty field where every now and then an airplane would land to deliver some freight or pick up a lone passenger. In the 1920s and 1930s several technologies were introduced aimed at coordinating the approach of airplanes and the handling of passengers and freight, but essentially it was still an airstrip with very basic facilities. Today, however, the airport comprises a host of high-tech airside and landside facilities. At the same time, the airport has evolved from a service provider to airlines such as KLM into an independent actor (Schiphol Airport Inc.).

Initially the spatial problems faced by this airport could be solved incrementally, but in the course of the 1930s more substantial interventions—tied to specific planning efforts—became necessary.

At first Schiphol closely followed airport developments in the United States—for instance, in building paved runways. After World War II, and especially after the emergence of jets, the operational challenges increased significantly. Airplanes grew larger, and heavier, leading to larger passenger loads but also to calls for stricter standards as to the length and wideness of the runways. Similarly, more negotiation on implementing inevitable changes was needed, and more actors became involved, as airports had to be integrated into national transportation infrastructures and aspects such as noise nuisance grew more prominent. Even though the dynamic of site-specific development and negotiation is different for each airport, there are clear patterns related to an airport's location and to how airport leaders respond to developments elsewhere. A case study of a single airport provides insight into the dynamic of innovation at airports as a specific type of location.

Ad Hoc Technology Development at Schiphol Airport, 1916–1940

When the first three airplanes landed at Schiphol, on September 19, 1916, all sorts of activities took place simultaneously: passengers walked across the platform while the grass was being mowed and another plane was landing nearby. Carts, pedestrians, trucks, and airplanes were all part of a single disorderly process. There was as yet no strict separation between landside and airside.[51] Passengers and airplanes belonged to a single spatial ensemble rather than to separate flows. The aircraft itself was at centerstage: the pilot of the incoming plane decided exactly where he wanted to land, and the lights were on the plane (as with automobiles) instead of the landing site. In its first years Schiphol was merely a grassy field with some wooden barracks, and it had not yet been determined where the infrastructure for landing and taking off should be. The "airport" had to be developed from scratch, absent a prior concept or vision of what an airport actually was. The first airports for civic aviation such as Croydon and Schiphol could look at military airports, but their history, too, was quite brief. Basically, those who built Schiphol had to find a technical solution for each problem that presented itself, without being able to fall back on models or handbooks.[52]

Initially KLM was the dominant actor in the airport's development, but in 1926 the city of Amsterdam formally gained authority over the airport, and in the same year appointed Jan Dellaert, a former KLM employee, as airport commissioner. Increasingly the city—and later on, after 1958, Schiphol Airport Inc.—developed new plans for the airport, but until around 1935 measures undertaken grew out of ad hoc technology development or

whatever was available. For example, when it came time to install a lighting system, airport staff visited airports in other countries to see what they had and then purchased a specific system. The optimal utilization of the available space became an increasingly important dimension tied to airport technologies. An early example is the hydrants system for storing fuel. Shell took the initiative to put in fifteen-thousand-liter underground gasoline storage tanks and a distribution system. Pumps made it possible to get the fuel from the storage tank to the six-meter-high tank on the ground. This system became operational in 1927 but functioned only to a degree because there was no fixed line-up of airplanes, so trucks continued to be necessary for fueling.[53] Use of space is also a factor in the linking of air traffic and ground traffic. A brick terminal had been put in place in 1928 for the 1928 Olympics in Amsterdam. In 1927 the State Highway Plan projected a highway connecting Amsterdam with The Hague and Rotterdam. Albert Plesman, director of KLM, elaborated on it and in 1934 proposed to the director of the city's Trade Agency, L. Boogerd, the construction of a new airline terminal, to be located along the projected highway between The Hague and Amsterdam so that passengers coming from the highway could conveniently enter the airport.[54]

The airside changed, too. As aircraft became heavier and faster, they severely damaged the grass surface, so Schiphol decided in 1938 to put in paved runways—the first European airport to do so. This meant that aircraft would have to take off and land on fixed runways, so it was important to consider the runways' exact location carefully in advance. The prevailing wind directions were one major factor; another was the linking of runways for landing and taking off and the taxiing trajectories to the piers and gates of the main terminal building. Paved runways had several other implications, too: the pilot no longer determined which direction he would land or take off in. Lighting became integrated into the runways through reflectors (sunken lighting units with

"Your air freight is also in good hands!" This KLM Royal Dutch Airlines promotional poster, from 1959, communicates that air freight would be as carefully handled as luggage and, indeed, passengers. In the early days, a delivery bicycle or car would take freight to the plane, and the loading was done piece by piece by manual labor. Now, the poster shows that conveyor belts had replaced stairs and that there was an appropriate vehicle for each type of air freight. At the old Schiphol Airport in Amsterdam, there was no integration between airplanes and the arrival and departure halls and the various transport connections. This led to a jumble of cars and buses on the platform for transporting travelers, freight, and fuel from and to the aircraft. Also, the aircraft had to be inspected and fueled. Over the years, activities such as loading of catered food and boarding crew and passengers were increasingly coordinated.

sodium lamps). Paving, lighting, and communication were thus increasingly integrated, gradually causing the airside to develop into a coherent ensemble.

An interesting further development involved the establishment of Air Traffic Control (a government agency) in 1939, which caused Schiphol's space to be expanded vertically into the sky.[55] After the introduction of radar systems in the 1950s this space became even more structured by the instructions from the air traffic control tower. It became possible to reduce traffic congestion in the air as well as to regulate the distance between two aircrafts by identifying air corridors and the various "blocks" into which they were divided. In the 1950s the landing procedure was formalized through the Instrument Landing System (ILS).[56] This system, which was tied to the geometry of the runways, made it possible to land in conditions of poor visibility.

Planning the Airport, 1940–1967

The developments of the late 1930s not only ensured the airport's technological stabilization for the time being but also supplied the building blocks for a more integrated vision. During World War II planning began for the layout of a central airport, and in 1949 an engineering firm, Netherlands Airport Consultants(NACO), was named to oversee the job.[57] In this planning period the concept of an ideal airport took shape. For instance, a plan proposed in 1949 by the airport commissioner, Jan Dellaert, foresaw integrated development of the airside, the landside, and the connection with national infrastructures.[58] As Dellaert put it in 1949:

> Runways, taxi runways, platforms, terminals, offices, hangars, workshops, roads, runway lighting, radio-technical facilities—these and countless other elements together constitute the enormous arrangement of a modern airport's operations….With the design of Schiphol its facilities are given shape once and for all; it will offer the Netherlands a future in international and intercontinental air traffic, which will largely determine its role in the world's airline system.[59]

Obviously, a wide range of technologies would have to be integrated. This quotation also contains an aspect of site-specific technology development that we have not yet encountered: the physical construction of the new Schiphol would define the airport's role far into the future. Given the enormous investment costs to build and operate the airport, things could not easily be altered once they were built. Recognition of this point led to some built-in flexibility. The Dellaert plan incorporated most of the airport's exist-

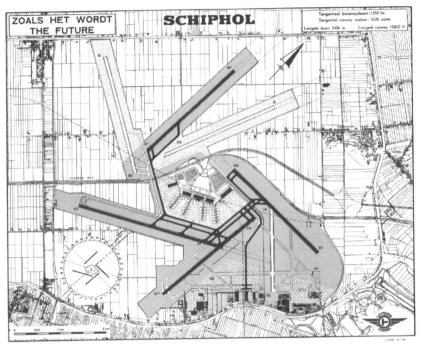

Designing specific locations was an innovation in its own right. This 1956 plan for Schiphol Airport, titled "Like it will be, the Future"(*Zoals het wordt)*, was largely the work of the airport's director, Jan Dellaert. This design of the space, similar to urban planning approaches, exemplifies "concerted disentanglement": there are separate runways for taking off and landing in the same wind direction. Takeoff runway 32 of the "old" Schiphol (shown in white, near the bottom of the plan) is integrated into the plan for the new airport.

ing infrastructure—later known as Schiphol-Oost—but also anticipated enormous capacity growth and possible future expansions.[60] Two runways located in Schiphol-Oost became part of the new system; overall, flexibility was a key notion.

At the center of the plan was the main terminal. Around the terminal was a road for vehicular traffic. The taxi runway was a circular road around this traffic area. The runways for take-off and landing were configured to form lines running off the circular taxi runway at a tangent. These runways did not cross each other—a major advantage of a tangential system. Access to the vehicular-traffic area was effected by linking a railway station and an exit of Highway 4 (The Hague–Amsterdam) to the traffic area, but to avoid intersecting with the circular road the train track and the highway would be put in tunnels at the crossing points. The national highway was constructed between and along the take off and landing runways. This plan, then, neatly separated the various flows. Dellaert's plan also separated landing and take-off in the same wind direction: one runway for landing and one for taking off. The tangential system ensured departing and arriving planes would not get in each other's way. Dellaert thus followed the creed of a "continuous flow of simultaneous traffic without individual control," formulated already before the war. This system also kept open the option of extending runways.[61]

Although Dellaert tried to base his plan on an estimate of the future volume of air traffic, he was unable to provide quantitative forecasts of the future number of travelers.[62] The weakness of the quantitative underpinning of his plan was a problem. The plan's basic premise was that planning for the airport's capacity had to be based not on "the total number of passengers per year" but on the highest volume at any one point in time: "the number of actions that have to be performed at the airport."[63] Various actors refused to accept the quantitative foundation of Dellaert's plan, and KLM asked NACO to do a critical analysis of it.[64] NACO opined that Dellaert's estimate of the capacity needed was too high and also that Dellaert underestimated technical developments in aviation concerning reduced vulnerability to crosswinds, which in turn reduced the number of runways needed for landing.

The smooth integration of airside and landside was no easy task. The separate flows had to function optimally on their own while their interferences had to be tackled as effectively as possible. However, analysts revealed contradictory design criteria in the plan.[65] The walking distances for passengers had to be minimized, but at the same time the new heavy jet planes, whose fumes and noise were a big problem, had to be kept at a distance from the terminal. One possible solution was to create great distance between terminal and airplane bays (the area adjacent to the gates, where planes taxi and "park"), and to connect them via tunnels. Moving floors would help passengers to walk from the terminal to the plane and vice versa. This option was rejected, however, because of the high construction cost of tunnels in the soggy soil of Schiphol. The airport management ultimately opted for a terminal with piers that protruded into the bay and a two-floor terminal: one floor for arriving passengers and one for departing passengers. In 1959 the working group that coordinated the planning agreed to a plan in which the terminal would have a two-story structure, integrated with the piers. Further efforts were necessary to bring airside and landside into line with each other. For example, ventilation and air conditioning were installed to keep the piers free of fumes, and space was created beneath the piers for facilitating aircraft handling in the bay.[66]

The Aviobridges at Schiphol are icons of modern airport technology. They form a perfect connection between airside and landside. Passengers can board and debark from the aircraft in a fully enclosed space and they are separated from baggage handling and all sorts of maintenance activities around the aircraft. The pier's roof is designed as a walking promenade and is a popular destination for visitors to the airport.

The Passenger Bridge (Aviobridge)

The passenger-boarding bridges between the plane and the arrival or departure gate was a major example of how the challenges of integrating the airside and the landside were met. Flexible passenger bridges connect the plane with the terminal. When Queen Juliana opened the new Schiphol Center in 1967, the "elephant trunks," as the bridges were called, immediately caught people's attention as a symbol of modernity, combining comfort and efficiency.[67]

The advent of jet planes and, later, wide body aircraft meant that the airport was forced to process larger numbers of passengers swiftly. The less time the planes had to be at the terminal, the more efficiently the airlines could operate. By now the planes were being met by a growing number of ground crew and their service carts for fuel, electricity, oil, water, water-methanol, defrosting fluid, air conditioning, toilet cleaning, catering, and baggage and freight handling.[68] As airlines sought to promote an aura of luxury and comfort, passengers had to be shielded from the chaos of this servicing operation with its noise and stench and also from dirt and the elements. The passenger bridge was sold as a solution that reduced both connection and turnaround time. Since that time enclosed passenger bridges have become a characteristic element of large modern airports. Initially the bridges were owned by Schiphol Airport and were operated by special ground handling crew of KLM, but today they are operated by an independent company.[69]

There were alternatives to passenger bridges, whereby various aspects played a role. From the airside perspective, as many airplanes as possible should have access to the terminal at the same time. From the landside perspective, walking distances for passengers should be as short as possible. One ideal solution, called "jet island," seemed to be to have airplanes "enter" the terminal with their "noses": a plane that had its nose turned right to the terminal would take up the least external space (an airside interest), while passengers could board and exit the plane via the plane's "nose" (a landside interest). But this solution was considered risky because of the danger of plane collisions while it also limited the space for aircraft handling at the gate.[70]

Another solution was to design a fully "mobile section" or "mobile lounge" that would take passengers from the terminal to the plane, where it would be hooked on to the plane so that passengers could enter. These units are still used in some airports; the advantage is that it creates flexibility regarding the exact position of planes at the gates, affording more handling room for planes. However, the mobile lounge actually added more traffic to an already busy gate, and late passengers had to be picked up and taken especially to the plane.[71]

Entangling Technology via Plans

After 1940 major airports such as Schiphol were increasingly constructed on the basis of specific plans, which themselves were based on interlinking land and air traffic flows, giving rise to specific innovations such as the passenger boarding bridge. Airport development evolved more toward the ideal of the planned, modern, and efficient airport. Schiphol's planning effort can be viewed as an effort at sociotechnical design of the site itself—the site itself becoming an object—but in practice this effort neither was comprehensive nor was it the only factor in the dynamics involved. Many companies took the initiative to design airport technologies in the course of Schiphol's development. Shell, together with other oil companies, developed the hydrants system; engineering agencies specializing in aviation developed baggage-handling systems; the passenger bridge was an initiative of KLM and was developed by a company, Aviolanda, that later became part of Fokker; the later

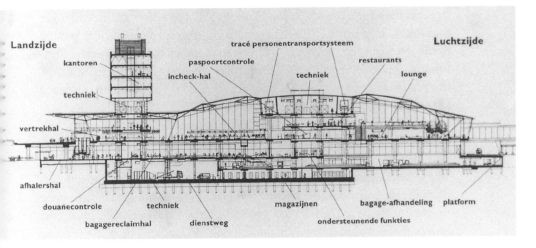

The arrival and departure hall of Terminal West at Schiphol Airport, built in the early 1990s, shows the application of spatial disentanglement. For example, arriving and departing passengers use separate floors, and their luggage is handled in separate locations. The hall simultaneously serves as the connecting link between landside (*landzijde*) and airside (*luchtzijde*).

integration of the airport with the national railroad network was an initiative of Netherlands Railroad. These sociotechnical innovations were developed by those with specific expertise and in interaction with projections of the airport's specific user context comprising many different types of users.

In the period after 1967 the airport's "self-planning" was further stimulated by a new emphasis on knowledge development within Schiphol Airport Inc. This was a development similar to the establishment of staff departments in large organizations. This was part of the effort to become a more business-minded company with specific expertise. The airport developed its own R&D department (which was abolished again in 1986) as well as an extensive network of design bureaus and contractors who worked for Schiphol (sometimes nearly exclusively so). The optimization of the airport became the prevailing goal, expressed and made measurable in the airplanes' maximum occupancy rate and minimal turn-around time. These concepts transpired in new innovations such as the baggage-handling system and the passenger-boarding bridge.[72]

By the 1980s Schiphol Airport had become an independent actor. Increasingly its leadership acted autonomously and determined how its location was used—the most striking example being the construction of the Schiphol Plaza shopping mall in the 1990s. These new initiatives were motivated by the fact that the airport, despite spectacularly growing passenger traffic, had never managed to earn profits. The airport faced ongoing operating deficits because of the need to keep making capital investments. In 1969, Berenschot Consultancy issued a report on how the airport company could be reorganized, the first in a series of reports that is still ongoing. Various reorganizations have already been undertaken.[73] Increasingly the Schiphol location was defined as company that had to operate efficiently and position itself in relevant contexts. At the close of the twentieth century another dynamic of airports became visible: their embeddedness in larger transportation networks. The airport's conceptualization as the central node of a transshipment network (in the context of the slogan "Nederland Distributieland") was a clear example at the conceptual level. By 1986 the airport had been fully integrated into national traffic systems, including railroad traffic. The connection of Schiphol to the new high-speed-rail link represents another step toward integration of an European transportation infrastructure where train traffic and air traffic complement each other. At the same time, the increase of its non-aviation-related activities turned Schiphol from a "service provider" into a city where one can do just about everything except live.

The Airport as Location: From Infrastructure to Planned Junction

Airports started out as empty fields at the beginning of the twentieth century, but they increasingly came to be regarded as junctions where land and air traffic and other flows intersect. Flows were disentangled so as to make more effective connections between them, on runways, at gates, and in terminals. To a large extent the airport as a location has been able to develop its own sociotechnical outlook—even though in practice this occurred as a result of plans that were already dated by the time they were executed because of rapid developments in aircraft technology and aviation. In the late 1930s, the technological development of Schiphol, just like that of many North American airports, reached a preliminary endpoint, which subsequently served as starting point for further planning and realization of the *perfect* airport—a model that optimizes landing and take off in interaction with flows of passengers and freight. Yet an airport's lengthy planning horizon, linked to the large number of actors involved and the various financial restrictions, accounts for the fact that perfect airports aren't possible. Not only technology but also contexts change, through resistance to further expansion and also because of competition and inclusion into overall transportation infrastructures. At the start of the twenty-first century the ideal airport integrates much more than airplanes and flows of freight and passengers. As a type of location, however, airports remain a localized infrastructure, even if they are dependent on the specific services the public demands for their operations and further development. Schiphol airport was able to develop its own plans, but could never realize them without permanent negotiation with other actors such as airlines, local and national government, and, increasingly, the people living next to the airport.

Cities: Linkage in Space and Time

In the introduction to this chapter we discussed examples of site-specific innovations that took place in cities. As we noted, the city is a type of location in its own right, one that comprises all sorts of locations that are subject to innovation(such as households and offices) and that develop in interaction with urban infrastructure and specific construction plans and regulations.

Early in the nineteenth century hygienists and engineers began to develop perspectives on urban planning. Partly because of the growth of the urban population and the increasingly active social role of cities and the national government, a need arose for systematic design and planning, but for a long time this approach could be taken only in new towns, districts, or subdivisions. After World War I a new group of professional urban developers entered the arena; they came from various backgrounds but shared ideals about planning cities or parts of them as a single integrated ensemble. The kind of planning they developed is the focus of this section because it reveals a specific feature of this type of location that is closely associated with its prevailing character.

Technology Linkage, Ad Hoc Solutions, and the Rise of the Welfare City

Between 1890 and 1914 the creation of a healthy city was the primary focus of city innovation. This development was a response to deplorable living conditions that had emerged because of rapid population growth since the mid-nineteenth century. (This was actually a pan-European pattern.) Increasingly, hygienists, engineers, and various government agencies began to view these conditions as untenable, but before interferences that occurred in cities could be restructured in a new and better way, first they had to be disentangled. This called for insight as well as larger municipal authorities.

This "coherent disentanglement" is visible in early views on urban problems, such as in a 1882 report by the city architect of Haarlem, H. W. Nachenius. This report, commissioned by the city's Association for the Promotion of Public Health, addressed the conditions in the "homes of those of little

Installation of new underground infra-structures also required innovations, reflected here in new means of transport and the characteristic construction method for putting in sewers. Shown here, the Nicolaas Beetsstraat in Utrecht, around 1910.

means."[74] After visiting actual homes, the staff documented their findings in a number of categories such as "drainage and sewers," "soil," and "privies." Under the heading "Privies," for instance, it was reported that "garbage is thrown into the water" and "the house comes with a dunghill that is also used as privy, while the actual privy is used as storage." Under the heading "Environment" the staff reported that "in some locations the home's waste-water and the rainwater run straight across the pavement into the canal" and also contain traces of offal or feces.[75] This combination, so the association claimed, led to unhealthy situations and indecency—undesirable conditions which at the time were much battled. In a publication that appeared two years earlier, Nachenius offered a remedy for the problems he had observed. In the introduction of his *Bijdrage tot de kennis van den stedebouw: Eene po-pulaire studie* (*Contribution to Knowledge of Town Planning: A Popular Study*), he claimed that "technical, sanitary, economic and aesthetic standards [are] interrelated in town planning to such degree that they cannot be separated from each other."[76] The issue could perhaps be solved only by a versatile and integral approach. Matters that had led to unwanted situations because of their entangling in practice had to be separated again by adequate design, so that problems related to public health and immorality could be solved.

Coherent disentanglement happened in practice. One example involved the filling in of canals in The Hague.[77] What was the problem? Bargemen and many of those who lived along the canals had increasingly used them as a dumpsite for feces and garbage. Furthermore, the canals used space: when mobility increased, this became a problem. The proposal to fill in canals was tied to notions of public health and the city's accessibility over land, even though this solution would also cause the loss of the canals' economic function as transportation channel. Nearby residents welcomed the proposal because they would no longer have to deal with the nuisance of the nearby markets served by canals. Thus various aspects and functions had become entangled in the course of time—the canals as commercial system, as sewer system, as garbage dump, as drinking water system. The city's public works director, J. A. Lindo, solved the problem through disentanglement: he designed an integrated sewer system for discharging sewage, wastewater, and rainwater, and he simultaneously designed channels (bigger than the canals that would be filled in) that could replace the economic function of the canals. Connections with the other waters that fell outside Lindo's managerial authority had to be arranged with the polder and dike board of Delfland. The export of The Hague's garbage beyond the city limits led to a new relationship between the city and its surroundings. Although the port designed by Lindo integrated a number of functions, it was not based on an integral vision on the city's overall town planning. Rather, it involved a series of pragmatic choices aimed at solving separate problems.

Urban Planning: "The Art of Synthesis"

At first the ad hoc problem-solving approach could also be seen in typical city-planning challenges such as urban expansion. A construction engineer, J. G. Watjes, claimed in 1911 that the "making of a town plan chiefly involves drawing a network of streets."[78] This changed after World War I. Initially, as at other types of locations, this effort involved consultants who were asked to design expansion plans based on new town planning visions (H. P. Berlage is a leading example). Increasingly, however, public works departments developed their own expertise for outlining such plans. As a result, the development of town-planning technologies increasingly occurred in cities. A new professional group emerged, consisting of professionals from heterogeneous backgrounds but with shared attention to and interest in urban development. Interaction among these professionals took place at conferences, in journals, and in professional organizations. Only after World War II did town planning become an engineering discipline within the TU Delft.

The town-planning effort's target was to achieve a "synthesis."[79] Initially

Different groups of residents used inner-city waterways for their own purposes. Some viewed canals mainly as sites for dumping garbage; others used them for transporting their merchandise, and others used them for drinking water. Various functions of the waterways had grown entangled through use, but the new "urban architects" and planners would increasingly disentangle these functions. The filling in of canals, as here at the Loosduinseweg in The Hague, was an example of this "concerted disentanglement." Children in their turn also appropriated the new space.

such a synthesis was defined in terms of construction and aesthetic design, with the goal of creating an urban space in which the road system and the city blocks were brought into line with each other. This is well illustrated by the perspective drawings made by Berlage, which showed a city in three dimensions. However, the notion of synthesis was soon broadened from purely city aesthetics to include more problems and challenges. Apart from hygiene and accessibility, a spontaneously developed urban pattern, such as ribbon development, the construction of houses along the roads radiating from a town, was increasingly considered to be undesirable. Ongoing ribbon development was caused by the construction of new roads along which the value of lots increased, which caused speculation. In 1928 the Standing Commission for Expansion Plans of North Holland, a precursor in the area of town planning at the provincial level, claimed that the star-shaped urban form that had thus came into being was unaesthetic and uneconomical:

If one imagines what would emerge [if ribbon development were to go on], the resulting picture is anything but attractive. But we should not resist this kind of development just because it destroys scenic beauty; it is also uneconomic. It raises the cost of building and maintaining roads, sewers, and other systems; needlessly it forces residents to cover long distances to reach the center with its shops and public utilities. The connection of the various buildings that line these roads and the empty lots in-between also leave much to be desired.... One should pursue concentrated development. The locations of ports, waters, railroad tracks, and attractive scenery should determine the direction of a particular development.[80]

As this quotation suggests, the cost of providing technological systems and the distances to be covered by city residents had meanwhile become the subjects of town planning. Town planning had to balance a variety of interests, namely those of business, public housing, traffic, and spiritual needs and relaxation. To do so effectively was the "art of synthesis."

Increasingly the conviction took hold that this discipline should be based on quantitative predictions. Initially planners relied strongly on demographic data and extrapolations from it produced by the city's departments, not just at the service of town planning but also of public housing. In the late 1920s, the civil engineer Th. K. van Lohuizen was a pioneer in conducting quantitative studies of urban traffic flows.[81] His work provided the foundation for Amsterdam's General Expansion Plan (Algemeen Uitbreidingsplan), formulated between 1928 and 1935. This plan was a synthesis of design and science. Van Lohuizen worked in particular on its scientific underpinning, for example, in the area of traffic. These studies provided the basis for the traffic system as proposed in the plan. It started at the level of the intercommunity road system; the next step was to identify "how the local road system needs to be designed in order to meet the standard that these areas [for living, working, and leisure] are connected with both the existing city and with each other by the shortest and easiest accessible roads.... An optimal road system design is fundamental to this goal, from an economic and social perspective."[82]

The development of town plans was an innovation in itself. The analyses and predictions that were part of the plans gave them quite a dynamic character, as they foresaw current and future flows and their interactions. Concrete innovations were also stimulated from the practice of looking at a city from the angle of town planning. For example, beltways led traffic around inner cities instead of through them.[83] From 1915, when the engineer Rückert designed the beltway around Tilburg, more people argued in favor of such circular road as solution for congested traffic in downtown areas. In many Dutch cities a beltway was put in, a development that culminated in the "diamond" around Rotterdam.[84] Town plans first served as basis for technical adjustments in cities in the 1920s and '30s. For example, the plan could determine the importance of a road (its status). This in turn determined its wideness and therefore the height of the buildings and whether or not a tram line would also run on the thoroughfare (trams obviously only ran on a selection of streets, and as they use space, the roadway needed to be wide enough). Roads were designed by cross sections (an engineer's drawing) that defined where the cars should drive, people should cycle, and people should walk; where trees would be placed and where the tram tracks would be installed. As a technical representation of a road, the cross section itself is an example of planned disentanglement on a flat surface: the traffic flows were neatly separated from each other, each flow having its own lane. In a similar way the plans provided a framework for solving layout problems: the boundary between public urban development and private living space was subject to debate among various actors. In the second half of the 1930s, specific modes of zoning appeared as part of town planning. Lots had to meet a host of different criteria, such as optimal access to light and air in homes and optimal property use.[85] A city's town-planning effort integrated the various urban flows and spatial functions into a coherent ensemble. Increasingly the city was seen as a human ensemble that could be planned. In practice this implied substantial pressure on all of the city's technical departments to coordinate their work.[86]

Innovation Below Sea Level and into the Air Space

Urban development plans contain proposals for the city's spatial layout. An interesting new insight was that spatial design had not just a horizontal dimension but also a vertical one. Technological systems such as the sewer system and the water system served as inputs in urban development plans in the sense that a specific plan should optimize the cost of building these systems.[87] Particularly in inner cities after 1890 the construction of such systems meant that streets had to be dug up regularly, which caused complaints about lack of coordination and made public works departments fear for their reputation.[88] The Amsterdam architect C. Hellingman developed an innovation, the "floor below sea level," as a solution to coordination problems between the various city companies working on the construction of the expansion for Plan Zuid in Amsterdam (around 1920). His innovation created an open space below sea level which could accommodate many functions. The area of Plan Zuid lay just under two meters below sea level. The entire terrain would had to be raised by two and a half meters, a necessity that could be obviated by Hellingman's invention. His patent application for the below-ground-floor level was extensively discussed in a leading technical journal.[89] This innovation would have made construction substantially less expensive because the below-ground-floor level made adding the layer of sand unnecessary. This level, which was an open space between the foundation and the street level, could solve multiple problems. First, the street would no longer need to be dug up to install new systems. Second, this level could serve as space for economic activity, implying the new notion of integrating living and working. Third, this level might perhaps even provide space to flows of traffic. Glass paving stones at street level were meant to ensure light below. Although this innovation was not realized, the issue of exploiting scarce space was significant enough to ignite a debate in this journal.

One of the principal themes of urban technology development in the twentieth century is the exploitation of the vertical dimension. Gradually, cities grew taller and also deeper. Urban streets were endlessly dug up in order to put in new technological systems. These systems were planned coherently and concurrently in new subdivisions. Cities became deeper and this vertical dimension was also exploited in the design of systems: sewage systems and water towers both made use of gravity.

The use of vertical space applied first to layered traffic systems such as realized in Arnhem around 1950. Later on, combinations were made with shopping centers and high-rises. Utopian views on this had already been articulated in the 1920s and 1930s, but now they were realized. The plans largely served as a solution to the increased volume of automobile traffic: by exploiting vertical space in layered facilities, automobiles could be accommodated and urban functions could be retained. In existing downtown areas of cities, where automobile traffic was hardest to integrate, vertical construction efforts had to be combined with the existing buildings. The Hoog Catharijne project in Utrecht—an altogether new and spatially multilayered ensemble of indoor shopping malls, office buildings, apartments, and a train station to the west of the city's historical downtown—was a unique (and contested) solution. Social functions were stacked instead of being horizontally organized. Offices, shops, and apartments were combined with accessibility: to facilitate automobile traffic nearby canals were filled in while passengers could cross the new highway via pedestrian walkways. But this new downtown utopia also proved to have its shadow sides, for instance, as a magnet for hobos and vandals. Meanwhile, our understanding of what constitutes a utopia has shifted as well. Currently, in the Netherlands planners try to preserve historical inner cities as much as possible, but building vertically, both into the ground and up into the air—has become a common feature of city innovation.

Contested Locations

My focus on what occurs in locations, which interferences take place and how various categories of actors involve themselves in improving things, has revealed various interesting aspects of technology development. A key is how on-site interference of practices and technologies, together with views of how things might be improved, lead to innovations, initially ad hoc but increasingly as part of an integral approach, supported by plans, blueprints, and schemas. This kind of approach has been a characteristic innovation of the twentieth century in its own right. The planning ideology that became visible in the interwar years and prevailed in the 1950s and 1960s is commonly seen as a driving force. The examples discussed suggest that this planning ideology was partly the product of experiences with on-site problem solving. This holds for all four types of sites we discussed, although we demonstrated it in more detail with respect to the city as a site.

The on-site linking of technologies, our starting point, has thus taken on an extra dimension. We discussed examples of horizontal linking, such as the railing lamp, the hatch, and the passenger bridge, and vertical linking, such as the Hoog Catharijne project. Specific artifacts or new technologies made it possible to combine diverse functions in homes and offices; conversely, in other contexts it was necessary first to disentangle flows in order to link them again in better or more productive ways: the passenger bridge created a specific way of linking flows of passengers and aircraft movements at the gate, while modern cities exist by virtue of processes of linking and disentangling, flows of traffic being a clear example.

Aside from horizontal and vertical linking we can also identify "reflexive" linking: based on what went wrong or right in specific situations, there is reflection on potential good combinations and on how these might be technologically designed or supported. As we have shown, blueprints and schemas are technologies used for this reflexivity. They refer, implicitly or explicitly, to ideal kitchens, offices, airports, or cities, but also, as became clear in particular in our discussion of households, to ideal users: handbooks and courses always feature some concept of the perfect housewife. This was part of a more general civilizing offensive in interaction with changes in home design and social patterns. Since the interwar period, planning, including its professionalization, have prevailed as a quasi-neutral version of this civilizing offensive. This neutrality could not be upheld in the face of contestation later on in the twentieth century, yet the notion of anticipation, and hence reflexive linking, would remain important.[90]

It is possible to identify a clear development of planning from the late

Hoog Catharijne in Utrecht is one of the earliest examples in the Netherlands of "three-dimensional" spatial innovation. Flows of traffic are disentangled not only in two but also in three dimensions. The overall design reflects a realization of urban concepts from the 1930s. The central urban space had to make room for functions such as shopping and traffic. All flows can cross without running into each other, which is of course intended to improve the flow. But the spaces that were byproducts of the process could be appropriated for idiosyncratic purposes and have unexpected effects: dark tunnels proved unsafe and drew crime, and the indoor shopping mall did not turn out to be a cozy environment.

nineteenth century onward. Social visions and social movements such as the hygienists and their spokespersons but also social democrats and various professionals (from domestic science experts to urban developers), became involved in planning efforts. These efforts were interactive, especially at first. Planning as a concept was still unstable, and it could not be implemented top-down. However, ongoing professionalization, World War II, and the ensuing reconstruction phase, including the interrelated top-down ideology, gave rise to a world in which everything became subject to plans and planning efforts. Still, as the story of Schiphol shows, actual practices were often unruly. It is perhaps inevitable that practices do not automatically adapt to planning because they always have a local dimension, and this will continue to be the case.

In each of the sites discussed, professionals were active who increasingly claimed their site as a domain of specific expertise. They redefined practices and formalized hitherto informal practices. To a greater or lesser degree this went hand in hand with a civilizing offensive vis-à-vis users. All these experts developed site blueprints, plans, schemas, designs, and the like that were intended to serve as guideline for the deployment of new technologies and their use. These professionals established their own firms, but they also founded journals to publish about their experiences, visited exhibitions and conferences and set up professional organizations. They promoted the circulation of knowledge in order to learn from each other. In some cases they developed new training facilities. Domestic science experts represent an early case, and after World War II town planning became a separate study. To this day, however, there is no specific training curriculum for office and airport planning professionals.

Initially, professionals were geared toward identifying bad situations and providing new solutions. Most involved disentanglement: separation of drinking water and wastewater to avoid unhygienic conditions, separate runways for landing and take off, spatial disentanglement of traffic flows to avoid traffic chaos and to separate flows of information processing. Later on, in each of our types of site, a plan or design started to serve as the basis for giving shape spatially to the planned linking of technologies. This plan then became the standard for the deployment and use of technology. In the case of offices and households, the use of technologies more or less coincided with specific tasks: housewives produced meals or did the dishes, and clerical office workers fed data into punch-card machines. The fact that kitchens and offices were labor-intensive locations may explain why all sorts of professionals mobilized scientific management, notably in times of economic crisis. The various plans, designs, and schemas were innovations in their own right. The two-dimensional representation of three-dimensional practices turned these practices into objects of manipulation in the hands of experts, managers, directors, producers, and users. Machines, closets, houses, runways—they could all be arranged differently as well.

In the case of airports and cities, plans and designs gave shape not only to actual space but also to uncertain futures. In the case of urban expansion plans it became increasingly evident that the plans were future-oriented, but they also determined what was or was not to be done in the present.[91] Predictions about the future volume of airplane passengers, and hence the number of airport activities, grew more important. Airport managers relied on agencies and institutions to give their planning a quantitative basis.

Significantly, planning never occurs in a vacuum. Aside from various interested parties and growing participation, there is the physical unruliness of actual sites. Once a road or housing block, including its visible and invisible infrastructure, has been constructed, such a site acquires obduracy. Except in unusual situations such as bomb explosions, technologies applied in given situations can be hard to remove or adapt to new needs. The financial investments may be huge, or various technological artifacts may be linked to specific interests and ideals of social actors. This certainly applies to airports, even though Schiphol after its "completion" in 1967 continued to be a permanent construction site. Obduracy is less of a factor in the case of offices and households. Removing a kitchen cabinet or adding a new layer of paint hardly involves a far-reaching physical adjustment, and office technologies tend to be written off for tax purposes and replaced after a few years. The future's uncertainties, however, cannot be filled in arbitrarily, and occasionally the frozen past has a determining influence, as can be seen especially clearly in cities.[92] Meanwhile the phenomenon of obduracy has led to an acknowledgment of the relevance of flexibility and its adoption, where possible, as an additional design criterion.

Besides this dialectics of the physical design of a specific location's future, which then limits further design, there are two other shared aspects of site-specific innovation. First, the innovation that results from optimizing space, of which we have seen many examples: in offices the foldable typewriter was introduced, and the Bruynzeel kitchen was designed to have all sorts of cabinets so that a limited floor space could still hold all the modern kitchen technologies.

In cities we have seen how city road systems transformed into multilayered (three-dimensional) systems, whereby especially in new towns such as Lelystad and Almere the flows of traffic were designed as multi layered structures. The increased number of high-rises should also be understood from this angle: functions are stacked rather than arranged side by side. Second, the ways in which innovations came into being were increasingly tied to coordination, which increasingly was actively pursued. How, exactly, this was done differed from one location to the next. For households it is possible to speak of a "discourse coalition" that emerged between producers, domestic science experts, women's organizations, and users. In the case of offices, producers, importers, and users were equally influential in the development of technologies at first, but starting in the 1920s the emerging groups of new professionals involved became very dominant.[93]In the case of Schiphol, influential actors such as KLM's director, Albert Plesman, and the airport commissioner, Jan Dellaert, developed new plans. Yet already during World War II a much more complex decision process evolved, whereby the national government, the city of Amsterdam, and various agencies began to exert a pow-

Bridges are very clear examples of innovation flowing from the effort to "bridge" conflicting design requirements of separate systems. For instance, land traffic has different requirements for bridges than water traffic: a bridge should allow trains to pass while not forming an obstacle to boats. These railroad bridges across the Westerkanaal in Amsterdam were designed as drawbridges to accommodate trains' need for a horizontal roadbed; "tall" barges and ships could only pass at fixed times when the bridge was up. This bridge no longer exists; the photo can't be dated.

erful influence on final plans. The plans' realization was in the hands of several engineering firms, and to this day this network continues to be involved in the technological design of Schiphol. Until World War II, Dutch cities were fairly autonomous in developing and implementing expansion plans, but this occasionally meant that city engineers in charge of expansion efforts acted as if their city was their fiefdom.[94] The developments of the 1950s and 1960s took place within these coordinative frames. Although they would continue to be important as power structure, new socioeconomic and cultural-political realities proved erratic and would occasionally challenge them. The trajectory set in motion in the 1920s and 1930s and pursued as an emphatically modernist trajectory (albeit with variations at different sites and always rife with heterogeneous elements) met with new challenges, partly through external developments such as the contestation of modernization.[95] One area of such contestation was the environment. The achieved unity of on-site activities was tested by individualization and by integration of sites into larger ensembles, and also by concrete issues such as traffic congestion, environmental pollution, and the effects of new information and communication technologies. To what degree did emerging innovation on-site patterns, including input from planners and other professionals, allow for solving new problems and incorporating new technologies? The integral approach, which came into its own in the 1950s and 1960s, has led to major accomplishments involving sociotechnical planning. This can always be improved on, of course, but it is a substantial achievement of modernity. The challenge is to avoid too large a gap between sociotechnical planning and what occurs on-site. The increased emphasis from the 1970s onward on the planning process—including citizens' participation—rather than just its final product can be seen as a way to minimize the gap.

It is hard to address this challenge effectively because of ongoing scale increases and the interrelated changing responsibilities. Cities continue to expand and fuse: some already think of a highly urbanized country such as the Netherlands as one city. What is currently motivating social and technological change or overall impulses toward change? The role of "agent" is increasingly taken over by supranational institutions such as the European Union. There is a tendency toward globalization, but also toward new regionalization ("city Europe"). Apart from the political aspect of separatism, new types of sites may come into being that are no longer defined as sociotechnical entities with a specific function like "office" or "household" or "airport," but as a regional junction of junctions. The city has always been such a junction of junctions, and new forms are emerging today. One indication is that urban and regional "innovation systems" have become the subject of policies and studies.[96] The dynamic of location-specific technology entanglement is bound to remain; this should have been but was not always taken into account by the integral planning that emerged after 1970. Now the issue is whether the various types of site have meanwhile become so interconnected that individual sites are no longer marked by a dynamic of their own. Although networks, new information and communication technologies, and new opportunities for linking technologies (orchestrated by various infrastructures) increasingly define our world, it will never be possible to ignore the local dimensions of technologies. Innovations may be motivated by large-scale global processes, but they are always implemented at a specific location.[97]

Acknowledgements

This text is the result of four years of work. In late 1998 I started out as a post-doc in the Department of Philosophy of Science and Technology at the University of Twente. My research proposal served as basis of this text. At that time, the concept of 'innovation junction' was new to me. From Adri Albert de la Bruhèze I had received the minutes of the editorial meetings about the series on 'Technology in the Netherlands in the twentieth century,' which revealed to me that one of the issues to be addressed was the 'densification' of technologies in a particular location. The first volume of the series also came out in late 1998, and after reading the book on office and information technology published in this first volume, I began to reflect on 'flows' in specific locations. Johan Schot subsequently advised me to investigate several specific locations addressed in several volumes. This gave rise to many conversations with Gijs Mom about Schiphol Amsterdam Airport as an innovation junction, which involved, among other things, the role of transportation in the processing of flows of passengers and freight. Together with Nil Disco and Hans Buiter I explored the city as an innovation junction. Nil and I had many talks, and this led to me concentrate on spatial dimensions. Ruth Oldenziel stressed the role of groups that claim locations as their domain. It was perhaps unfair to some people, such as Arie Rip, that I did nothing with regional-economic ideas. In addition, the views reflected in this chapter evolved in conversations with my fellow researchers who worked on the other chapters. To me our meetings were always highly stimulating and I feel honored to have been part of this group. The concept of innovation junction developed into one of the 'contested ideas' within the overall project after publications by Nil Disco and me in *Technikgeschichte* in 2001 and by Onno de Wit, Jan van den Ende, Johan Schot, and Ellen van Oost in *Technology and Culture* in 2002. This triggered rather harsh comments from referees that were nevertheless a source of inspiration to me. I also would like to thank everyone who has helped to structure and clarify my thoughts. Furthermore, there is also a quite practical dimension to a four-year process of text production. At the Technical University of Delft, Karel Mulder created the space for me to continue my research: historical research in a department devoted to a sustainable future. I am indebted to him for being a facilitator. Next, Harry Lintsen, in the Department of History of Technology at Delft, created room for me as well. In the final stages of the project Arie Rip was an outstanding editor. Frank Veraart did a fine job of finding images to go with the text. Finally a word of thanks to Johan Schot, who for four years has always been available as supervisor in the background for all sorts of advice and practical support. There is always more a person can learn from him.

Notes

1 C. Disco, H. Buiter, and A. van den Boogaard, "Inleiding," in Schot et al., *Techniek in Neder-land*, vol. 6, 13–23.

2 I am indebted to C. Disco for the notion of a "physical construction with characteristic tools" for a site that is transformed into an innovation junction.

3 On the social construction of the Dutch kitchens and housewives see A. van Otterloo and M. Berendsen, "The Family Laboratory: The Contested Kitchen and the Making of the Modern Housewife," in A. A. de la Bruhèze and Ruth Oldenziel, *Manufacturing Technology, Manufacturing Consumers: The Making of Dutch Consumer Society* (Amsterdam: Aksant, 2009) 115–138; see also R. Oldenziel and M. Berendsen, "De uitbouw van technische systemen en het huishouden: Een kwestie van onderhandelen 1919-1940," in Schot et al, *Techniek in Nederland*, vol. 4, 37-61.

4 Ruth Schwartz Cowan, *More Work for Mother: The Ironies of Household Technology from the Open Hearth to the Microwave* (New York: Basic Books, 1983); Ruth Schwartz Cowan, "The Consumption Junction: A Proposal for Research Strategies in the Sociology of Technology," in Wiebe E. Bijker, Thomas Hughes, and Trevor Pinch, eds., *The Social Construction of Technological Systems: New Directions in the Sociology and History of Technology* (Cambridge, Mass.: MIT Press, 1987). For an interesting analysis of overlapping networks regarding the microwave, see C. Cockburn and S. Ormrod, *Gender and Technology in the Making* (London: Sage, 1993).

5 For the articulation of the innovation junction as a research topic, see J. W. Schot, H. W. Lintsen, and A. Rip, "Techniek in ontwikkeling," in Schot et al., *Techniek in Nederland*, vol. 1, 15–51; Onno de Wit et al., "Innovation Junctions: Office Technologies in the Netherlands, 1880–1980," *Technology and Culture* 43, no. 1 (2002): 51: "[An] innovation junction . . . is a space in which different sets of heterogeneous technologies are mobilized in support of social and economic activities and in which, as a result of their collocation, interactions and exchanges among these technologies occur. These interactions and exchanges lead to location-specific innovation patterns." See also Adrienne van den Bogaard and Cornelis Disco, "The City as Innovation Junction," *Technikgeschichte: Zeitschrift des Vereins Deutscher Ingenieure* 68, no. 2 (2001): 107: "Innovation junctions result from actor strategies shifting from the design of singular technologies to the design of technological ensembles specific for the location. Innovation junctions utilize the inevitable interferences and linkages among discrete technologies that occur in technologically dense locations. Hence the location and its diverse technological aspects become reflexively (and coherently) coproduced." In this chapter the following aspects are added: reflexivity at locations generally does lead not only to integrated, entangled design but also to "conscious disentanglement." Furthermore, we devote more attention in this chapter to spatiality. Space is always a factor in designs, for instance, in shaping spatial flows or optimizing space.

6 Lindy Biggs, *The Rational Factory: Architecture, Technology and Work in America's Age of Mass Production* (Baltimore: John Hopkins University Press 1996); Philip B. Scranton, *Endless Novelty: Specialty Production and American Industrialization, 1865–1925* (Princeton, N.J.: Princeton University Press, 1997).

7 Frederick W. Taylor, *The Principles of Scientific Management* (New York: 1915), 142–143.

8 The integration of conscious site-specific innovation and the rationalization of labor progressed in the Netherlands in the 1920s. In 1924, in the context of the founding of the Royal Institute of Engineers' new section on Technical Economics, A. E. C. van Saarloos, a government tax accountant in the Dutch East Indies, argued that the modern engineer had to be an "efficiency engineer," which he defined as follows: "The efficiency engineer has to be a very skilled engineer who knows the entire company and its parts. He has to

be technician, statistician, economist, etc. etc., but because he also, and primarily so, has to do with human workers he has to be a judge of human nature, a psychologist as well." See A. E. C. van Saarloos, "De betekenis van het bedrijfshuishoudkundig element in de ingenieursopleiding," *De Ingenieur* 39, no. 33 (1924): 622–627. Modern engineers had to integrate two sorts of problems: the rational processing of materials and the rational organization of the work. In his introductory lecture for the establishment of the Technical Economics section, I. P. de Vooys argued that engineers should receive specific training from the view that innovation of technology mostly means "the creating of a higher-level degree of organization." See I. P. de Vooys, "De economische beteekenis van den vooruitgang der techniek," *De Ingenieur* 39, no. 25 (1924): 464–466. On the application of scientific management in the design of coal mines, see B. P. A. Gales, J. P. Smits, and R. Bisscheroux, "Steenkolen," in Schot et al., *Techniek in Nederland*, vol. 2, 45–65; for a more general history of scientific management in the Netherlands, see E. S. A. Bloemen, *Scientific Management in Nederland, 1900–1930* (Amsterdam: Nederlandsch Economisch-Historisch Archief, 1988); on scientific management in offices and households, see Francina Maria Hartveld, *Moderne zakelijkheid: Efficiency in wonen en werken in Nederland, 1918–1945* (Amsterdam: Spinhuis, 1994).

9 Hettie Pott-Buter and Kea Tijdens, eds., *Vrouwen: Leven en werk in de twintigse eeuw* (Amsterdam: Amsterdam University Press, 1998).

10 Some household tasks were done by servants in wealthier families in large cities and others such as Arnhem. It is estimated that 16 percent of households had one or more servants.

11 Centraal Bureau voor Statistiek, "Woningen en woningkenmerken per hoofdbewoner en huishouden, periode 2000"; see also "7,2 miljoen huishoudens," www.cbs.nl/nl-NL/menu/themas/bevolking/publicaties/artikelen/archief/2007/2007-90079-wk.htm.

12 H. Baudet, *Een vertrouwde wereld* (Amsterdam: Uitgeverij Bert Bakker, 1986), 95.

13 R. Oldenziel, "Het ontstaan van het moderne huishouden: Toevalstreffers en valse starts, 1890–1918," in Schot et al., *Techniek in Nederland*, vol. 4, 22.

14 On the Housing Act, see Schot et al., *Techniek in Nederland*, vol. 6, in particular A. van den Bogaard, "De geplande stad, 1914–1945," 51–73; E. M. L. Bervoets, "Betwiste deskundigheid: De volkswoning, 1870–1930," 119–141; and E. M. L. Bervoets, "Woningbouwverenigingen als tussenschakel in de modernisering van de woningbouw, 1900–1940," 143–159.

15 Nederlandsch Instituut voor Volkshuisvesting en Stedebouw, *De Woningwet 1902–1929: Gedenkboek samengesteld ter gelegenheid van de tentoonstelling gehouden te Amsterdam, 18–27 October 1930, bij het 12-jarig bestaan van het Nederlandsch Instituut voor Volkshuisvesting en Stedebouw* (Amsterdam: NIVS, ca. 1930); P. van Overbeeke and G. P. J. Verbong, "De strijd om het huishouden," in Schot et al., *Techniek in Nederland*, vol. 2, 180–182; R. Oldenziel, ed., "Huishouden," in Schot et al., *Techniek in Nederland*, vol. 4, 11–151.

16 Margrith Wilke, "Kennis en kunde," in R. Oldenziel and Carolien Bouw, eds., *Schoon genoeg: Huisvrouwen en huishoudtechnologie in Nederland, 1898–1998* (Nijmegen: SUN, 1998), 61.

17 C. J. Wannée, *Handleiding voor de huisvrouw: Eenige wenken op het gebied van de gezondheidsleer, voedingsleer en warenkennis* (Amsterdam: Maatschappij voor Goede en goedkope Literatuur, 1919), 34.

18 Ibid., 37.

19 Ibid., 17–18.

20 Oldenziel, "Ontstaan van het moderne huishouden," 26.

21 Anneke H. van Otterloo, *Eten en eetlust in Nederland, 1840–1900: Een historisch-sociologische studie* (Amsterdam: Uitgeverij Bert Bakker, 1990), 116–119.

22 Despite the fading of the "feminist" ideals associated with 1920s efficiency thinking, these views continued to have guiding force. See Hartveld, *Moderne zakelijkheid*, 214. This study has a chapter on collective ideals versus individual strategies. R. Oldenziel refers to

it as "the debate on strategies," in Oldenziel and Berendsen, "Uitbouw van technische systemen en het huishouden," 37–43.

23 L. Zwiers, *Ons huis: Hygiëne en gerieflijkheid* (Haarlem: Ruijgrok, 1924), 37.

24 Ibid., 41.

25 Ibid., 44.

26 Marja Berendsen, "Het 'gezinslaboratorium': De betwiste keuken en de wording van de moderne "huisvrouw,'" *Tijdschrift voor Sociale Geschiedenis* 28, no. 3 (2002): 301–322; Oldenziel and Berendsen, "Uitbouw van technische systemen en het huishouden," 51–61

27 Hartveld, *Moderne zakelijkheid*, 198–200; Oldenziel and Berendsen, "Uitbouw van technische systemen en het huishouden," 51–61 (on the kitchen see p. 57 and following pages.

28 Quoted in Hartveld, *Moderne zakelijkheid*, 199.

29 K. Limperg, G. J. Meyers, and R. Lotgering-Hillebrand, *Keukens* (Rotterdam: Nijgh en Van Ditmar, 1935).

30 Ibid., 52.

31 Ibid., 51.

32 Ibid., 56.

33 Ibid., 57.

34 Wilke, "Kennis en kunde," 61.

35 A. H. van Otterloo, "Prelude op de consumptiemaatschappij in voor- en tegenspoed, 1920–1960," in Schot et al., *Techniek in Nederland,* vol. 3, 270.

36 *Bruynzeel-keuken* (Zaandam: Bruynzeel's Door Company, Kitchen Department, 1944), 2.

37 The analysis here owes much to De Wit, "Innovation Junctions," 50–72.

38 There was still some interest in office techniques that did not involve machines. For example, the *Maandblad voor het Boekhouden en Aanverwante Vakken* explained the use of one's ten fingers in multiplying large numbers. See, for example, Anonymus, "Hoe anderen vermenigvuldigen," *Maandblad voor het Boekhouden en Aanverwante Vakken* 23, no. 270 (February 1, 1917): 116–117.

39 W. O. de Wit, "De opkomst van de moderne administatie, 1880–1914," in Schot et al., *Techniek in Nederland,* vol. 1, 229.

40 E. van Oost et al., eds., *De opkomst van de informatietechnologie in Nederland* (The Hague: Netherlands Society of Informatics and Foundation for the History of Technology, 1998), 35.

41 J. G. Ch. Volmer, "Het Taylorsysteem en de boekhouding," *Maandblad voor het Boekhouden en Aanverwante Vakken* 22, no. 258 (February 1, 1916): 127.

42 See J. Goudriaan, "Over organisatie-schemas," *Administratieve arbeid* 2, nos. 4–7 (April–July 1924): 101, 104, 135, 167, 195.

43 Goudriaan, "Over organisatie-schemas," 173–174 (author's translation).

44 J. G. de Jongh. "De z.g.n. mechaniseering der boekhouding," *Administratieve arbeid* 1, no. 5 (May 1923): 121, 122; *Administratieve arbeid* 1, no. 6 (June 1923): 153.

45 A.C. [A. Cohen], "Korte aanteekeningen over technische hulpmiddelen in de administratie," *Adminstratieve arbeid* 3, no. 3 (March 1925): 85; see also *Adminstratieve arbeid* 3, no. 8 (August 1925): 225.

46 Van Oost, *Opkomst van de informatietechnologie in Nederland*, 42, column 1.

47 H. Koster, "Toepassing van het ponskaartensysteem bij de centrale boekhouding van de Dienst der Publieke Werken te Amsterdam," *Publieke Werken* 2, no. 4 (1932): 58 (H. Koster was head accountant of the Public Works Department).

48 Van Oost, *Opkomst van de informatietechnologie in Nederland*, 41.

49 S. C. Bakkenist and H. C. King, *De interne stedelijke bestuursorganisatie van de gemeente Amsterdam*, part 1, *Rapport uitgebracht aan het College van Burgemeester en Wethouders en Voorstudie A: Groei en ontwikkeling van de Gemeentesecretarie sinds 1851* (Amsterdam: Stadsdrukkerij, 1959), 83.

50 J. van den Ende, "Kantoortechnologie in de twintigste eeuw," in Schot et al., *Techniek in Nederland,* vol. 1, 332.

51 For the history of Schiphol Airport, see G. Mom et al., *Schiphol: Haven, station, knooppunt sinds 1916* (Zutphen: Schiphol Group and Foundation for the History of Technology, 1999), and M. L. J. Dierikx, J. W. Schot, and A. Vlot, "Van uithoek tot knooppunt: Schiphol," in Schot et al., *Techniek in Nederland,* vol. 5, 117–143.

52 Mom et al., *Schiphol,* 92.

53 Ibid., 19.

54 A. M. C. M. Bouwens and M. L. J. Dierikx, *Op de drempel van de lucht: Tachtig jaar Schiphol* (The Hague: , 1996), 76–78.

55 Mom et al., *Schiphol,* 24, 32.

56 Ibid., 36.

57 Its name was Netherlands Airport Consulting Office N.V. (NACO). The first director was G. C. van Wageningen.

58 L. Boogerd, [?] van Heemskerk, and U. F. M. Dellaert, eds., *Plan voor uitbreiding van de Luchthaven Schiphol* (Amsterdam: Amsterdam Department of Public Works, 1949).

59 Ibid., 59.

60 Ibid.; see, for example, 42.

61 M. P. Blaauw, "Het plan voor uitbreiding van de luchthaven Schiphol," *De Ingenieur* 61, no. 15 (1949): A150–A153.

62 Boogerd, Van Heemskerk, and Dellaert, *Plan voor uitbreiding van de Luchthaven Schiphol,* 27.

63 Ibid., 28.

64 Other scientific institutes, such as the Foundation for Economic Research of the University of Amsterdam and the Netherlands Economic Institute of Erasmus University, Rotterdam, also did extensive calculations to project the passenger, freight, and transit traffic volumes.

65 Mom et al., *Schiphol,* 58.

66 Ibid., 59.

67 A. Vlot, *Aviobrug tussen luchthaven en luchtvaartmaatschappij: De ontwikkeling van de Aviobruggen voor Schiphol, 1958–1971* (Eindhoven: Stichting Historie der Techniek, 2000), 1.

68 Ibid., 8–9.

69 KLM took the initiative for the development of the Aviobridge.

70 Vlot, *Aviobrug tussen luchthaven en luchtvaartmaatschappij,* 11–16.

71 Ibid.

72 Mom et al., *Schiphol,* 96.

73 Bouwens and Dierikx, *Op de drempel van de lucht,* 314–319.

74 H. W. Nachenius, *Verslag van het onderzoek naar den toestand van woningen voor min-vermogenden* (Haarlem: Society for the Promotion of Public Health in Haarlem, 1882).

75 Ibid., 12.

76 H. W. Nachenius, *Bijdrage tot de kennis van den stedebouw: Eene populaire studie* (Haarlem: De Graaff, 1880), iv. This study was intended as be a popular guide on urban improvement and expansion for municipalities.

77 This case is given extensive treatment in H. Buiter, "Werken aan sanitaire en bereikbare steden, 1880–1914," in Schot et al., *Techniek in Nederland,* vol. 6, 29–30.

78 J. G. Watjes, "Stedebouw," *Bouwkundig Weekblad* 31, no. 9 (1911): 95.

79 For this formulation, see [Van der Pek-Went], *De woningwet 1902–1929,* 103.

81 Standing Commission for Expansion Plans in North Holland, *Leidraad bij de samenstelling van uitbreidingsplannen* (Haarlem: 1928), 13–14.

81 Th. K. van Lohuizen was a member of the Social-Technical Association of Engineers from 1910 to 1935. In this context he became acquainted with P. Bakker Schut, who was active

in the "Technical Economics" section of the Royal Institute of Engineers and was head of the city-planning department of The Hague. He later became director of the State Agency for the National Plan. Since 1912 this association had been advocating the introduction of urban planning as a separate field at the Technical College Delft, where Van Lohuizen became the first professor in urban development in 1948. He was also one of the first proponents of a quantitative approach in urban development. Two of his contributions were counting vehicles and developing flow charts of traffic. He also devoted much attention to the development of business. His research was characterized as "urban development studies of an economic-technical nature." See Arnold van der Valk, *Het levenswerk van Th. K. van Lohuizen: De eenheid van het stedebouwkundig werk* (Delft: Delft University Press, 1990), 76.

82 Amsterdam Public Works Department, *Algemeen Uitbreidingsplan Amsterdam: Nota van toelichting en bijlagen* (Amsterdam: 1934), 127.

83 Standing Commission for Expansion Plans in North Holland, *Leidraad bij de samenstelling van uitbreidingsplannen*; Amsterdam Public Works Department, *Algemeen Uitbreidingsplan Amsterdam*.

84 Michelle Provoost, *Asfalt: Automobiliteit in de Rotterdamse stedbouw* (Rotterdam: NAi, 1996). This suggestion came from Ed Taverne.

85 See, for example [J. de Graaf], *Het plan voor een woonwijk te Amsterdam: Uitbreidingsplan Bosch en Lommer* (Amsterdam: Amsterdam Public Works Department, 1937).

86 This pressure is visible in accounts by urban engineers in the 1915 volume of *Technisch Gemeenteblad*.

87 See also Amsterdam Public Works Department, *Algemeen Uitbreidingsplan Amsterdam*, 11.

88 See, for example, W. J. Groot, "Welke plaats behooren de gemeentebedrijven in de gemeentelijke huishouding in te nemen?" *Technisch Gemeenteblad* 1, no. 4 (1915): 69–76.

89 C. Hellingman, "Nieuwe werkwijze voor de uitbreiding Zuid van Amsterdam en dergelijke op poldergrond zich uitbreidende gemeenten," *Technisch Gemeenteblad* 7, no. 6–7 (1921): 247–255 and 293–300.

90 See also chapter 2 in this volume.

91 H. P. Berlage tried to ground his designs in statistics, and from the 1930s the use of statistics in planning increased markedly.

92 Urban planning was geared especially to suburbs, as inner cities were hard to change; see also Van den Bogaard, "De geplande stad, 1914–1945," 55.

93 See De Wit, "Innovation Junctions: Office Technologies in the Netherlands, 1880–1980," 71.

94 The number of engineers working for cities increased. A 1931 amendment to the Housing Act recommended coordination with neighboring municipalities, but only after 1942 were cities required to coordinate their plans in collaboration with higher levels of government. See Van den Bogaard, "De geplande stad, 1914–1945," 63.

95 See chapters 1 and 7 in this volume. Such contestation is not a new phenomenon, as evidenced by events in the late nineteenth and early twentieth centuries.

96 See John De la Moth and Gilles Pacquet, eds., *Local and Regional Systems of Innovation* (Boston: Kluwer Academic Publishers, 1998).

97 For the importance of local knowledge in planning and innovation see also James C. Scott, *Seeing Like a State. How Certain Schemes to Improve the Human Condition Have Failed* (New Haven: Yale University Press, 1998) especially chapter 9.

In the 1950s and 1960s the NORIT peat-processing company managed to lower the price of its product, pulverized coal, through mechanization and economies of scale. This 1961 photo of the factory's exterior shows the result of expansion and modernization. Scale increase, especially in industry, was one of the most visible developments of the twentieth century (though the trend did not affect all sectors to the same degree). In some industrial branches companies would start growing in size or output only after years, be it gradually or in spurts. Heavy-industrial companies such as the steel producer Hoogovens, however, were established on a grand scale from the beginning.

4 Scale Increase and Its Dynamic

Rienk Vermij

Around 1900 the emergence of larger markets in a variety of areas made it possible for a number of Dutch manufacturers to expand and, ultimately, to mass-produce. Instead of producing what individual consumers wanted them to make, they became geared to producing standard products in large quantities that they subsequently tried to market. Although this was not an altogether new phenomenon, earlier this approach had been adopted only in certain specific sectors such as ceramic stoneware (pipes) and print (almanacs, playing cards). Gradually, however, companies began to mass-produce and then market other products, too, such as tin, enamelware kitchen utensils, and clothes irons made of cast iron.[1]

From the late nineteenth century onward, series production of more complicated mechanical devices grew common as well. Factories for such products first emerged in the United States with its large home market. Well-known early examples include sewing machines by Singer, Samuel Colt's revolvers, and—perhaps the most famous example of all—the Model T Ford, launched in the early twentieth century. Its prominence flowed not only from the enormous volume of the production and the immense wealth it brought to Henry Ford but also from the way he managed to organize his factory optimally for the production of a single product. This example shows that optimal organization of production alone was not enough to turn a mass-market company into a success, for despite Ford's superiority in this area his company ran up against serious problems later on. Companies active in mass markets also had to solve problems with respect to marketing and advertising, planning and anticipating future developments. If successful manufacturers of mass goods responded to increased sales potential, they also tried to keep up and improve the facilities that enabled mass markets.

Throughout the twentieth century, this overall development has stimulated the rise of large-scale companies. In much economic and business history, the need for scale increase is presented as inevitable. By enlarging outputs, the unit manufacturing cost decreases—so-called economies of scale. Assuming there is enough sales potential, this will offer large-scale companies a competitive edge and in time, based on a survival-of-the-fittest dynamic, they will start dominating the market. The story of these large-scale structures should be regarded as one of the main themes of twentieth-century history in general, and this chapter will explore how this trend toward scale increase in manufacturing prevailed in Dutch economic life during that era. Our major assumption is that such expansion is not a matter of blind forces that operate automatically or irresistibly. Efforts aimed at scale increase also met with obstacles that in some cases made such pursuits impractical or unfeasible, and in many cases there were viable alternatives. In other words, efforts toward scale increase might involve processes that for various reasons were challenged or contested. This chapter deals with actually realized scale increase in Dutch manufacturing during the twentieth century, its underlying ideals and motives, and the discrepancy between these ideals and the real world of commerce.

Scale Increase as a Historical Problem

The American business historian Alfred Chandler has outlined an influential model of historical development in which the tendency toward scale increases plays an important role.[2] In his view, the years between 1890 and 1930 are the crucial period, one that some even refer to as the second industrial revolution. During this time, all sorts of factors—such as better communications, improved technologies and the rise of mass markets—created the preconditions for large-scale production. Entrepreneurs capitalized on this by creating large corporations with multiple branches. This was not a matter of simply raising production, but of making targeted investments in three domains: production, marketing, and management. The latter in particular caused more old-fashioned individual entrepreneurs to be replaced by a class of professional managers. Soon these expanding companies opened such a lead over their competitors that it became very difficult for the latter to reach the same level of output. Many of the companies that grew large in the period around 1900 would continue to dominate economic life throughout the twentieth century, whereas only a small number of challengers managed to join their ranks, and when this did occur it was because of very large investments, government subsidies and other forms of support, or special circumstances.

Chandler's model is quite enlightening. Compared to earlier business economics analyses, which tend to be one-sided in their highlighting of the benefits of scale, Chandler's close attention to management is a notable major improvement. He argues that large companies do not just emerge simply through raising and expanding their production effort, but also through

The Netherlands Yeast and Spirits Factory (Nederlandsche Gist- en Spiritusfabriek), founded in 1869, was one of the first modern companies in the Netherlands. The company was founded to pursue large-scale production based on scientifically defined procedures. In 1882 a separate pilot plant was established, followed by a bacteriological laboratory three years later. The founder and first director, Jacques van Marken, was also known for his progressive social views. Here an exterior view of the complex in Delft, from 1908.

managerial investments. Yet his model, like any model, has its limitations. Its main shortcoming, we feel, is that Chandler hardly looks beyond large-scale enterprise, thus revealing himself as a prophet of the big-is-beautiful axiom. In his approach, the emergence of big corporations basically coincides with economic and social progress. This big-is-beautiful praise has met with much criticism. It is argued, for instance, that Chandler presents developments in the United States as a universal standard. Moreover, not all large twentieth-century companies focus exclusively on mass production; in some fields, making very specific or unique products has played a vital role in their success—one that Chandler systematically ignores.[3]

Another basic limitation is that Chandler examines only *industrial* companies. Trade, transportation, and agriculture remain largely beyond his scope. He ignores all that takes place outside of the world of business proper. Yet there are various modern professional sectors that gained prominence outside that world, including government, health care, and international organizations. Furthermore, Chandler's argument that large corporations can only exist by virtue of modern management does not mean that, conversely, organizational modernization should only be geared to realizing growth. After all, smaller companies and organizations may also rely on modern management, and the rise of professional management is not just a response to the emergence of large companies but also one of its causes.

A much more comprehensive approach to these developments is taken by J. R. Beniger, who focuses specifically on the notion of control.[4] In his view the trend toward ever larger-scale and faster production, facilitated by technological progress, has resulted in a "crisis of control": since these developments were realized so quickly, control and guiding mechanisms failed to keep pace, giving rise to other management problems. Production increases, for instance, caused company managers to lose track of the overall production effort, notably the many varied activities of the now much more anonymously laboring masses. In individual cases this would ultimately lead to a turning point at which the various control mechanisms were adapted to the new situation and company management would again take the initiative by implementing a new command structure, a rational distribution of tasks, and a systematic study of the production process. If Beniger, like Chandler, emphasizes the importance of management, he moves beyond a mere consideration of big industry by zooming in on the emergence of new forms of control. However, in his account, too, those forms of control are a response to problems associated with scale increase.

Multiple Forms of Scale Increase

In seeking to analyze the role and significance of large-scale economic structures in the twentieth century, it is important to be aware of the multiple meanings of the term "scale increase." This concept is commonly used in discussions at the level of individual companies or institutions: farms, offices, chemical plants, power plants, hospitals, and so on. But exactly what one is measuring to determine the size or scale level of such companies or institutions is not fixed: in offices it frequently applies to the number of employees or the quantity of data processed; in farms it may apply to the number of cattle (but not the number of employees); in hospitals to the number of beds; in other cases to sales or outputs.

Furthermore, frequently scale enlargement in companies is described in terms of problems of yield or productivity, or by focusing on the interaction

"Big is beautiful!" Manufacturers tried to impress the public by emphasizing the size of their plants—the aerial view was a favorite strategy. Here an impressive panorama of the merged textile companies J. A. Raymakers and G. W. Kaulen, in Helmond, in the early twentieth century. The image was a contribution to a "tribute by trade and industry" to Prince Hendrik, the new consort of Queen Wilhelmina. The company employed two hundred fifty to three hundred people at that time.

between scale increase and rationalization (or management innovations). For instance, scale increase is accompanied in offices with the introduction of new office technologies and Taylorism, in coal mining with scientific management, and in chemical industries with automation or the transition to continuous processing. The authors of the introduction to the seven-volume *Techniek in Nederland in de twintigste eeuw* (*Technology in the Netherlands in the twentieth century*) highlight this interrelation between scale increase and organizational innovation by discussing the much more diffuse process involving the flow of goods in the port of Rotterdam.[5] As such, scale increase does not primarily involve a technical problem in that it transcends the issue of productivity or rationalization within individual companies. This wider interpretation of "scale increase" underscores the extent to which it comes with control-related problems.

In this chapter we will discuss "scale increase" both in the strict sense of enlarging the size of individual production units and as a more differentiated phenomenon. Because the concept of scale increase is largely linked up with individual companies, the international discussion on its implications mainly pertains to this specific context. Accordingly, we first address the extent to which the situation in the Netherlands with respect to company size can be compared to that in other countries. Next, from a more basic perspective we discuss the nature of this country's large-scale process industries such as oil refining and mining, structures that came into being in the course of the twentieth century. Familiar models of development toward large-scale production strongly emphasize growth of individual companies. However, our discussion also emphasizes an element that is somewhat underexposed in overviews of business history: cooperation between companies. Also when companies remain fairly small, collaborative efforts may contribute to creating large-scale structures. In other words, there are various potential responses to challenges posed by mass-market structures, and the preferred answer in a given context will always depend on specific circumstances.

Second, the existing literature on scale increase as a phenomenon is strongly tied to the problem of management. In many cases, realizing production on a larger scale involves a complicated process. A rapidly-growing company or institution has to face all sorts of new problems and its existing small units do not automatically fit into larger ensembles. The literature suggests that anticipating or responding to these various concerns has been the key challenge for company leadership. Business historians argue that the many new tools and technologies developed during the first half of the twentieth century were mainly meant to make scale increases manageable. In this chapter we demonstrate that in the Netherlands this proved to be a slightly

One of the first industrial complexes—whose emergence was linked directly to World War I and the increased demand for equipment—was the steel producer Hoogovens, in IJmuiden, pictured here on a photo from 1982.

more complicated matter in concrete situations. Views about business organization were fed by many sources, not just sources tied to business economics. Even a tool such as standardization, which is preeminently geared to enabling larger connections, was sometimes applied in individual practices with quite different objectives in mind. We develop this argument with reference to the rationalization ideal in general and on the basis of the history of standardization and certification.

Big Industry in a Small Country

The authors of a 1993 study of the extent to which Chandler's model also applied to the Dutch context—E. Bloemen, J. Kok, and J. L. van Zanden—considered the one hundred largest Dutch companies. They used 1913, 1930, 1950, 1973, and 1990 as sample years and included industrial companies but also retail firms such as Albert Heijn and De Gruyter. Company assets served as measure of size, implying the study focused on companies officially in Dutch hands. Foreign-owned companies with branches in the Netherlands were not taken into account, whereas a company such as Philips, with many foreign branches, was part of the overview based on its total assets (while

Unilever and Shell only counted for the part that is formally Dutch: 50 percent and 60 percent, respectively).[6]

The authors largely saw Chandler's assumptions corroborated: around the start of the twentieth century a fairly small group of Dutch companies managed to acquire a dominant position through expansion. Based on international standards, half a dozen major corporations set the pace: Royal Petroleum (Shell), Philips, Unilever, AKU/Akzo, DSM (State Mines Company), and, possibly, Hoogovens. This leading group has been remarkably stable over the years. Although in 1913 Philips and AKU (Algemene Kunstzijde Unie, now named Enka, for Eerste Nederlandse Kunstzijdefabriek Arnhem) ranked much lower, this half dozen companies has been in the top rank since 1930, occasionally switching positions. Lower on the list, however, there is much less stability. Although the top thirty still comprises a more or less stable group of companies, among the lower-ranking companies it is still hard to notice clear historical patterns.

The study by Bloemen and his coauthors, apart from largely confirming Chandler's assumptions, identified two striking differences between the Dutch situation and that in other countries. First, major Dutch companies came into being fairly late, which perhaps should come as no surprise. Whereas in the United States the birth of many large companies dates back to the late nineteenth century, in the Netherlands such companies became a major factor only after World War I. A second striking difference applies to their level of concentration, or their contribution to the Dutch national income. For both 1950 and 1973, the total assets of the 100 largest Dutch companies as part of the national income is proportionally far more than double that in England, Germany, or the United States. Of course, it is only natural that in bigger countries with more companies the weight of a random sample makes up a smaller share of the total. However, if we consider only the 100 largest companies, the concentration level proves to be substantially higher in the Netherlands. The total assets of numbers 21 to 100 on the Dutch list together account for some 12 percent of the total assets of the 100 leading companies; in the other countries mentioned this is more than 40 percent. In terms of assets, Royal Petroleum/Shell is the unchallenged leader for the entire period. If this company is excluded from the calculation, the difference between the Netherlands and the other countries is less marked, albeit still significant. In the Netherlands, then, a small number of companies have acquired a disproportionate piece of the cake. There are 5 or 6 large corporations that count internationally. There is a wide gap between them and the large number of much smaller companies, a gap that is especially striking in the period before World War II. Afterward, a specific group of companies has gradually drawn nearer to the six leaders; they managed to gain access to international markets without developing into leading global players.

The study by Bloemen and his colleagues pertains to the size of companies as a whole, rather than the size of their separate businesses or branches. From this second angle, the Netherlands situation is much less of an exception. In 1971 P. A.V. Janssen published a quantitative analysis of the growth of Dutch businesses for the years 1953, 1968, and 1980, in which he took the number of employees in the Netherlands as measure of the businesses' size. Janssen only considers industrial enterprises, including those in the construction and dairy industries, and basically takes into account all Dutch businesses with more than ten employees. His international comparison of the distribution of businesses in terms of their category and on the basis of their size does not suggest any large differences between Dutch industry and that in other countries.[7] Further indications are found in a European Community study quoted by Janssen. For the period 1962-1963 it offers more detailed data on company size categorized by sector in several European countries (Germany, the Netherlands, Belgium, France, and Italy), as well as Japan and the United States. Although this material is rather limited for our purposes, the numbers do not indicate that big companies in the Netherlands are less common than in larger countries.[8]

Above-average business concentration, then, mainly occurs at the level of corporations, but barely at the level of businesses or companies within corporations. It is not the size of the businesses that is essentially different, nor their distribution across size categories, but the ownership structure of which they are part. This, then, is about the organization of assets, not technology. What is necessarily absent in the study by Janssen is the international dimension; after all, the biggest companies in the Netherlands are multinational corporations. Their assets are largely reflected in their (partial) ownership of foreign branches and only to a limited extent in companies that operate in the Netherlands.

Later on Jan Luiten van Zanden made estimates of employment among large Dutch companies. It turns out that employment grew strongly with both the six leading companies and in the top one hundred companies in the period up to 1973, after which it began to drop. This applied to their employment worldwide. Only for the six top-seeded companies did Van Zanden make employment estimates of their Dutch branches as well. This revealed the same trend, with the qualification that the growth before 1973 was weaker and the drop after 1973, very obvious. In this respect companies below the top-ranking ones did much better after 1973. The share of the six leading companies in Dutch employment dropped from 17.9 percent in 1973 to 12.5 percent twenty years later.[9]

In the Netherlands a small number of large companies contribute a disproportionate share to the national income, as Bloemen and his coauthors suggest. Since this situation cannot be derived from Dutch employment structures, it may also be interpreted to suggest that the few Dutch companies that operate internationally do so quite powerfully. The five top-ranking companies have mainly earned their position by means of internationalization. Strikingly, three of these companies acquired their position by merging with a foreign partner: Royal Dutch/Shell is the product of a merger of Royal Petroleum with the British Shell Group; Unilever is the result of a merger in 1929 of the British Lever Brothers with the Margarine Unie (the latter was itself the result of a merger of companies owned by Jurgens and Van den Bergh, both of which were among the five leading Dutch companies in 1913); AKU (synthetic fibers) came into being in 1929 after a merger of Enka with the much larger German Glanzstoff. In order to properly grasp, then, the emergence of large Dutch industries it is relevant to take into account the various ramifications of international commerce and how they could be dealt with.

Chandler's model predicts that the companies that around 1900 evolved into large corporations are mostly found in new, capital-intensive sectors. The study by Bloemen and colleagues supports this conclusion for the Netherlands. In this country, corporate development mainly evolved in the process industry and the electro-technical industry. Other sectors, however, such as steel, heavy machinery, and transportation, have clearly been underrepresented in the Dutch top one hundred, at least when compared to other countries, and in the course of the twentieth century their significance decreased even more. (Notably the disappearance of Dutch shipbuilding has been a major factor; this sector's problems were hardly unique to the Netherlands.) In this respect, Bloemen, Kok, and Van Zanden point to a clear pattern: industry in the Netherlands proves to be complementary with that of Germany to a high degree. Whereas Germany mainly relies on steel industries, the chemical sector, and industries tied to transportation, in the Netherlands oil and food industries prevail. This aligns well with our observation that Dutch companies have been able to grow into big companies exclusively through internationalization, whereby they were not solely dependent on domestic factors, of course, but also on international power relations.

Dutch companies never managed to develop into genuinely global players in sectors in which German companies already prevailed. For example, Van Berkel's Patent, in 1930 ninth on the Dutch list of major companies, is a company that mass-produced mechanical products such as cutting devices, using the most advanced methods, and the Begemann machine company in Helmond had already modernized before World War I, mainly specializing in centrifugal pumps. At a time when most Dutch construction workshops and machine factories had long vanished, both managed to develop into major industries, albeit without becoming major players on a global scale. Although companies such as Begemann or Van Berkel's Patent managed to acquire an international position with several specialized products, they did not become sector leaders overall. What they lacked, in Chandler's terms, was scope, or product diversity.

Because international industrial relationships clearly disfavored the growth of the Dutch steel and chemical industries, these companies' size remained limited. This is not to suggest that companies in the food sector, which suffered much less from such relationships, saw unrestrained growth. Major companies in this sector that innovated early on, such as NGSF (Nederlandse Gist- en Spiritusfabrieken, Dutch yeast and spirits factory, today Gist-Brocades) and Calvé Delft never became genuine leaders. Although in 1913 they still ranked ninth and eighth, respectively, in later years their ranking fell significantly on the list of Dutch major companies.

The application of Chandler's model thus points to several features that are specific to the Netherlands. Compared to Germany and especially the United States, the most striking difference is the small domestic market in the Netherlands. This made it hard for a company to grow simply by producing more, or more inexpensively. A Dutch company that aspired to go international ran up against serious obstacles, and as such it is no wonder that initially but a small number of companies succeeded. A Dutch company that wanted to grow preferably had to internationalize through mergers, takeovers, or in some other way. The small domestic market thus constituted some sort of glass ceiling when it came to expansion. Only companies that managed to break it could achieve real growth, which was only possible under favorable circumstances.

Finally it should be pointed out that although several large multinational companies represented—and represent—a disproportionate share of the total assets of all Dutch companies, there were also a large number of small and medium-sized companies. Some of them were quite advanced, as in the case of Van Berkel's Patent and Begemann, two flourishing companies with modern management and clear ideas on marketing and production. Many of these companies were established because around 1900 entrepreneurs capitalized on the profit potential of growing mass markets. Still, these companies remained small. A question we will not address here is the extent to

In 1917 the Utrecht chapter of the Association for Dutch Products took the initiative to organize an annual fair (*jaarbeurs*) in Utrecht. The fair was open from February 26 until March 10. Its huge success—it drew 690 participants and over 150,000 visitors—led to the fair's becoming an annual event and laid the groundwork for today's international company, Utrecht Fair.

which the small domestic market in the Netherlands represented not just an impediment to growth, but perhaps also provided a favorable context and protection to small-sized companies. After all, at home they may have had an advantage over large foreign competitors that otherwise would have pushed them out of the market (sooner).

Economy of Scale: Ideals and Limitations

The question how to respond to the new economic opportunities and difficulties was in particular pressing at the time of World War I. The study by Bloemen, Kok, and Van Zanden and more recent work by H. J. de Jong reveal that this period was crucial to the development of Dutch industry.[10] Moreover, entrepreneurs and others were quite aware that they were living in an era of new opportunities but also one of great risks for those who backed the wrong horse. From the outset, the various implications for Dutch competitiveness were a relevant topic. In 1915 the secretary of the Industrial Association (Maatschappij van Nijverheid), G. de Clercq, wrote to Kipp, a firm in Delft:

> Concerning the future of our industry, one party is afraid that after the war the Netherlands will be flooded by German brands. This party is already beginning to urge raising import duties. By contrast, another party, and in confidence I can convey to you he is on the board of the Industrial Association, feels that after the war Dutch industry will be stronger than German industry. The latter will have to face the aftermath of crisis, a lack of skilled leaders and workers, expensive money, and disorganization because of the many company arrangements geared toward ammunition as well as changing market circumstances. This group, then, feels that after the war the Netherlands will be able to compete with Germany very well.

Despite De Clercq's anticipation of expensive money after the war, at the moment the money market was still favorable. Thus, those eager to expand should act swiftly and not wait.[11]

The question was no longer if industry had to be modernized but how. In various cases, scale increase was somehow seen as the proper way to go. Precisely in areas in which the Dutch had operated clearly in the shadow of Germany, the chemical and steel industries, new opportunities for industrial expansion seemed to present themselves. It is true, however, that high-flown efforts did not always produce the best results. One of the grandest efforts in this respect was the attempt by G. Hondius Boldingh to set up an integrated chemical-industrial complex in the Netherlands, a project in which he received strong support from the Industrial Association. The starting point was the creation of a Dutch dye-stuffs industry, but this also called for setting up production of a whole chain of raw materials and intermediary products. In theory this plan could have been realizable, but all sorts of practical problems caused so much delay that the favorable moment passed and the entire project went up in smoke.[12]

In this period a firm plea for uniform mass production reverberated also in the Dutch heavy machinery industry. In 1915, in the Industrial Association's journal, the industrialist and engineer E. A. du Croo, who manufactured hoisting cranes, railway materials, and dredging equipment, claimed that Dutch industry had to switch to production of standard products. His ideal was mass production as practiced in the Ford factories in Detroit.[13] Another advocate was the engineer and industrialist E. P. Haverkamp Begemann. He headed up a manufacturing company that specialized in centrifugal pumps. Through smart design and construction he managed to produce a large variety of pump designs that used a limited number of standard parts. He denounced the tendency of Dutch companies (he was mainly thinking of machinery manufacturers) to produce a jumble of products. Because they had insufficiently specialized, their products fell short compared to those of more specialized foreign competitors: "The difficult economic struggle that lies ahead should teach us that today these companies have to switch promptly to production of only one or just a few products, because otherwise the competition will be unbeatable after the war."[14]

More cautious voices could be heard as well; they suggested there were also good reasons for Dutch machine companies to be flexible. Ir. P. Lugt, director of the Conrad Dockyard (Wharf Conrad) in Haarlem, responded to Begemann's argument by referring to a trip he had made to the United States twenty years before, in part to study the opportunities for purchasing modern equipment for a planned factory. He reported that each of his requests to manufacturers to make a few adjustments to account for specific circumstances triggered the very same response: "Ask my neighbor!" Perhaps one could afford such response in the American market, but this was certainly not the case in the European market. On this point German manufacturers proved much more accommodating and this was one of the reasons, Lugt explained, that German industry had managed to outdo the Americans in the European market.[15]

Although the Dutch failed to pose a threat to German dominance in

The growing influence of large companies was an established topic in literature and the arts. Around 1900 the power of big capital was of course a major concern in socialist circles. The opening in 1903 of a new store in Amsterdam by a Belgian department store chain occasioned this cartoon, *The Large and the Small,* by the well-known political cartoonist Albert Hahn, showing a large department smashing small stores that sell textiles, perfumes, leather goods, chinaware, toys, and other products.

chemical industries and heavy machinery, much was accomplished in this period, and the pleas by Du Croo and Begemann did not fall on deaf ears entirely. The Industrial Association, in collaboration with the Royal Institute of Engineers (Koninklijk Instituut van Ingenieurs, KIvI), took the initiative to establish a commission tasked to find out where Dutch industry had a need for product standardization and that also had to initiate its implementation. (The work of this "Main Commission for Standardization in the Netherlands' is discussed in more detail later in this chapter.)

Caution in the Interwar Period

Aside from the penchant for large-scale mass production, as felt particularly during World War I, other considerations were relevant. Although Dutch industry increasingly aimed at organizing production as efficiently as possible, in most cases scale increase was approached with ambivalence rather than being actively pursued. Still, people were quite aware of the business-eco-

nomic benefits of economy of scale, which were formulated mathematically
in 1910 by the German economist Karl Bücher in his "Law of Mass Production."[16] In the Netherlands, however, scale benefits were mostly couched in
terms of trade rather than production. After all, the benefits of scale do not
occur only in production but also in transportation and all sorts of services.
In the context of expansion or scale increase, one should not only think of
the emergence of mass production, but also of the rise of big business in general—such as retail business, with its supermarkets, department stores, and
store chains.[17] Emile Zola, in his novel *Au bonheur des dames* (*The Ladies' Par-
adise*), described how a trader could earn a profit by lowering his prices if he
simultaneously managed to raise his turnover rate. Jacob Blokker, founder
of a Dutch retail chain for household products, prided himself in his ads in
1907 that because of his larger sales volume he could offer articles at prices
that no one could beat.[18]

In industry this was much less an issue. Although around 1900 there was
certainly a sense that big industry was modern and represented the future, for
a long time most people were hardly aware of scale benefits as such, it seems.
The effort aimed at unifying production primarily targeted reducing chaos—
a lack of organization and transparancy in the production process, which
thereby tends to become less manageable—not to increasing production volume. Although the various Dutch handbooks on business economics from
the interwar period discuss the potential positive cost price effect of expansion, they mainly point to its risks. When a factory is organized and managed
in a specific way, it has a certain optimal production whereby the cost per unit
is lowest. Further expansion is only possible through large investments; this
puts the optimal production at a higher level, but because of the investments
the cost per unit will first appear to go up. Expansion is certainly not recommended as a general business strategy. Expansion contains "a *speculative*
element, namely anticipation of the product's future sales."[19] And:

> Because of expansion the cost per unit is lowered either directly or
> after some time: the *company optimum has gone up* and will continue
> to do so in later expansions, until the constant costs are basically no
> more part of the cost per unit or until the company has grown to a size
> at which the organization deteriorates again and its employees begin
> to act more like being part of a bureaucratic system. Their sense of
> tedium or routine will push economic motives into the background.[20]

The economists from this period agree that business expansion should not
be simply explained by economic factors such as the pursuit of higher prof-

its or lower cost prices. A. A. D. Bouwhof and J. C. Lagerwerff refer instead to the "inner urge that reveals itself in each energetic leader to make the company grow."[21] C. Huijsman submits that the industrialist does not primarily pursue profits, but rather the success of his company: "The craving for expansion thus carries a sporty element."[22] O. Bakker, in the wake of A. S. Dewing, largely agrees, but adds that the pursuit of higher profits may be subordinate as a motivation, yet it is essential as a justification when trying to find financiers and allies. Because psychological factors play a leading role, there is quite possibly a risk, Bakker argues, that the urge for expansion will contribute to exceeding the optimal company size.[23] A. B. A. van Ketel, finally, claims that in a company's evolution expansion is "a "natural" affair, and that its seed is in the company's organization: the company becomes ever more perfect, its employees become better trained, its market position improves, capital investment is more constantly profitable, capital providers have more trust in the company, and so on—this explains the company's development."[24]

Scientific Management

This leads to the question of whether striving for economies of scale was an element of the modernization of Dutch industries in the interwar period. Although not all revolutionary dreams were instantly realized, this should not blind us to the fact that World War I provided a powerful incentive to Dutch business and industry, and that in this period the foundation was laid for large-scale modern industry. To trace this process, however, we will have to look less to grand schemes for answers than to concrete changes in production and management in companies. Which changes, exactly, can be identified and what was their background? And in particular: Did striving for scale enlargement act as a direct or indirect engine behind these developments?

The modernization of industries in the interwar period took place largely under the banner of the scientific management theory developed by the American F. Taylor, who felt that industry had to be led by specialist managers. A careful analysis of all production activities, based for instance on length of time needed for each process, would allow one to arrive at detailed protocols as to how and how fast each task should be performed. Scientific management was inspired by the American practice of mass production and it only made sense at a certain level of production volume. Even so, the link between Taylorism and economies of scale was indirect at best. Although Taylorism was rarely applied in its "pure" form (if such term makes sense at all), many in Dutch business were guided or inspired by the basic notion of

scientific organization. In a lot of companies—and not exclusively mass-pro-
ducing ones—some kind of "scientific" ideas were tried. One of the first
Dutch firms that successfully practiced these methods was the construction
workshop of De Vries Robbé in Gorinchem, which made bridges, roof
frameworks for railway stations, and such.[25]

Rather than being a panacea for the entire industry, scientific management was geared toward a certain category of problems and therefore had
limited validity. The problems it aimed to solve had to do in particular with
manipulation of labor. In the manufacturing industry, workers make or put
together various products. Production is a matter of getting workers going and
keeping them going, and this is what the overall management has to focus
on. Its task is primarily a disciplining one. The worker has to do what he is
told to do, instead of hanging around much of the day. Second, management should not lose track of the situation when production is rising. If a
product is basically made by one worker from start to finish, as in a crafts-
man's workshop, it is easy to determine how much labor went into it. However, this is much harder to determine for products that consist of many parts
and involve as much as a dozen workers, who work on other products and
perform other tasks to boot. Or, to give another example: employees of a
Dutch machine factory, one that was technologically advanced, would just
write down the number of hours spent on a product at the end of the week
from memory, and the manager subsequently used this information to calculate the cost price. When the company suffered serious losses, in 1909, an
accountancy office had to tell the company leadership that it had sold its
products below cost price.[26]

Because of the technological developments at the start of the century, in
many branches wage calculation had become problematic. Management
would lose sight of the amount of labor that went into a product, and hence
of its cost and the wages to be paid. Even before World War I, then, much
experimenting was going on with all sorts of new wage systems, such as the
premium systems of Halsey and Rowan. Inasmuch as Taylorism made
inroads in the Netherlands, it chiefly applied to determining the correct wage
rates by means of time recording.[27] However, it is quite common that a focus
on innovating one system, in this case wage systems, will also lead to other,
partly interrelated, improvements. For example, other organizational aspects
were changed, such as the proper position of machines, the care of tools, and
so forth. Taylorism, then, was also attractive to company managements
because it provided opportunities for getting a handle on *how* workers actually worked and realizing operational modes that were most efficient and
profitable. This also implied disciplining workers by means of strict rules.

In the interwar period, assembly line production had an aura of modernity and was associated with large-scale outputs. This photo shows one of the many lines for assembling radio receivers at the Netherlands Seintoestellen Fabriek in Hilversum, which worked closely with Philips. In late 1928 it employed over three hundred people. The slowly moving conveyor belt to which work tables were linked triggered quite some amazement in those days. The whole process would be perfected over time. The arrangement shown on the photo, with the work tables at right angles to the moving conveyer belt, was called the "school system." Mostly female workers were hired for this work, which was repetitive and monotonous yet required careful attention.

Scientific management was a way to render work manageable and as such it was mainly deployed where labor was the bottleneck of the organization's productivity. This applied to industrial sectors such as textiles and steel, to coal mining and to some extent offices.

One of the most visible or defining elements of industry's modernization in this era was the assembly line. For a long time historians considered this tool the preeminent instrument of uniform mass production. Assembly lines were mostly associated with large factories such as those of Ford or Renault, where a new automobile rolled off the line every few minutes. Undoubtedly the assembly line has functioned well for this kind of work. In the Netherlands, however, the assembly line was mainly deployed in smaller factories and for changing series of products. About half of all Dutch companies that had assembly line production in 1940 were clothing workshops. This included a few large companies, with a few hundred or even more than one thousand employees, but many were much smaller, having a few dozen employees at most. In companies that mainly produced underwear or shirts, styles and specifications were more or less constant, but assembly lines were also used for making dresses, coats, pants, and other garments. These products were highly dependent on fashion. The orders came in at unpredictable times and maintaining stock was risky. So manufacturers had to go back and forth between specific series, and sometimes they had to switch from one product to another every few days, if not more often. In this context the assembly line was not a tool for mass production, but a means for disciplining the (female) workers. Some companies mainly implemented assembly lines to have workers work more, or produce more, for the same wages, but this was often bound to fail. The advantages to assembly lines lay chiefly in improved flow rates, production control, and in the fact that employees needed to have less education because of the more limited tasks they had to perform.[28]

Probably the largest-scale use of assembly lines in the Netherlands in this period was the radio assembly line of Philips. Here too, however, uniform mass production was not the reason for its use. Radios came in all kinds of models, so that on a limited number of production lines many different series were made. From a marketing angle it was illogical to start with a new series when the old one was finished; instead, assembly line workers constantly had to switch between ongoing series. Apart from being used as a means of transportation, the assembly line mainly served "to force a specific pace upon workers."[29]

Assembly lines should be viewed as an expression of the era's new scientific management. Their implementation was of a piece with the age's general trend, whereby the responsibility for production was taken as much as possible out of the hands of workers and put into those of management. One contemporary consultancy bureau claimed that the benefit of an assembly line was mainly that it allowed "control of overall performance to shift from workers to management," offering it new opportunities for raising performance through further measures.[30] Thus, the assembly line served in particular as a perfected means for disciplining human labor.

Operational Management in the Process Industries

These considerations do not apply to industry as a whole. Industrial sectors differed; notably, the process industry encountered quite different problems of production control than the manufacturing industry. When Taylorism started attracting attention, after World War I, G. de Clercq, in the 1920 volume of the *Chemisch Weekblad* (*Chemical Weekly*), devoted an essay to the question of whether Taylorism could also be applied to the chemical industry. He did not believe it could. Chemical processes could not be improved by having workers work either longer hours or more efficiently: "Whereas a worker in a machine factory contributes a specific amount of labor that can be expressed in concrete units, the worker in the chemical industry merely serves as a faithful watchman who is standing by as chemical reactions are taking place according to the laws of nature."[31] Likewise, other commentators argued that workers in the chemical industries had to meet other standards than workers in manufacturing industries. In a plea for separate training for worker in the chemical industry in 1955, Professor J. G. Hoogland claimed that steel workers "walking around with a wrench in their hand, always hunting for something that needed repair," had the wrong attitude for monitoring chemical processes. By contrast, the chemical industry needed workers who possessed farmers' quiet power of observation: "Farmers know that a ruminating cow should be left alone."[32]

Because work in the chemical industry was based on the "laws of nature," De Clercq argued that increased "capacity or output is only possible along the lines of scientific research."[33] Company management should not be preoccupied primarily with disciplining workers, but with finding the best procedures, designing the most efficient installations to facilitate the various processes, and searching for the proper settings and measuring values. In chemical industries, in other words, rationalization was a matter of technology rather than management.

One way of realizing this rationalization involved scale enlargement. In many cases a single large installation proved much more efficient than a number of smaller ones with the same total capacity. Having a larger plant did not necessarily require more workers for process monitoring, whereas in the case of more (small) plants each plant of course needed to have a full staff. This explains why chemical plants, at least those for bulk products such as salt, fertilizer, or sulfuric acid, were inclined to strong growth and why opting for larger production units was prominent in this field in particular. During the crisis of the 1930s, DSM sometimes responded to collapsing markets precisely by production capacity increases and raising production, in order to be able to put products on the market for lower prices.[34] This concentration and scale increase of production units was applied in the power supply sector, also an area where efficiency was determined much more by technology than by labor.

A second major aspect of rationalization involved the design of installations that could be operated around the clock. P. J. H. van Ginneken, in a 1923 article in *Chemisch Weekblad*, put forward that processes have to be altered and adapted until a "*continuous* mode" is achieved, one that ensures "stability" at each point in the process. He felt that "it must be possible to prove scientifically that this particular mode is also the most economic."[35]

One concrete advantage of continuity in process industries was more cost-effective use of fuel because installations could be kept at the same steady temperature.[36] Although in many cases it was not easy to find methods for having processes take place on a larger scale as well as continuously, these challenges were mainly of a technological-scientific nature.

Our consideration of the role of scale increase in rationalizing production during the "second industrial revolution" warrants the conclusion that this strategy was of great significance in Dutch process industries and some related sectors, such as energy, but that its potential benefits in manufacturing were overshadowed by problems related to other aspects of business activity. This is not to say that scale advantages played no role in production industries; but such advantages were mainly found in areas such as sales and purchasing.

Opting for Scale Increase or Cooperation?

To take advantage of growing mass markets and other tendencies toward economic integration, there were options beyond straightforward expansion, either of production units or companies. Although specialization by focusing on manufacturing a single product could be attractive from a production-technical angle, from the angle of the market it was not without risk, especially if this product was sensitive to market trends. This concern was all the more urgent because of the large investments that inevitably accompanied industrial mechanization. A common reaction to these risks involved efforts aimed at reducing competition, its being one of the largest risk factors. Open competition, long a dogma in entrepreneurial circles, fell into discredit. Already in the late nineteenth century the view that competition was not sacrosanct gained terrain in Germany.[37] Mechanization in Europe, then, generally led to joint efforts and cooperation rather than to specialization and efforts aimed at economy of scale. This collaboration could take on many forms: price fixing, dividing the market among individual companies, takeovers, and horizontal and vertical integration. As a result, large conglomerates such as IG Farben and VDF (Vereinigte Drehbank-Fabriken, United Lathe Factories, founded in 1927) emerged in Germany.[38]

Similarly, Dutch companies pursued collaboration rather than a monopoly in the years prior to World War II. Especially in the 1930s, because of the economic crisis, one could hardly find people anymore who believed that society could still be left to the game of free market competition.[39] In reality, however, skepticism on this point was much older. For example, in 1923 the economist J. Goudriaan argued that in many cases the risks of specializing production—whereby a product is mass- or series-produced and the company offers it for a lower price in a struggle to achieve the sales it needs—were too large: "It is either bound to fail or one does not even give it a try." He felt, however, that "collaborating business enterprises in the same field might be able to negotiate specialization."[40]

Prewar manuals of business economics expend few words on scale benefits, but they cover joint ventures and cartels or trusts extensively:

> The struggle to retain their markets and conquer new ones is very tough for industrial companies, which are forced by the operation of the law of increasing returns to maximize production…. For many companies it is vital to eliminate undue competition, and hence thereby to solve their sales problem.[41]

Business economists who studied profitability challenges in times of crisis notably advocated joint efforts among companies, be they various forms of functional cooperation or something else.[42]

During this era the rise of large-scale companies occurred most often as a result of agreements and takeovers rather than from exploiting specific production benefits vis-à-vis other companies. NGSF (yeast and alcohol) was quite adamant about this. Its 1930 commemorative book proudly mentions that the company "did not pursue expansion of the already marginal market otherwise than by taking over distilleries it was offered. Rather we tried to set up collaborative efforts in order to save this early national industry from its total demise."[43] Scale increases in the period prior to World War II mainly involved cartel formation rather than the rise of mass production companies. Before World War II the Dutch government tolerated cartels or even encouraged them. The Entrepreneurs Agreements Act (Ondernemers-overeenkomstenwet) of 1935 made it possible to declare agreements between companies formally binding for the entire sector.[44]

Of course, efforts aimed at circumventing market risks took on various forms. An original tactic was that of ENCK (Eerste Nederlandse Coöperatieve Kunstmestfabriek, First Dutch Cooperative Fertilizer Factory), based in Vlaardingen. Fertilizer was a market controlled by several big international firms. Via all sorts of cartels, which often broke up and then re-formed, these companies managed to shape the market to their will. ENCK's establishment in 1916 was a response to this. It was an initiative of several agricultural cooperatives that wanted to control supplies of the fertilizer they needed, rather than continue to be the playthings of other fertilizer manufacturers. But the man they subsequently asked to build and lead the factory, F. W. Bakema, saw quite different opportunities. Without the board's realizing it, he built a factory not with a capacity of 50,000 tons per year (the projected need), but 200,000 tons, making it one of the largest fertilizer factories in the world. He could do so because the market risks were slight: the cooperatives' members were forced to purchase the fertilizer for a fixed price, while surpluses were put on the market. After several years, the cooperatives' members learned, much to their surprise, that they were making quite a tidy profit. Their joy, however, was slightly blunted by the fact that shortly beforehand they had unknowingly consented to a substantial bonus scheme for director Bakema and the board. The adopted plan also gave rise to other troubles, particularly because members found out that outsider customers sometimes had to pay less for their fertilizer than the cooperative members. Ultimately, however, the company proved to have a bright future, which underscores that a host of underlying factors and motivations apart from technological and business-economic considerations could be involved in scale-enlargement efforts.[45]

Cooperation in Agriculture

Efforts at cooperation played a role not only in Dutch industry but also in Dutch agriculture, even though not always for the same reasons. On the one hand, there was a striving in agriculture for greater efficiency that went hand in hand with that of the industry and partly derived inspiration from the same sources. Around 1918, S. Koenen, a professor in rural economy at the agricultural college of Wageningen, held a lecture in which he compared industry and agriculture. Although he felt that agriculture was below industry when it came to optimizing the economic use of human labor through mechanization and labor division, it was agricultural science that facilitated agriculture's rationalization and modernization. Earlier crises, he argued, were in fact defeated by such rationalization and specialization.[46] This approach was mainly promoted by the agricultural engineers and other experts from Wageningen and agricultural experimental stations.

On the other hand, the countryside and farming represented specific values that contrasted with those underpinning modern urban society. Notably the people who were the most aware of the shadowy sides of progress stood up as advocates for the interests of farming. In 1890 a government commission found that more than others, "farmers preserve popular tradition and the old morals and customs that unite the various generations and add cohesion to people's history. In this sense, one has good reason to claim that the rural population constitutes the nation's core, which contrasts with the cosmopolitan character of large cities."[47] This quote appears to espouse a rather romantic vision of the farmer or peasant.

The brand-new dairy factory in Weidum, near Leeuwarden, along with (presumably) the cooperative's members. The cooperative movement was especially widespread in the agrarian sector, and around 1900, cooperative dairy factories were enormously successful. Farmers remained independent but tasks such as purchasing, processing, and sales were organized collectively.

In the history of Dutch dairy production, both tendencies—romantic versus modern, pastoral agriculture versus industry—have been interlaced. Traditionally, the processing of butter and cheese took place on the farm. In these circumstances, it was not possible to maintain a constant quality. In response to the improvement in the quality of foreign dairy products, the first Dutch milk factories were set up in the late nineteenth century. One of the first advocates of this development was J. Rinkes Borger, who wrote in 1878: "Dairy production has to be performed: systematically that is, according to fixed laws and rules, established by science and experience." This is clearly the voice of progress. Agriculture, like industry, had to be modernized by means of a scientific approach. For Rinkes Borger this also implied scale increase. Rationalization, he felt, was only possible on large landholdings or by moving dairy production to factories.[48]

In the ensuing period the processing of milk increasingly did take place in factories. This was realized in two ways: farmers either set up a dairy cooperative for factory processing of their milk or sold it to a private milk factory. Both solutions existed side by side for decades in the Netherlands. Historians who studied what defined the nature (private or cooperative) of a specific region's dairy industry mainly point to economic factors, suggesting that cooperatives only came into play where private factories failed to jump in.[49]

But J. C. Dekker, writing on the southern provinces of North Brabant and Limburg, has found that in this region cultural-ideological factors were actually of great importance: "The strength of cooperative dairy production lies in its connection to the traditional pattern of life of the farming population." Relevant here is that the farming communities find leaders who actively promote this cooperative ideology, such as agricultural instructors and dairy extension officers hired by the agricultural associations. In the largely Catholic countryside of Brabant and Limburg this role was played by the influential clergy, who pursued the establishment of a Catholic farming organization based on cooperative ideals and Christian charity. Nominally the clerics took a stance against socialism as well as against economic liberalism, meaning free trade and capitalist relationships. But they deployed cooperative ideas as a tool for spreading the Catholic faith, and in this approach modernization was inextricably bound up with social activism.[50]

This hardly meant that economic factors were unimportant. Regardless of the social aspirations of those who led cooperatives, if they had not been economically successful, they would never have been in business as long as they were. In the Netherlands the cooperative movement, especially in agriculture, evolved into a prominent socioeconomic factor. In addition to processing cooperatives there were credit cooperatives, purchasing cooperatives, and sales cooperatives.[51] They were not confined to rural areas and less-developed economic sectors. Cooperative auction halls gained a very strong position in factorylike, internationally competitive sectors such as floriculture and truck farming. Many of the first auction organizations were associations, such as the vegetable growers association of Rotterdam, which was established in 1903 and was restructured into a cooperative in 1926, because of substantial new commitments.[52]

Another relevant factor was the power relation between farmers or producers and their customers, who mostly were traders or wholesalers. Precisely in the agricultural sector the contradictions between these two groups were quite strong. Because agricultural businesses were quite small and had fairly few growth options, to gain market access they depended on the intermediary of wholesale trade when markets no longer had a strictly local basis. Small producers thus ran the risk of becoming a plaything of their cash-rich buyers. Furthermore, for purchasing as well, farmers were often dependent on very few wholesalers, who sometimes supplied them with credit as well. Collaboration and shared purchasing allowed farmers to bargain for lower prices and to ensure quality.[53]

At first the establishment of cooperatives was frequently accompanied by fierce conflicts between farmers and traders. For example, in 1904 the establishment of dairy cooperatives in North Brabant gave rise to the so-called "butter war." The various farming associations had set up cooperative butter markets, where they sold cooperative factories' products by Dutch auction. A group of traders decided to go boycott the cooperative butter market in Eindhoven and set up a butter market of their own in the same town, called the North Brabant–Limburg butter market. Each time one of their members bought at the cooperative market, he would be obliged to pay a fine of 5,000 guilders. Because butter was a perishable, any boycott posed a significant threat, but the farmers closed ranks and managed to find new sales channels for their butter. Their North Brabant Dairy Union established an export association and a central butter-kneading facility. Thus they brilliantly won the butter war.[54]

Collaboration was a strategy for creating one's own sales, purchasing, or processing channels, which reduced dependency on third parties. The power of cooperatives lay largely in their providing farmers with the opportunity to carry on as fairly small-scale businesses in a world that was increasingly dominated by large-scale structures.

After World War II: Competition versus Cooperation

The end of World War II had a reinvigorating effect on Dutch business and industry. If the prewar period was marked by economic crisis and an awareness of risks, the immediate postwar years were years of sustained growth and economic optimism. The collapse of Germany and the indisputable political and economic supremacy of the United States also caused a spiritual reorientation. Even more than before, Dutch industrialists and business economists viewed America as the country that represented the future. American techniques and methods were eagerly studied and practiced, which was promoted by the American government through the allocation of Marshall Aid funds and technological aid. Prominent Dutch businessmen and government officials made study trips to the United States to observe the achievements of American capitalism with their own eyes.

After the war there was a change in emphasis from collaboration to competition. This led to a new politics regarding scale enlargement in industry. Before the war the government had allowed or even encouraged the formation of trusts. The Entrepreneurs Agreement Act of 1935 in principle made it possible for government both to declare agreements between individual companies – on prices, market shares, etc. - void, and to make them binding for the whole branch. In practice, it was used only for the latter purpose. In 1941, however, the German occupation leadership ordered to render the Entrepreneurs Agreements Act inoperative based on the Anticartel Decree, which in 1943 was amended to include the option of taking steps against economic power positions in general. These measures were aimed to prevent the formation of trusts and agreements rather than at supporting them. Although after the war voices argued in favor of rescinding this measure again, this did not happen, implying that the Cartel Decision remained in force, which served as an impediment for companies' collaboration. Whereas the Entrepreneurs Agreement Act, as said, had been used only to declare agreements between companies as generally binding, all fifty trust decisions from the period between 1949 and 1958, the year the Antitrust Law (Wet Economische Mededinging) came into force, were precisely nonbinding declarations. This antitrust policy stimulated the growth of individual companies. Dutch economic relations thus acquired a slightly "more American" flavor, which perhaps also fitted better with the gradually emerging European market.[55]

In many areas more or less secret agreements between industrialists existed, but the nature of the public debate had clearly changed. Competition, rather than collaboration, became the new motto. During the interwar period and also in the war years, economists basically had viewed economic collaboration as either a healthy development or an inevitable one. Some even considered it as a transitional phase toward even larger changes—toward a new era marked by planned production and controlled price setting, imposed by changes in the structure of production and distribution.[56] After World War II, however, such idealist views quickly went out of fashion. A survey of Dutch industry, published in 1951, looked back at the prewar business relationships with a certain patronizing sense of superiority. During the interwar period, notably in the years of crisis, the older individualist economic mindset had been replaced by "trusts and other forms of collaboration among entrepreneurs.... Unlike before, there was less confidence in individual strength and people began to rely more on the power of collectivity." As a counterbalance, the (quite romanticized) model of the industrialist from the era of World War I was embraced: "He was "willing to compete and fight and firmly believed that first and foremost this would solve all practical problems." [57]

Large-scale automated production made significant inroads in the processing industries. In this photo, taken around 1955, the operator's main task is to check the meters and, where needed, adjust the settings. Some activities still are done using an old and tested technology: pencil and paper. Later on this would also be largely automated.

Scale Increase in Industry

In the light of the distinction between manufacturing industry and process industry, it seems only natural that when it came to expansion after World War II the Dutch process industry took the lead. The interwar period had already seen substantial growth of production facilities and factories, and this development accelerated. The average output capacity per factory in the petrochemical industry—for instance, in production of ethylene or ammonia—saw exponential growth into the 1970s. In 1948 DSM decided to build a factory with a daily output of 15 tons of urea, a basic material used in some plastics and fertilizer for tropical agriculture. This factory began operations in 1952. Four years later ongoing research led to a new factory with three units each having a capacity of 50 tons per day. In the late 1960s this factory, after being adapted to a new process, achieved its maximal capacity with 375,000 tons per year or over 1,000 tons per day. DSM not only built factories, it also sold its knowledge to others via Stamicarbon, a company especially set up for this purpose. By 1960 this company had sold six urea factories with a total capacity of 270,000 tons per year, while seven other factories (830,000 tons per year) were being built. In the ten years afterward Stamicarbon sold another sixty-six licenses.[58]

Large factories had a significantly lower production cost per tonnage. This meant that companies had to join in so as not to price themselves out of the market. Of course, this worked only as long as there was economic prosperity. Ultimately, it turned out that things had been exaggerated a little. The supersize installations that were in operation in the petrochemical industry in the mid-1970s hardly offered any more scale benefits over the slightly smaller installations. Moreover, they were efficient only when used at full capacity. When the demand declined or the cost of raw materials rose temporarily, smaller installations were more cost-effective. The ease with which giant factories could be built (the technology needed, after all, could simply be bought from companies such as Stamicarbon) resulted in worldwide overproduction. After the 1973 oil crisis the scale of the installations leveled off abruptly.[59]

The issue of scale increase also became prominent in the manufacturing industry in the wake of World War II. Just as had occurred during and shortly after World War I, many advocated large-scale mass production on the American model as a necessary condition to survive the international competition. The magic word this time was "type limitation": the active reduction of production to one or few standard product models instead of allowing customers to choose from a whole range. Essentially, this ideal had already been promoted by people like Du Croo at the time of World War I, when it served as a basis for the emergence of the standardization movement. The fact that now a new concept was needed suggests how much the original ideal either had gradually fallen into oblivion or had changed meaning.

It is clear, however, that expansion involved more challenges in product industries than in process industries. In 1948 the Netherlands Institute for Efficiency (Nederlands Insituut voor Efficiency, NIVE), which always had a good nose for the latest trends in business, organized its annual "Efficiency days' around the theme of "the transition of Dutch industry from one-off production to series and mass production." Instead of embracing the new opportunities uncritically, the preliminary recommendations, formulated as a basis for discussion during the event, unambiguously cautioned against high-flying expectations. The engineer and business management consultant Ernst Hijmans emphasized that mass production was anything but a panacea and that it had major drawbacks; for instance, the usability of products diminished through uniformity, a manufacturer's flexibility decreased, or transportation and communication problems came up. In addition, there were other ways than mass production to lower the transitional cost. He also found it questionable that producers tried to substitute mass production for professional skill: "Inasmuch as the urge toward mass production derives from a striving to be able to work scot-free and 'in a slapdash way,' giving in to that urge is not in the interest of our prosperity."[60]

Scale Increase in Agriculture

Before the war Dutch farmers had solved problems of their operations' size not by making internal changes within their businesses but by participating in cooperatives. After World War II, however, these organizations were themselves increasingly faced with a dynamic of economic expansion. In the first half of the century social considerations were often decisive. The cooperative format had not been chosen on purely economic grounds but also had major social significance, notably at the village level. Starting in the 1920s and 1930s, however, government officials urged for more joint efforts from existing organizations as a way to cut back costs, but they were nearly always blocked by cooperatives boards and members. They were too attached to their autonomy. As long as cooperatives suffered no losses, considerations other than economic or efficiency might gain the upper hand.[61]

After World War II, however, economic arguments began to prevail. The growing technological disadvantage vis-à-vis other countries and the need for large investments more or less compelled cooperatives to expand and restructure. In 1947 a study commission of the Southern Dutch Dairy Union (Zuid-Nederlandsche Zuivelbond) recognized the concerns of farmers, but it argued that the circumstances of the time forced one "to choose what cold pragmatism and hard necessity demanded."[62] The various smaller cooperatives were subsumed under larger regional units. Whereas some of the cooperatives set up at the start of the century had had no more than thirty members, now several thousand dairy farmers were members of such "cooperative circles" (kring-coöperatie). The inevitable result was that the farmers' attachment to their cooperative grew weaker. Under the new circumstances, their involvement was mainly based on economic grounds, and after several years this made it easier to restructure the smaller and less profitable milk factories.[63]

Scale increase and business concentration, then, proved possible within the cooperative frame. In this respect, cooperative opportunities for expansion, which in Dutch agriculture and agricultural industry have been of great importance, constitute a supplement to monopoly capitalism on the model of Chandler. The cooperative system allowed many small agrarian producers to participate and hold out in a mass market, which in fact was dominated by wholesale trade. It should be added that the modern cooperatives with their fully professional management began to look increasingly like regular businesses, albeit with members instead of stockholders. If cooperatives started off as collectivist organizations, they ended up as modern capitalist companies.

In dairy farming itself the postwar period was preeminently one of scale increases as well. This was stimulated by technological innovations such as cubicle stalls and milk tanks. A major factor was also the realization of the European Economic Community (EEC). The EEC and its subsequent embodiments not only offered a large and protected market for agricultural products but also for years supported the development of the European agricultural sector with large subsidies. Gradually, Dutch farmers became more businesslike, also in the sense that the old rural solidarity declined. They now sometimes acknowledged that they were competing with each other.

As a phenomenon, the process of scale increases in Dutch agriculture should primarily be viewed as involving the restructuring of smaller businesses. Real growth was mainly possible where technological innovations made production less dependent on the available acreage, as in hog raising or poultry, a sector that introduced battery cages after the war and was strongly backed by the feed industry.[64] These sectors took on a nearly industrial character. Through specialization and technological improvements the output of individual companies kept going up.

Although the postwar expansion of Dutch agriculture was very successful, eventually it did come up against limits. In the 1970s it was determined that further scale increases would not lower a farmer's cost price anymore, while it would raise his financial risks. Agricultural businesses became hard to control, both technologically and economically. Starting in 1975 the new slogan "Not more, but better" reflected that dairy extension officers had left behind the belief in further growth, but this new view did not immediately catch on and it proved hard to realize in practice. One contemporary observer put it this way: "Extension officers told farmers how to walk. But now that the latter have started to run, the former no longer keep up with them."[65]

Concerns about developments in agriculture were, again, in part motivated by social considerations: people tried to protect the family business and resist industrial conditions. In addition, large-scale economic forces were soon felt as well. High outputs were partly based on European subsidies. As production continued to rise, however, it was clear that these subsidies could no longer keep pace and that the system became too expensive. From the 1980s, support was trimmed down again while new policies specifically aimed at limiting production.

Scale Increase and Rationalization

In common models, the notion of scale increase is linked not only to new production technologies and sales channels but also to new forms of technological and organizational control. To lead a large company, one needs professional management, which should offer specific possibilities for control and intervention at all levels. Reform of this nature in industrial or other organizations was basically couched in terms of rationalization, rather than scale increase.

Rationalization and professional management, according to authors such as Chandler and Beniger, follow from the demands of business management as seen from a business economics perspective. It is questionable, though, whether these elements are so straightforwardly interlinked. Of course, rationalization had to provide a solution for practical problems. A company that sold its products below cost price had to do something about it. Rationalization, however, was more than a series of ad hoc solutions for problems that presented themselves. It was an active movement promoted by people whose interests were not limited to business. Often it involved specialists who had their own professional ethos and agendas, and who also responded to broader social problems. Ideas about rationalization, and thus about scale increase, were not determined exclusively by purely economic considerations.

Rationalization, Experts, and the Social Issue

At the start of the twentieth century it was hard to consider matters of technology and management in isolation from political and social activism. In many cases the pursuit of changes in these areas was prompted by what were seen to be social wrongs: unhealthy food, unsanitary living conditions, and so on—wrongs that were mostly a direct effect of sudden population increase and urbanization in the second half of the nineteenth century and the social problems related to it. Furthermore, growing awareness of social issues was also tied to the rise of the socialist movement and political mass movements and parties in general. Indeed, rationalization was meant to provide an answer to the various problems of a new and increasingly more integrated society, but the public mainly viewed these problems as the "social issue."

This pursuit of reform or rationalizations did not come out of the blue, but its source was in specific persons and groups. One of the most significant developments involves the emergence of groups of specialists that appropriated the problems of industry or society, the social issue thereby constituting a major catalyst. Basing his study on the public housing effort, P. de Ruijter has shown how a grassroots movement initiated by socially involved reformers eventually led to new technological expertise. The movement for better public housing started with outsiders, most often physicians who got involved in this issue based on their interest in problems concerning hygiene. Public housing was thus only one of the topics that drew their attention.

In essence they wanted to elevate the masses and improve society. As a result of their activism, local governments began to formulate and impose

Dutch apples being sold at auction in Rotterdam, around 1959. After the Second World War Dutch auctions, generally run by cooperatives, were still an adequate marketing tool for small producers in an expanding and increasingly large-scale market.

regulations, which led to the establishment of agencies that monitored building and housing. This evolved into a nationwide effort, and in 1901 the Housing Act (the Woningwet) was adopted.

The implementation of these government regulations required the practical input of architects, city officials, and contractors. City agencies had to expand their internal expertise or rely on outside knowledge. Public housing became a field of professional specialists who gradually took over the role of the earlier reformers. This specialty became institutionalized in 1918 with the founding of the Dutch Institute for Public Housing and Urban Development (Nederlands Instituut voor Volkshuisvesting en Stedebouw, NIVS).[66] It had grown out of this new professional group whose members gradually started to define their tasks more broadly. In addition to housing construction and urban development, they began to address more general problems of spatial planning. In due time, town planning became a separate science and a major policy instrument for central government.[67] This professionalization, however, did not resolve ideological differences right away. There were quite basic differences of opinion on the social task of architects, the role of modern industry, and, consequently, on the standards that modern-day cities and buildings would have to meet. As early as the late nineteenth century, H. P. Berlage had advocated large-scale, uniform building. His views were very popular among young modernist architects such as J. J. P. Oud and K. P. C. de Bazel. By contrast, M. J. Granpré Molière, a professor at the technical college in Delft, favored a historicizing mode of thinking and strongly renounced urbanization and industry.

Such differences of opinion did not keep urban developers and planners from steering policies into new directions. This is evidenced most clearly by the planning of the large land reclamation projects. In the nineteenth century, the Dutch government addressed the reclaiming of the Haarlemmermeer strictly as a public works project: the new land was simply subdivided and sold. In the twentieth century, however, government wanted to control things much more. This was not because circumstances in the new polders were so different, but because government itself reflected more on its role and actions and set new goals. Simply selling the land, as before, would stimulate land speculation and mainly a few wealthy farmers would profit. This was deemed no longer permissible.[68] The government's stance was thus inspired and motivated by a larger social sensitivity. Moreover, an army of experts energetically tried to persuade government of the need to be more active, while these same people also provided it with the instruments for getting a larger handle on planning efforts. For example, in 1926 the government appointed an aesthetic consultant to supervise the spatial planning of the first drained polder, the Wieringermeer. After Berlage declined the position, the choice fell on Granpré Molière. Furthermore, the government called in the help of the Dutch National Forest Service to oversee the polder's revegetation; serving on the Wieringermeer Directorate, established in 1930, were agricultural engineers from Wageningen.[69]

Urban development and planning were not the only technical specialties established in the first decades of the twentieth century. Social reformers also managed to turn all sorts of regulations into legislation with respect to labor sanitation, safety, working hours, social insurance, and so on. As a result, companies had to meet ever more formal requirements. This called for organizational adjustments and implied that, in addition to government monitoring of the implementation of regulations and guidelines, companies themselves needed to have expertise in all these areas as well. In reference to such organizational adjustments in companies and organizations in the twentieth century, the emergence of this new and broad field of professional expertise carries perhaps even more weight as a defining phenomenon than the dynamic of economic expansion and scale increase.

Agriculture

In the course of the twentieth century Dutch agriculture increasingly relied on professional experts. Because most entrepreneurs lacked the means to innovate on their own, in many cases the first efforts at innovation came from the outside. For example, the rationalization of land ownership by means of exchanges of small, dispersed parcels was possible only because the government actively facilitated it. In general the government has felt much more inclined to interfere in and regulate agriculture than industry. This was largely motivated by the serious crisis of agriculture at the end of the nineteenth century. In the last quarter of the nineteenth century cheap grain from oversees (mainly the United States) flooded European markets. This caused a sharp drop in the prices of agrarian products. Many agrarian enterprises struggled for survival and government was called upon for help. One should add that agriculture had a strong lobby. Dutch landowners and political elites were closely interrelated, and in the nineteenth century the former group was already well organized in agricultural associations at the provincial level.

It was mainly on the initiative of these associations that the State Agricultural College in Wageningen was established in 1876. Although this college received little funding, it still developed into a knowledge center for Dutch (and East Indian) agriculture.[70] A year later, in 1877, a state experimental station was set up by this school to conduct testing and analysis of samples of fertilizer, commercial feed, and seed. This station was modeled

The Food Inspectorate in action. The inspector samples the milk being sold by a milk man of Frisia. The appearance of professional, semigovernmental bodies with a public supervisory role in areas such as safety, nutrition, and hygiene led to a gradual long-term raising of standards in such areas.

on German examples. The first international conference on seed control in 1875 in Kassel served as the direct occasion for its establishment, because leaving the Dutch delegate's seat at the Kassel conference empty would have been undesirable.[71] At this station individuals (mostly wholesalers) could get samples of seed, feed, or fertilizer analyzed and certified for a fee. The agricultural associations also organized general or specialized agricultural exhibitions, first locally, but soon also on a larger scale. In 1884 Amsterdam hosted an international agricultural exhibition. According to contemporaries, it revealed that in many areas agriculture in the Netherlands was far less advanced than in other countries, implying a crucial need for further measures.

Pressured by agrarian interests and particularly the large crisis in agriculture of the final quarter of the nineteenth century, the Dutch government felt forced to study how it could contribute to improving the nation's agriculture, and in 1886 it set up, albeit reluctantly, an advisory commission. This commission emphasized the need for more government regulation, more support of education, more testing stations, government supervision of the butter trade and cattle imports, as well as regulations in the area of management, taxes, granting of credits, and so on.[72]

The significance of these measures was less in their immediate effects than in their indirect outcome: more agricultural training and experience gained in experimental stations gave rise to a new class of experts who advocated a scientific approach of agricultural issues and tried to get modern, rational methods accepted. The group of dairy extension officers, appointed in the 1880s and 1890s on the initiative of the agricultural associations with support from government, provided a major basis for such an approach. The same applies to the emergence of a context of expertise. For example, in 1886 the Association of Agricultural Engineers (Vereniging van Landbouwkundig Ingenieurs) was established and throughout the twentieth century its members played a major role in all sorts of farming organizations, while also contributing to shaping the practice of the celebrated so-called "OVO triptych," a project aimed at research (*onderzoek*), information (*voorlichting*), and education (*onderwijs*) relating to Dutch agriculture.

Industry

Social concerns or motivations did not always play a central role in efforts involving technology development. As mentioned previously, in the chemical industry the problems mainly centered on installations and instruments, which is why chemical technologists contributed little to social debates. The generation of electricity was a novelty of which few had any experience, despite extensive scientific understanding of the phenomenon. The issues involved in electricity were largely technological, and especially because of the related dangers, addressing these issues could not be entrusted to just anyone. Almost from the start, power generating stations were designed and managed by the new professional group of electrical engineers.

In the mechanical industries, by contrast, work and workers took center stage. At a time when the relationship between capital and labor was on edge, triggering much debate, it was hard to ignore the social issue. The problems of management in production industries were tackled in particular by engineers who had socialist sympathies (but also a strong belief in the blessings of technology and industry). For example, during and right after World War I the engineer and industrialist P. Haverkamp Begemann made an eloquent plea for standardization and mass production. It is not clear to what extent business concerns were uppermost in his mind; his real motivation was geared to other issues. Specifically, his ideal was universal international standardization, which, as he claimed, would "only be possible after our much-praised system of open competition has been done away with and we no longer produce for profits, but to meet the needs of the community. But that is left for the future."[73] The Russian Revolution elicited as much enthusiasm in this Dutch businessman as the "American way of manufacturing." In 1919 he stepped down as company director and sometime later moved to Russia to help build the young nation.

Engineers in the manufacturing industries were no longer exclusively concerned with nuts and bolts, but also with problems of labor and organization. Companies that did not have their own technical expertise increasingly called in external help. For example, in 1910 Thomassen, a motor company, hired the mechanical engineer Karel Gustaaf Simon, founder of the First Dutch Consultancy for Factory Management K. G. Simon. Among other things, Simon introduced transparent cost-price accounting, as well as a premium system (the Rowan system) for calculating employees' compensation.[74] After World War I the number of management consulting firms grew rapidly. They were mostly founded and run by engineers, but some other professional groups also became active in this market, such as psychologists who viewed their "psycho-technics" as a useful instrument for selecting employees.

The tool these engineers deployed to deal with problems was usually some version of Taylor's scientific management. Initially, in fact, scientific management was not exclusively regarded as an instrument to make workers work more efficiently, but as a way to bridge the contradiction between capital and labor. Business management developed into a separate discipline from its partially socially motivated origins. Despite professionalization, ideological contradictions continued to exist here as well. Some advocated rationalization, precisely because they mistrusted modern-day achievements. The Dutch Jesuit Jac. van Ginneken viewed Taylorism as a reprehensible system, but in line with his motto—"In many respects the children of darkness are smarter than the children of light"—he called on others to study this system and learn from it. He highly favored psychological tests as a way to get the right employee in the right place. His objection to existing tests was that they exclusively pertained to partial tasks, not to general character traits. Specifically he argued for the establishment of a psychological office to be initiated by the workers' cooperatives as a way to counterbalance the psycho-technicians hired by the employers.[75] In general, however, those who voiced such views in the world of Dutch industry remained outsiders. Leading ideas on how rationalization should be implemented were influenced or determined in particular by engineers who embraced a modernist ethos.

In this respect, World War II caused a change of socioeconomic climate as well. At first sight it seemed that social components were hardly addressed, while business strategies became almost exclusively dictated by anticipated profit margins. Dissatisfaction with the prewar political and social relations had been replaced with great confidence in democratic society (if not self-complacency). In the interest of prosperity and reconstruction, industry had to be able to develop freely, which caused noneconomic arguments to be sidelined.

At the same time, however, many specialties from the prewar era had meanwhile become firmly established and their practitioners were working toward realizing the welfare state. This led to ongoing expansion of the various regulating and monitoring agencies. While business was encouraged to earn profits, it was also subjected to new regulations in social care, safety, participation, and so on. The 1959 Industrial Medicine Act, passed in 1959, and the Occupational Health and Safety Act (called Arbowet, short for Arbeidsomstandighedenwet), in 1980, are just two examples. Moreover, regulations became ever more detailed. After the war, the scope of the regulations became wider. Initially they had existed to counter direct risks. Later, less urgent matters (say, access to adequate daylight) were also subject to regulation. As a result, a clear separation in competencies and responsibilities emerged. While social aspects were increasingly supervised by external agencies, company leadership saw its task as limited to the purely economic aspects of the business.

All in all, rationalization was a complex notion that not only implied a striving for maximizing profits but also was tied to the ambitions and ideals of the many new groups of experts, which also led to heterogeneous ideas on how rationalization should be implemented. Generally one can say that the new experts pointed to social needs while also referring to their scientific expertise. Unmistakably, then, the ideal of rationalization had a major technocratic dimension.

Scale Increase and Standardization

Our large-scale society was made possible through cooperation and economic expansion and by leaving all sorts of matters up to specialists, but this society also called for technological adaptations. Standardization was a major device that allowed or forced institutions and individuals to link up with developments in society as a whole. The concept may be defined as the articulation of standards regarding units, methods, product qualities (such as form, color, material), testing procedures, and so forth, with the aim of achieving larger uniformity. Uniformity makes it possible to achieve scale benefits in production or purchasing, or, in other cases, larger interchangeability or comparability and mechanized handling. The uniformity pursued is important in particular from the perspective of the system as a whole. Because setting standards relies on negotiation on all sorts of fragmented interests, standards will generally provide less than optimal solutions in any specific case. This is why, in concrete situations, standardization often meets with resistance, even though almost everyone will subscribe to its overall relevance. What is more, standards restrict the liberties of constructors and designers. No wonder, then, that the history of standardization has described a trajectory strewn with obstacles and setbacks. Moreover, most official standards merely have the status of being recommended, rather than being legally enforceable.

Standardization has been a familiar concept since the late nineteenth century. In England the Engineering Standards Association was established in 1901.[76] For the time being, however, standardization remained limited to just some areas. An early standardization project involved screw threads. Manufacturers used to make their own screws and nuts, which as a rule did not match those made by other companies. The nineteenth-century standards in this area, first formulated in England, made screws manufactured by different companies interchangeable. Other early projects involved rolled sections (the form in which pig iron was made available to the market) and some areas of electro-technology. Standardization also became a major theme in emerging large technical systems. In the Netherlands the various railway companies were working to standardize track sections and overhead wires. Around 1900 the Dutch Association of Gas Manufacturers was busy standardizing gas meters. The manufacturers wanted to fix their standards by producing a metal model showing the right sizes, but this turned out to be too expensive.[77] At the start of the twentieth century standards were commonly communicated by means of a "standard sheet," a sheet or booklet that concisely summarized the standards involved for specific cases, preferably including diagrams.

The Interwar Period: Standardization and the Efficiency Ideal

Standardization developed into a major concept during World War I. Notably the German engineers involved in rationalizing the war industry pushed for the standardization of products and methods. In 1917 their effort led to the establishment of the German Standards Committee (Deutsche Normenausschuss, DNA), later on the German Insitute for Standardization

(Deutsches Institut für Normung, DIN).[78] At this time, standardization was also beginning to receive more sustained attention from Dutch industrial reformers. In 1916, on the initiative of the Industrial Association and in collaboration with the Royal Institute of Engineers (Koninklijk Instituut van Ingenieurs, KIvI), the High Commission for Standardization in the Netherlands (Hoofdcommissie voor de Normalisatie in Nederland, HCNN) was established. This became the Netherlands Standardization Institute (Nederlands Normalisatie-Instituut, NNI, today better known as NEN, for Nederlandse Norm). This was (and is) a private foundation whose contributing members include several ministries and municipalities and also many companies, at first especially those in construction, the heavy equipment industry, and shipbuilding.

Individuals such as the aforementioned Du Croo and Begemann played a major role in the establishment of the HCNN. The establishment of this entity was inspired in particular by the notion that in the short run the industry needed to move on to mass production. The motives for establishing the HCNN are reflected in the words with which the chairman of the Industrial Association, J. van Hasselt, defended the proposal to create the HCNN before the board of KIvI (of which he was a member):

> In the Industrial Association various manufacturers have expressed the concern that there is little unity in the dimensions of many products. This is an obstacle to mass production…. The various standards make it difficult for Dutch industry to compete with other countries. Usually a Dutch engineer, when putting in an order abroad, will comply with the types that are common there. If he puts in an order in this country, however, he will be the one to set the standards. This renders it impossible for our industry to produce inexpensively.

His words met with agreement from the board. His fellow board member H. J. E. Wenckebach, former general director of DSM, commented, "Industry is the victim, and the engineers are the culprits."[79]

How, exactly, various issues could be resolved through standardization was barely considered when the HCNN was founded. The first project it engaged in, therefore, involved making a survey to gain more insight into the problems at hand. Although this organization did not yet have a clear brief, in its early years it formulated several major standards, such as for fittings and tolerances, which in fact largely followed German standards, as well as for technical drawings.[80] One may wonder, however, to what extent all standards were aimed at mass production. The first standard involved

The poster announces the forthcoming 1932 Efficiency Exhibition in Amsterdam. Visitors could get acquainted with administrative tools to improve the set up of both factory and office. In the interwar period the efficiency movement in the Netherlands saw strong growth. Expositions, conferences, and other gatherings were organized, and the Netherlands Institute for Efficiency issued its own publications. Ultimately, however, the movement's effect on business management was not always decisive.

screw bolts and rivets—products that had long been mass-produced. In this case it was not so much the manufacturers who had a stake in standardization, but their clients who could now buy the same product made by various companies. That this was of little concern to these manufacturers is illustrated by the fact that in 1920 one could hear the complaint that they were only willing to make and supply the standard bolts if clients in fact wanted to pay *more* for them than their regular trade price.[81]

The High Commission soon staked out its terrain and also began to address topics that were outside the scope of the manufacturing industry. It aspired to become the central standardization actor in the Netherlands and therefore tried to link up with existing standardization efforts in other areas, including railroads and electro-technology. The Dutch Society of Electricity Company Managers (Vereeniging van Directeuren van Electriciteitsbedrijven in Nederland) had already set up a standardization commission for cable fittings. In 1919 this work was taken over by the standardization commission.

Two years later its involvement in this field led to formal collaboration with the Netherlands Electro-technical Commission (the national branch of the International Electro-technical Commission), which would function as a subcommittee of the High Commission for Standardization, for electro-technology.

The actual standardization effort took place in the various subcommittees for individual fields, staffed by experts from these fields: technicians and others who worked with the manufacturers or technical colleges involved and who set aside time for this work as part of their regular job. The High Commission and its secretariat, the Central Standardization Bureau (Centraal Normalisatie Bureau), played mainly a coordinating and supporting role. In general the High Commission only got involved when some industry or other interested party put in a request. Not all requests, however, were honored. For instance, in 1920 the executive committee of the First Dutch Roads Conference asked the HCNN to provide a contribution to the standardization of roads nomenclature. This was not in line with its mission, the High Commission felt.[82] When C. F. Laurillard, director of Dutch Glazed Stoneware Pipes Industry, asked the HCNN to standardize firestone, the response was that for the time being it saw no way to do it "partly because of insufficient financial support from the interested parties."[83]

Indeed, financial support was a major consideration. For example, the HCNN responded in quite a reserved manner when the Central Statistics Office (Centraal Bureau voor de Statistiek, CBS) called in its help regarding standardization of the terminology for production statistics. The HCNN chairman, J. H. Hulswit, called to mind "that the Standardization Bureau was made possible through contributions from companies in the metal industry and that therefore it should be primarily concerned with this particular branch." Putting it even more unambiguously, vice chairman Wenckebach said that the High Commission only worked "for a specific sector within the industry."[84] He probably mentioned this in the middle of a heated debate because ultimately some form of collaboration with the CBS materialized. At the same time, however, his partiality seemed no isolated case. Upon formalizing collaboration with the Netherlands Electro-Technical Committee, E. Hijmans, director of the High Commission secretariat, cautioned: "Since the manufacturing branch currently provides proportionally much more support to us than the electro-technical industry, we handle requests coming from the manufacturing industry more readily, of course."[85]

Thus, the subsequent direction of the standardization effort was less determined by its original ideals than by the question of who was willing to pay for it. In addition to steel, machine, and electricity companies, initial support mainly came from government, both national and local. In 1920 this led to a standardization commission for sewer parts. The High Commission warned manufacturers involved that this implied extra work for the commission and that it had decided to expand, "trusting those interested were prepared to carry the cost for the work involved."[86] In 1923 a committee was set up for standardizing formats and testing specific paper standards. The latter implied moving beyond the terrain of industrial production. The use of standard paper formats was mainly advocated regarding the rationalization of office work—for instance, by the Association of Dutch Municipalities. This topic was hardly considered marginal. On the contrary, A4, the standard European letter-size sheet of paper, had major symbolic value for the standardization movement.

Resistance to expanding the tasks the High Commission set itself was motivated by the assessment of what could be accomplished in a given context and not by fundamental objections to the idea itself. The HCNN attributed very general significance to standardization, but its ambitions tended to be far-reaching. This is why, whenever possible, the High Commission moved beyond the industrial realm without having misgivings about doing so. Standardization was part of the interwar period's overall rationalization and efficiency ideal, whereby increasingly, sight was lost of the connection with mass production, and more and more attention was paid to the interests of the user. For instance, much effort went into standardizing firefighting equipment, in part because the equipment parts from different firefighting crews did not fit, which could pose a serious problem in providing emergency help in another district. A similar problem occurred in ambulances: stretchers of one ambulance did not fit in another one, which in case of calamities such as train wrecks could lead to unfortunate situations.

Thus, standardization is closely linked with the rising class of independent experts in the interwar period. Much work had social rather than economic importance. For instance, at this time the Dutch Labor Inspectorate (Arbeidsinspectie) used standardization in its effort to subject all sorts of installations to various safety rules. In 1935, after a number of accidents with cranes, it asked the HCNN to formulate rules for lifting equipment. In 1936 and 1937 there were requests concerning electro-technical regulations for working underground in mines and for devices in spaces where there was a risk of explosion. Electrical installations in factories and workshops were also standardized after a request from the Labor Inspectorate. The HCNN handed over such requests to one of its subcommittees, which sometimes were expanded with a few new members.[87] However, it did not become active only in cases of bottlenecks or risky contexts. For example, experts who in the interwar period sought to rationalize household work embraced standardization as well. The Institute for Household Work took various initiatives aimed at standardizing household articles and these were taken over by the HCNN.[88] Standardization, in other words, was no longer primarily an instrument for factory production; it had become a vehicle for giving shape to a more comprehensive and also quite vague modernization ideal—one that the HCNN eagerly capitalized on, of course.

Another example may further illustrate the role of HCNN. Around 1930, Dutch hospitals knocked on its door as well. In the 1920s a movement for modernizing hospitals had emerged, mainly led by new groups of experts such as hospital managers. They turned their attention first toward bookkeeping, patient records, and the uniformity of annual reports, but in time they also began to address standardization of patient data and hospital environments. These concerns were motivated in part by developments in the United States and the doctrine of scientific management. Remarkably, however, the need for this effort was not defended on purely economic grounds. Although the subsequent economic crises made it hard to evade cost control, it was barely acceptable to manage care institutions such as hospitals solely on the basis of economic considerations. But a pragmatic and businesslike approach could be justified with reference to Taylor's scientific methods: on the basis of "scientific insight" in the functioning of hospital sections, "it is easier to make decisions that are legitimized with the help of "objective arguments" without losing sight of the hospital's social dimension."[89]

This pursuit of rationalization also took in standardization. One of its advocates was P. C. Cleijndert, medical director of the Van Iterson Municipal Hospital in Gouda and one of the acknowledged proponents of rationalizing hospitals. He served on the Commission of Hospital Efficiency (Commissie voor Ziekenhuis-efficiëncy), established in 1922.[90] Standardization would raise efficiency because it facilitated collective purchasing. In 1929 the HCNN formed a special commission that in the ensuing years issued various standards, such as for hospital beds and linens. A 1940 survey revealed that the standards were poorly applied, however; fewer than half of the facilities studied made use of standardized products or central purchasing.[91]

Although standardization may be viewed as a strictly utilitarian and economic affair, in the interwar period several authors discussed it as more than just a way of raising profit margins. Tellingly, in a 1930 essay on "the evaluation of technology," K. F. Proost treats standardization in a section on ethics; he argues that life "becomes simpler" and that "this simplification involves a leveling we do not admit right away…. Technology leads to larger unity and solidarity—a unity and connectedness that used to be unthinkable before."[92] The standardization ideal of the first half of the twentieth century, related to ideals such as modernization or efficiency, involved a broad cultural movement. Modernist artists eagerly relied on the standardization idiom as well. In 1918 the architect H. P. Berlage argued for standardization in housing construction as a basic demand of the modern era. Other Dutch artists who promoted standardization were J. J. P. Oud, Gerrit Rietveld, and Piet Zwart.

Company Standardization

The economic usefulness of standardization was explored also at lower levels. Various companies set up standardization departments and defined their own "company standards." Some companies made mutual agreements on standards outside the HCNN. Such company standards tended to be more modest in scope, serving the requirements of warehouse and stock management rather than aiming to rationalize production. The standardization department of the Netherlands Railroad Company, for example, was mainly concerned with generic issues unrelated to railroads as such: office furniture, bicycle stands, safety glasses, and so on. (Standardizing railroad equipment was a matter of international committees.)[93]

From a business-management angle this effort was anything but irrelevant. In fact, it was of great importance to control the rising flows of goods and information within companies at least somewhat. In this area standardization often went hand in hand with assigning specific codes to various products. To counter the threat of disorganized warehouses full of unmarketable items, companies produced or stocked a limited assortment of standardized products. This also allowed for larger-scale and cheaper purchasing. It should be added that until World War II, only half a dozen Dutch companies set up their own standardization departments: Werkspoor, DSM, and several dockyards. Only after the war did it become a real trend, which is undoubtedly related to huge production increases. In 1962 a wholesaler in the metal sector admitted that his colleagues did not welcome the standardization effort. Initially the fear existed that a significant reduction of the number of product lines would undermine the role of wholesalers. This objection, however, had grown outdated after World War II; the number of different products one needed to have in stock had skyrocketed, which is why more basic and cost-saving ways of handling them were eagerly adopted.[94]

In addition, standardization could have major ramifications for companies that wanted to participate in the international market. Deviating standards, or just the use of different formats and specifications, actually constitute major trade barriers. The post–World War I establishment of national

In mass production, interchangeability of parts was of central importance, and this was possible only when items were made according to exact standards. Among the first products to be standardized were screws, nuts, and bolts, which earlier varied widely. In this advertisement, from around 1915, P. van Thiel & Sons, of Beek en Donk, showed a modest selection of their product lines.

standardization committees throughout Europe certainly had to do with these factors. Large industrial nations such as Germany and England tried to impose their own standards onto smaller nations.[95] The HCNN, in its founding document, explicitly warned that Dutch standard forms should not coincide one-sidedly with "standard forms of only one foreign economic unit. The HCNN should determine its choices in such way that the most extensive competition among non-Dutch economic units will be possible."[96] National standardization was also meant as a tool for joining discussions on realizing international standards and as such for protecting one's own commercial interests.

One Dutch company that has always been strongly committed to standardization is Philips. In 1923 it first formulated specific company standards for its mechanical engineering plant. Five years later a standardization department was set up to serve the entire company. W. H. Tromp, the director of the Central Standardization Bureau, was specifically hired to lead the department. His budget was substantial (41,400 guilders in 1933). Later on Philips continued to devote much attention to standardization. In 1954 an "International Conference on Standardization" was convened in Eindhoven to decide on a new international collaborative structure. A "Company Standardization Board" was set up to define the overall parameters. Whereas in most other companies the standardization department, if it existed at all, was rarely taken serious by the company leadership, in the case of Philips standardization constituted an integral element of executive board policies.[97]

The central standardization department of Philips also focused on central purchasing. One of the first issues addressed was chairs for factory workers. From the start, however, there was also a concern with standardization as a tool for enlarging markets. One example involves light bulb fittings. When all light bulb manufacturers worked with the same measures for fittings, it was possible in theory to put a Philips bulb in each lamp. This meant that Philips could sell its light bulbs worldwide. This also implied of course that in its home market it had to compete with foreign companies such as Osram, but the Dutch company accepted the competition as part of the deal. From the outset Philips has pursued internationalization. Standardization was an active business strategy aimed at raising sales, not by directly rationalizing production but by stimulating uniformity of the market itself. Philips played a leading role in international standardization of many electric products—for instance, batteries, which the company itself did not manufacture but which were used in many of its products.

Especially after World War II the company played a major— possibly leading—role in the High Commission on Standardization at the national level. It is telling that when, in 1968, P. van Zuuren wanted to accept the post

Wholesalers who stocked products played a major role in the development of business and technology in the Netherlands. Some of the products they sold were standardized. Shown here is Econosto, a supplier of both valves and fittings, around 1957. Econosto supplied both standardized and custom-designed products.

of publicity director with the Netherlands Standardization Institute, he first had to drop by at a Philips director's office in Eindhoven and introduce himself. Only if he passed muster with this director could the appointment go forward.[98] Philips's interference with the HCNN, and its successor, the Nederlands Normalisatie-Instituut, however, clearly served the company's own business strategy. International standards were made by negotiations between delegates from the various national standardization committees, so in order to have a say in the making of international standards, Philips needed to participate in the HCNN in the first place. In 1954 the head of the Philips standardization department, N. A. J. Voorhoeve, more or less admitted this: Philips made use of Dutch standards and even helped the HCNN effort, but only as a way to gain access to the international standardization effort.[99]

Philips's attempts to determine international standards figures prominently in a 1948 report that Voorhoeve wrote for the Philips executive leadership in response to a request for more support for national standardization. He summed up the advantages of standardization for Philips and also indicated that in the context of the International Electro-technical Com-

The importance of standardization was realized early on in military technology. By the end of the medieval period standards were defined for the calibers of canons. This photo shows the production of cartridge cases at Hembrug, a state-owned munitions company, in the late 1920s.

mission he had been appointed international technological secretary of a commission for standardization in radio: "Thus we can exert major influence on realizing international standardization in this area. That this may tremendously serve our interests needs no further argument." Voorhoeve also cited several examples of foreign requirements that had been dropped for certain devices (such as receivers and tubular lighting devices), which made it possible for Philips to produce them more cheaply. Although he first underscored the *general* importance of standardization, he was quite clear about who should be in charge: "By supporting standardization in the Netherlands we promote its private nature. Too much government interference in this area would be very damaging to us in various respects."[100]

Standardization and Business After World War II

After World War II, the HCNN was much interested in the reconstruction effort and it actively sought to link up with the Marshall Aid program. For instance, a working group on the restriction of product models was set up. In addition, government stressed the relevance of standardization for trade, with respect to both raising exports and countering the import of inferior products, and that of international standardization at large.[101] This is not to deny that over time national standardization had become a matter of special experts, who followed their own agenda. In the interwar period the original aim of enhancing industrial production through standardization was replaced by a more general ideal of rationalization and raising efficiency. The practical relevance of this was not always obvious, as is clear in the case of hospitals, mentioned above, where the standards thus made were implemented only to a limited extent.

Moreover, there was the risk that a focus on achieving a certain product quality would be at the expense of business-economic considerations. Before World War II this was hardly addressed, although dissident voices had been heard at the HCNN meeting on October 26, 1939:

> From certain quarters of the industry one can occasionally hear the remark that some standards are developed in too much detail. [Speaker] is thinking, for instance, of products for which the Nenorm label [a hallmark that the HCNN could grant to products to indicate that they followed its standards] is applied, because in these products, with a view to testing, many specifics were standardized. This may lead to reduced opportunities for useful competition in sales and may also work toward somewhat less easy adoption of technological improvements.

At the time, this kind of objection was brushed aside, even though it had to be recognized that the character of standardization had gradually changed: "It is odd, to be sure, that at first standardization of sizes and connecting sizes was given most attention, but in later years safety and effectiveness for use by the public have gradually become key factors in standard formulation."[102]

After the war the standardization effort began to be criticized more aggressively. Business was given a more pronounced voice in national standardization when in December 1957 the HCNN established the Committee in Business Standardization (Commissie voor Bedrijfsnormalisatie, called Cobeno). One of the first tasks it undertook was the formation of a study group to investigate the value of national standards for business. It sent questionairies to representatives of ten large companies, which revealed that some eighty national standards were unusable. "It turned out that a large number of standards have become dated and contain errors. Frequently, standards that should coincide with foreign standards come with incomprehensible deviations that render their application impossible."[103]

Evidently, this was a painful conclusion. The study group—soon referred to as the "rebel club"—thus rather rudely shook up the slightly drowsing Dutch standardization world. In the ensuing years, Cobeno also raised the issue of the usefulness of standards for companies. In this respect the 1962 collection *Bedrijf en norm* (*Business and Standards*) put forward several critical comments, including the following:

> It has become abundantly clear that national standards are unsuitable as company standards because of their comprehensiveness. The national standards seem to include everything on which experts tend to agree. But the individual engineer fails to see that also part of what the standard covers may meet the client's needs, while this is even desirable for economic reasons. The standards do not show that a product can be cheaper if it does not have to meet all requirements at once.... In the current national standardization effort, which is geared toward standard uses and regulations rather than limitation of types, commercial and technical factors outweigh business management aspects.[104]

Another voice in this volume argued: "As insiders will know, limitation of types as such ... does not necessarily lead to optimal results in business management terms."[105] In 1967 A. T. Hens challenged the interrelationship of standardization and mass production. He claimed that one-off or small series production would precisely benefit most from (national) standardization, because in large series and mass production "one can afford to take larger individual liberties in specific circumstances."[106]

It would be somewhat premature to view this as an attack by the business world on the standardization effort. It was primarily a discussion within the standardization movement. The Cobeno did not represent business but was, rather, an association of those involved in business or company standardization. It mainly involved a discussion, then, between two groups of specialists. In most Dutch companies standardization was not a high priority for the company leadership (Philips being an exception). This meant that those in charge of standardization within companies had to defend the usefulness of their specialty and thus paid much more attention to business management aspects than those mainly engaged in national or international standardization. Moreover, a possible factor is that the social task embraced by engineers in the interwar period gradually eroded after the war. Engineers developed more into specialists in their specific fields.

The economist P. van Zuuren, who in 1968 became a publicist with the Netherlands Standardization Institute (into which the HCNN had merged in 1959) and who was in charge of the Cobeno, later recalled that the section directors of the institute were exclusively interested in technical issues and ignored the business management aspects of standardization:

> By far most people in the NNI organization failed to see the contro-
> versy between the significance of standardization as efficiency problem
> and as economic problem—as rationalization problem. They reveled
> in doing technical things, making drawings, formulating standards,
> but they had no interest whatsoever in other aspects…. NNI's focus
> was technical, technical and again technical. This I tremendously
> regretted.[107]

The Cobeno was concerned in particular with organizing meetings and courses for those in business standardization. Its influence on policies remained minimal, and after growing dissatisfaction on both sides Van Zuuren left the institute in 1973. The institute meanwhile continued to focus on international collaboration, not on cultivating contact with the Dutch business world.

Innovation became a concern only in the closing decades of the twentieth century. Dutch industry became increasingly aware of the importance of standardization, mainly as a result of increasing European integration. Because of the struggle against trade barriers, it was no longer tolerated that something was approved in one country while not in another. All sorts of standards with respect to safety, health and so on had to be brought into line with each other. The European Community (later, European Union) resolutely stepped up its standardization effort. In addition to national and worldwide standards, an entire system of European standards emerged as well. In the process, standards became ever more coercive, in part as a result of increasingly strict demands applied to products and sanctions for noncompliance. Manufacturers of standardized products generally could pass on part of their product liability to higher echelons, whereas a manufacturer of a nonstandardized product had to carry the entire liability burden alone. Even where the use of existing standards was formally not mandatory, it proved difficult to evade. In this situation it was increasingly difficult for business to stomach that the Netherlands Standardization Institute seemed so uncooperative. Pressured by the Ministry of Economic Affairs, the institute underwent several reorganizations. As a result, in the 1990s it came to be more open to the wishes of businesses and more aware of the significance of market interests.[108]

What in all likelihood contributed most to the rather unexpected embracing of standardization by the upper echelons of Dutch trade and industry is the increasing attention to quality and the strong rise of international ISO 9000 quality standards since the 1980s. These standards had major ramifications for the relationships between companies. The key issue became certification.

Scale Increase and Certification

The rise of largely anonymous mass markets caused the disappearance of the personal relationship between manufacturers and users. It became much harder for consumers to decide whether a product would meet their expectations. Still, companies that produced for mass markets somehow had to win consumers' confidence. The same applied to a supplier who delivered parts to a company at the other end of the world. Economic scale increases could only be realized if these problems were solved. In this respect, one option was to invest in an extensive network of sales outlets. Another was to try to infuse one's company name with an aura of reliability, an approach that became the basis of the rise of brand products. A third option was to have an independent agency ensure a product's reliability by issuing a quality mark for it. In this case we speak of certification.

Safety
Certification is certainly not the only way to gain access to the market, nor is it the easiest way. Until the final quarter of the twentieth century in most areas certification only played a modest role, with the exception of products where concerns related to safety, health, or nuisance potential were at issue. Here it was less the market than government—as guardian of the public interest—that set standards. The government set safety standards for some products early on, mainly electro-technical products. To this end, in 1927 the Inspectorate of Electro-Technical Materials, known as KEMA (Keuring van Electrotechnische Materialen), was established as a certification agency that was formally recognized by the Dutch government (KEMA is now an international consultancy). Electro-technical products were allowed to be sold or installed in the Netherlands only after the Inspectorate had tested and approved them. Also in sensitive sectors such as aviation, aircraft construction and the defense industry, inspection and certification were applied early on.

Agriculture and Nutrition
Certification has also long been important in the areas of food and agricultural products. This had less to do with their importance for public health (which was seen by the Dutch government as a matter of individual respon-

Leiden's municipal seal of quality for meat products for 1890. If buyers or consumers knew nothing about a meat product's origin, the quality mark assured them that it met the quality standards. On the left the "Approved" seal, in the middle the "Rejected" seal, and on the right a seal stating that the meat was of second quality because tainted by bovine distemper (*parelziekte*) but could be consumed after proper cooking.

sibility) than with the structure of the market. Certification in Dutch agriculture is linked to the cooperative movement and has a similar organizational structure. Farmers looked for collaborative ways to protect their sales and purchasing. As indicated, farmers had good reasons for mistrusting those who traded their products. This caused them to undertake collective action, both in sales and purchasing, an effort that might also include certification.

A good example is dairy inspection. The tampering with milk and dairy products was a serious problem, but for a long time the government refused to take measures to protect consumers. (Some municipalities were more willing.) According to the prevailing liberal view, users simply had to take charge of things on their own. What made government decide to take action in the end, however, was that tampering with dairy products had negative effects on trade. Dutch butter exports suffered serious losses though large-scale fraud, in particular by wholesalers, who often sold inferior butter as if it were high quality. Foreign customers eschewed Dutch products as they could not be sure they would not get inferior ware. This was grounds for a nationwide intervention by the national government. In 1889 (thirty years before the Food and Drugs Act) the first Butter Law was enacted, recommended by the 1886 Agriculture Commission, with the purpose of ensuring quality.

However, neither this law nor several later amendments managed effectively to prevent tampering. Finally the agricultural associations took the initiative to establish butter inspection stations; producers could become members only if they met stringent conditions and opened their production to inspection at all stages of the production process. This inspection was paid for by the members but it took place under the auspices of government, which introduced a special butter mark.[109] This mark's acceptance by the co-operative butter market in Eindhoven was in fact the direct occasion for the 1904 butter war: traders hoped to weaken this mark when they set up their alternative market, but thanks to the support of farmers' organizations it quickly gained recognition. For a long time this state dairy mark met the needs of the cooperative dairy factories. Not until the 1920s there was a debate on the use of trademarks, as an alternative way to gain customer's confidence.[110]

Dairy companies also began to negotiate with each other so as to reach agreements about their product. The establishment of the Association for Dairy Industry and Milk Hygiene (Vereeniging voor Zuivelindustrie en Melkhygiëne) in 1908 was initiated by the producers. Probably they preferred to keep inspection and management in their own hands, rather than waiting to see whether or not the government would intervene further. Another relevant factor was the opposition between private and cooperative factories, or between traders and farmers. The association emerged out of a merger of two earlier industrialists' associations and comprised only the private dairy companies; the cooperative companies had their own club, the General Dutch Dairy Union (Algemeene Nederlandsche Zuivelbond), a federation of unions of cooperative dairy companies established in 1900.

The inspection of sowing seed and seed potatoes also goes back to the century's beginnings. Here, too, it was relevant that major trade products were involved. Particularly seed potatoes developed into a major export product. Strictly speaking the quality of sowing seed can only be assessed if one knows its exact origin, and otherwise only when it sprouts. A farmer who bought sowing seed from a wholesaler had to do so in good faith, which proved not always to be justified. In particular in the light of the growth and increasing complexity of the markets and the flows of trade, the need cropped up for an independent agency that could issue a quality guarantee. A first step was the establishment of the state testing station for seed control in Wageningen in 1877. On a voluntary basis, wholesalers participated in inspections at this station, meaning that they permitted their clients to have a seed sample tested.

Initially the provincial agricultural associations played a major role in this area, too. Especially Groningen hosted many exhibitions of sowing seed in the closing decades of the nineteenth century. To determine the sound origin of the seed, in around 1900 inspection of crops in the field started. From the start there was a focus on collaboration, such as in provincial inspection commissions set up after 1910 whereby the actual leadership increasingly shifted to the new professional group of agricultural engineers. The Central Committee on Crops Inspection (Centraal Comité inzake Keuringen van

Gewassen) was established in 1919, on the initiative of a number of agricultural associations, agricultural extension officers, and the agricultural institutes at Wageningen; it formally became part of the Institute of Plant Improvement (Instituut voor Plantenveredeling) at this college. This commission was also intended as a means to be less dependent upon wholesalers. The wholesalers did not want to give up their advantage in information, which allowed them to keep the upper hand over their customers, and in response set up a facility of their own, the General Testing Institute for Sowing Seed and Seed Potatoes (Algemene Keuringsinstituut voor Zaaizaad en Pootgoed). This gave rise to prolonged and unpalatable bickering. The presence of two different institutions did not benefit exports. This was the main reason the government intervened, but it didn't manage to restore unity of inspection through legal regulation until 1941.[111]

Industry

Industry had much more direct access to the market than agriculture, which explains why certification had always played a less prominent role for industrial products than it did for agriculture. Industrial products were mainly sold on the basis of their brand or manufacturer's name. Occasionally, external certification was relevant for some special reasons, such as safety. The first organization to engage in certification of manufactured products was the Dutch Product Association (Vereeniging Nederlandsch Fabrikaat), founded in 1915. Its certificate was meant to guarantee that a product was made in the Netherlands, by a Dutch company, and its objective was to raise domestic sales. The association reasoned that the Dutch public ought to prefer domestic products for patriotic reasons, especially in times of war and economic hardship. It also undertook other activities geared toward promoting Dutch products that had a less nationalist focus.

The High Commission on Standardization also issued a quality mark for some time: its Nenorm mark was one of the ways in which during the interwar period it tried to promote the standardization effort. This mark was first used in 1934 and guaranteed that the products bearing it met Dutch standards. It said little about a product's quality. The names of products that received the Nenorm mark were published in the HCNN's journal. Another well-known certificate was the Dutch Association of Housewives' quality mark, which dates back to 1928.[112] This example is in some respects similar to the situation in agriculture, involving as it did a large number of individual customers (housewives) who could fall through the cracks because of the many new products and suppliers, but who manage to become a serious party by combining forces.

Certification of industrial products thus became used mainly in special areas or for specific goals, and a more general certification in Dutch industry seemed to have little relevance. In 1946 the Dutch Product Association also tried to introduce a quality mark. This initiative, however, was abandoned within a year after objections were raised by its main section, on industry, which feared that companies that somehow would not manage to get such quality mark would suffer.[113]

But companies eagerly welcomed external certifications if they could obtain them. After World War II the Shell laboratory in Amsterdam coordinated the purchase of laboratory equipment for all Shell branches worldwide (the United States excepted). At first Shell tested all equipment in Amsterdam before buying it. Later on, some products of several manufacturers were bought without prior testing. Still, Shell's standards were high. It not only tested products but its staff also inspected the factories where the products were made. A list was made of approved manufacturers and equipment, and purchasers basically adhered to this list quite strictly. Listed companies could mention their being on the list to potential customers, thus, Shell's testing lab functioned more or less unintentionally as an unofficial yet authoritative certification bureau.[114]

ISO 9000

In this respect the international introduction of ISO 9000 standards in the 1990s meant a definite turn. The ISO 9000 Standards, first published in 1987, are a set of guidelines created by the International Organization for Standardization that assure that businesses meet certain quality control and management standards. ISO 9000 standards define the requirements of quality management systems. Companies can get a certificate, issued by an outside party, to attest that they in fact answer to these standards. A quality certificate based on ISO 9000 standards certifies the production process and subsequent handling, including design, supply, and so on. Very soon these standards became authoritative. Shell, for one, gave up its testing of lab equipment and lab materials it planned to purchase, as well as its own list of approved manufacturers because all firms certified according to ISO 9000 standards were eligible suppliers. By relying on ISO standards, Shell could subcontract work much more often and much more easily and hence concentrate on its core activities. Many other large companies and governments besides Shell began to require a quality certificate from their suppliers, having such a certificate became a must in many business branches.[115]

Thus the introduction of quality certificates in industry was interrelated not only with increasing world market integration but also with a change in

Starting in the 1980s, standardization was geared not only to products and technical specifications but also to issues such as operational management and quality control. Companies could acquire certification if they met specific conditions, and this increasingly applied to nontechnical companies and institutions as well. In early 2003 the Antonius Binnenweg Nursing Home in Rotterdam received the Model Internal Quality System for Nursing Homes seal of quality from the former minister of health, Hans Simons.

the politics of large companies. They could subcontract many more activities, with the result that the various production stages of a specific product became less transparent, also to the actual manufacturer. Quality certificates ensured one that the parts used in a product were made according to a basic standard. Another factor was the trend toward "just in time" production. If a company could be sure that parts suppliers would deliver on time, this implied one did not have to stock a large quantity of these parts. An ISO 9000 certificate guaranteed that suppliers could in fact meet such obligations.

A certificate is not an absolute guarantee of quality. A spectacular example of this was the Sleipner A oil platform, built by Norwegian subcontractors for Norwegian Statoil. In the case of earlier oil platforms built by this successful and experienced company, quality control had mainly been an internal affair, but in the case of the Sleipner A platform Statoil insisted on following the new ISO 9000 procedures. Soon, in 1991, a construction error caused the oil platform to sink to the ocean floor even before it was completed.[116] Of course, the quality standards cannot be held directly responsible for this outcome, but putting too much faith in the procedures themselves without paying much attention to the stress their introduction caused, did cause rather than prevent trouble.

Quality certification is of recent date, which makes it hard to evaluate its efficacy reliably. Still it is safe to suggest that quality certification, too, has been a contested innovation to some extent. As occurred with standardization, certification proved to develop a dynamic of its own whereby business management aspects could be lost sight of. The importance of good quality is acknowledged by everyone, but is hard to express in money. This is all the more a significant factor because improving an organization takes time and money, and companies have to pay for certificates. The journal *Machine Design* ran a series on the "quality myth," in which one author mentioned the example of the American pipes and valves manufacturer Wallace, which a few months after it had won the prestigious Baldridge National Quality Award was on the verge of bankruptcy. Such a debacle is often price-related: customers are not willing to pay more for better quality.[117] Every now and then critical voices could be heard in the Netherlands as well. For example, an article in 1990 argued that, contrary to the facts, quality care had come to be regarded as a panacea.[118]

Oddly, however, such criticism has mainly come from quality experts themselves. Unlike standardization, quality certification did not gradually evolve as a specialty, but was mainly imposed from outside in an abrupt manner, as a result of policies initiated by governments and large customers. Suddenly there was much demand for quality courses, quality auditors, and quality certificates. From a discipline-specific angle, this rapid growth undoubtedly caused unwanted side effects. The question is of course to what extent such angle is decisive. The real reason for quality certificates may be different from the demands of good quality standards, as defined by experts.

Certification follows a general pattern, whereby specific issues in an industry increasingly become the domain of specialists with their own rules and ethos. Possibly in some cases companies rely on certification because it relocates responsibility for the process, rather than because of a real concern about quality itself.[119] Meanwhile, however, it is hard to imagine our economic life without certification processes. It is telling that in recent years certification processes have also been adopted by exactly those groups that are quite critical of the world of business. For example, nowadays certificates are sought to indicate that a specific product is ecologically sound, made without exploiting child labor, supports developing world economies, and so on.

Consumer representatives were influential in establishing more or less defined standards and warranties for consumer goods. One well-known seal of quality for consumers and producers alike was that of the Dutch Association of Housewives. Testing and inspecting products was a central activity of this association, which had its own testing facility. Its executive board is shown on this photo from the early 1970s.

Conclusion

In the twentieth century distances became smaller, outputs rose, and sales soared. The unprecedented economic growth has changed our society unrecognizably and deeply influenced the nature and qualities of its various institutions. Obviously, a town is not just a village of some size, while it is hardly productive to compare the functioning of a village store to that of an international company. These qualitative differences are couched in terms of scale increases: processes that occur at other levels are no longer mutually comparable. This also means that older technologies or modes of organization have become unusable in the current dynamic. Scale increases influence societies not only through what they generate—in the sense of more, larger, and faster—but also through what they demand.

One of the most defining elements of scale increase as a phenomenon is mass production, which is only possible by virtue of giant companies with an extensive organization. Such companies have contributed inordinately to the image and feel of the twentieth century. They receive undue attention precisely because we are so constantly made aware of their significance. For this reason it is difficult to determine where reality stops and our imagination takes off, and the same can be said of historians who try to define the essential characteristics and motivations of this period.

The most prominent historian of economic expansion in the twentieth century is Alfred Chandler. In his view a tendency to scale increase reveals itself in particular in business, which is more or les coerced by the functioning of economic laws that allow larger units to operate more cost-efficiently. The principle of economy of scale functions as the engine of many other social developments. In order to grow companies have to adapt their organization, in particular their management structure, which leads to the rise of modern methods such as scientific management. The growth of industrial production thus causes organizational changes and these in turn eventually lead to other far-reaching social changes.

This perspective, however, is hardly self-explanatory, and one might as well opt for the opposite approach: the process of scale increase is not primarily located in individual companies or installations but in the economic system as a whole. Thus there has been an intensification of global trade and the gradual expansion of flows of goods and information. This increases society's complexities and interconnections and puts new demands on the existing institutions to grow, to cooperate, and so forth. The changes in companies, as in modes of production and organization, are a response to this development, rather than a cause of it.

It's a classic chicken-and-egg problem, and of course the two factors—the economy's pressure on companies and society, and companies' pressure on the economy—are interlinked. The emergence of large-scale industry was tied to how material and immaterial flows were generated and regulated in the twentieth-century world. Large companies could only exist by virtue of large potential markets, offering sufficient opportunities for sales, transport, and so on. Still, it cannot be denied that large mass production companies have been co-responsible for the growth of production and distribution of goods, and thus for the effective realization of the mass markets and flows of goods on which they depend.

It is an oversimplification to say that the twentieth-century intensification of production could only be realized via mass production in large-scale industries. This was neither the sole cause of scale increases, nor the only possible reaction to them. Likewise, one cannot say that the rise of modern management techniques in companies was exclusively, or even chiefly, a reaction to the problems that followed from raising production efforts. We have shown that company size and company organization, on the one hand, and the size and regulation of total production and distribution, on the other, can be interrelated in various ways.

Of the three examples we discussed, the process industry, inasmuch it produces bulk goods, was the only sector where the intensification of production was mainly a result of enlarging the size of factories and installations.

The account management department of the Postcheque- en Girodienst in The Hague in the 1930s. This banking service, set up in 1918, had 365,000 accounts by 1939. Processing the growing volume of data was initially managed by hiring more employees.

Scale increases in this sector, as long as the markets proved cooperative, had clear economic benefits, which is why the available options to do so were maximally exploited. Yet this sector certainly did not perform a leading role in implementing modern management methods. In the process industry, "scientific management" could be and was applied only to a limited degree.

Although in the manufacturing industry the new methods attracted much attention, it turned out that such rationalization was largely unrelated to size. Instruments that are traditionally seen as belonging to big industry, such as production lines or scientific management, also appear in smaller companies, or in variable production. Their purpose was not mass production per se (even though in that context they were used as well), but rather disciplining workers—a much more immediate problem in the manufacturing industry than in the process industry. Scientific management was largely a social tool, rather than strictly (as some have suggested) a response to the size of factories.

An industrial company that aspired to grow into a large-scale company first had to pay attention to sales opportunities. This factor is precisely relevant for Dutch companies because of their limited domestic market.

Europe's national borders have always been quite effective trade barriers. Companies that wanted to serve mass markets first had to create them, as it were. Various Dutch companies decided to merge with a foreign partner. A more creative option was chosen by Philips. Although this company set up branches abroad, it also tried to reduce the effect of Dutch national borders by forcefully promoting international standardization in the areas directly relevant to the company. This led to a more uniform market. The company's larger size, then, was made possible in part by regulating flows of goods in the outside world.

Getting a handle on these flows was not just of relevance to large companies, however. Here, too, any automatic correlation with mass production is absent. Initially, modernization in agriculture was mainly driven by trade and processing rather than by production. For the most part farm businesses remained small-scale operations, even though they modernized, of course, in a variety of other ways. The markets for agrarian products, by contrast, had meanwhile evolved into mass markets. Farm businesses responded by adapting production but especially by adapting the format of their connection to commerce via cooperative factories and Dutch auctions. These fairly small businesses managed to function well for a long time.

Significantly, whereas the modernization of the Dutch industry basically took off at the time of World War I, only after World War II has there been a clear trend toward scale increase among companies in which economic motivations prevail. In agriculture a similar thing happened: after World War II economic considerations more or less forced restructuring and scale increases, both of individual farm businesses and of the cooperatives in which they collaborated. In other words, for Dutch companies the realization of the European market seems to have been a more important impulse to growth than the "second industrial revolution."

The large and ever-growing output, mobility, and information supply underpinning modern society, then, are only marginally caused or sustained by big industry, understood as large corporations that conquer markets via mass production and marketing and manage to keep going thanks to their modern organization. More important, at least in the Dutch context, are measures that regulate the interactions between manufacturers, clients, or consumers and other players. Coordination between the various parts of the system is more important than the size of its individual units.

This coordination took place partly on the basis of strictly economic measures—price fixing, dividing markets, and so forth. Other measures, standardization and certification in particular, clearly had a technological component. What mattered was not jut the sheer existence of standards or testing criteria but also their content. This is why technicians often had a major say in their formulation. Still, standardization and certification have mainly been important as factors improving coordination and facilitating integration. That standardization would be a means to mass production, as its Dutch advocates claimed when HCNN was established, has not been supported by actual developments. It is precisely large companies that could afford to ignore others and go their own way; thus genuine mass production can do without standardization. But for fairly small countries such as the Netherlands or for smaller companies that sought to connect to the world market it was a major tool. The same applied in fact to countries or companies that wanted to shield themselves from the global market: by using deviating standards they could isolate themselves from others. As the twentieth century progressed, however, this option became increasingly less viable.

Large companies with their mass production represent but one aspect of scale increase in the twentieth century, and so the question presents itself as to why they could have dominated its image to such an extent. It cannot be denied that big companies have played a major role in society at large. Their significance to the economy and employment is evident, as shown at the start of this chapter, and they generally had and have a lot of political clout. Still, one can wonder whether the image of the century has been shaped primarily through such facts. Prior to being significant in the Netherlands, large-scale production methods already stirred up people's imagination. As early as the World War I era, Ford's automobile production counted as the foremost model of what people imagined the industry of the future would be like.

One must consider the role played by the spectacular character of the successes of the American industry, which rendered them usable as a rhetorical ploy for people who wanted to promote social, technological, or management improvements. Such reformers voiced their concerns clearly in the first half of the twentieth century. All sorts of groups of experts organized in response to the social problems tied to industrialization, urbanization, and modernization. These experts derived their legitimacy in part from their view of a utopian future, which they claimed was within reach. They thus had a need for a communal ideal to express their mutual interrelatedness and present themselves as trail blazers on the path toward tomorrow. In this situation large-scale American society became the preeminent ideal to many—a future that was inevitable if one wanted to count in the modern world and make a difference.

This image was determined by industry, rather than agriculture, and perhaps not just industry in general but large-scale industry. Those working in industry had no qualms about arrogating to themselves this exemplary role as their due. In industry, too, modernization was largely taken on by new groups of experts, who viewed mass production, scientific methods, and large organizations as main features of a modern society. As a result, the measures of coordination and integration, which in many cases were quite important with respect to scale increase processes in the Dutch context, remained underexposed. Although in practice experts were mainly preoccupied by precisely such problems, they were less usable for propaganda. The ways in which standardization commissions forged often hard-won compromises did not result in a clear and appealing image of the future. In other words, the promoted image of mass production and scale increase was distorted in a way that raised its propagandist value, but this rendered genuine *technology* largely invisible.

Acknowledgments

Of the many people who contributed to this chapter through their advice and commentary, I mention Ernst Homburg first. Particularly Johan Schot, Harry Lintsen, and Ruth Oldenziel provided major suggestions on the final text. Many postdocs and other staff have offered useful comments within the context of the *Techniek in Nederland in de twintigste eeuw* project; I am especially grateful to Peter Baggen for our pleasant collaboration. On the topic of standardization and certification, I received valuable support from Henk de Vries and Piet Vos. For permission to consult the archive of the Nederlands Normalisatie-Instituut I am grateful to its director, J. A. Wesseldijk.

Notes

1 Cf. R. Oldenziel, "Het ontstaan van het moderne huishouden: Toevalstreffers en valse starts, 1890–1918," in Schot et al., eds., *Techniek in Nederland*, vol. 4, 22, 25.

2 See A. D. Chandler, *The Visible Hand: The Managerial Revolution in American Business* (Cambridge, Mass.: Harvard University Press, 1977), and Alfred D. Chandler, *Scale and Scope: The Dynamics of Industrial Capitalism* (Cambridge, Mass.: Belknap Press, 1990).

3 Philip Scranton, *Endless Novelty: Specialty Production and American Industrialization, 1865– 1925* (Princeton: Princeton University Press, 1997). For critical discussions of Chandler's work, see, for instance, R. H. Tilly, "Grossunternehmen: Schlüssel zur Wirtschafts- und Sozialgeschichte der Industrieländer?" in *Geschichte und Gesellschaft* 19 (1993), 530–548; C. S. Maier, "Accounting for the Achievements of Capitalism: Alfred Chandler's Business History," in *Journal of Modern History* 65, no. 4 (1993): 771–782; B. W. E. Alford, "Chandlerism, the New Orthodoxy of US and European Corporate Development," in *Journal of European Economic History* 23 (1994): 631–643.

4 J. R. Beniger, *The Control Revolution: Technological and Economic Origins of the Information Society* (Cambridge, Mass.: Harvard University Press, 1986).

5 J. W. Schot, H. Lintsen, and A. Rip, "Methode en opzet van het onderzoek," in Schot et al., *Techniek in Nederland*, vol. 1, 48.

6 E. Bloemen, J. Kok, and J. L. van Zanden, *De top 100 van industriële bedrijven in Nederland 1913–1990* (The Hague: Adviesraad voor het Wetenschaps- en Technologiebeleid, 1993).

7 P. A. V. Janssen, *Groot en klein in de Nederlandse industrie, 1953–1968–1980: Een poging tot kwantificering van het concentratieverschijnsel en van de verschuivingen binnen het industriële patroon* (The Hague: Nederlandsche Centrale Organisatie voor Toegepast-Natuurwetenschappelijk Onderzoek, 1971).

8 Commission of the European Communities, *De industriepolitiek van de gemeenschap: Memorandum van de Commissie aan de Raad.* (Brussels: Commissie van de Europese Gemeenschappen, 1970), 71–87, notably 81–82.

9 Jan Luiten van Zanden, *The Economic History of the Netherlands 1914-1995* (London: Routledge, 1998), 36-38.

10 H. J. de Jong, *De Nederlandse industrie 1913–1965: Een vergelijkende analyse op basis van de productiestatistieken* (, Amsterdam: Nederlandsch Economisch-Historisch Archief, 1999).

11 State Archive North-Holland, Haarlem, Archive Nederlandsche Maatschappij voor Nijverheid en Handel in Haarlem; Letter of G. de Clercq to Kipp, 22 December 1915.

12 E. Homburg, "De Eerste Wereldoorlog: Samenwerking en concentratie binnen de Nederlandse industrie," in Schot et al., *Techniek in Nederland*, vol. 2, 322–329.

13 E. A. du Croo, "Normalisatie!," *Tijdschrift der Maatschappij van Nijverheid* 83 (15 October 1915): 504–507.

14 E. P. Haverkamp Begemann, "Iets over normalisering in machineconstructie: Voordracht gehouden in de vergadering van de Afdeeling voor Werktuig- en Scheepsbouw op 16 December 1916," *De ingenieur* 32 (1917): 220.

15 W. F. Boterhoven de Haan, "Normalisatie als integreerend onderdeel van wetenschappelijk beheer: Voordracht gehouden in de vergadering van de Afdeeling voor Werktuig- en Scheepsbouw on 12 mei 1917," *De ingenieur* 32 (1917): 764. Regarding the case of Siemens, Schmidt could in fact establish that this company sought to meet the wishes of individual clients. This turned standardization into a sluggish effort. "Das grösste Hindernis stellte aber zweifellos die Orientierung des Werkes dar, eine möglichst breite und sich laufend ändernde Produktpalette anzubieten." See Dorothea Schmidt, *Massenhafte Produktion? Produkte, Produktion und Beschäftigte im Stammwerk von Siemens vor 1914* (Münster: Westfälisches Dampfboot, 1993), 94.

16 Wolfgang König, *Geschichte der Konsumgesellschaft* (Stuttgart: Steiner, 2000), 48, with

reference to Harald Winkel, *Die deutsche Nationalökonomie im 19. Jahrhundert* (Darmstadt: Wissenschaftliche Buchgesellschaft, 1977), 107.

17 In addition to mass production, mass distribution is addressed by König, *Geschichte der Konsumgesellschaft*, 91–107.

18 Henk Povée, *De eeuw van Blokker: Honderd jaar huishoudbranche in Nederland* (Bussum: Thot, 1996), 24–25.

19 A. A. D. Bouwhof and J. C. Lagerwerff, *Elementair leerboek van de bedrijfshuishoudkunde*, 5th ed. (Groningen and Batavia: Noordhoff, 1939), 313.

20 C. Huijsman, *Leerboek der bedrijfshuishoudkunde*, 2nd ed. (Amsterdam: Brinkman, 1937), 93; see also 259 and the elaborate section on limits of expansion, 262–269.

21 Bouwhof and Lagerwerff, *Elementair leerboek van de bedrijfshuishoudkunde*, 312.

22 Huijsman, *Leerboek der bedrijfshuishoudkunde*, 268.

23 O. Bakker, *Bedrijfshuishoudkunde*, 4 volumes (Purmerend: Muusses, 1947–1954), "Expansie der onderneming," volume 4, part 1, 7–10.

24 Albertus B. A. van Ketel, *Bedrijfseconomische schetsen omvattende leerstof van doctoraalcolleges, 1935–1938: "Financiewezen der onderneming"* and *"Toegepaste bedrijfshuishoudkunde"* (Tilburg: Katholieke Economische Hogeschool, [1941]), 113.

25 R. B. Hartevelt and H. R. Wortmann, *De Vries Robbé & Co 1881–1956: Tijdsbeeld van driekwart eeuw technische en maatschappelijke vooruitgang* (Gorinchem: De Vries Robbé & Co. N.V., 1956), 139–141. On the history of the "scientific" approach in Dutch manufacturing in general, see R. Vermij, E. Nijhof and F.C.A. Veraart, "Verwetenschappelijking van productie en organisatie," in Schot et al., eds., *Techniek in Nederland*, vol 6, 284-301.

26 W. Visser, *Van een half tot drieduizend pk—1906–30 mei 1956: Uitgegeven ter gelegenheid van het 50–jarig bestaan der N.V. motorenfabriek Thomassen, De Steeg* (The Hague: N.V. Motorenfabriek Thomassen, 1956), 91.

27 On the reception of Taylorism in the Netherlands, see E. S. A. Bloemen, *Scientific management in Nederland 1900–1930* (Amsterdam: Nederlandsch Economisch-Historisch Archief, 1988).

28 R. H. Vermij, "Gedwongen tempo: De lopende band in Nederland tot de Tweede Wereldoorlog," *NEHA-Jaarboek voor economische, bedrijfs- en techniekgeschiedenis* 64 (2001): 227–257. See also R. H. Vermij, "Het gaat vanzelf - of niet? Industriële automatisering in Nederland," in Schot et al., eds., *Techniek in Nederland*, vol. 6, 303–317.

29 Philips Company Archives (Eindhoven) 644.21, Syllabi Interne technische opleidingen-radioapparaten, box 162, ITO (Internal technical training), course RA 42 1939–1940, 29.

30 National Archive, The Hague, 2.18.24, Archive Adviesbureau voor bedrijfsorganisatie van J. M. Louwerse, (1923), 1925–1953 (inventory no. 37), report 43–6, Report on Maatschappij tot Exploitatie van Bendien's confectiefabrieken, Almelo, 18 December 1935.

31 G. de Clercq, "Het taylorstelsel in de chemische industrie," *Chemisch weekblad* 17 (1920), 173.

32 Minutes of meeting, Vereniging Nederlandse chemische industrie, 14 December 1955, 9 (in the collection Ernst Homburg).

33 De Clercq, "Taylorstelsel in de chemische industrie," 173.

34 Ernst Homburg, *Groeien door kunstmest: DSM Agro 1929–2004* (Hilversum: Verloren, 2004), 69–72, 202–206.

35 P. J. H. van Ginneken, "Suikerindustrie en chemische wetenschap," *Chemisch weekblad* 20 (1923): 522.

36 For example, on the production of sulfuric acid, see Ernst Homburg and H. van Zon, "Grootschalig produceren: Superfosfaat en zwavelzuur, 1890–1940," in Schot et al., *Techniek in Nederland*, vol. 2, 287.

37 H. H. Vleesenbeek, "Overheid, parlement en economische mededinging: Analyse van de parlementaire diskussie n.a.v. de wet tot het algemeen verbindend en onverbindend verklaren van de ondernemersovereenkomsten, 1934/1935," in J. van Herwaarden, ed., *Lof der*

historie: Opstellen over geschiedenis en maatschappij (Rotterdam: Universitaire Pers Rotterdam, 1973), 379–380.

38 T. Siegel and Th. von Freyberg, *Industrielle Rationalisierung unter dem Nationalsozialismus* (Frankfurt am Main: Campus, 1990), 203–205.

39 Vleesenbeek, "Overheid, parlement en economische mededinging," 383.

40 J. Goudriaan, *Samenwerking van bedrijven als middel ter verhooging van de efficiency* (Purmerend: Nederlands Instituut voor Efficiency, 1923).

41 Bouwhof and Lagerwerff, *Elementair leerboek van de bedrijfshuishouding*, 54, see also 54–81; Huijsman, *Leerboek der bedrijfshuishouding*, 401–433.

42 For example, E. Hijmans, "Productievraagstukken in verband met crisis," *De naamloze vennootschap* 1 (1922–1923): 186–188.

43 N.V. Nederlandsche Gist- en Spiritusfabriek te Delft, *De ontwikkeling der onderneming in zestig jaren, 1870–1930* (Delft: 1930), 111.

44 Cf. D. J. G. Arnoldus, "Nederlandse kartelvorming in de olien- en vettenindustrie in de jaren dertig," *NEHA-jaarboek* 60 (1997): 226–257.

45 Dirk de Wit, *Windmill, wieken naar de wind gekeerd: Van boerencoöperatie naar internationale organisatie* (Vlaardingen: Hydro Agri Rotterdam, 1990). In the Vlaardingen Municipal Archive, 357, Archive ENCK, 1971–1981 (inventory no. 99), see the following: P. Lindenbergh, "Geschiedenis van de ENCK tot en met 1933," unpublished manuscript, n.d., (inventory no. 216); C. H. Buschman, "Veertig jaren ENCK [1903–1961]," unpublished manuscript, n.d. (inventory no. 264); Minutes, Board ENCK (inventory no. 266); Minutes, Board Zeeuwse Coöperatieve Kunstmestfabriek, papers concerning the General meeting (inventory no. 301). Homburg and Van Zon, "Grootschalig produceren," 283, offer a slightly different interpretation, however.

46 S. Koenen, *Inleiding tot de landhuishoudkunde (Wat iedere Nederlander omtrent den vaderlandschen landbouw dient te weten)*, ed. H. W. C. Bordewijk (Haarlem: Erven Bohn, 1924), 30–37, 56–59, 77–80.

47 Quoted by W. H. Vermeulen, *Den Haag en de landbouw: Keerpunten in het negentiende-eeuwse landbouwbeleid* (Assen: Van Gorcum, 1966), 71–72.

48 J. Rinkes Borger quoted in V. R. IJ. Croesen, *Geschiedenis van de ontwikkeling van de Nederlandsche zuivelbereiding in het laatst van de negentiende en het begin van de twintigste eeuw* (The Hague: Mouton, 1932), 85–86.

49 J. C. Dekker, *Zuivelcoöperaties op de zandgronden van Noord-Brabant en Limburg, 1829–1950: Overleven door samenwerking en modernisering—Een mentaliteitsstudie* (Middelburg: s.n., 1996), 26; for a survey of theories, see 26–32.

50 Dekker, *Zuivelcoöperaties*, 77–78, 83–86.

51 Cf. J. H. van Stuijvenberg, *Het Centraal Bureau: Een coöperatief krachtveld in de Nederlandse landbouw 1899–1949* (Rotterdam: Stichting het Economisch Instituut, 1949).

52 B. Havenaar, ed., *Gedenkboek veiling "Berkel en Rodenrijs" G.A. 1903–1953* (Amsterdam: s.n., 1953), 65.

53 Van Stuijvenberg, *Centraal Bureau*, 96–103.

54 Dekker, *Zuivelcoöperaties*, 317–322.

55 For more on the history of this legislation, see C. H. Schouten, *Opstellen over economisch ordeningsrecht*, (Rotterdam, 1973), part 2; Dutch Employers Association, *Wet economische mededinging, waarin tevens opgenomen de kartelbepalingen in het E.E.G.-verdrag: Tekst en commentaar* (The Hague: Verbond van Nederlandsche Werkgevers, 1958), 7–16. On postwar Dutch antitrust policies, see, for instance, B. Baardman, "La notion d'entreprise(s) en position dominante et les méthodes de contrôle en droit antitrust néerlandais," in J. A. van Damme, ed., *La réglementation de comportements des monopoles et entreprises dominantes en droit communautaire / Regulating the behaviour of monopolies and dominant undertakings in community law* (Bruges: De Tempel, 1977), 277–287; Corwin D.

Edwards, *Control of Cartels and Monopolies: An International Comparison* (New York: Oceana, 1967), 355–358.

56 W. T. Kroese, *Vormen van samenwerking in de Nederlandsche katoenindustrie 1929–1939* (Leiden: Stenfert Kroese, 1946). The study was completed in 1944.

57 A. de Graaff, *De industrie* (Utrecht: Het Spectrum, 1951), 31–33.

58 The data on DSM urea production were generously provided by Arjan van Rooij. See *Building Plants: Markets for Technology and Internal Capabilities in DSM's Fertilizer Business, 1925-1970* (Amsterdam: Aksant, 2004). On Stamicarbon, see Arjan van Rooij and Ernst Homburg, *Building the Plant: A History of Engineering Contracting in the Netherlands* (Zutphen: Walburg Pers, 2002), 77, 96.

59 P. H. Spitz, *Petrochemicals: The Rise of an Industry* (New York: Wiley, 1988), 418–458, in particular, 453–458.

60 E. Hijmans, "Serievergroting en massaproductie gezien vanuit bedrijfstechnisch standpunt," in T. J. Bezemer, E. Hijmans, and H. van Mourik Broekman, eds., *De overgang van de Nederlandse industrie van stukfabricage op serie- en massaproductie* (The Hague: Nederlandse Insituut voor Efficiency, 1948), 17–25.

61 Dekker, *Zuivelcoöperaties*, 113, 126.

62 Ibid., 129.

63 Ibid., 124–134.

64 J. Bieleman, "De legkippenhouderij," Schot et al., *Techniek in Nederland*, vol. 3, 170–71.

65 A. H. Crijns, *Van overgang naar omwenteling in de Brabantse land- en tuinbouw 1950–1985: Schaalvergroting en specialisatie* (Tilburg: Stichting Zuidelijk Historisch Contact Tilburg, 1998), 103–105.

66 P. de Ruijter, *Voor volkshuisvesting en stedebouw* (Utrecht: Matrijs, 1987).

67 P. de Ruijter, *De rijksplanologische dienst: Instelling en ontwikkeling* (Delft: Technische Hogeschool Delft, Planning Theorie Groep, Afdeling Bouwkunde, 1975).

68 A. J. Geurts, *De "groene" IJsselmeerpolders: Inrichting van het landschap in Wieringermeer Noordoostpolder Oostelijk en Zuidelijk Flevoland* (Lelystad: Stichting Uitgeverij De Twaalfde Provincie, 1997), 18.

69 Ibid., 19–25.

70 Dutch Ministry of Agriculture, Industry and Trade, Directorate Agriculture, *Een en ander betreffende regeeringsbemoeiingen in zake den landbouw* (The Hague: 1907), 9–10; Harro Maat, *Science Cultivating Practice: A History of Agricultural Science in the Netherlands and Its Colonies, 1863–1986* (Dordrecht: Kluwer Academic Publishers, 2001), 101–103.

71 N. H. H Addens, *Zaaizaad en pootgoed in de Nederlandse landbouw* (Wageningen: Veenman, 1952), 52, 56–60; Maat, *Science Cultivating Practice*, 58.

72 Vermeulen, *Den Haag en de landbouw*, 69–74.

73 Boterhoven de Haan, "Normalisatie als integreerend onderdeel van wetenschappelijk beheer," 766. I am grateful to Giel van Hooff for providing me with data on Paul Begemann.

74 Visser, *Van een half tot drieduizend pk, 1906–30 mei 1956*, 96–97.

75 Jac. van Ginneken, *Zielkunde en Taylor-systeem: Rede, voorgedragen op de gecombineerde hoofdbestuursvergadering der R.K. vakbonden, 27 nov. 1917, te Utrecht* (Amsterdam: De R.K. Boekcentrale, 1918). On Van Ginneken, see Bloemen, *Scientific management in Nederland*.

76 It is now the British Standards Institution. Its history is told in C. Douglas Woodward, *BSI: The Story of Standards* (London: British Standards Institution, 1972).

77 The attempts at standardization of gas-meters and the difficulties they met were recorded in *Het gas* 17 (1897): 128, 269; *Het gas* 19 (1899): 121–130, 201; *Het gas* 20 (1900): 195; *Het gas* 21 (1901): 223; *Het gas* 22 (1902): 281; *Het gas* 23 (1903): 291; *Het gas* 24 (1904): 292.

78 Th. Wölker, "Entstehung und Entwicklung des deutschen Normenausschusses," Ph.D. dissertation, Freie Universität Berlin, 1991.

79 National Archive The Hague, 2.19.047.01, Archive of the Royal Institute of Engineers, 1847–1960 (henceforth: Archive KIvI) (inventory no. 15), Minutes Board of Directors, 1913–1915, part 1915; Meeting held 21 December 1915.

80 Vermij, Nijhof and Veraart, "Verwetenschappelijking", 292-293.

81 E. Hijmans, "Normalisatie [Lecture at the meeting of the Royal Institute of Engineers on 2 February 1920]," *De Ingenieur* 35, no. 24 (1920): 427.

82 Sociaal Historisch Centrum Limburg, Maastricht, Mega-Archive, report on meeting of the HCNN, 14 January 1920, 2.

83 Sociaal Historisch Centrum Limburg, Mega-Archive, report meeting of the HCNN, held on 18 January 1921, 2.

84 State Archive in North-Holland, Haarlem 446, Archive Nederlands Normalisatie-Instituut, box 1, report of the Centraal Normalisatie Bureau on discussion of Wednesday, June 12, 1918, between HCNN members and the directors of, respectively, the Central Standardization Bureau and the Central Statistics Bureau. In the final minutes things are articulated a little more carefully: "Mr. Hulswit pointed out that eventually the HCNN would not want to limit its effort to the metal industry. But since it had received financial contributions needed for its future operations and for the establishment of the CNB principally from this industry, the HCNN felt it should at least ensure that standardization of the metal industry would not be slowed down as a result of more general efforts it undertakes."

85 Sociaal Historisch Centrum Limburg, Mega-Archive, Minutes of the Meeting of the Electro-Technical Advisory Commission, HCNN, February 9, 1921, 6.

86 State Archive North-Holland, 446, box 1, draft letter of the HCNN to Dutch Manufacturers of Parts of Sewer Systems, July 1920.

87 State Archive Limburg, Maastricht 17.26/17A, DSM Company Archives: All Coal Mines, Limburg, 1923–1939 (inventory no. 11), CNB to All Coal Mines, 25 June 1935; State Archive Limburg 17.26/17A (inventory no. 11), Board HCNN to Vereeniging tot behartiging van de belangen der Limburgsche mijnindustrie, December 18, 1936, and April 27, 1937.

88 R. Oldenziel and M. Berendsen, "De uitbouw van technische systemen en het huishouden: Een kwestie van onderhandelen, 1919–1940," in Schot et al., *Techniek in Nederland*, vol. 4, 55–56.

89 H. van der Boom, "De geschiedenis van het ziekenhuiswezen in Nederland, 1900–1940: Het ontstaan van het moderne ziekenhuis vanuit sociaal constructivistisch perspectief," master's thesis, University of Maastricht, 1997, 80.

90 Ibid., 66, 79.

91 Ibid., 110–111.

92 K. F. Proost, *De waardering der techniek* (Arnhem: Van Loghum Slaterus, 1930), 62.

93 F. J. van Lith, "4 voet 82 duim," *Normalisatie* 39 (1963): 156.

94 P. P. A. E. van Heeswijk, "Materiaalnormalisatie bezien uit het standpunt van de groothandel in walsprodukten: Voorraadshandel en normalisatie," in *Preadviezen materiaalnormalisatie op het gebied van ferrometalen walsprodukten (platen, profielijzer, enz.): Kwaliteit en afmetingen—Benamingen: Landelijke bijeenkomst van de regionale contactgroepen bedrijfsnormalisatie 8 mei 1962* (The Hague: Nederlands Normalisatie-Instituut, 1962), 9–15.

95 Cf. Th. Wölker, "Der Wettlauf um die Verbreitung nationaler Normen im Ausland nach dem Ersten Weltkrieg und die Gründung der ISA aus der Sicht deutscher Quellen," *Vierteljahrschrift für Sozial- und Wirtschaftsgeschichte* 80 (1993): 487–509.

96 High Commission on Standardization in the Netherlands, *Grondbeginselen: Uitgegeven door Hoofdcommissie voor de normalisatie in Nederland ingesteld door de Maatschappij van Nijverheid en het Koninklijk Instituut van Ingenieurs en de Raad voor de Normalisatie in Nederlandsch-Indië* (The Hague: 1931), points 13–15.

97 N. Nagtegaal, "De eerste 25 jaar normalisatie bij Philips: De geschiedenis van een af-deling," unpublished manuscript, 1982, Philips Company Archives (Eindhoven). I am grate-ful to L. van Rooy, who provided me with various data on standardization relating to Philips Electronics. See also N. A. J. Voorhoeve, "De normalisatie bij het Philips concern," *De ingenieur* 62, no. 17 (1950): A 213–222.

98 P. van Zuuren, author interview, The Hague, July 2, 2001.

99 State Archive Limburg, 17.26/21C, DSM Company Archives; Organization, 1950–1969; inventory no. 366: data on the Centraal Normalisatie Bureau, 1952–1966; notes of fact-finding visits on February 11 and 25, 1954, by a DSM delegation to Philips Electronics, where talks were held with, among others, N. A. J. Voorhoeve on Philips's standardization effort.

100 Philips Company Archives (Eindhoven), 727.6, Normalisatie, note by N. A. J. Voorhoeve, Jan-uary 8, 1948.

101 In line with Hoofdcommissie voor de Normalisatie in Nederland, *Nota normalisatie in Nederland* (The Hague: HCNN, 1947).

102 Archive KIvI (inventory no. 25), Papers Concerning Stichting Fonds voor de Normalisatie and HCNN 1912–1941; summary of discussion in the meeting of HCNN, held October 26, 1939.

103 State Archive Limburg, 17.26/28 C, DSM Company Archives: Membership and Subsidy, 1959–1975; inventory no. 243: Membership of the Nederlands Normalisatie-Instituut, 1950–1969; summary of discussion in the meeting of the General Board of Nederlands Normalisatie-Instituut, held November 27, 1959. On Cobeno, see, for instance, G. M. de Beer, "Tien jaar cobeno," *Normalisatie* 43 (1967): 237–241; G. A. Rosenthal, interview, *Normalisatie* 49 (1972): 34–37.

104 F. Tollenaar, "Normalisatie in het bedrijf," in Commissie Bedrijfsnormalisatie, ed., *Bedrijf en norm* (The Hague: Nederlands Normalisatie-Instituut, 1962), 117–113, 130.

105 H. Vijgenboom, "Normtechniek" in de praktijk," in Commissie Bedrijfsnormalisatie, *Bedrijf en norm*, 161.

106 A. T. Hens, "Normalisatie: juist nu!," in *Normalisatie* 43 (1967): 3.

107 P. van Zuuren, author interview.

108 P. Vos, author interview (telephone), October 18, 2001.

109 Dutch Ministry of Agriculture, Industry and Trade, *Een en ander betreffende regeerings-bemoeiingen in zake den landbouw*, 77–83; J. H. van Stuijvenberg, "Aspecten van overheids-ingrijpen," in *Honderd jaar margarine 1869–1969* (The Hague: Martinus Nijhoff, 1969), 303–307; Dekker, *Zuivelcoöperaties*, 449–455.

110 Dekker, *Zuivelcoöperaties*, 302.

111 J. Bieleman, "Tarweteelt en tarweveredeling," in Schot et al., *Techniek in Nederland*, vol. 3, 187; Addens, *Zaaizaad en pootgoed in de Nederlandse landbouw*, 115–168.

112 Oldenziel and Berendsen, "De uitbouw van technische systemen en het huishouden," 51–55.

113 See "Een kwaliteitsmerk," *Sigma* 9 (1963): 115–119.

114 P. Holleman, author interview (telephone), February 1, 2002.

115 See also Vermij, Nijhof and Veraart, "Verwetenschappelijking," 298-301.

116 G. L. Wackers, "The Sleipner A GBS Loss (1991), or: Why Did a Highly Competent Company Fail in What They Were So Good At?," paper presented at BOTS colloquium, University of Maastricht, September 19, 2001. The case is also dicussed in: G. Wackers, "Offshore Vul-nerability: The Limits of Design and the Ubiquity of the Recursive Process," in: C. Owen, P. Béguin and G. Wackers, eds., *Risky Work Environments. Reappraising Human Work within Fallible Systems* (Farnham, Surrey: Ashgate, 2009), 81-98.

117 J. Smith and M. Oliver, "The Baldridge Boondoggle," *Machine Design* 92 (1992): 25–29.

118 M. Gruisen, "Help, de kwaliteitszorg beheerst zichzelf niet meer," *Sigma* 36 (1990): 16–19.

119 See the analysis of the meaning of accounting in Michael Power, *The Audit Society: Ritu-als of Verification* (Oxford: Oxford University Press, 1997).

The midlevel technical education system often responded more quickly to new developments than the Technical University Delft. In 1934 the Haarlem midlevel system began a specialized curriculum in aircraft engineering; five years later a similar curriculum was instituted at Delft.

5

The Rise of a Knowledge Society

Peter Baggen, Jasper Faber, and Ernst Homburg

Although nearly everyone agrees that computers and cellular phones contain more knowledge, so to speak, than firestone axes and wheels, it is less clear what the knowledge intensity of technology amounts to. Knowledge intensity is the "amount" or "degree of sophistication" of the knowledge needed to produce a certain product, or run a particular process. It is a truism to say that in the course of time the level of know-how embodied by products and artifacts has increased. But what, exactly, is covered by this umbrella concept, knowledge intensity, and what, as a consequence, may remain hidden from view?

Consider the impending shortages of nitrogen compounds at the start of the twentieth century. This gave many chemists an incentive to look for processes geared to fixing atmospheric nitrogen. One of them was Fritz Haber, a German physical chemist who worked at the Technische Hochschule in Karlsruhe. Haber, an expert in thermodynamics, had carried out numerous experiments at extremely high temperatures and pressures. One day he hit upon the idea of combining atmospheric nitrogen with hydrogen in a high-pressure reactor, which resulted in ammonia. In 1908, after having found a suitable catalyst for this reaction, Haber knocked on the door of the Badische Anilin- und Soda-Fabrik (BASF), where he met Chief Engineer Carl Bosch.

In trying to apply Haber's discovery of nitrogen fixation on an industrial scale, Bosch ran into countless difficulties, tied in particular to the extreme conditions involved such as pressures of more than 200 atmospheres and temperatures of over 500°C. Under these conditions hydrogen reacted with the carbon that was present in the steel of the reactor wall, which regularly caused leaks or even explosions. To solve these and other problems, BASF hired a large number of scientists from various disciplines, not only physical chemists and mechanical engineers but also metallurgists, physicists, and electrical engineers. They managed to achieve industrial nitrogen fixation by designing a double-walled reactor in which a strongly improved catalyst did what it was meant to do. In 1913 the first plant to exploit this principle was ready for operation. When World War I broke out the next year and the demand for nitrogenous explosives increased, BASF strongly expanded its production capacity and developed a host of new processes. By the war's end, in 1918, the German company had gained a major edge over its competitors.[1]

In several respects the knowledge intensity of this new technology is obvious. Without a background in thermodynamics, Fritz Haber would never have been capable of discovering the basic principles of nitrogen fixation, and without the deployment of countless physical-chemists, mechanical engineers, metallurgists, physicists, and electrical engineers, BASF would never have managed to develop a reliable reactor, let alone build a plant within five years that produced ammonia on an industrial scale. Is it possible to generalize this example and argue convincingly that in the twentieth century scientific knowledge has become an indispensable ingredient in the development of new technologies? Or, put in economic terminology: is it safe to suggest that manufacturers who after 1890 refrained from investing in building technical expertise were likely to lose their market share to competitors who were highly committed to investing in new knowledge?[2]

Knowledge-Intensive Technology?

A careful analysis of late-nineteenth-century changes in Western science, industry, and technology reveals that the dynamic of the unfolding transformations is anything but clear. Several historians who study this era have referred to it as a "second industrial revolution." They suggest that a new kind of economy emerged around science-based industries such as the dyestuffs and the electro-technical industry in which multinational companies played first fiddle, while the rising intervention state actually promoted this development. Partly as a result of these changes, these historians argue, global economic power shifted from England to Germany and, especially, the United States in the course of the twentieth century. Yet the very concept of a "second industrial revolution" invokes a dividing line that is quite arbitrary. After all, regardless of the growth of overall knowledge intensity, traditional knowledge and established ways of knowledge acquisition remained and still remains important in industry. Moreover, the level of increase strongly differed from one economic sector to the next, and it is altogether unwarranted to speak of a "revolution" in all sectors.[3]

Still, many governments were quick to embrace the story of a more knowledge-intensive technology and also adapted their policies to it. Already in the first decades of the twentieth century, the Dutch government began to invest in the development of science with the aim of stimulating technological development and economic growth. In the interwar period, the government was quite sensitive to the argument that "our current technology is entirely science-[based]."[4] In the 1950s the Dutch government actively began to pursue a policy of supporting basic research as a route to increased prosperity, with technology serving as intermediary.[5] At the same time, new industrialization plans stressed the significance of putting "more brains" into Dutch products. Because government felt that those directly involved in the industry did too little to foster such development, the country's political leadership defined it as one of its own tasks.

In this chapter we recount the story of how this happened, fill in many of its details, and put it into a broader perspective. After all, a more sophisticated understanding of knowledge intensity involves not just the knowledge embodied in a specific technology, such as the mere fact that the Haber-Bosch process, a laser, or a computer was based on more, and more sophisticated, knowledge than a stepladder or a bicycle. Knowledge intensity also pertained to what happens in a sector as a whole or what has to be done on the work floor, including the implications for the education and training of those involved. For instance, the rise of large-scale industrial sugar production in the Netherlands and the Dutch East Indies caused a growing need for competent employees. When it turned out they were not available in sufficient numbers, the sector had no alternative but to set up schools that could meet their need for educated employees.

Technical Education

Knowledge intensity is a dimension not only of products of technologies but also of individuals and organizations. And in that respect more is at stake than setting up training facilities for technicians. Company managers would travel to learn more about the current level of knowledge in specific technological sectors; technicians in the textiles industry such as colorists sought to gain more knowledge by working for different companies; and ever higher standards were applied to on-the-job training. Moreover, for the correct application of techniques, specific tests and analytic methods became important, notably from around 1900. This led to the establishment of laboratories, which were either linked to companies or set up as autonomous operations. Concern for certainty and advance reliability, which is especially relevant in construction (think, for example, of large bridges or structures made of reinforced concrete), stimulated further knowledge development. For example, it was found that the structures designed by Roman and medieval architects were often more sturdy than needed, at least from a technical angle. In the nineteenth century, the calculations for the strength of

Young students at the basic technical school in Dordrecht, around 1900, who were training to become blacksmiths. After drawing schools (evening schools), basic vocational schools constituted the largest segment of the technical education system set up at the start of the twentieth century.

constructions, such as bridges, were substantially refined, as shown by Eda Kranakis in an interesting article. These calculations were part of the negotiations among engineers, contractors, and clients, which raised the need for making knowledge explicit and codifying it. Thus the significance of what was called "formal knowledge" increased.[6]

In the course of the nineteenth and twentieth centuries all sorts of practices met with new technological challenges: in factories, in construction, on farms, and in hospitals. New and more knowledge-intensive competencies had to be developed, whether for newly invented technologies or imported ones. One relevant perspective for studying this dynamic involves the system of technical education, which engaged in an ongoing effort of responding to new developments, adapting courses, and anticipating the level of knowledge expected from its graduates. Between 1870 and 1964, the number of schools for junior, intermediate, and higher technical education went from 9 to 430 in the Netherlands. Major growth took place after 1945, when the number of technical schools tripled at all levels. This strong rise of formal technical (daytime) education was accompanied by a similar increase of the number of technical courses, from 31 in 1870 to 389 in 1964, which mostly involved evening education, aimed at those who worked during the day. After 1968, when technical education had become fully incorporated into the overall education system, the importance of evening courses sharply decreased.[7]

The growth of the Dutch technical education system (discussed later in this chapter) was marked by a double transition: the system evolved from providing basic technical training to offering elaborate technical schooling and from practice-oriented training to theory-oriented instruction. Moreover, once the overall framework of the system was in place, it took on a dynamic of its own and no longer operated exclusively in response to the specific challenges of the multifarious technological practices involved. By generating scores of well-trained technicians each year, the Dutch technical education system itself became a driving force in raising the knowledge intensity of technology.

Technological Research

From the late nineteenth century onward, Dutch companies increasingly faced the challenge of having to be proactive in technological knowledge development, which gave rise to industrial research efforts and their institutionalization. In the years between 1875 and 1885 this process occurred first in the German synthetic dyestuffs industry, then in German and American electro-technical companies. The success of these pioneers caused many other companies and sectors to adopt a similar strategy. We will map these developments for the Netherlands until World War II, whereby it will become clear that specific circumstances persuaded the Dutch pioneers, such as Royal Dutch/Shell and Philips, to engage in research before this had become part of the company's strategy.[8] Together with the state labs that were set up in the same period, this led to a substantial increase of the volume of technological-scientific research in the Netherlands.

The development of Dutch research laboratories in the twentieth century has not been extensively studied. Much is still unknown with regard to the establishment and expansion of company laboratories in particular. Still it is evident that the number of labs grew steadily. This trend started in the commercial lab sector that emerged after 1865 in response to the growing demand for physical and chemical analyses in commerce, industry, and public health. Initially their number grew slowly, but after 1890 there was exponential growth, leading to a total of fifty-two labs by 1909. This phase was followed by a period of decline that reached a low of thirty-three commercial labs in 1925. This proportional decline, however, was compensated for

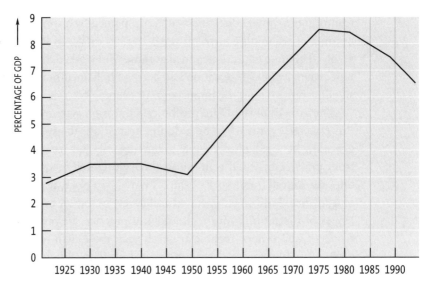

Graph 5-1. Combined investments in education and R&D as percentage of GDP
(For sources see page 603)

(and caused by) the emergence of company labs and state-funded labs. Between 1905 and 1930, the government decided to set up a large number of labs within and outside higher education, both on its own initiative and in collaboration with third parties. More than half of the partially government-controlled labs established between 1860 and 1940 were set up during this period. Although no exact figures are available yet for the period after 1945, the overall trend was for more and more companies to start their own labs.[9]

To give an impression of the sheer size of the changes in the development of technological education and research in the Netherlands, several figures illustrate the enormous growth of institutions that produced or spread formal knowledge.

This first of all applied to investments in education and research and development. Graph 5.1 shows that between 1921 and 1973 these fields saw an increase of over 300 percent, expressed as percentage of the gross domestic product, with the main growth taking place between 1947 and 1960, the period of industrialization. After peaking in 1973, the relative size of these investments began to decrease again, which is perhaps ironic because precisely in the years following the early 1970s oil crisis it became common to refer to Western societies as "knowledge societies."

Coordination of Research and Education

In the course of the twentieth century government began to play an ever larger role in technological education and research in the Netherlands. On the one hand, government acted as one party among many. It was faced by technological challenges, for example, in the area of water management, and it also developed a need for doing scientific-technological research, such as for defense, carried out by the State Defense Organization (Rijksverdedigingsorganisatie), a division of the Organization for Applied Scientific Research (Toegepast Natuurwetenschappelijk Onderzoek, or TNO) or road construction, research performed by the State Highway Laboratory (Rijkswegenlaboratorium). On the other hand, the Dutch government increasingly took charge of general services in technological research and the coordination of the (national) technological knowledge system and its various sectors. If in the course of the twentieth century technologies grew significantly more complicated, the same is true of policies relating to technology and technological research. Increasingly, the new challenges called for a response at the "systems level."

World War II constituted a decisive break in this development, mainly as a result of developments in the United States that involved massive government funding, such as for the Manhattan Project and the development of synthetic rubber, radar, and insecticides. An altogether different, much more scientific and technological world came into being after the war. Scientific research had played a major role in the victory of the allied forces, the power of nuclear energy had become quite tangible, and Europe's reconstruction took off on a massive scale, which in turn affected the role and functioning of Europe's various governments and national systems.[10]

Given these new circumstances, it only seemed natural that in the 1950s the Dutch government emphasized that Dutch products and industries should reflect "more brains" and that government itself had a role to play in achieving this goal. Reports on industrialization expressed a sense of enhanced knowledge intensity, even if the term was not used. In addition to economic rationales, strategic considerations and motives related to national status played a role. No country could allow itself to ignore nuclear energy research and development.

The Reactor Centrum Nederland was established; later it was renamed Energie Centrum Nederland. In addition to the postwar problem-solving approach, which started from the premise that more knowledge leads to better products, the importance of an infrastructural or facilitating approach gained ground. Dutch companies and educational institutions increasingly saw that facilitation as a task of government, and at various levels the gov-

The Dutch Reactor Center Foundation was established in 1955, at the instigation of the Ministry of Economic Affairs, with the aim of exploring the possible peaceful applications of nuclear energy in the Netherlands. Its research facility, comprising three research reactors, was set up in Petten.

ernment also addressed the issues involved, even though it did so somewhat slowly and hesitatingly, especially when compared to other countries. The various Dutch science and technology policies underscored the increasing role of knowledge intensification at the "systems level," notably in the second half of the twentieth century.

By the 1980s the contours of a genuine knowledge-intensive and technology-based society—rooted in concerted actions of government, industry, and universities—had become visible in the Netherlands.[11] Concern about the assumed innovation gap between America and Europe since the 1960s was mitigated by evidence of an increasing number of areas of innovation competition. In the mid-1980s, in an effort to diminish the American and

Japanese edge in integrated circuits ("chips"), Philips-Siemens launched a mega-chip-project with substantial support from the Dutch and German governments. The project only partially realized its objective (with input from Japan's Toshiba), because American and Japanese competitors beefed up their research efforts as well.

This example highlights not only the globalization of technological development but also the strategic importance of global competition for future technology design. In addition to being a dominant production factor, knowledge itself has become a product that can be sold and strategically deployed in all sorts of ways and contexts. At the close of this chapter we return to a discussion of these recent developments.

Using this outline of crucial twentieth-century developments, in this chapter we examine three topics in more detail. Our discussion focuses on the Dutch reaction to the increase in knowledge intensity; each of the three aspects we have selected for closer examination marks a transition in how enhanced knowledge intensity was dealt with. First, we focus on the Dutch technical education system, which initially had to respond to the new challenges but subsequently also tried to anticipate them through curriculum renewal. Second, we address the emergence of industrial research, which turned knowledge intensity into a proactive issue. This was a factor not only for industrial companies, but also in agriculture and in the medical sector; government support also played a role, for instance, in setting up agencies for various sectors. Finally, we consider various concerns expressed about the functioning of the system as a whole, which clearly revealed a new level of proactive response to knowledge intensity. The national government is the most natural focus here, but our interest is less in how it specifically acted and intervened than in how this focus allows us to expose the various reactions and responses at the systems level.

The Growing Importance of Technical Education, 1890–1930

In the nineteenth century the transfer of technical skill and knowledge usually took place on the work floor, possibly supplemented by evening classes.[12] From 1850 on, a variety of factors—specialization, scale increases, the expansion of technical education abroad, specific technical professional pursuits, and the invention of new technologies—contributed to a growing need for institutions geared to technical knowledge diffusion. Technical schools were set up for training technical managers in different sectors.[13] In examining these developments we concentrate on the emergence of schools at the intermediate level because the technical education sector was the one that saw the greatest expansion in the Netherlands in the first decades of the twentieth century. A major reason for this development was that in a growing number of commercial branches company management felt it could no longer entrust production supervision to individuals with limited technical training because the processes and machinery involved had become too complex. More advanced technical education at an intermediate level met that need, even though in several other sectors, traditional or more basic technical education remained dominant way into the twentieth century.

In the early decades of this century, one hardly ever encountered an academically trained engineer in a production unit. Engineers either served as laboratory staff or were on the executive board. This changed after World War II, when process industries and several other industrial branches had grown so complex that the skills and knowledge of midlevel technicians no longer sufficed. In September 1956, for instance, the manager of the DSM fertilizer company wrote to his company's executive officers:

> We reached the conclusion that because of rapid developments in the [ammonia] sector, the tasks of production managers ... have become too complex ... [not just because of] the higher demands put on management, coordination, and maintenance efficiency in recent years ... but mainly because of the nature of the [process], marked as it is by very high pressures accompanied by both high and low temperatures and with large quantities of poisonous and inflammable gasses. The standards with respect to knowledge of applied materials, safety and lay-out have substantially increased, and a machinist or someone with intermediate technical training does not meet them anymore.[14]

This manager proposed to replace the technician who left the company with an academically trained engineer.

Our example not only underscores the relevance of the interrelationship between the "knowledge intensity" of industrial technology and the educational level of technical staff but also suggests that this interrelation is not straightforward: the technical standards and considerations that played a role in the decision to hire an academically trained person cannot simply be linked to one specific technical discipline. Moreover, technical knowledge may fulfill various roles regarding one specific production technology. In addition to the technical knowledge needed by a production manager of an ammonia plant to supervise the production process and coordinate maintenance, more in-depth technical knowledge is called for if research labs and pilot plants are to find more profitable or efficient ammonia production processes. In addition, a certain level of technical knowledge among the plant's operational staff is crucial, even though it may be less extensive than that of its production manager. This makes it complicated to link a technology's "knowledge intensity" unequivocally to the educational level of the staff involved. For industrial sectors such as textiles, where for a long time production technologies were simply purchased abroad and maintenance was done in part by machine manufacturers and contractors, this applies even more strongly.[15]

However, there are at least two major reasons for paying attention to the development of Dutch technical education here. First, its growth may allow us to trace the broad outlines of technology's increasing knowledge intensity. Although there is no straightforward interrelation, this drawback is more than compensated for by the advantage that educational developments potentially reveal the general patterns that apply to multiple sectors and technologies. Second, the growth of Dutch technical education is not only a reflection of an increasing demand for knowledge on the part of industry and government-provided technology-based services but also is itself one of the driving forces of technology's increased knowledge intensity. Technical school teachers played a large role in theorizing and mathematizing technology. Moreover, in technical education there was a cross-fertilization between disciplines, whereby a domain's specific theoretical and practical standards exerted influence on other fields.[16]

In the next section we deal with the rise of technical education and the decline of informal knowledge transfer in the twentieth century. The sugar industry and the construction industry function as two exemplary business sectors in this respect. In both sectors technical day schools were established, but for quite different reasons. The specific need for knowledge was met in different ways, depending on the conditions in the sectors involved. Next, we address in more detail the emergence of a national regulation for technical education at the intermediate level.[17]

Schools for the Sugar Industry

In the second half of the nineteenth century, the Dutch sugar industry developed into a large-scale, mechanized process industry that made use of steam power and vacuum pans. This development took place in each of the sector's three major areas: sugar refining, sugarcane cultivation and processing, and beet sugar cultivation and processing.

Of these three, sugar refining was the oldest, yet because of the strong competition from neighboring countries, its importance for the sector as a whole rapidly declined after 1850.[18] The other two sectors in fact flourished

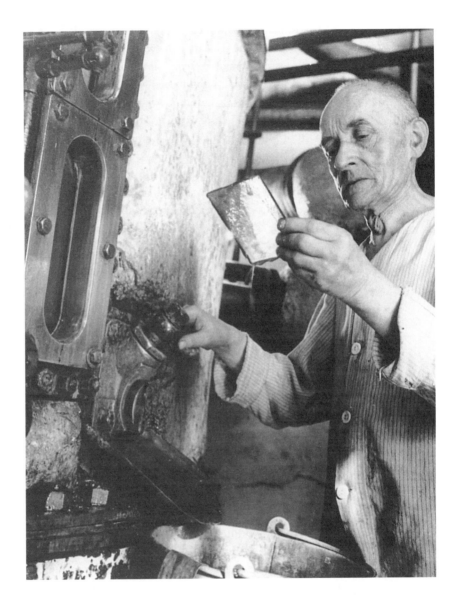

Process control was—and is—based primarily on simple tools and use of the senses. Here, a worker determines the syrupiness of the sugar solution in the boiling pan by letting the fluid flow across a glass plate. He could adjust the production process if necessary.

in the second part of the nineteenth century. Beet sugar cultivation and processing was completely new in the Netherlands. The first Dutch beet sugar factory dates back to 1859 and was founded by the Amsterdam sugar refiners, who countered the worsening conditions in the market for refined sugar by pursuing "reverse" integration, so to speak. For their factory equipment they relied just like their competitors on French and German technology.[19] This same technology was implemented in sugarcane cultivation and processing on Java (then part of the Dutch colonial empire in the East Indies).[20]

As a consequence, the Dutch and Dutch East Indian sugar producers had to find workers capable of operating the systems, which was a challenge. It also proved difficult to train workers on the work floor because of the limited experience gained with the new equipment, and there was no school for sugar technicians. This problem mainly applied to the middle management level. Those in lower positions hardly needed specific experience or education; the work largely consisted of simple actions such as crushing and pressing beets or cane. It was also fairly easy to fill a sugar house's upper management positions: owners or their relatives mostly ended up in these positions, regardless of whether they had any specific knowledge of the sugar business. However, midlevel technical skills such as sugar boiler and machinist posed a major problem. Both had to work with the new equipment. Sugar boilers monitored the functioning of the vacuum pans and machinists were responsible for servicing the equipment and operating the steam engines. They were essential to the sugar industry in the Netherlands as well as on Java.[21]

To find good sugar boilers and machinists, sugar manufacturers had three options: they could hire experienced workers from abroad, send inexperienced personnel to foreign schools, or set up their own schools. In choosing between these options, there was a substantial difference between the Dutch and the Java-based sugar producers. The former made more use of informal educational trajectories, while the latter clearly manifested more faith in formal training, the decisive factor being the distance to the centers from which the sugar technology was disseminated.

Dutch beet sugar manufacturers could hire experienced sugar boilers and machinists in Germany, France, and Belgium. Some of them stayed in the Netherlands for a long period, either working in one sugar house or moving from one factory to the next, whereas others disappeared again after several working seasons. Upon leaving a particular employer, they mostly received a testimonial.[22] By contrast, Javanese sugar manufacturers hired staff on this basis much less often because the large distance to the German and French centers came with high costs and risks to both manufacturers and prospective employees.

The second option for sugar manufacturers to get qualified personnel was to send people to foreign training schools. German schools were popular in particular, such as the *Berlin Institut für die Zuckerindustrie*, set up in 1866 by Dr. C. Schreiber. This school taught the general chemical and mechanical aspects of sugar production and also devoted attention to analysis methods for process control. Between 1883 and 1909 this institute attracted at least fifty-one Dutch students.[23] Similarly, the *Schule für die Zuckerindustrie* in Brunswick, established in 1876 by Dr. R. Frühling and Dr. J. Schulz, attracted many Dutch students. Unlike the Berlin institute, this school put less emphasis on chemical analysis and also offered classes in business management.[24] Many Dutch students from Berlin and Brunswick found employment in Javanese sugar mills, which suggests that their managers saw these schools as a reliable source of qualified personnel.

A third option for sugar producers was to set up their own schools. The first Dutch school of this type was the Training Facility for Machinists (Kweekschool voor Machinisten), set up in 1878 in Amsterdam. This school sought to train machinists for ocean-going steam vessels, sugar mills, and other large-scale facilities. After 1895, only graduates from three-year high schools were admitted to the two-year and, later on, three-year curriculum in mechanical engineering. From the start close ties existed between this training facility and the Javanese sugar industry. For instance, many sugar-cane producers and sugar refiners were among the school's founders,[25] and the curriculum also underlined its ties to the sugar industry. In 1885 a chemistry instructor was hired to bring the chemical knowledge of prospective machinists up to standard, an effort that twelve years later became a separate sugar course. Dr. J. J. Docters van Leeuwen taught chemistry, sugar chemistry, and sugar production. His sugar course soon developed into a six-month specialty, in which students could enroll after completing courses for mechanical engineering. Whether they had taken this sugar course or not, many actually gained practical experience in a sugar mill. Of all students who did an internship between 1878 and 1894, one out of seven ended up working in a Dutch beet sugar mill.[26] Nearly a third of all graduates from the Training Facility for Machinists found a job in the sugarcane industry.[27]

A second Dutch school for the sugar industry started up in 1894 in Amsterdam on the initiative of two chemists, G. Hondius Boldingh and J. K. van der Heide. This school was part of their commercial laboratory, set up one year before, where those interested could gain practical chemistry skills.[28] Initially this school offered a preparatory course of two months, participation in a internship in a sugar mill, and a core four-month course. In 1898 it also began to offer a two-month follow-up course on sugar production topics,

In 1911, a lobby of local entrepreneurs led to the establishment of the second midlevel technical school in The Netherlands, in Dordrecht. The school started with a Department of Sugar Technology, which after 1917 was renamed the Department of Chemical Technology. Chemistry courses were a central element in the curriculum of this department.

but also on chemical analysis of petroleum, gold, coal, and foods. This school, in which the Javanese sugar industry seems to have taken an interest, had an alumni association whose membership comprised a third of all graduates. More than half of them had worked in the Dutch East Indies and about 10 percent in the Dutch beet sugar industry. Several alumni rose to the ranks of manager or ended up on a sugar mill's executive board.[29]

The Intermediate Technical School (Middelbare Technische School) in Dordrecht, established with support from local industrialists in 1911, had a section called Sugar Technology. In 1917 this school added a section on Chemical Technology, which had several subjects in common with Sugar Technology. The curriculum was quite similar to that of the Training Facility for Machinists in Amsterdam, with much emphasis on the sciences and mathematics. In the first year all students had to take courses in algebra, goniometry, and trigonometry, descriptive geometry and stereometry. This was followed in the second year by mandatory classes in analytic geometry, mechanics, physics, and chemistry, and in all years there were classes in Dutch, German, and English. By 1934 the Sugar Technology section had 140 graduates, or some 20 percent of the school's total. Later on, its courses in sugar technology became part of the Chemical Technology department.[30]

Together these three examples underscore the emergence of a need for technically trained personnel arising from increasing demands of the Dutch industry. Local industrial elites and inventive academics responded to this need by setting up schools. Most graduates from these schools found employment in the sugar industry, contributing to the competitiveness of the Javanese sugarcane industry and the Dutch beet sugar industry, which were both active in international markets.

The Construction Industry and Technical Daytime Education

The educational needs of locally operating construction businesses looked quite different. Whereas in the sugar sector the establishment of schools more or less preceded the formation of a Dutch professional group of trained sugar technicians, in construction the evolving professional groups and professions themselves largely defined the nature of daytime schools.[31]

Like the sugar industry, the Dutch construction industry developed into a large-scale sector after 1850. Before, construction largely relied on craftsmen and a limited degree of division of labor. An entire construction project was mostly in the hands of a single person, the master builder. He not only made designs but was also responsible for drawings, specifications, budgeting, and supervision. He hired craftsmen for the actual building effort and also worked as foreman or overseer.[32] After the 1850s, however, the division of labor in construction increased, mainly because more and more building projects were public contracts. The aim was to save costs by having several potential contractors put in a bid on a project. Together with scale increases this led to a gradual separation of design and construction. This in turn gave rise to new professional groups, such as architects, contractors, supervisors, and draftsmen, each of whom specialized in one aspect of the overall building process.[33]

This emergence of new professional groups had major consequences for construction education. In the first half of the nineteenth century craftsmen and master builders were mainly trained on the job, and attended evening schools for additional knowledge in drawing and mathematics. Prospective craftsmen could be trained in one of many technical drawing schools while future overseers could enroll in drawing academies in Rotterdam and The Hague.[34] In addition, the Royal Academy of Fine Arts in Amsterdam provided courses in architecture (even though as a daytime facility its significance for construction was limited).[35] Only after the 1850s did daytime education become more important for this sector. Many of the newly emerging professions established their own daytime schools. Although they were not all successful, competition among them played a major role in this sector's educational reform.[36]

In pursuing the establishment of a training facility of their own, architects found that competing professional groups constituted an obstacle. Their main opponents, the civil engineers, had already been indirectly involved in construction since 1808, when the Dutch government formally assigned the management of its building efforts to the Rijkswaterstaat (the State Corps of Civil Engineers). To bring to standard the architectural knowledge of civil engineers, their school, the Royal Military Academy (Koninklijke Militaire Academie) at Breda, had offered courses in architecture ever since 1828.[37] When architects first articulated their desire to have their own daytime school, in 1841, they sought to link up with this facility. Three architects presented a petition in which they proposed to develop architecture into a full curriculum, with courses not only in mathematics, physics, mechanics, construction theory, and materials, but also in art theory and the history of architecture.[38] The government, however, decided otherwise. In 1842 the Royal Military Academy's civil engineering unit was moved to the Royal Academy of Civil Engineers (Koninklijke Academie voor Burgerlijke Ingenieurs) in Delft, but its curriculum dealt with architecture merely as a technical minor.[39]

In the ensuing years the involvement of civil engineers in civil architecture increased further. Technological developments played a key role in this shift of educational needs, notably regarding the use of new materials: after 1865 cast-iron construction became common and later on reinforced concrete was introduced.[40] Frequently, civil engineers were hired who knew these materials well from port and railroad construction. The architects, forced onto the defensive, responded by presenting themselves as a professional group that knew more about design, an issue with which engineers were assumed to be unfamiliar.[41] The contradictions came to the fore in particular when, in 1863, the Royal Academy for civil engineers was reorganized into a polytechnic. Unlike its predecessor, the new school had a separate curriculum in architecture, but its character was strictly technical. When the architects asked the minister to expand architecture courses at the Royal Academy of Fine Arts in Amsterdam into a full curriculum, the opposite happened: in 1870, following a reorganization, this academy eliminated its courses in architecture instead of expanding them.[42]

This letdown caused Dutch architects to bolster their professional image even more energetically. They regularly met in Architectura et Amicitia, an Amsterdam-based society set up in 1855, and the Union of Dutch Architects, established in 1908.[43] In this same year several of the society's members founded their own training facility for architects, named Secondary and Higher Architecture Education, with a three-year curriculum organized by leading architects such as J. Th. J. Cuypers and H. P. Berlage. Major subjects taught were art history, design, and ornamentation. Unlike the plans of 1842 and 1863, this school offered evening classes, rather than daytime education, and it was open to anyone with enough hands-on experience in construction to profit from it.[44]

The establishment of daytime schools for contractors, foremen, and draftsmen generated fewer problems. The main difference was that these groups did not compete with each other within the construction business, which allowed them to act as one party against the architects' interests. Contractors took the lead in this movement. In the nineteenth century they were held responsible for all the risks inherent in construction, which they claimed was unfair. The architects were seen as merely representing their client and thus stayed out of range, even if they were actually responsible for many problems during the construction process.[45] The main objective of the Netherlands Contractors Union (Nederlansche Aannemersbond), set up in 1895, was to do away with this disadvantage. This union's main feat was the formation in 1907 of a permanent Council of Arbitration (Raad van Arbitrage), whose brief was to mediate in conflicts between contractors, architects, and clients.[46] Another union, the General Dutch Overseers and Draftsmen Union (Algemeene Nederlandsche Opzichters en Teekenaarsbond), was set up in 1906 by overseers and draftsmen in construction who no longer accepted the power of architects and architectural bureaus.[47]

The contractors, foremen, and draftsmen joined forces in their concern for professional education. In 1907 the Dutch Contractors Union issued a report in which it concluded that contractors felt a need for senior secondary technical drawing education.[48] By increasing their drawing ability, they could improve their position and options in negotiations vis-à-vis architects. This proposal received support from the General Dutch Overseers and Draftsmen Union. In 1910 the Intermediate Technical School for construction opened in Utrecht to train contractors, overseers, and draftsmen. It admitted students with enough experience in construction and some knowledge of architectural drawing. It involved a three-year curriculum comprising subjects such as grapho-statics, surveying, and leveling, reinforced-concrete construction, architectural drawing, building materials, plans, budgeting, and lists of materials. In the first ten years of its existence 160 students graduated. Strikingly, only thirteen of them started a contracting business; many others joined government agencies, but also architectural bureaus and railroad companies, while contractors willingly used the services of these new middle managers.[49]

These various developments illustrate the complex dynamic between technology's knowledge intensity and the sector's level of education. On the one hand, the educational level went up in response to the development of new construction technologies and the introduction of new materials. On the other hand, the rivalry among professional groups in this sector helped to increase its level of training, leading in turn to the adoption of new and ever more complex technologies. As suggested earlier, it underscores the relevance of professional education's development in raising the intensity of technological knowledge.

Harmonizing Intermediate Technical Education

In the first decades of the twentieth century the number of technical daytime schools strongly increased in the Netherlands, especially at the intermediate level. Between 1911 and 1922 eight intermediate technical schools were established, including in Sneek, The Hague, and Heerlen.[50] Some of them were based on existing educational facilities and as a rule private initiative—from industrial sectors or professional groups—played a role.[51] As at the end of the nineteenth century, this ongoing development of intermediate technical education was stimulated by the demand for educated labor and by specific aspirations from professions and groups of teachers.

After 1922 the rapid expansion of the intermediate technical school system ended. Until 1940 only one new school was set up, the Intermediate Technical School for Architecture in Amsterdam. Even though, in the 1930s,

Over time, training in technical fields became firmly embedded in secondary education, especially at the advanced *Hogere Burgerschool* (high schools), first established in 1863. Most students in higher technical education were graduates of such schools. The photo, from The Hague in 1933, shows a physics lab for *Hogere Burgerschool* seniors.

the number of students hardly grew (see table 5.1), after 1922 Dutch technical education did not come to a standstill. On the contrary, it would see drastic changes, but the Ministry of Education, the Education Inspection Authority, and the school boards now took the lead from private actors. They conceived of the intermediate technical schools as a single coherent system of educational facilities, rather than as a collection of separate schools, each serving a specific professional segment.

This approach was not entirely new, for already at the end of the nineteenth century there had been a debate on the content of midlevel technical training.[52] This discussion pitted two groups against each other, "pragmatists" and "theorists." The pragmatists argued that technical education was mainly about learning informal, practical knowledge. They wanted to keep entry requirements and exit standards of the intermediate system as modest as possible. By contrast, the "theorists" felt that technical education should be mainly about formal knowledge. This group advocated raising the system's entry and exit standards as high as possible. The pragmatists were chiefly from architecture, while the theorists mainly came from mechanical engineering and chemical technology.

The kick-off for the discussion came from A. Huet, a prominent member of the Society of Civil Engineers. In 1891, as a member of a commission that had to report on the future of higher technical education, Huet published a minority paper in which he revealed himself to be a pragmatist. He considered intermediate technical education a curriculum for "overseers and foremen, for managing activities in all sorts of factories and the implementation of public works."[53] Such a curriculum needed to be more theoretical than that of the existing schools, but it should not develop into scientific treatment of issues: elementary mathematics but no infinitesimal calculus, land surveying but no geodesy, experimental physics but no mathematical physics. It should be a two-year curriculum, or at most a three-year one. Huet felt that students with enough job experience were eligible for this kind of training. After primary education until age fourteen, they ought "to work in a workshop or a public works project until age seventeen in combination with preparatory evening school courses, to be followed by a two-year or at most three-year course at a technical school, leading to a diploma as overseer or work master and, finally, a life-long career in practice."[54]

Huet's proposal touched on a sore spot with H. Enno van Gelder, an engineer who worked as a technician in a sugar mill and not long afterward became director of the Training Facility for Machinists in Amsterdam. In 1895, on the request of the Dutch Society of Mechanical Engineers and Marine Architects, he considered the issue of what intermediate technical education should be like, whereby he revealed himself to be a theorist. He disagreed in particular with Huet's claim that this level of technical education was appropriate for training overseers. Enno van Gelder argued that "bosses such as foremen in factories or at work" did not need any specific education because they came from the working class anyhow. Such training he deemed necessary, however, for technicians who stood above the foremen but below the engineers. For this group, he felt, a curriculum had to be created that was hardly different from that of the Training Facility for Machinists: a three-year curriculum for which three-year high school served as preparation.[55]

The two reports underscored the contrast between pragmatists and theorists.[56] These groups had different views of the specific positions for which intermediate technical education should train students: Should it train prospective overseers and foremen or well-educated technicians? There was also the interrelated difference of opinion on prior training: should incoming students have extended primary education with some years of practice or should they have a three-year high school diploma? Moreover, the two parties disagreed on the role of practice as part of the curriculum: should practice come before or after enrolling in the curriculum? And there was another ideological issue: in general the pragmatists viewed intermediate technical education as an option for working-class youths to rise on the social ladder, whereas theorists viewed that same education as the natural domain of the middle classes.

Quite soon this debate on intermediate technical education drew attention from politicians. In response to the debate between Huet and Enno van Gelder, in 1904 the minister of internal affairs, A. Kuyper, presented a proposal for the establishment of a general intermediate technical school at Haarlem.[57] This plan was very similar to Huet's earlier proposals, which is why Enno van Gelder in 1906 opposed Kuyper's proposal.[58] One year later Kuyper's successor, J. Rink, tried to reconcile his predecessor's view with that of Enno van Gelder by launching a plan to expand the latter's Training Facility for Machinists into a general school. This met with criticism from the pragmatists. They believed that such an advanced technical school would be difficult for lower-class students to get into.[59] In part because of these protests, the formulation of a law on intermediate technical education was delayed. It finally went through only after the adoption of the 1919 Technical Education Act. Unlike the proposals from Kuyper and Rink, the new law did not choose sides in the conflict between pragmatists and theorists. To ensure maximal flexibility of technical education, the legislator decided to omit provisions on curriculums, entry standards, and exams from the law.[60] By and large the existing situation was endorsed and for the time being schools could do as they pleased.

In the long run, however, the Technical Education Act had several specific consequences. In exchange for the larger financial contributions required by the law, national government claimed more influence on the content of curriculums. The effect was that the differences between the various intermediate technical schools became smaller after 1921, notably in the area of entry standards and curriculums. This trajectory's first step was made at a 1924 meeting with representatives from the ministry, the schools, the employers, the Royal Institute of Engineers and the organizations of midlevel technicians. At this meeting the participants soon agreed that access to intermediate technical education had to be tied to an entry exam at the level of the three-year high school. This decision was a victory for the theorists. The pragmatists' attempt to make practical experience mandatory failed as well. The biggest stumbling block for this regulation was its implementation. Candidates with theoretical knowledge but no experience would have to compensate for their poor prior training by learning the ropes in actual practice, but companies had no interest in inexperienced and untrained workers. Theoretical knowledge became mandatory for everyone, but not practical experience.[61]

The second step toward harmonizing technical education at this level came about at the National Conference for Trade Training Issues in 1931. This conference was organized to find out whether technical education in fact met the demand for educated workers. At this conference theorists and pragmatists clashed on two points: the role of higher mathematics and the fine details of practice in the curriculum. The theorists felt that technicians at the intermediate level, just like engineers, had to have knowledge of infinitesimal calculus. The pragmatists countered this with the argument that higher mathematics was not mandatory for construction engineers. The pragmatists lost out yet again: a proposal in favor of higher mathematics in intermediate technical education was adopted. Regarding practice, the groups reached a compromise. Where the theorists argued that practice had to follow training, the pragmatists held that it should precede training. This led to the adoption of a proposal saying that the exact role and timing of practice in the curriculum should depend on specific circumstances: in construction mostly before gaining theoretical knowledge, but in mechanical engineering, chemical technology, and electro-engineering generally after having acquired such knowledge.[62]

The National Conference's recommendations prompted the Ministry of Education to set up a commission to raise uniformity in midlevel technical

The technical education system paid great attention to practical skills such as technical drawing. The Amsterdam midlevel school, which opened in 1878, had four drawing halls by the mid-1920s.

training. In 1933 the Hofstede Commission, led by the then inspector-general of industrial education, G. Hofstede, presented curriculum plans to which all schools had to adhere. Basically these proposals marked a victory for the theorists. The commission emphasized, for instance, that the normal way of getting into intermediate technical schools was via three-year high school; practical experience was not seen as either necessary or useful. Moreover, the plans for all sections (including architecture) foresaw a sizable number of specific subject areas in the first year that a majority on the commission deemed necessary ("a firm basis in mechanics, mathematics, physics and chemistry is indispensable for a sound treatment of technical subjects"), and the second-year students had to spend a number of hours each week on higher mathematics. Finally, the report stipulated that practice and theory be offered side by side.[63]

The Hofstede Commission report met with substantial criticism. For example, several teachers unions felt the commission was biased against junior technical school students, for whom "it rather preferred to keep the doors of intermediate technical schools shut."[64] Such criticism made little impres-

Table 5.1: Number of students in technical daytime education

	Junior Technical Schools, Craft Schools	Extended Junior Technical Schools or Extended Craft Schools	Intermediate Technical Schools (after 1957, Technical Colleges)	Polytechnics (After 1905, Technical Universities)	Total	Evening or Part-Time Courses
1870	422		175	171	768	2,692
1875	803		350	263	1,416	6,506
1880	757		700	224	1,681	6,625
1885	718		1,050	319	2,087	6,265
1890	1,145		1,400	255	2,800	7,116
1895	2,543		1,200	420	4,163	10,050
1900	3,218		1,245	784	5,247	13,772
1905	4,902		2,434	1,123	8,459	17,327
1910	6,924		2,592	1,179	10,695	22,328
1915	10,506		3,496	1,371	15,373	25,171
1920	11,808		3,213	2,393	17,414	30,292
1925	17,791		2,511	1,675	21,977	34,273
1930	20,940		3,268	1,743	25,951	39,928
1935	29,699		3,188	1,842	34,729	33,238
1938	31,143		4,105	1,838	37,086	43,501
1945	36,722	600	7,442	4,072	48,836	31,280
1950	52,995	1,374	9,505	5,615	69,489	61,015
1955	63,693	3,636	11,003	5,062	83,394	67,985
1960	111,703	9,945	12,543	7,916	142,107	83,733
1964	129,297	16,394	15,082	9,803	170,576	103,338

(For sources see page 602)

sion, though, for soon the government announced it would adopt the recommendations. In 1938, after a brief transition period, the curriculum plans of the Hofstede Commission became mandatory for all intermediate technical schools.[65]

Proliferation of Technical Education

After World War II the growth of technical daytime education accelerated (see table 5.1). In 1.900 there were 5,247 students in Dutch technical daytime education; in 1930 this number went up to 25,951 and in 1960 to 142,107. Table 5.1 shows the strong growth of technical daytime education between 1870 and 1964. In these years the number of students increased annually over 6 percent. In 1870, 0.1 percent of all Dutch youths age twelve to twenty-five participated in technical daytime education, but by 1900 this had gone up to 0.4 percent. At this time many boys went to work after primary school. In 1930 1.3 percent of the age group mentioned went to technical school, and in 1960 the figure was 5 percent. In 1957 the intermediate technical schools were changed into technical colleges (*hogere technische scholen*, literally, higher technical schools). Even though the technical education system's large growth came after the 1930s, its basic curriculum, with an emphasis on theoretical subjects, dates from before this era, which is why in this section we especially concentrate on this episode.

The Dutch technical daytime education system saw growth at all levels, but the rate of growth was highest in junior technical schools. The growth of intermediate versus higher technical education was less even. Between 1875 and 1895, for instance, the number of students in intermediate technical education increased more rapidly than the total number of students in technical daytime education, while in this same respect higher technical education was lagging. This was tied to the demand for educated labor: employers favored technicians with midlevel training rather than engineers.[66] In its short-lived existence, the extended junior technical education system finally saw solid growth as well: between 1945 and 1964 its student population rose from 600 to 16,394.

The growth of technical daytime education was accompanied by a similar increase in part-time technical education. For the most part this education consisted of evening classes, taken by students who worked during the day. No wonder, then, that in 1870 the part-time education system exceeded daytime education (2,692 students versus 768). In the ensuing years the number of participants in part-time education continued to go up, but the pace of growth lagged behind that of daytime education. In 1900, for instance, 13,772 students enrolled in part-time training, and this figure increased to 39,928 in 1930; for daytime education the numbers were 5,247 and 25,951, respectively. Already in 1935 daytime education attracted more students than the part-time system, which, after a brief resurgence in 1938, would systematically lag in terms of student numbers. In 1960 technical daytime education attracted more than one and half times as many students than the part-time system.

Our discussion of the historical developments in Dutch midlevel technical education—as one case in the country's overall technical education history—suggests that this specific sector displayed an interesting dynamic. Initially schools were set up in response to a field's immediate practical needs, either because imported machinery or equipment put higher demands on the competence of workers, as in the case of the sugar industry, or because professional groups wanted to enhance their expertise to strengthen their position, as in construction. Once intermediate technical schools were established, however, the system partly took on a dynamic of its own, whereby the increasing level of formal education became more removed from the practical or professional needs that had provided the original impetus. The variety

Around 1950 greater demands were being made on the operating staff of chemical installations, such as here at the Servo Company, in Delden. Several chemical companies, in collaboration with the Royal Dutch Chemical Association, set up a training facility for chemical workers (*chemiciens*), which gave rise to a professional training curriculum for the processing industries, called VAPRO, a Dutch organization for training and educating process operators.

of technical training facilities gradually gave rise to what was eventually a fairly uniform type of advanced technical training that was determined in particular by the views and wishes of the "theorists." Sound education in mathematics and the natural sciences constituted the cornerstone of midlevel technical curriculums.

Relatively little is known about the effects of this fairly one-sided theoretical orientation on professional practice and technology development in general. The growth of a well-educated middle management technical staff in companies probably made it easier to appropriate foreign technologies or implement improvements. DSM, to mention one example, hired graduates from intermediate technical schools as research assistants in its research laboratories, as assistants for staff engineers in its factories, and as managers of production units. In these positions they played a role in quantitative research, measuring and regulation efforts, process automation, and other activities.

Through their training in mathematics and physics they contributed to a specific type of knowledge intensification of technology. DSM in turn exerted a strong influence on the curriculum of the intermediate technical school in nearby Heerlen, which was the second school in the Netherlands—after Dordrecht—to introduce chemical and physical process engineering (modeled on the American example), even before this subject was taught at the Technical University in Delft (Technische Hogeschool Delf, earlier the Royal Academy for Civil Engineers). In the next section we discuss in more detail the activities of academically trained engineers and natural scientists, who could increasingly be found in industrial research labs, notably after World War I.[67]

The Rise of Technoscientific Research, 1910–1950

The knowledge intensity of technology changed not only because the formation of a nationwide system of technical schools enhanced the *spread* of technical and scientific knowledge at all levels, but also, and perhaps more important, because in the first half of the twentieth century many institutions were established for the *creation* of new technical knowledge, including industrial research laboratories and design departments, university laboratories, and the Technical University in Delft. By hiring academically trained scientists and engineers and maintaining contacts with higher education facilities, Dutch companies managed to import knowledge of complex technologies and develop or invent new technologies on their own.

From Technology Import to New Technology Design

Both aspects, the acquisition of foreign technology and the design of new technologies, are nicely illustrated by the career of a physical chemist from Leiden University, G. Berkhoff. When, in 1928, DSM was planning the establishment of a nitrogen-fixing plant (*Stikstofbindingsbedrijf*, or SBB) for which it had bought technology in Italy and Belgium, the DSM leadership realized it had to recruit well-educated staff to master this complex technology. The company lacked experience in the area of cryogenics, the cooling technique for extracting the nitrogen from the air at very low temperatures. On September 20, 1928, DSM's director, Frits van Iterson wrote a letter to Professor W. Keesom, of the famous cryogenic lab at the University of Leiden, asking him whether there might be a young physicist in his lab who wanted to work for DSM's new plant. Unfortunately, Keesom replied, he had no such physicist, but he did know a young chemist who was working on his Ph.D. and had done a minor in physics, G. Berkhoff. On April 1, 1929, Berkhoff began to work for DSM as the company's first academically trained chemist. He also continued to work in Keesom's lab for some time, after which he moved to the ammonia plant of the Belgian-Italian company that supplied the nitrogen-fixing technology, so as to gain experience in production practice.[68]

Berkhoff's first feat in industrial research involved a quite prosaic production problem. When SBB began its operations in 1930, it soon turned out that the crystals of its main product, ammonium sulfate, were too small and were shaped too irregularly. The fine, powderlike salt was difficult to scatter and tended to cake, which caused the bags of fertilizer to become as hard as "gravestones" during overseas transport. In the same period several competitors, relying on another production process, began to market so-called "coarse-crystalline" ammonium sulfate, and DSM was faced with a huge problem.[69]

The company assigned Berkhof to study the crystallization of ammonium sulfate in-depth in his nitrogen-fixing lab, where he got to the bottom of the matter. His approach is a perfect example of the role fundamental scientific insights can play in solving very practical problems. Having studied with Professor Schreinemakers from Leiden, an internationally recognized expert in the area of physical chemistry, Berkhoff was well equipped to study aspects of crystallization processes. Starting from theoretical insights about crystal growth, he determined experimentally in the SBB lab that the oversaturation had to be as limited and uniform as possible to obtain large crystals. In addition, he discovered that pollutants in the sulfuric acid used had a negative influence on the crystallization process. Through a simple trick—adding small quantities of superphosphate to which the pollutants would bind—Berkhoff managed to solve this problem effectively forever. DSM patented the invention immediately.

By systematically calculating and mapping the influence of countless parameters, Berkhoff managed to control the crystal growth process within a year, and in his lab he could produce whatever crystal size he wanted. Thus, the theoretical and experimental foundation was laid for new process conditions and for reconstructing the reactors. Mechanical engineers from SBB, on the basis of on Berkhoff's specifications, designed new reactors that ensured better mixing. These improvements of existing devices were also patented by DSM. The results were astounding: the coarse-crystalline product it marketed from 1932 onward soon earned this company a great reputation.

These improvements in the production process far exceeded the direct economic interests tied to sales of ammonium sulfate. The product's sym-

bolic meaning and its influence on the future of SBB as a knowledge-intensive company were significant for three reasons. First, the improvements
demonstrated that the company, after only three years, had become a main
player on the global stage of nitrogen producers. Second, the discovery of
coarse-crystalline ammonium sulfate soon led to the sale of licenses to third
parties, which gained the company substantial earnings. Finally, the accomplishments underlined the potential value of scientific research for practical
production problems. Berkhoff subsequently proved to be one of the moving forces behind the establishment of DSM's Central Laboratory, completed
in 1940. Three years later he took charge of all its research and development,
a leadership that lasted until 1961, when he left the company after a most
fruitful career.[70]

The First Industrial Research Laboratories

The example of Berkhoff and SBB perfectly illustrates how, within a few
years, a company could change from a technology importer into a supplier
of new technology by choosing to pursue scientific laboratory research. Scientific networks played a role in such processes as well. In our description of
when and why the first industrial research labs were set up in the Netherlands, we concentrate on how these labs established close ties with professors from technical academic institutions, notably the Technical University
of Delft. Many of the products generated through these collaborative efforts
were "knowledge-intensive" in the sense that developing these products
according to accurate specifications called for cutting-edge scientific knowledge and sustained lab research. Even a seemingly simple matter like the
crystal size of a fertilizer salt took almost a year of lab research to produce a
good result.

Between about 1910 and 1950 new patterns of knowledge acquisition and
knowledge production emerged in the Netherlands that strongly deviated
from nineteenth-century practices, when knowledge acquisition often meant
bringing in knowledge from elsewhere. Industrialists imported this knowledge in several ways: they paid visits to companies abroad that were active
in the same sector, organized an internship in a foreign company, hired foreign technicians or engineers, took out a license on a foreign product or
process, or bought equipment abroad.[71] However, if Dutch manufacturers
wanted to develop an invention of their own or apply a foreign invention in
their company, they could hardly call on institutions or persons outside their
company. They had to start working on it by themselves, and many in fact did
so. For example, Meindert Honig experimented for twenty years in his starch
factory, hoping one day to produce quality corn starch. Because he failed to

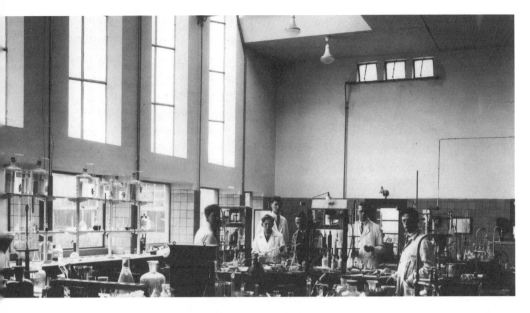

A laboratory of the Nitrogen Binding Company, part of DSM, led to the establishment of the Central Laboratory in 1940. Its research effort was significantly stepped up. Partly through the activities of this laboratory, DSM made a successful transition from mining company to chemical corporation.

get a handle on the product's quality, he had his son Klaas study chemistry and in 1887 he sent him to the Chemical State Laboratory in Hamburg. Back at the factory Klaas outdid his father and succeeded in improving the quality of starch while also developing several new applications. Likewise, other industrialists' sons and prospective industrialists increasingly opted for a technical or scientific education before starting their careers. A well-known example is J. C. van Marken, who after graduating from the Polytechnical School and taking a trip abroad established the Netherlands Yeast and Spirit Factory (Nederlandsche Gist- en Spiritusfabriek, NGSF).[72]

Scientific training would provide industrialists the advantage of potentially applicable knowledge in production processes, but it was not always enough to solve production problems. In his yeast company Van Marken ran into the problem of yeast's fluctuating quality, a problem that he could not solve, despite his chemistry background. Convinced that deeper scientific understanding could bring the solution nearer, he hired a technologist from the engineering school at Delft to do research on the growth of yeast. This too failed to solve the riddle. In the annual report of 1884 Van Marken concluded that chemistry alone could not spread "enough light in the darkness that still enshrouded the yeast business," which is why he announced the establishment of a microbiological research laboratory. He hired M. W. Beijerinck, who had a doctorate in biology, to improve products and processes. Beijerinck studied, among other things, the infection of yeast with micro-organisms in the air and assisted the factory in switching to a new production procedure.[73]

The Netherlands Yeast and Spirit Factories laboratory, established in 1885, is generally considered to be the first industrial research lab in the Netherlands. Several decades later other companies followed suit, two of the country's largest companies, Royal Dutch/Shell and Philips, leading the pack.[74] Royal Dutch/Shell had entered into a fierce competitive battle with Standard Oil in the early twentieth century. It was crucial for the Dutch company to raise the quality of its products. Accordingly, its technologist contacted a professor at Delft who put one of his assistants to work on the problem involved. The successful approach of this chemical-technologist caused the company leadership to hire him full-time in 1906 and also to set up a special lab for him in Schiedam. This was the first laboratory of Royal Dutch/Shell; its main task was research aimed at solving specific production problems (much in the manner of Beijerinck's Netherlands Yeast and Spirits lab). In the ensuing years the oil company's lab focused on an increasing number of production problems. After it was moved to Amsterdam in 1914 the number of employees quickly rose and by 1925 this lab was the largest Dutch industrial lab.[75]

After establishing the company that developed into Philips Electronics, Gerard Philips was personally involved in technological development and solving production problems. This is why for a long time Philips, unlike NGSF or Royal Dutch/Shell, had no separate laboratory organization. New technological developments, however, twice threatened Philips's competitive position and twice the company responded by establishing a special lab. The first effort followed in the wake of the 1905 invention of filament lamps, when Siemens, AEG (Allgemeine Elektrizitäts Gesellschaft), and General Electric succeeded in building a strong patent position, which posed a serious threat to Philips sales in Germany. In 1908 Gerard Philips therefore decided to set up a chemical lab to refine the art of drawing metal wire and use this knowledge to make a dent in the American-German patent cartel. Not much is known about the results of this "chemical laboratory"; probably its research allowed Philips to draw good metal wire. When in 1913 General Electric (G.E) marketed the so-called half-watt lamp, the situation was even more threatening to Philips. Its leadership realized that the company needed to build a good patent position to keep up the competition with GE and the German companies. This is why in 1914 Philips set up its Physics

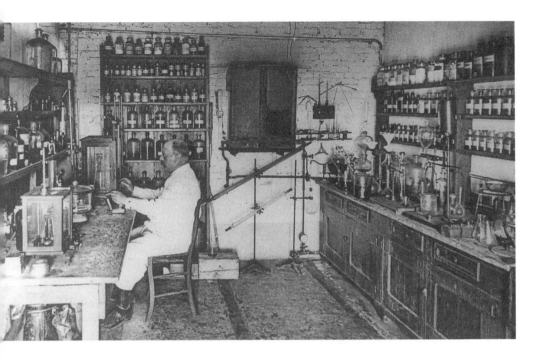

In the first half of the twentieth century the number of laboratories geared to inspection and research increased markedly. By around 1914, larger soap and detergent companies usually had their own research labs. Here, at the De Duif soap factory in Den Dolder, there was a laboratory equipped to carry out product and production control.

Laboratory (Natuurkundig Laboratorium, or NatLab for short). In the first years it mainly concentrated on improving light bulbs and studying the related physical phenomena. Later on, research efforts also concentrated on other kinds of lighting and radio systems.[76]

In the case of smaller Dutch companies, international competition was less of a factor in establishing research departments. For example, Noury & Van der Lande, a flour and oil company in Deventer, set up a lab in 1916. Earlier the company had asked an independent researcher to do a chemical study of the bleaching of flour and paint oil, and decided to hire him after his first positive results. In the first years he and his staff mainly concentrated on improving flour bleaching, developing a production process for a bleaching agent and patenting other bleaching methods, so as to hamstring the competition. In the 1930s, when the company's profits dropped, the management called on the lab to search for new products that might allow the company to enter more profitable markets.[77]

Between 1910 and 1930 Dutch companies also set up research labs in sectors where innovations came from the design table rather than the lab. In 1914, the electro-technical company Heemaf in Hengelo hired an electro-technical engineer to develop new devices. Here, the introduction of a Patent Act in 1912 was a major factor. Between 1869 and 1912 there had been no patent law in the Netherlands, so a company could freely use technologies invented by others. Like many companies in the electro-technical industry, Heemaf usually copied foreign products, but after 1912 this was no longer a valid option, because foreign companies registered their patents in the Netherlands. Even with its small research department the company managed to develop a new type of engine, which became a worldwide success.[78]

The exact number of Dutch industrial research labs established in the first half of the twentieth century is hard to determine. There is a dearth of historical sources, and it is not always easy to distinguish between a research lab and an inspection lab or between a design department and a research department. Still, it is clear that the number of labs markedly increased. In 1900 there was only one, the NGSF lab, but around 1940 one could find companies in each sector that structurally generated scientific and technological knowledge to improve or renew products and processes.[79]

Well-Educated Industrial Employees

The research departments that were set up in the first decades of the twentieth century were chiefly staffed by engineers from the Technical University in Delft (TU Delft) and by academically trained scientists. This caused a sharp rise in the number of those with a higher-education degree who worked in industry. In 1879 only 6.7 percent of the graduates of Delft were industrial engineers, but in 1900 their share had gone up to 29.7 percent, to reach 37.9 percent by 1917. A similar increase can be seen with respect to the role of scientists in industry. Around 1900 there were only a few chemists and pharmacists who worked for industrial companies, but by 1930 11.6 percent of all academics in salaried positions had a job in industry, agriculture, commerce, or banking. The largest contingent of academics in salaried positions could be found among scientists: 26.7 percent of them worked in the sectors mentioned.[80]

The growth of the number of well-educated employees in Dutch industry during the first half of the twentieth century is depicted in graph 5.2. A distinction is made between chemists, physicists, pharmacists, and biologists (all university graduates) and chemical, electro-technical, mechanical, and physical engineers (all graduates from a technical university or its forerunner). The graph shows that in 1900 there were about 100 college gradu-

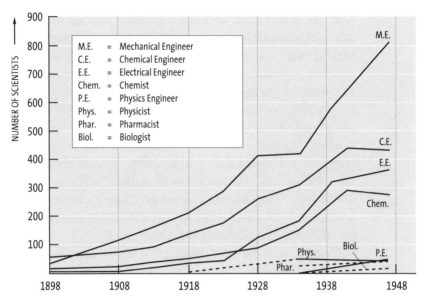

Graph 5-2. Number of scientists in Dutch industry
(For sources see page 603)

ates active in the Dutch industry. In 1915 their number had gone up to about 350 and by 1940 to about 1,800. Most growth, then, took place in the inter-war period.[81]

In engineering, mechanical and chemical engineers in particular found their way to industry, while in the sciences it was especially chemists who found industrial employment. For mechanical engineers this situation can be explained from the fact that their profession had a practical origin anyway. Before 1890 Dutch mechanical engineering had developed in the context of the machine industry. The rise of mechanical engineers after this date was for the most part caused by the replacement of practically trained mechanical engineers by theoretically trained engineers.[82]

After 1880 university-trained chemists were attractive to the industry because chemistry as a discipline developed rapidly. The emergence of new fields such as physical chemistry and organic chemistry held great promise for the further expansion of chemical industries. The fact that after 1900 large numbers of chemists found employment in the industry can be explained by their substantial theoretical qualifications.[83]

Chemical engineers, finally, occupied an intermediary position. In the nineteenth century many already had ties to industry. Furthermore, around

1900 the training for chemical engineers became more scientific. This improved scientific basis in combination with their image of being practical-minded made chemical engineers even more valuable to the industry.[84]

The engineers and scientists who populated the new research departments were trained to do scientific research. In the second half of the nineteenth century, college and university students were increasingly trained as researchers, and research experience in a university lab became integrated into the curriculum. The research methods taught to students were more and more geared toward solving industrial problems. The training for engineers, too, was quite scientific. Students at the Technical University in Delft first had to take many courses in mathematics and sciences before actually receiving design training. Consequently they were inclined to approach problems from a mathematical, theoretical perspective.[85]

The scientific relevance of industrial labs is suggested in part by the fact that industrial research might serve as groundwork for an academic title or career. It is telling, for example, that several researchers from the lab of Noury & van der Lande earned their doctorates on the basis of their company research.[86] Several researchers also earned their doctorates while working at Philips, and between 1929 and 1946 nine of its NatLab staff members became professors.[87] Whether it was possible to earn a Ph.D. on the basis of industrial research depended not only on the nature of the research but also on the extent to which a company allowed lab data to be made public. This probably indicates that many more researchers could have earned their Ph.D.s for industrial research according to the prevailing academic standards than actually did so. Still, that some of them managed to do so underscores the scientific nature of industrial research and the validity of industrial problems for science.

The nature of research in process industries and product industries differed substantially. In the latter, notably in the machine and equipment industry, research tended to be closely tied to design, thus retaining a crafts-man-like character. Aside from calculations and scientific theory, designers started from their visual memory of other designs, from basic rules, and from practical examples of what worked, whether or not it could be scientifically explained. Yet scientific knowledge did begin to play an ever larger role in designs, partly as a result of the curriculum at the TU Delft.[88]

Industrial-Academic Networks

Industrial researchers often stayed in touch with their former university teachers. In the early lab years, researchers with Royal Dutch/Shell regularly asked their professor at the TU Delft, S. Hoogewerff, for advice, as was true of researchers with Noury & van der Lande, who had been trained by

The Physical Laboratory of Philips maintained close ties with the academic world. For several years its director, the physicist Dr. G. Holst (left) held a special chair at the University of Leiden. In this photo from 1935, a member of his staff, Dr. B. van der Pol (who also held a special chair at Delft), explains the operation of a radio tube.

Professor J. Böeseken. Sometimes such contacts became more formal, as in the case of H. R. Kruyt, a professor of physical chemistry at Utrecht, who in the 1930s organized a monthly colloquium for all his former students. The NatLab of Philips had formal ties with the professor in experimental physics at Utrecht, L. S. Ornstein. He offered advice to NatLab staff, and, more important, he also supplied this lab with new researchers. Of Ornstein's ninety-four Ph.D. graduates between 1920 and 1941, eighteen found employment with Philips.[89]

Like Philips, many other companies recruited their researchers from specific training facilities. For example, NGSF had a preference for chemical engineers from Delft, where the company founder, Van Marken, the factory manager, F. G. Waller, and the microbiologist, Beijerinck, had studied chemical technology. In 1885 Waller succeeded Van Marken as director, to make room in 1923 for his son, who was also a chemical engineer. A similar situation applied to Royal Dutch/Shell and Dordtsche Petroleum Maatschappij. More than half of the forty-two college graduates hired by Royal Dutch/Shell between 1890 and 1914 had studied chemical engineering at Delft, and as many as 75 percent of the research staff hired by Dordtsche came from this department.[90]

The example of Ornstein's ties with Philips is hardly an isolated case. Many Dutch companies initiated long-term relationships with specific professors or academic labs. In other ways, too, contacts were established between companies and technical colleges or universities, contributing to the emergence of a network of relationships that in the course of the century grew denser and ever more complex. Companies hired professors as board members or outsourced research to universities. A growing number of companies appointed professors as advisers, many of whom had supervised doctoral projects of students who were already on the laboratory staff of the same company. However, companies that had no research department of their own might also ask professors for particular advice; for instance, the Glue and Gelatin Factory in Delft asked Kruyt to serve as its adviser.[91]

In the 1930s the Amsterdam lab of Royal Dutch/Shell hired at least four chemical professors as advisers. Two professors from Delft, F. E. C. Scheffer and H. I. Waterman, offered advice to the lab on physical chemistry and chemical technology, respectively, and two professors from Amsterdam, J. P. Wibaut and J. Westerdijk, on organic chemistry and biocides. Smaller companies also appointed professors as advisers or board members. In addition to Professor Böeseken, W. Reinders, a professor at Delft, was involved in the lab of Noury & Van der Lande, where some of his students were employed. Kruyt not only had direct ties with the Delft Glue and Gelatin Factory but also with Albatros Superphosphate Factories and the Netherlands Linoleum Factory.[92]

Many professor-advisers were approached by companies' leadership or their research staff. Some professors were so open to collaboration with industry that mutual contacts were established more or less automatically. A case in point is A. M. Sprenger, a professor of horticulture at Wageningen who succeeded in getting many business sectors interested in his research. In the late 1920s he began studying the deep-freezing of vegetables and fruit, for which he received a grant of 9,000 guilders from the Dutch Association of Refrigeration Technology (Nederlandsche Vereeniging voor Koeltechniek) and Zeeland's Proeftuin. Next, Sprenger became involved in a NatLab study on the influence of electric light on hothouse plants. Fruit-processing companies provided funding for his research of sweet most, a nonalcoholic fruit juice preserved exclusively through filtration. In 1935 he also succeeded in persuading the Amsterdam Superphosphate Factory to fund his research on the influence of fertilizer on fruit trees.[93]

Another scientist who showed entrepreneurial initiative during the interwar period was the physicist A .M. J. F. Michels. He was much interested in measurements at high pressure, a topic that not only was crucial for thermodynamics but also had major industrial applications. In 1925, with support from leading physicists such as H. Kamerlingh Onnes and P. Zeeman, Michels managed to attract nearly 50,000 guilders from, among others, Stork and Werkspoor. The equipment that could be built or purchased from this money soon turned the Amsterdam lab into a recognized center of high-pressure measurements. It received constant requests for advice and help, also from abroad. For instance, the English chemist R. O. Gibson visited Michels's lab in 1925. Via Gibson, Michels came into contact with the English company Imperial Chemical Industries (ICI), which invested heavily in the Amsterdam lab. From 1928 on, each year Michels received 2,000 (or about 24,000 guilders) to buy instruments and train company personnel. ICI certainly benefited from this collaboration: in 1933 Gibson, who meanwhile worked for ICI, discovered polyethylene with help from Michels; after 1939 this plastic became one of the company's major profit makers. The war disrupted Michels's further contact with ICI and in 1942 he became scientific adviser to DSM, where he contributed to polyethylene production after the war.[94]

Professors had several reasons for establishing close ties with companies. For one thing, being an adviser or board member was lucrative; it could substantially raise their income or, as in the case of Sprenger, contribute to the expansion of their institutes. For other professors, such as Kruyt, dissatisfaction with their role in the ivory tower of science was a major factor; they wanted to contribute to society from a general sense of social commitment. A final reason for actively developing contacts with industry had to do with the possibility of securing jobs for students. Given Ornstein's many contacts with companies, it is no coincidence that over 40 percent of his Ph.D. students continued their careers in industry. Similarly, Waterman managed to arrange employment for several of his students in the lab of Royal Dutch/ Shell.[95]

The density of the Dutch industry and higher education network was further enhanced by the fact that many companies had research done in university labs or at Delft. Initiatives in this respect did not come exclusively from the side of industry, but also from other parties, such as the Industrial Association (Maatschappij van Nijverheid). In 1912 it argued for appointing "industrial fellows" at universities, an idea derived from "The Chemistry of

Commerce," a study by the American chemist R. K. Duncan. These fellows were supposed to research specific technical problems of companies with the aim of solving them. All research costs would have to be paid for by the companies involved and in exchange the universities would promise to keep the research results secret for three years, giving the companies a chance to exploit them. The Industrial Association mediated in placing industrial fellows, but despite repeated calls, few companies were interested in the project.[96]

This motivated some academics to pursue a more active approach. For instance, W. Reinders, professor of inorganic and physical chemistry at the TU Delft, approached Philips in 1913 to ask whether the firm was interested in using the services of an industrial fellow, for which he already had in mind one his students, the chemical engineer L. Hamburger. Instead, Philips *hired* Hamburger, so as to add him to the staff of its chemical laboratory, which he agreed to. This did not end Hamburger's dealings with Delft, though: in 1917 he earned his doctorate with Reinders on the basis of a study that was linked to his work for Philips. Later on, Hamburger proved a fierce advocate for the establishment of state laboratories that perform research for industry.[97]

Although initially the experiment with industrial fellows was hardly successful, this changed in the 1920s, when research and development efforts expanded and became more comprehensive. In 1928, for example, the firm of photocopying materials of Van der Grinten paid for research by a young chemist, G. Elsen, who joined the laboratory of Professor Böeseken at Delft. Elsen conducted research on the possibility of producing a specific colorant. His contract ended in 1930, but in 1932 Van der Grinten hired the services of another chemist who joined Böeseken's lab. A similar solution was chosen by NGSF, which in 1933 paid for two researchers on the staff of Albert Kluyver, professor of microbiology at Delft. Between October 1933 and February 1935 the two worked on problems related to waste fluids, writing biweekly progress reports. In December 1936 the agreement between NGSF and Kluyver was renewed, which made it possible for another researcher to join the project.[98]

In the 1920s, several research organizations were set up to coordinate the industrial research at Delft and the various universities. One of the largest organizations was the "electro-technical industry fund for the enhancement of natural-scientific research," an initiative of the Dutch Society of Directors of Electricity Companies (Vereeniging van Directeuren van Electriciteitsbedrijven, VDEN). Starting with its establishment in 1913, the association had been concerned with technical research, such as into the quality of electrical devices and power lines. In 1925 this research was allocated to

the Inspectorate for Electro-Technical Materials (Keuring van Elektrotechnische Materialen Arnhem, KEMA) in Arnhem. But not all the research the
association initiated was done by its organization. Since 1920 there had been
contacts with Professor H. S. Hallo at Delft, who led the research effort of
several "company engineers" who worked on calibrating electricity meters.
Later on, contacts were established with the group around the physicist L. S.
Ornstein at Utrecht, which carried out studies of the quality of transformer
oil, focusing on its insulating effect, aging, and dielectric losses. This effort
was based on close collaboration with Royal Dutch/Shell.[99]

The collaboration with Ornstein's institute proved so successful that in
1927 the chairman of the Association of Electricity Company Directors, J. G.
Bellaar Spruyt, took the initiative to set up the electro-technical industry
fund. The financial basis was laid by companies such as Dutch Railroads,
Hollandsche Draad- en Kabelfabriek, Stork, Heemaf, Royal Dutch/Shell,
and Philips. In the first ten years of the fund's existence, the total input of
these companies amounted to 65,171 guilders, which was used for research
bridging technological and scientific practice. In 1930 it was decided to
broaden the reach of the fund to include all Dutch industry and its name
was changed to Fund for Natural-Scientific Research for the Furthering of
Dutch Industry (Fonds voor Natuurwetenschappelijk Onderzoek ter
Bevordering van de Nederlandse Industrie).[100]

Another research organization set up in the 1920s was the Heat Research
Foundation (Warmtestichting, 1927). It was an offshoot of the Technical-
Economic Society (Technisch-Economisch Genootschap, TEG). In the early
twenties the Technical-Economic Society approached Ornstein and asked
him whether he was interested in heat-technological research. Ornstein rather
liked the idea and set aside one of the rooms of his lab for this research. The
heat technology research conducted at Ornstein's lab was coordinated by the
Heat Research Foundation. It involved, among other things, the research
into the insulating effect of building materials and constructions, the buildup
of heat in walls, the insulation of pipelines, and the heat emission of pipelines
and radiators.[101]

Higher Education for Industry

As industrial companies began to carry out more research and needed more
researchers to secure their competitive position and sustain diversification,
they also had greater stakes in the quality of higher education. Several large
companies, notably Philips and Royal Dutch/Shell, actively engaged in shaping the content of curriculums and study facilities at Delft and other Dutch
universities.

Developing their own research equipment is an essential activity of many laboratories. The glass and instrument shop was an important component of the Central Laboratory of Dutch State Mines, shown here in 1952.

A clear example involves the foundational history of the technical physics curriculum at Delft. In 1922 the university's General Sciences section proposed setting up a new curriculum called "technical physics." Graduates would find employment in public works agencies, hospitals, and industry. With the exception of the Mechanical Engineering Department, none of the departments at Delft welcomed this plan, however. The Mining Engineering and Chemical Technology departments were actually very negative, fearing for the quality of the first-year physics courses for the other students, as well as for competition on the job market. Philips, by contrast, was extremely enthusiastic. In a letter of November 14, 1924, the company informed the Delft administrators that it welcomed the new study program because of the developments in radio and X-rays. To underline its support, two months later it invited the administrators of the Technical University to the NatLab. Two years later the Delft initiative also received support from Royal Dutch/Shell, which was hoping the new curriculum would be launched soon. This convinced the Dutch government, and despite resistance from several universities, the new program started up in 1929. Two of its three professors came from the Philips NatLab.[102]

In the first ten years of its existence, the new curriculum produced seventy-nine physical engineers. A substantial number of them ended up working with companies that had actively supported the Delft initiative: nineteen joined Philips and eleven were hired by Royal Dutch/Shell. In this period the technological-physical research at the Technical University saw enormous growth as well. Upon the establishment of the new curriculum, Philips and Royal Dutch/Shell had offered to pay the costs for a new laboratory for technical physics, which began operations in 1930. It was primarily intended for teaching, but it also served as a facility for doing research for third parties. To give a sampling of its research activities, they pertained to heat conduction of building materials, acoustic insulation measurements, inspection standards for yellow automobile lights, gauging anemometers and pyrometers, and repair of oscillographs and transmitting lamps for Post, Telegraphy, and Telephony. The lab also allowed third parties to do their own projects, with support and advice from its staff. For instance, the AVRO (Algemene Vereniging Radio Omroep) performed an acoustic study, the Hollandsche Draad- en Kabelfabriek studied heat loss of television cables, and Royal Dutch/Shell studied the behavior of catalysts.[103]

After World War II the oil company continued to support the technical physics and other scientific curriculums. In 1946 it donated 1,000,000 guilders for the construction and design of a pilot plant for chemical engineering research at Delft, in which students could be familiarized, both theoretically and in practice, with physical processes and operations of the petroleum industry and the chemical industries. Obviously, Shell had a need for students with such knowledge. The chemical technology faculty received a similar donation and the University of Leiden also got 1,000,000 guilders for expanding its physics lab.[104]

After this investment in facilities, Philips, Royal Dutch/Shell and AKU (Algemene Kunstzijde Unie—synthetic fibers) also supplied staff to universities and colleges. From 1946 Philips funded a large number of special chairs at universities. Many of those who held these positions were also on the staff of its NatLab. Thus Philips could influence the research of specific university departments, track talented students, and build formal and informal ties with the academic community. Similar considerations may have been at the basis of appointments of AKU employees to major positions in the knowledge infrastructure pertaining to the textiles industry. In the 1950s one of its employees became professor in mechanical engineering at Delft, with special attention for textiles technologies, while another employee became director of a partly government-funded foundation that mapped new developments in textiles technology.[105]

Scientific Research in Companies

In the first decades of the twentieth century, Dutch companies increasingly replaced subjective and random approaches to knowledge acquisition, often tied to the abilities of specific individuals who happened to be on the scene, with more structural approaches. Knowledge acquisition became an integral part of business practices through the establishment of research departments. The reasons for this change in the pattern of knowledge acquisition differed from one company to the next. Some companies were interested in solving production problems or securing their international competitive position, but in others issues related to innovation and diversification or the introduction of the 1912 Patent Act played a major role.

More than previously, the technologies developed in the research departments were based on scientific knowledge or obtained through scientific methods. Having a research staff with sound training in science commonly led to novel approaches to tackling specific problems. This effect became even more pronounced where professors were hired as advisers or part of the research was contracted out to university laboratories.

The institutionalization of research and the advisory role of professors gave rise to ever closer contacts between industry and higher education. This affected the character of the knowledge used in both industry and higher education. By providing financial, political, and employment support to colleges and universities, companies ensured that the scientific knowledge gained by students matched the knowledge needed by the industry.

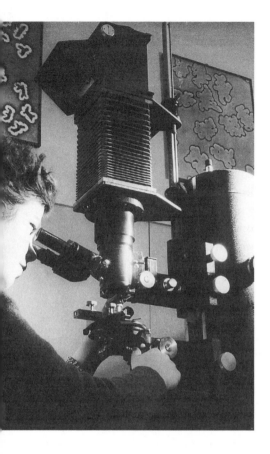

The synthetic fiber industry developed into one of the most successful industries in the postwar Netherlands, partly as a result of extensive research departments and pilot plants where industrial practices were tested on a small scale. The largest company, AKU, based in Arnhem, set up a new synthetic fiber pilot plant between 1948 and 1950. Here an employee studies a cross-section of the fibers with the help of a phase-difference photomicroscope.

The Emergence of a Science and Technology Policy, 1930–1970

The previous sections of this chapter have examined how Dutch companies and other actors variously responded to external threats and challenges. In the nineteenth century, schools were set up to meet the need for educated personnel in sectors such as the sugar industry, or because specific groups of professionals wanted to reinforce their position vis-à-vis other professions. The resulting technical education system subsequently took on a dynamic of its own, which in turn inspired further educational improvements.

In the twentieth century, partly parallel to the developments already described, specific new challenges—production problems, international competition, and patent legislation—caused companies to embrace a more proactive approach to knowledge acquisition by setting up their own research departments. In addition to institutions for spreading technological knowledge, institutions for creating technological knowledge came into being.

The scientific-technological laboratories gradually evolved into a more autonomous system as well. Particularly after World War II the industrial research labs of large Dutch companies concentrated more and more on basic scientific research that was unrelated to direct production problems.[106]

Meanwhile various sectors and the Dutch economy at large also faced challenges that exceeded the competencies or possibilities of individual actors. A need developed for mutual coordination and orchestration of the various initiatives and for institutional renewal at the systems level. This provided space for government in particular to take on new roles. The sector in which this occurred first was agriculture.[107]

Experiment Stations and Information Services

In the final quarter of the nineteenth century, Dutch agriculture was faced with declining prices for its products, which caused a substantial decrease in income for a significant portion of the population. The government responded by supporting agriculture technologically and organizationally. In 1890 it set up an information service and three experiment stations, which performed soil analyses, seed experiments, and research on issues such as soil fertility. The agricultural experiment stations were the first Dutch facilities that engaged in government-funded scientific research for the benefit of groups outside government.[108]

The organization of agricultural experiment stations—geared to research, inspection, and disseminating information—soon became a model for other sectors in society. This is best reflected in the founding of the State Industrial Agency (Rijksnijverheidsdienst). In the first decade of the twentieth century the Shopkeepers' Association (Middenstandsbond) and the Industrial Association (Maatschappij van Nijverheid) both pleaded for "a state agency ... to help small industries to develop, more or less in the same spirit as done already in agriculture." In 1910 the government responded by appointing a state industrial information officer and establishing the State Industrial Agency, which gave technological advice to small and medium-size industry. There was so much demand for such advice that after three years two more officials were appointed. At the same time, the State Industrial Laboratory was set up in Delft; its objective was "to gather technical data and address technical questions in order to support the state industrial consultants" that visited the companies.[109]

The creation of the State Industrial Agency was followed by the establishment of a large number of agencies linked to specific sectors. For instance, the State Leather Agency (Rijkslederdienst), an experiment station and information service for the leather industry, was set up in 1911, followed in 1913 by

Agricultural experiment stations, three of which were founded in 1890, were the first research facilities funded by the Dutch government. Well into the twentieth century these institutions continued to perform diverse research on agricultural products, such as a study of thistle suppression by the Plant Disease Service. The field on the left has not been treated with a certain herbicide, whereas the treated field on the right is free of thistles.

the State Rubber Agency (Rijksrubberdienst), for the rubber trade and industry, the State Fiber Agency (Rijksvezeldienst in 1919 and the State Peat Experiment Station (Rijksturfproefstation) in 1925. Most of these agencies had their origins in private initiatives, by either individuals or manufacturers associations.[110]

These various agencies took quite different approaches. The State Leather Agency, associated with the State School for Tanners and Shoemakers in Waalwijk, mainly concentrated on inspection and analysis of raw and auxiliary materials on demand. It did very little research aimed at finding new or improved products and processes.[111] By contrast, the State Rubber Agency, founded at the initiative of Professor G. van Iterson Jr. at the TU Delft, engaged in large-scale research of, among other things, vulcanization methods and latex applications.[112] The difference between the two agencies can be largely attributed to a difference in institutional embedding. The State Leather Agency was linked to a technical school, where research was not a tradition. The State Rubber Agency, by contrast, was housed in Van Iterson's lab at the TU Delft. Van Iterson was outspoken on the importance of science for commerce and industry. He felt that companies and countries could gain a competitive edge by basing their technology on scientific insights. The State Rubber Agency put these views into practice.[113]

The founding of the various state agencies shows that in this period the Ministry of Trade and Industry (Ministerie van Handel en Nijverheid), which oversaw this effort, saw research, inspection, and technical information as a way to promote fair trade and industrial development, which was a task of government. Unlike the case in agriculture, however, there was no umbrella organization for industry in general. Each agency was a separate unit and could develop on its own.

Organized Applied Research

The shortages and privations of World War I were the initial impetus for the creation of an umbrella organization to coordinate technical and scientific research for industry and other sectors of the economy. Faced with scarcity in countless areas, members of the Royal Netherlands Academy of Sciences (Koninklijke Nederlandse Akademie van Wetenschappen, KNAW) created an organization geared to ensuring "that all scientific competence and experience the country has at its disposal should search for means and ways to extract maximal use from scarce available raw materials and means of production." In February 1918 this resulted in the creation of the Scientific Commission for Advice and Research in the Interest of Public Welfare and Resilience, also named Lorentz Commission, after its chair, Antoon Lorentz, professor of physics at Leiden. It received ample funding from government to do research. Various subcommittees formulated proposals for solving specific problems, such as communications with the East Indies (the subcommittee decided that a radio communication was possible) and the shortage of manure (the subcommittee suggested using the contents of public toilets). Because World War I ended sooner than foreseen, the government eventually implemented none of the proposals in fact. Having mixed feelings about its limited influence, the commission discontinued its activities in 1919.[114]

But the Royal Netherlands Academy of Sciences remained convinced that there was a real need in the Netherlands for an organization for scientific research in the service of actual practice, partly because of World War I experiences and presumably also because of similar developments abroad, such as the founding of the Kaiser Wilhelm Gesellschaft in Germany and the Department of Scientific and Industrial Research in Great Britain, to promote the sciences by means of research. What, exactly, such organization should look like became the subject of a debate that saw the gradual emergence of two parties pitted against each other: academic scientists versus industrialists and the steadily growing number of scientists employed by industry.[115] C. J. van Nieuwenburg, a chemical engineer who worked for Glasfabriek Leerdam, clearly represented the second group. He drew a sharp distinction between "scientific-technological work" and "pure science," the former aiming to develop specific products or processes that might generate profits for a company whereas the latter pursued the discovery of general laws. Given this difference, academic scientists were not suitable for scientific-technological work. Van Nieuwenburg blamed the limited influence of the Lorentz Commission on the fact that it had mainly consisted of academic scientists. A new organization, to be established, should be led by engineers and scientists with knowledge of the "commercial side of companies." Van Nieuwenburg felt that such an organization should focus on contract research for the industry and research of subjects of general importance, such as minerals and alternative energy sources. In industrial circles his plea met with strong support, for instance, from the Industrial Association.[116] However, academic scientists, united in the KNAW, held very different views on these issues. If their standpoints were clearly articulated only later on in the discussion, the Lorentz Commission had already shown that its members believed their competence to be exactly in what Van Nieuwenburg called scientific-technological work.

Because of the specific wishes that circulated within the Academy of Sciences and the powerful intervention by Van Nieuwenburg, the government decided to ask for advice from I. P. de Vooys, a professor at the TU Delft and an adviser to several companies. He proposed not to establish new institutions, but, in part given the government's financial situation, to organize the scientific-technological work more effectively by merging the separate industrial agencies that had been created into one organization, on the model of the agricultural experiment stations that had been restructured in 1915. This was the only way that "powerful expansion and strengthening" would be possible in the long run.[117]

In 1923 the advice from De Vooys was considered again when the head of the Ministry for Agriculture and Internal Affairs together with his colleague from the Ministry of Education, Arts, and Sciences established a commission "to study by which measures and in which form applied natural-scientific research in this country can be made to function more in the public interest."[118] This commission was directed by F. A. F. C. Went, a professor of biology at Utrecht University and the chairman of the Academy of Sciences. Strikingly, the academy had urged a change from the term "scientific-technological work," used by Van Nieuwenburg and De Vooys, to "applied scientific research" (*toegepast natuurwetenschappelijk onderzoek*). This was more than just an editorial adjustment. The term "applied scientific research" implies that there is something to be applied. The Went Commission's report made it crystal-clear that "applied scientific research" presupposed "pure scientific research." To improve "overall prosperity," "resilience" and the "public interest," academy scientists felt that both applied research and pure research had to be promoted. This suggests the beginning of a systems approach that after World War II became even more influential: to secure a basic level of prosperity and competitiveness, science would have to be stimulated at all levels. Obviously, this view departed from the "guided" stimulation of scientific-technological work as advocated by Van Nieuwenburg.[119]

According to the Went Commission, government should foster research and stated, "One of the major means for raising prosperity is encouraging and systematically guiding scientific research." Researchers, however, should be able to work in full freedom, rather than being controlled, as occurred in industry. The only research guidance the commission viewed as acceptable was that researchers should not lose sight of the objectives pursued through their continuous contact with actual practice.[120]

The Went Commission proposed an organizational model in which all existing government institutes in the area of applied scientific research would combine flexibility with researchers' freedom. Government subsidized this organization, but scientists, engineers, and representatives of industry, trade, and agriculture would manage it. Following the commission's recommendations, in 1932 the Central Organization for Applied Scientific Research (Centrale Organisatie voor Toegepast Natuurwetenschappelijk Onderzoek, or TNO) was founded.[121]

In the event, the Central Organization's beginnings were anything but auspicious. The industrial agencies initially did not join the new organization. The Ministry of Trade and Industry had taken no responsibility for the Went Commission; and continued to set up its own experiment stations and information services, which operated outside the Central Organization's framework and could therefore be controlled directly by the ministry .

Divergent Views on the Social Role of Science

The Dutch Ministry of Trade and Industry approached the scientific research for which it was responsible from the perspective of direct utility: science should be deployed to strengthen industry and guarantee fair trade. In the 1930s this was emphasized further when the ministry actively began to promote industrialization. Although this policy had been initiated earlier, the ministry's activist approach intensified when H. C. J. H. Gelissen became head of the ministry in 1935. At that time, Gelissen, a chemist with some years of experience in the chemical industry, was director of the Provincial Limburg Electricity Company, as well as of the Economic and Technological Institute Limburg (Economisch Technologisch Instituut Limburg, ETIL). Gelissen continued the Economic and Technological Institute's approach when he became a member of the national cabinet.[122]

The Limburg institute deployed science and technology for an active industrialization politics. Founded in 1931 by the province of Limburg, on Gelissen's initiative, this organization was a good example of the orchestration of economic and technological activities at the systems level. In response to rising unemployment and loss of prosperity in the 1930s, the Economic and Technical Institute wanted to pursue an "active regional welfare policy" in order to "provide to as many people as possible a living that was as decent as possible." This called for the establishment of new industries and these industries required technological and economic knowledge. ETIL supplied this knowledge. Its economic department studied conditions for setting up business and cost-price calculations, while its technological department provided information.[123]

Not only did organizations such as the Economic and Technological Institute Limburg contribute to transforming the diffusion and creation of new technologies into more knowledge-intensive processes, but as a result of the new focus, policies regarding education and research became based more on well-considered processes grounded in social, scientific, and economic knowledge. Not long after the establishment of ETIL, other provinces followed suit, setting up similar Institutes.

As minister, Gelissen further pursued the ETIL focus and relied on science and technology to promote industrialization. For instance, he had lists made of products that were not produced in the Netherlands, or only on a marginal scale. Based on these lists, the provincial economic-technical institutes and the new Central Economic-Technological Institute (Centraal Economisch-Technologisch Instituut, not to be confused with the Central Organization for Applied Scientific Research, set up in 1935, could develop industrialization plans.[124] In its "Nota inzake de industrialisatie" (industrialization memo), the Trade and Industry ministry directed that Dutch industry should focus on producing quality products for export. In textiles there was room for new companies, but also in machine building and the chemical sector, which were knowledge-intensive sectors that would not survive without scientific and technological support.[125]

The Ministry of Education, Arts, and Science (Onderwijs, Kunsten en Wetenschappen, OK&W) approached scientific research from an angle that was much more in line with the Went Commission's. Its main assumption was that the enhancement of science automatically leads to increased prosperity. Meeting this objective required an organization that guaranteed scientific freedom. Research would be most useful if conducted in full freedom; direct guidance would actually be at the expense of the specific value of scientific research for actual practices.

In part because of the different ministries' approaches, TNO was off to a difficult start. According to the original plan, this organization, which fell under the Ministry of Education, Arts, and Science, would take over the existing experiment stations and state agencies, but the Ministries of Agriculture and of Trade and Industry opposed this. The main obstacle was their loss of authority over these institutes.[126]

An unusual circumstance made it possible to transfer the state agencies to TNO after all, namely World War II. Like World War I, this war resulted in a scarcity of raw materials, which increased the need for research. Concerns about budget cuts concerning these agencies reduction and their inclusion in the German war economy provided an additional incentive to hand them over to TNO, which had no formal ties to the government and so the German occupation authorities had less control over it. The new chairman of TNO also managed to solve a lingering concern of the Ministry of Trade and Industry: all research institutes would become foundations, and the ministry would be allowed to appoint several of their board members. Thus this ministry retained at least some say in the agencies activities. All the agricultural experiment stations except for four of them, however, remained outside of TNO. The Ministry of Agriculture had built a solid organization of research, education, and information that it refused to give up to another organization. Between 1940 and 1948 Ministry of Agriculture even set up new research institutions that largely remained outside the scope of TNO. In 1957 the Central Organization for Applied Scientific Research completely lost its mandate over agricultural research, after which the National Council for Agricultural Research (Nationale Raad voor Landbouwkundig Onderzoek, NRLO) took over the task of coordination.[127]

Industrialization and Pure Scientific Research

After World War II, the government's involvement in science and technology was strongly determined by reconstruction and industrialization. The ministry of Education, Arts, and Science spent extra money on basic research, in response to initiatives from university scientists impressed by U.S. scientific progress, which they had been unable to monitor during the war. One key example: the atomic bomb revealed that within a short time span fundamental scientific discoveries could be converted into technological appli-

This photo from 1963 shows a member of the working group headed by Professor Dr. H. Brinkman and his colleague H. de Waard at the University of Groningen, engaged in special study of the beta spectra and the so-called Curie diagrams. After the war, basic scientific research was undertaken at academic and interrelated institutions, and research in nuclear physics was pursued on a larger scale in the Netherlands than before. Under the auspices of the Netherlands Organization for Pure Scientific Research (Organisatie voor Zuiver Wetenschappelijk Onderzoek), nuclear physicists were active in communities of the Foundation for Fundamental Research on Matter, and others participated at the European level.

cations. Moreover, in these years of reconstruction Dutch scientists sought to apply their knowledge to solving social problems. With government support mathematicians established the Mathematical Center, which combined fundamental research with contract research and data processing for the industry. Physicists set up the Foundation for Fundamental Research on Matter (Stichting Fundamenteel Onderzoek der Materie, FOM), to study peaceful applications of atomic energy.[128]

Next, the Education Ministry established a commission aimed at designing an organization whose purpose was to stimulate basic scientific research in the Netherlands in all branches of science. This commission subscribed to a view already articulated by the Went Commission: "Without fundamental research in this country, we may perhaps imperfectly imitate foreign accomplishments, but it will be impossible to set the pace, which is urgently called for in light of our need to export." In two ways basic research contributed fundamentally to prosperity: it produced well-trained scientists who could work in industrial laboratories, and the insights generated through research could give rise to innovations. This was the view of the American presidential adviser Vannevar Bush, who in *Science, the Endless Frontier* argued that new basic scientific discoveries would ensure a continuous flow of innovation.[129]

This advice led to the establishment, in 1950, of the Organization for Pure Scientific Research (Organisatie voor Zuiver-Wetenschappelijk Onderzoek, ZWO). In addition to the role of the sciences in encouraging the expansion of prosperity, the government also considered pure science to be a major cultural product that contributed to the country's international prestige and the "spiritual well-being of our people." Even more than TNO, ZWO was an organization in which scientists had a great deal of freedom as to how funds were spent. They apportioned government subsidies on the basis of the scientific importance of specific research.[130]

The Ministry of Economic Affairs (Ministerie van Economische Zaken) meanwhile tried to help industry to benefit from science more directly. In the early postwar years this ministry adopted a fairly directive approach in the area of industrialization. It mapped the gaps in Dutch production and developed ideas on how Dutch industry could optimally profit from the collapse of its German competitors. Initially this took place in product commissions consisting of company representatives, scientists, and ministry officials. A characteristic example is the Plastics Commission, with representatives from DSM, the Bataafsche Petroleum Maatschappij (Shell), Philips, AKU, Sikkens, and the head of the TNO Plastics Institute. Led by G. A. Kohnstamm, a ministry official, this commission tried to bring the supply of raw materials into line with the Dutch processing industry's needs; tried to identify what knowledge the industry was most in need of; discussed what know-ledge was available; considered the information about the German plastics industry that became available; and reflected on the options for developing new knowledge.[131]

Later on Kohnstamm wrote a general industrialization plan that drew much attention nationally. Its point of departure was that the Netherlands, especially after the possible independence of the Dutch East Indies, did not have enough foreign exchange to pay for its imports. Moreover, the population was growing rapidly and mechanization of agriculture had reduced the number of jobs in this sector. Industrialization was the answer to these problems: people would find work in industry and the export of industrial products might improve the country's stock of foreign currency, which during the early postwar years was essential for buying foreign scientific instruments and machinery because the Dutch government prohibited the use of the florin for foreign trade. Kohnstamm claimed that industrialization would have to be geared mainly toward producing scarce goods and quality goods that "required much thinking and handwork." On the basis of these premises he formulated a "national research program" in league with the TNO.[132]

Kohnstamm's plan was a precursor of the more substantial yet less directive industrialization policy initiated by the Ministry of Economic Affairs after 1949 and developed and clarified in a series of eight memorandums that appeared at irregular intervals between 1949 and 1963. In contrast to Kohnstamm's views, the first memo argued that the Netherlands had to bet in particular on expansion of basic industry because given its favorable location it was easy to import raw materials. Promoting top-quality, labor-intensive industry was deemed impossible because Dutch workers lacked proper training, and the "designers and constructers" were lacking in experience. Only in the long run would the country be able to move toward "production of top-quality products." Yet first it was necessary to do away with the "educational disadvantage of [Dutch] workers vis-à-vis those in other countries, and scientific research would also have to be expanded."[133]

In the context of this industrialization policy the number of technical schools was enlarged, at the junior as well as the intermediate level. The number of students grew even faster (see table 5.1). The expansion of research, the minister reasoned, would mainly have to be driven by the companies themselves, but for small and medium-sized companies he recommended that TNO should help them. The cabinet increased the funding of TNO's technological organization from 3.5 million guilders in 1949 to 28 million guilders in 1963, while calling on TNO to improve its contacts with industry. So-called "contact men" were appointed to inform companies about TNO's services and to promote TNO's findings with companies. Moreover, Economic Affairs expanded its State Industrial Agency by establishing offices in each province and it instituted a Technological Development Credit to support the development of innovations in companies.[134]

Later industrialization plans incorporated the view that it was doubtful that Dutch industry and the national research facilities could generate enough knowledge to close the gap with the United States. The minister felt that knowledge imported from America was needed. This was done by stimulating American companies to establish branches in the Netherlands and by appointing a technological-scientific attaché to a Dutch embassy post in Washington. His task was to "provide information to industry and research institutions in the Netherlands about research developments in the United States and Canada." Dutch companies were encouraged to make use of his services.[135]

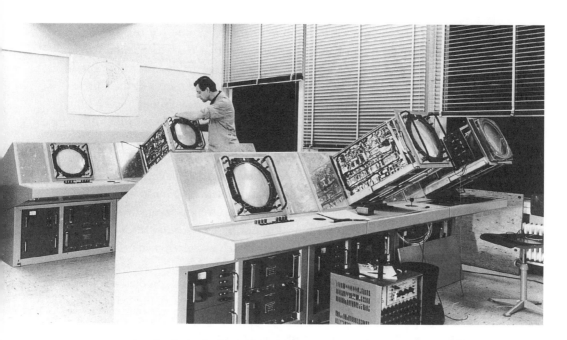

A 1969 photo of a laboratory of the Organization for Applied Scientific Research (Toegepast Natuurwetenschappelijk Onderzoek, TNO). In 1932, the national government set up a separate central organization to meet the research needs not only of business and society but also its own needs. For instance, the TNO State Defense Organization, set up in 1946, worked for the Ministry of Defense and the armed forces. The Physical Laboratory of the TNO, in The Hague, undertook research relating to radar images.

Economic Affairs stimulated research in areas it felt were important for Dutch prosperity but received too little or no attention from the industry. The clearest example of this was nuclear energy. After the atomic bomb had demonstrated that knowledge of nuclear fission could have practical applications, research into peaceful applications started on a worldwide basis. From 1946 onward, the Dutch government funded basic nuclear energy research, both within and outside the context of the Organization for Pure Scientific Research, the ZWO. Five years later, the Netherlands collaborated with Norway to set up an experimental reactor; the government considered the knowledge thus obtained might make it possible to apply nuclear power industrially. To this end, in 1955 it founded the Reactor Centrum Nederland, a collaborative effort of scientists and industry led from with the Ministry of Economic Affairs. This joining of science and industry is exemplary of this ministry's approach in the area of industrialization.[136]

Science Policy

In the 1960s and 1970s, the Education, Arts, and Sciences Ministry (OK&W) began to pursue a policy aimed at focusing science and technology more on solving social problems. This partial policy shift, which threatened to restrict researchers' freedom, was partly a result of international influences and took quite some time. In 1963 the ministry's director-general for sciences, A. J. Piekaar, first articulated the reasons for embarking on such "science policy." In a lecture at the University of Leiden he described how the costs of scientific research were skyrocketing and pointed out that this pace of growth could no longer be sustained. Moreover, he expressed the concern that science and technology were developing so rapidly that humanity as a whole threatened to lose its control over the direction of their development. This made it necessary to make informed choices as to which field should be allowed to grow and which not, so that science could be guided into a direction favored by society.[137]

Piekaar's lecture did not go into the actual policies, as the ministry's officials were still too busy thinking about how guidance was possible without infringing too much on scientists' freedom. And in any case, it was still unclear what direction guidance ought to take. This reflective process had been going on for some years, also in a European context. In 1959, J. M. L. Th. Cals, the minister of OK&W, informally discussed the issue with his European colleagues; the talks continued during the following years.[138]

Meanwhile, in 1962, the Organization for Economic Co-operation and Development (OECD) organized a ministerial conference on science policy. In this the OECD was mainly concerned about the economic relevance of science. One of the reasons for organizing the conference was the observation that Europe profited less from basic scientific research than the United States. The economic perspective met with criticism from OK&W and was also strongly disapproved by Cals. Science, he felt, could not be discussed in an economic organization such as the OECD because science, apart from being a production factor, was a major cultural product with non-economic, intrinsic value. But his protests against this intervention in his policy domain failed to stop the conference.[139]

The OECD conference gave rise to a ranking of countries that had a national science policy. The Netherlands proved to be quite low on the list, which led the new cabinet that had come into office try to put the country on the map by pursuing a "forceful national science policy."[140] It took quite some time, however, before an actual science policy could be implemented. In 1966 the Advisory Council for Science Policy (Raad voor Advies van het Wetenschapsbeleid, RAWB) was set up to provide advice to the minister of

Education, Arts, and Science in this area and one year later the first Science Budget, an inventory of money spent on research, was made public. In the Science Budget the minister of Education, Arts, and Science presents her view on the state of the science system in general and puts forward her plans for the coming years. This first Science Budget provided an overview of all government expenditures for science but did not offer any sense of direction, nor did it articulate any specific priorities. Precisely on this crucial science policy aspect, the Advisory Council for Science Policy failed to formulate consensus recommendations.[141]

In 1971, when M. L. de Brauw was appointed minister without portfolio for higher education and science policy, the objectives of Dutch science policy became more focused. The 1972 Science Budget stated, "Science and technology" should be put "more at the service of the major problems that will need to be solved in this decade. This means that science policy applies not only to the production of goods, education and defense; communication, public health, housing, spatial planning, recreation and so on have come to be within its reach as well."[142]

De Brauw's successor, F. H. P. Trip, was the first minister to develop specific views on steering and coordination in this field. Plans were made to abolish ZWO and replace it with an organization that would not only allocate money but also steer the development of science, taking into account specific governmental aspirations. But these plans were fought successfully by the Royal Netherlands Academy of Sciences, the Academic Council, and the Organization for Pure Scientific Research. After the Den Uyl cabinet stepped down, however, Trip's ideas soon disappeared from the policy agenda.[143]

In the late 1970s, in the wake of the 1973 oil crisis and as the economy stagnated, technological innovation was increasingly put on the government's agenda. In 1979 the ministers for science policy, economic affairs, and education and science together issued the so-called *Innovatienota*, in which they outlined a policy aimed at improving the innovativeness of businesses and industries. This meant the introduction of a special technology policy, in addition to the already existing Dutch science policy. When in 1981 a new cabinet was formed, this technology policy became a responsibility of the Ministry of Economic Affairs, while science policy remained a responsibility of the Education, Arts, and Science Ministry. The Ministry of Economic Affairs initially continued the research programs set up by Education, Arts, and Science in the field of innovation enhancement. In this period its policy was mainly geared toward increasing the supply of knowledge.

In the course of the 1980s and 1990s this policy increasingly shifted toward promotion of the transfer and diffusion of knowledge, notably from universities and research institutes to small and medium-sized businesses (and, in practice, also to large companies). In the eighties the Organization for Applied Scientific Research was restructured into a more market-oriented and flexible organization. In the 1990s the Netherlands Organization for Scientific Research (Nederlandse Organisatie voor Wetenschappelijk Onderzoek, NWO) replaced the Organization for Pure Scientific Research, and the new entity slowly but surely moved toward a strategic and more client-oriented approach to funding research. Its new instruments or projects focused on innovation and technology ("innovation-oriented research programs" and "first-rate technological institutes"), and this seemed only natural. The concept of a "national system of innovation," or of "research and innovation," guided diagnoses of how well the system functioned, with the OECD (again) serving as trailblazer.[144]

Orchestrating Scientific-Technological Research

In retrospect, we may conclude that the coordination of applied scientific research first occurred in sectors dominated by small companies in sectors that could not undertake their own research: agriculture and the leather industry. From the start, government has played a large role in those areas. After 1910, the Ministry of Trade and Industry began to establish services aimed at research and inspection for specific sectors, thus showing it valued research for industrial development and trade enhancement. Although these services were formally part of the ministry, they operated as autonomous bodies that evolved independently. After 1932 further coordination of applied research was also influenced by the establishment and slow initial development of the Central Organization for Applied Scientific Research. This research organization differed from the Trade and Industry agencies: although TNO was also geared toward industrial development, the ministries involved only had indirect influence on its policies.

After World War II, when the Mathematical Center, Fundamental Research on Matter, and the Organization for Pure Scientific Research were founded, the knowledge reservoir for technological research also began to encompass fundamental research. However, the various elements of the national knowledge system—technical and scientific education, industrial research, TNO university research (partly funded via the Fundamental Research of Matter Foundation and the Organization for Pure Research)—largely existed side by side. For a long time, coordination was marginal at best.

After 1960 this changed only gradually. The science policy implemented by the Education, Arts, and Sciences Ministry evolved first, and initially its concrete effects for technological development were modest. After 1980, when these policy approaches and those of Economic Affairs were increasingly brought into line with each other, the coordination of applied research increased strongly, mostly through stimulation programs.

The effort to orchestrate technological-scientific research thus evolved from various ad hoc and sector-oriented activities to responsibility at the level of the entire national system. This is reflected in the interest in science and technology policy inventories and in the initiation of a so-called "technology radar," which tracks new technological developments world wide and puts policy priorities on the agenda. This did not mean that all technological-scientific research became nationally orchestrated. The demand side was also exerting a force on the direction of scientific research. In this, the responsibilities of the national science organizations and the government were overshadowed by the growing importance of developments on the European level and by globalization trends in business and technology development.

Toward a Global Knowledge Economy?

Starting in the 1980s, everything seemed properly in place in the Netherlands to support a knowledge-intensive technology in a knowledge-intensive society. Technical education was organized at various levels, industry-specific research and development—whether or not performed by the industrial sector itself—was broadly accepted, and science policies and innovation policies were implemented in a more or less integrated manner. That the infrastructure was in place did not imply, however, that all was going well. For one thing, the interest in technical education suddenly began to wane, not only in the Netherlands but throughout Europe and North America. The stress on short-term relevancy in most industrial research and development efforts had reformulated the goals of academic research in terms of its relevance to the industrial and service sectors. Moreover, Dutch policies had to respond to international developments, which entailed the risk of renewed fragmentation. At the same time, technology's knowledge intensity increasingly led to high-tech rivalry on a worldwide basis, while environmental problems and development issues raised challenges for new technological developments with their own kind of knowledge intensity.

How will these various interrelated issues evolve further? And what can be learned from the past, from the vantage point of a contextual technology history? A major lesson from this chapter's analysis is that knowledge intensity per se is not a driving factor in technological development. Knowledge intensity resulted from actions and interactions by various actors who had a variety of motivations and increasingly made use of the appeal associated with this notion. In the period before World War II we saw how increasing scientific and technical knowledge provided industrialists with the opportunity to raise profits through innovation. The government wanted to expand prosperity by stimulating overall knowledge intensity. Professors who became advisers not only secured higher incomes for themselves but also wielded their influence in companies to secure employment for their students. Scientists and engineers, finally, acquired a whole new field of activity as a result of the increasing knowledge intensity, precisely at a time when employment in traditional fields such as secondary education was dropping. Yet at the same time, actors reflected on knowledge developments and how to respond to them. The various standpoints advanced are still relevant, even if they are pushed to the background because of the rhetoric involved in the knowledge society concept.

The effort to harmonize intermediate technical education was marked by a division between pragmatists, who viewed education as supplementary to and a strengthening of practical knowledge, and theorists, who felt that a solid base of formal knowledge was necessary for fulfilling a midlevel technical function. In the legislative regulation of technical education, the views of the theorists gained the upper hand because entry standards were prior education rather than practical experience and because in the first years of the curriculum ample space was set aside for mathematics and sciences. To this day the tension between practice and theory remains visible at all levels of Dutch technical education.

In the debate on practice-oriented research, industrialists and industrial researchers were pitted against academic scientists. The former argued for close collaboration between research and practice, as well as for a commercial orientation of technological research. The academic scientists stressed that only research conducted in conditions of maximum freedom would bear fruit. After World War II a temporary social contract was established in which the mere promise that automonous research would almost automatically have a technical and commercial relevancy was already enough to guarantee a high level of financing. The title and content of *Science, the Endless Frontier* (1945) by Vannevar Bush on the deployment of scientific research as a U.S. national interest expressed the nature of such compromise.[145] In industrial research the freedom of researchers was also a major element in developing new technologies, but starting in the early 1970s the tensions involved once again made themselves more strongly felt.[146] A new compromise was found in the notion of strategic research, which starts from assumptions concerning the long-term social relevancy of scientific research, while at the same time, contrary to the 1950s, holding the researchers accountable for the realization of concrete social and economic objectives.[147]

For the Dutch government the issue was how proactive it should and could be in promoting knowledge intensity and orchestrating research. Partly through actions from professors and company directors (as in the TNO's prior history), governmental organizations took on more responsibilities, even though they thus risked being blamed for having insufficient expertise to orchestrate techno-scientific research. The problem of how to weigh government decisions or the impact of all sorts of technology actors (including special interest groups, such as the environmental movement) continues to be a major concern. Only in the case of military technology do government and technology actors seem to have shared interests—notably in the United States.

Presumably, the knowledge intensity of technology will partly shift from (exclusive) embedding of knowledge in artifacts and individuals to strategic deployment of new and sometimes marketable forms of knowledge. The rise of information and communication technologies has significantly changed the nature of knowledge intensity, largely because today's knowledge reservoirs are organized differently. A last major development concerns military technology: because of the emphasis on targets rather than costs and the need to be assured in advance that specific objectives are met, the knowledge intensity in this sector has always been high and it continues to be a major booster of scientific-technological research and development at large.[148]

These developments are now playing out at a global level, an arena the Netherlands has fully belonged to all along, of course. Although this country was always one of technology import and diffusion, at the start of the twenty-first century national differences have become less pronounced in this context. As the twentieth century unfolded, the Netherlands developed into more of a producer of new knowledge and technology than it had been in the nineteenth century. In the twenty-first-century global knowledge economy, scientific and commercial success will largely remain dependent on the skill or talent to, as the nineteenth-century German chemist Justus Liebig put it, "drill with a saw and saw with a drill."[149]

Acknowledgments

The authors have greatly benefited from comments on earlier drafts from editors and postdocs associated with the project "Technology of the Netherlands in the Twentieth Century," in particular Arie Rip, Geert Verbong, Aad Bogers, Giel van Hooff, Erik van der Vleuten, Harry Lintsen, Johan Schot, and Dick van Lente. We appreciate especially the contributions of the participants in an informal discussion group on the history of R&D in the Netherlands that met in Eindhoven between 1997 and 2002. We are most grateful to Harro Maat, Kees Boersma, Marc de Vries, Rienk Vermij, Arie Rip, Ernst Homburg, and Arjan van Rooij.[150]

Notes

1 Ernst Homburg, James Small, and Piet Vincken, "Van carbo- naar petrochemie, 1910–1940," in Schot et al., *Techniek in Nederland*, vol. 2, 340–341.

2 J. W. Schot, H. W. Lintsen, and A. Rip, "Methode en opzet van het onderzoek," in Schot et al., *Techniek in Nederland*, vol. 1, 46–48.

3 See E. Homburg, "De tweede industriële revolutie: Een problematisch historisch concept," *Theoretische Geschiedenis* 13, no. 3 (1986): 367–385.

4 F. A. F. C. Went (commission chairman), *Rapport der commissie, ingesteld bij beschikking van zijne excellentie, den Minister van Onderwijs, Kunsten en Wetenschappen en zijne excellentie, den Minister van Binnenlandsche Zaken en Landbouw, dd. 30 juni 1923, met opdracht: Te onderzoeken, door welke maatregelen en in welke vorm het toegepast-natuurwetenschappelijk onderzoek hier te lande in hooger mate dienstbaar kan worden gemaakt aan het algemeen belang* (The Hague: Algemeene Landsdrukkerij, 1925), 12.

5 Albert E. Kersten, *Een organisatie van en voor onderzoekers: De Nederlandse Organisatie voor Zuiver-Wetenschappelijk Onderzoek (Z.W.O.), 1947–1988* (Assen: Van Gorcum, 1996), 14–15.

6 Ernst Homburg, "From Colour Maker to Chemist: Episodes from the Rise of the Colourist, 1670–1800," in Robert Fox and Agustí Nieto-Galan, *Natural Dyestuffs and Industrial Culture in Europe, 1750–1880* (Canton, Mass.: Science History Publications, 1999), 219–257; Ingrid Vledder, Eddy Houwaart, and Ernst Homburg, "Particuliere laboratoria in Nederland," part 1: "Opkomst en bloei, 1865–1914," *NEHA-Jaarboek voor economische, bedrijfs- en techniekgeschiedenis* 62 (1999): 249–290; Cornelis Disco, "Made in Delft: Professional Engineering in the Netherlands, 1880–1940," Ph.D. diss., University of Amsterdam, 1990; Eda Kranakis, "The Affair of the Invalid Bridges," *Jaarboek voor de geschiedenis van bedrijf en techniek* 4 (1987): 106–130.

7 Jan Wolthuis, *Lower Technical Education in the Netherlands 1798–1993: The Rise and Fall of a Subsystem* (Leuven: Garant, 1999).

8 Ernst Homburg, "The Emergence of Research Laboratories in the Dyestuffs Industry, 1870–1900," *British Journal for the History of Science* 25 (1992): 91–111.

9 Vledder, Houwaart, and Homburg, "Particuliere laboratoria in Nederland," 254; J. J. Hutter, "Nederlandse laboratoria 1860–1940: Een kwantitatief overzicht," *Tijdschrift voor de Geschiedenis der Geneeskunde, Natuurwetenschappen, Wiskunde en Techniek* 9, no. 4 (1986): 150–174; Jasper Faber, *Kennisverwerving in de Nederlandse industrie 1870–1970* (Amsterdam: Aksant, 2001), 40–60.

10 James Phinney Baxter, *Scientists Against Time* (1946; Cambridge, Mass.: M.I.T. Press, 1968).

11 Cf. Gernot Böhme and Nico Stehr, eds., *The Knowledge Society: The Growing Impact of Scientific Knowledge on Social Relations* (Dordrecht: Reidel, 1986); Nico Stehr, *Knowledge Societies* (London: Sage, 1994).

12 For the concepts "shop culture" and "school culture," see M. A. Calvert, *The Mechanical Engineer in America, 1813–1910: Professional Cultures in Conflict* (Baltimore: Johns Hopkins University Press, 1967).

13 For a more general treatment of the transition from "shop" to "school," see P. Lundgreen, "Engineering Education in Europe and the USA 1750–1930: The Rise to Dominance of School Culture and the Engineering Professions," *Annals of Science* 47, no. 1 (1990): 33–75.

14 Ernst Homburg, *Groeien door kunstmest: DSM Agro 1929–2004* (Hilversum: Verloren, 2004), 136–37.

15 Cf. Keith Pavitt, "Sectoral Patterns of Technical Change: Towards a Taxonomy and a Theory," *Research Policy* 13 (1984): 343–73; Faber, *Kennisverwerving in de Nederlandse industrie*, 61–107.

16 G. P. J. Verbong, "Techniek, beroep en praktijk in Nederland," in H. W. Lintsen et al., eds.,

Geschiedenis van de techniek in Nederland: De wording van een moderne samenleving 1800–1890, 6 vols. (Zutphen: Walburg Pers, 1992–1995), vol. 5, 290–303; Ernst Homburg, *Van beroep "Chemiker": De opkomst van de industriële chemicus en het polytechnische onderwijs in Duitsland (1790–1850)* (Delft: Delft University Press, 1993), 287–339.

17 On the emergence of basic and higher technical education in the nineteenth century, see G. P. J. Verbong, "Ingenieurs en het technisch onderwijs 1863–1890," in Lintsen, *Geschiedenis van de techniek*, vol. 5, 116–155. For an early history of intermediate technical schools (*middelbare technische scholen*) see also P. Baggen, "Technischer Vollzeitunterricht in den Niederlanden am Beispiel der Mittelstufe des technischen Unterrichts," *Zeitschrift für Technikgeschichte* 68, no. 2 (2001): 157–179.

18 M. S. C. Bakker, "Suiker," in Lintsen et al., *Geschiedenis van de techniek in Nederland*, vol. 1, 214–251.

19 Ibid.

20 Margaret Leidelmeijer, *Van suikermolen tot grootbedrijf: Technische vernieuwing in de Java-suikerindustrie in de negentiende eeuw* (Amsterdam: Nederlandsch Economisch-Historisch Archief, 1997).

21 M. S. C. Bakker, *Ondernemerschap en vernieuwing: De Nederlandse bietsuikerindustrie 1858–1919* (Amsterdam: Nederlandsch Economisch-Historisch Archief, 1989), 193–211.

22 Ibid., 199–203.

23 Ibid., 206–208.

24 M. S. C. Bakker, "Industrieel onderwijs en de Nederlandse suikerindustrie," *Jaarboek voor de Geschiedenis van Bedrijf en Techniek* 2 (1985): 151–172.

25 W. J. Heijdeman, *Gedenkschrift ter gelegenheid van het vijftigjarig bestaan der Middelbare Technische School Amsterdam (tevens kweekschool voor machinisten), 1878–19 October 1928* (Amsterdam: 1928).

26 Bakker, "Industrieel onderwijs en de Nederlandse suikerindustrie."

27 Hans Schippers, *Van tusschenlieden tot ingenieurs: De geschiedenis van het hoger technisch onderwijs in Nederland* (Hilversum: Verloren, 1989), 79–80.

28 For more on this laboratory, see Vledder, Houwaart, and Homburg, "Particuliere laboratoria in Nederland."

29 Bakker, "Industrieel onderwijs en de Nederlandse suikerindustrie."

30 Dordrecht Middelbare Technische School, *Gedenkboek uitgegeven ter gelegenheid van het 25-jarig bestaan der Middelbare Technische School te Dordrecht in september 1936* (Dordrecht: 1936); F. Goudriaan et al., eds., *Vijftig jaren Middelbaar Technisch Onderwijs te Dordrecht 1911–1961: Gedenkboek uitgegeven ter gelegenheid van het vijftig jarig bestaan van de Hogere Technische School te Dordrecht door het bestuur van de Vereeniging voor Middelbaar Technisch Onderwijs* (Dordrecht: Vereeniging voor Middelbaar Technisch Onderwijs, 1961).

31 In the sociology of professions the distinction is made between "profession construction" by external groups (as occurred in the sugar industry) and "professionalization" by those directly involved (as occurred in the construction sector). Cf. L. Burchardt, "Professionalisierung oder Berufskonstruktion? Das Beispiel des Chemikers im Wilhelminischen Deutschland," *Geschichte und Gesellschaft* 6 (1980): 326–348.

32 Coert Peter Krabbe, *Ambacht, kunst en wetenschap: De bevordering van de bouwkunst in Nederland 1750–1880* (Zwolle: Waanders, 1998), 17–23.

33 W. R. F. van Leeuwen, "Woning- en utiliteitsbouw," in Lintsen et al., *Geschiedenis van de techniek in Nederland*, vol. 3, 199–200; Krabbe, *Ambacht, kunst en wetenschap*, 138–139.

34 Jan Alexander Martis, "Voor de kunst en voor de nijverheid: Het ontstaan van het kunstnijverheidsonderwijs in Nederland," Ph.D. dissertation, University of Amsterdam, 1990, 21–44; Nicolaas Bastiaan Goudswaard, *Vijfenzestig jaren nijverheidsonderwijs: Een onderzoek betreffende het nijverheidsonderwijs zoals het in Nederland aan jongens en ouderen is

gegeven vanaf de officiële opheffing van de gilden in 1798 tot de inwerkingtreding van de Wet op het Middelbaar Onderwijs in 1863 (Assen: Van Gorcum, 1981), 29–32, 112–119; Hendrikus Peter Meppelink, *Technisch vakonderwijs voor jongens in Nederland in de 19e eeuw: Een sociografisch onderzoek inzake structuur en uitbouw* (Utrecht: Elinkwijk, 1961), 18–23.

35 Krabbe, *Ambacht, kunst en wetenschap*, 69–83; Martis, *Voor de kunst en voor de nijverheid*, 93–102, 288–292.

36 Cf. V. Clark, "A Struggle for Existence: The Professionalization of German Architects," in Geoffrey Cocks and Konrad H. Jarausch, eds., *German Professions 1800–1950* (Oxford: Oxford University Press, 1990), 143–160.

37 Krabbe, *Ambacht, kunst en wetenschap*, 86–104; Engelina Blanca Faustinia Pey, "Herstel in nieuwe luister: Ideeën en praktijk van overheid, kerk en architecten bij de restauratie van het middeleeuwse katholieke kerkgebouw in Zuid-Nederland (1796–1940)," Ph.D. dissertation, Catholic University Nijmegen, 1993, 220–221; Van Leeuwen, "Woning- en utiliteitsbouw," 205–206.

38 A. W. van Dam, E. S. Heynincx, and Is. Warnsinck, "Memorie van toelichting behoorende bij het adres aan Zijne Excellentie der minister van binnenlandsche zaken betrekkelijk de middelen ter verbetering van de staat der burgerlijke bouwkunst hier te lande," in H. R. R. Roelofs Heyrmans, ed., *Gedenkschrift van de Koninklijke Akademie en van de Polytechnische School 1842–1905: Samengesteld ter gelegenheid van de oprichting der Technische Hoogeschool* (Delft: Waltman, 1906), appendix 7.

39 Krabbe, *Ambacht, kunst en wetenschap*, 115–241. In 1863 the Koninklijke Academie voor Burgerlijke Ingenieurs was reorganized and its name changed to Polytechnische School. In 1905 the Polytechnische School was elevated to the rank of a Technical University (*Technische Hogeschool*).

40 H. Schippers, "IJzerconstructies," in Lintsen et al., *Geschiedenis van de techniek in Nederland*, vol. 3, 272–300.

41 A. de Groot, "Rationeel en functioneel bouwen 1840–1920," in D. van Woerkom, A. de Groot, and M. Bock, *Het nieuwe bouwen: Voorgeschiedenis* (Delft: Delft University Press, 1982), 25–30; Krabbe, *Ambacht, kunst en wetenschap*, 165–173.

42 Martis, *Voor de kunst en voor de nijverheid*, 93–102, 152–155; Krabbe, *Ambacht, kunst en wetenschap*, 232–234; De Groot, "Rationeel en functioneel bouwen," 46–47.

43 Jeroen Schilt and Jouke van der Werf, *Genootschap architectura et amicitia* (Rotterdam: Uitgeverij 010, 1992); Dion Kooijman, *Wortels van het architectuuronderwijs* (Delft: Delft University Press, 1995), 23–28.

44 "Afdeeling Voortgezet en hooger bouwkunst-onderricht," *Architectura* 16, no. 14 (1908): 114–119.

45 Eric M. Fontein, *De rechtspositie van de architect 1850–1985: Een onderzoek naar de relatie tussen ontwikkelingen in de maatschappelijke positie en rechtspositie* (Delft: Delftse Universitaire Pers, 1988), 18–29 and 74–91; F. de Herder et al., eds., *Gedenkboek uitgegeven door den Nederlandschen Aannemersbond ter herdenking van zijn vijf en twintig-jarig bestaan 1895–November-1920* (Bussum: Brand, 1920), xi-xlviii.

46 Henk Reinders, *Een eeuw verenigd bouwen in Nederland* (Bunnik: Ned. Verbond van Ondernemers i.d. Bouwnijverheid, 1995), 16–22.

47 J. L. B. Keurschot, C. Boenk, and J. G. van den Berg, "Vereeniging van Nederlandsche bouwkundige opzichters en tekenaars," *Vademecum der Bouwvakken* 18 (1903), 65–66.

48 J. N. Hendrix et al., *Rapport der Commissie in zake het Middelbaar Technisch Onderwijs* (Amsterdam: Nederlandsche Aannemersbond, 1907).

49 Herder et al., *Gedenkboek uigegeven door den Nederlandschen Aannemersbond*, 206–230.

50 Schippers, *Van tusschenlieden tot ingenieurs*, 32–41.

51 J. J. Beljon, *300 jaar Koninklijke Academie van Beeldende Kunsten 's-Gravenhage 1682–1982: Een beknopt overzicht* (The Hague: Koninkl ke Academie van Beeldende Kunsten, 1982);

F. N. Maas, *Honderd jaren Academie van Beeldende Kunsten en Technische Wetenschappen* (Rotterdam: Donker, 1951); Loes A. Peeperkorn-Van Donselaar, *Een eeuw beroep op onderwijs: De geschiedenis van de Vereniging voor Beroepsonderwijs Haarlem, 1891–1991* (Haarlem: Vereniging voor Beroepsonderwijs Haarlem, 1991); *Gedenkboek van de Middelbare Technische School te Heerlen bij gelegenheid van haar twaalf en een half jarig bestaan, 4 December 1922–4 juni 1935* (Heerlen: Middelbare Technische School te Heerlen, 1935).

52 A similar discussion emerged in Germany. See Kees Gispen, *New Profession, Old Order: Engineers and German Society, 1815–1914* (Cambridge: Cambridge University Press, 1989).

53 A. Huet, "Nota over de regeling van het hoger en middelbaar technisch onderwijs," in R. A. I. Snethlage, *Verslag der commissie in zake het technisch onderwijs, benoemd ingevolge het besluit van de algemene vergadering der Vereeniging van Burgerlijk Ingenieurs op 18 juli 1891 te 's-Gravenhage* (The Hague: Belinfante, 1895), 46.

54 Huet, "Nota over de regeling van het hoger en middelbaar technisch onderwijs."

55 H. Enno van Gelder, *Overzicht van het technies onderwijs in Nederland* (Zutphen: Thieme, 1919), 50–53.

56 Cf. H. Enno van Gelder, "Technisch onderwijs," *Tijdschrift der Maatschappij van Nijverheid* 70, no. 10 (1903): 595–606; A. Borgman, "Middelbaar vakonderwijs," *Tijdschrift der Maatschappij van Nijverheid* 70, no. 12 (1903): 691–711.

57 Memorie van toelichting bij het "Ontwerp van wet tot verhooging en wijziging van hoofdstuk V der Staabegrooting voor het dienstjaar 1904," *Bijlagen van het verslag der Handelingen van de Tweede Kamer der Staten-Generaal* 1903–1904, no. 176.3.

58 H. Enno van Gelder, "Het Middelbaar technisch onderwijs: Wat nu?," *De ingenieur* 21, no. 11 (1906): 198–199.

59 Sociaal-Technische Vereeniging van Demokratische Ingenieurs en Architecten en Algemeene Nederlansche Teekenaars- en Opzichtersbond, *Adres inzake het middelbaar-technisch onderwijs* (The Hague: 1907); J. N. Hendrix, *Rapport der Commissie in zake het Middelbaar Technisch Onderwijs; Enquête ingesteld door de Commissie Middelbaar Technische Onderwijs (M.T.O.) van de Bond van Technici* (Amsterdam: Nederlandsche Aannemersbond, 1907); G. Homan van der Heide, *Het middelbaar technisch onderwijs, een verwaarloosd volksbelang: Hoe tot eene regeling te geraken in de quaestie betreffende dit M.O.* (Leiden: Brill, 1907); H. J. C. Haver et al., *Congres ter bespreking van het middelbaar technisch onderwijs* (Amsterdam: 1908).

60 Schippers, *Van tusschenlieden tot ingenieurs,* 36–38; Udo Teunis, *De geschiedenis van het MTO (HTO), 1990–1940* (Amsterdam: University of Amsterdam, Pedagogical-Didactic Institute, 1980), 40–44.

61 Schippers, *Van tusschenlieden tot ingenieurs,* 41–47; Teunis, *De geschiedenis van het MTO (HTO),* 45–51.

62 Vereniging tot Bevordering van de Vakopleiding voor Handwerkslieden in Nederland, *Congres-verslag: Nationaal congres voor vakopleidingsvraagstukken, gehouden op 28, 29 en 30 december 1931 te 's-Gravenhage in de Grafelijke zalen op het Binnenhof* (Leiden: Vereeniging tot Bevordering van de Vakopleiding voor Handwerkslieden in Nederland, [1932]).

63 G. Hofstede (chairman), *Rapport van de Commissie inzake het Middelbaar Technisch Onderwijs, ingesteld bij beschikking van den Minister van Onderwijs, Kunsten en Wetenschappen van 21 December 1931, no. 20642, Afdeeling Nijverheidsonderwijs, aangevuld bij beschikking van 4 Januari 1932, no. 20761, Afdeeling Nijverheidsonderwijs* (The Hague: Algemeene Landsdrukker , 1932), 44.

64 L. Lammerse (chairman), *Rapport van de Commissie in zake het Voorbereidend Middelbaar Technisch Onderwijs* (Alkmaar: 1934), 28.

65 Schippers, *Van tusschenlieden tot ingenieurs,* 54–56.

66 G. P. J. Verbong, "Delftse ingenieurs tussen wetenschap en industrie (1875–1900)," *Gewina* 16, no. 3 (1993): 248–260.

67 Ernst Homburg, *Groeien door kunstmest: DSM Agro 1929–2004* (Hilversum: Verloren 2004), 141. On the German midlevel technicians, see Gispen, *New Profession, Old Order*.

68 Ernst Homburg, *Groeien door kunstmest: DSM Agro 1929–2004* (Hilversum: Verloren 2004), 147–149.

69 G. Berkhoff, "De bereiding van grofkristallijn ammoniumsulfaat," *Chemisch Weekblad* 32, no. 13 (1935): 186–197; G. Berkhoff, "Het kristalliseeren van technische producten en van ammoniumsulfaat in het bijzonder," *Chemisch Weekblad* 35, no. 51 (1938): 868–872.

70 Arjan van Rooij, "Aangekochte technologie en industriële research bij het Stikstofbindingsbedrijf van de Staatsmijnen in de jaren 1930," *NEHA Jaarboek* 66 (2003), 263–286; Arjan van Rooij, *Building Plants: Markets for Technology and Internal Capabilities in DSM's Fertiliser Business, 1925–1970* (Amsterdam: Aksant, 2004). For the history of the Centraal Laboratorium, see H. W. Lintsen, ed., *Research tussen vetkool en zoetstof: Zestig jaar DSM Research 1940–2000* (Zutphen: Stichting Historie der Techniek, 2000).

71 Ernst Homburg and Arjan van Rooij, "Die Vor- und Nachteile enger Nachbarschaft. Der Transfer Deutscher chemischer Technologie in die Niederlande bis 1952," in Rolf Petri, ed.), *Technologietransfer aus der deutschen Chemieindustrie (1925–1960)* (Berlin: Duncker & Humblot, 2004), 183–226. On the role of purchasing equipment, see Arjan van Rooij and Ernst Homburg, *Building the Plant: A History of Engineering Contracting in the Netherlands* (Zutphen: Walburg Pers, 2002), 18–59.

72 H. W. Lintsen, "Kennisverwerving in de Nederlandse industrie in de 19e eeuw," *Tijdschrift voor de Geschiedenis der Geneeskunde, Natuurwetenschappen, Wiskunde en Techniek* 9, no. 4 (1986): 175–189; Chantal M. C. Vancoppenolle, *Tussen paternalistische zorg en zakelijk management: C. J. Honig als eindpunt van persoonsgericht sociaal ondernemersgedrag in een Zaans familiebedrijf (1930–1957)* (Amsterdam: Nederlandsch Economisch-Historisch Archief, 1993), 28–31.

73 B. Elema, *Opkomst, evolutie en betekenis van research gedurende honderd jaren gistfabriek* (Delft: Koninklijke Gist- en Spiritusfabriek N.V., 1970), 6.

74 G. P. J. Verbong and E. Homburg, "Theorie en praktijk: Chemische kennis en de chemische industrie," in Lintsen et al., *Geschiedenis van de techniek in Nederland*, vol. 5, 243–269; E. Homburg, "Speuren op de tast: Een historische kijk op industriële en universitaire research," inaugural lecture, University of Maastricht, October 31, 2003 (Maastricht: Universiteit Maastricht, 2003).

75 Ernst Homburg, Arie Rip, and James Small, "Chemici, hun kennis en de industrie," in Schot et al., *Techniek in Nederland*, vol. 2, 299–305.

76 A. Heerding, *Geschiedenis van de N.V. Philips' Gloeilampenfabrieken*, vol. 2: *Een onderneming van vele markten thuis* (Leiden: Nijhoff, 1986), 27–38, 147–182, 384–391; F. K. Boersma, *Inventing Structures for Industrial Research: A History of the Philips NatLab, 1914–1946* (Amsterdam: Aksant, 2002), 31–74; Marc J. Vries, *Eighty Years of Research at the Philips Natuurkundig Laboratorium (1914–1994)* (Amsterdam: Pallas Publications, 2005); I. J. Blanken, *De geschiedenis van Philips Electronics N.V.*, vol. 3: *De ontwikkeling van de N.V. Philips' Gloeilampenfabrieken tot elektrotechnisch concern* (Leiden: Nijhoff, 1992), 1–26.

77 Faber, *Kennisverwerving in de Nederlandse industrie*, 202–204, 214–222.

78 Ibid., 155–156, 161–163; A. N. Hesselmans, "Tentoonstelling van nationale electrotechnische industrie," *Jaarboek Electrotechnische Vereeniging* 30 (1989): 203–207.

79 Faber, *Kennisverwerving in de Nederlandse industrie*, 46.

80 H. W. Lintsen, *Ingenieurs in Nederland in de negentiende eeuw: Een streven naar erkenning en macht* (The Hague: Nijhoff, 1980), 353; J. Limburg (chairman), *De toekomst der academisch gegradueerden: Rapport van de commissie ter bestudering van de toenemende bevolking van universiteiten en hoogescholen en de werkgelegenheid voor academisch gevormden* (Groningen and Batavia: Wolters 1936), 31.

81 Hutter, "Nederlandse laboratoria," 162.

82 H. W. Lintsen, G. van Hooff, and G. Verbong, "Werktuigbouwkunde in Nederland in de negentiende eeuw: Een techniek zonder beroepsgemeenshap," *Jaarboek voor geschiedenis van bedrijf en techniek* 8 (1991), 81–100 C. Disco and H. W. Lintsen, "De vervlechting van ingenieursberoep en industrie," *Tijdschrift voor Sociale Geschiedenis* 9, no. 3 (1983): 343–369; Frida de Jong, *Tussen tandwiel en turbulentie: De opleiding tot werktuigkundig ingenieur aan de TU Delft* (Delft: Technical University Delft, Faculty of Mechanical Engineering and Maritime Technology, 1992).

83 Geert Jan Somsen, *"Wetenschappelijk onderzoek en algemeen belang": De chemie van H. R. Kruyt (1882–1959)* (Delft: Delft University Press, 1998), 9–35, 57–91; Homburg, Rip, and Small, "Chemici, hun kennis en de industrie," 307–313.

84 Homburg, Rip and Small, "Chemici, hun kennis en de industrie," 305–307.

85 Peter Baggen, *Vorming door wetenschap: Universitair onderwijs in Nederland 1815–1960* (Delft: Eburon, 1998), 117–137; Disco, *Made in Delft*, 149.

86 H. C. J. H. Gelissen earned his doctorate in 1925, H. Ebbinge in 1940, and P. R. A. Maltha in 1946, all at Delft.

87 De Vries, *Eighty Years of Research at the Philips Natuurkundig Laboratorium*, 47.

88 Eugene S. Ferguson, *Engineering and the Mind's Eye* (Cambridge: MIT Press, 1992); Walter G. Vincenti, *What Engineers Know and How They Know It: Analytical Studies from Aeronautical History* (Baltimore: Johns Hopkins University Press, 1990).

89 Homburg, Rip, and Small, "Chemici, hun kennis en de industrie," 299–305; Faber, *Kennisverwerving in de Nederlandse industrie*, 207; G. J. Somsen, "Selling Science: Dutch Debates on the Industrial Significance of University Chemistry, 1903–1932," in A. S. Travis, H. G. Schröter, E. Homburg, and P. J. T. Morris, eds., *Determinants in the Evolution of the European Chemical Industry, 1900–1939: New Technologies, Political Frameworks, Markets and Companies* (Dordrecht: Kluwer Academic Publishers, 1998), 144; H. G. Heijmans, *Wetenschap tussen universiteit en industrie: De experimentele natuurkunde in Utrecht onder W. H. Julius en L. S. Ornstein 1896–1940* (Rotterdam: Erasmus, 1994), 160–161, 176–177; J. J. Hutter, *Toepassingsgericht onderzoek in de industrie: De ontwikkeling van kwikdamplampen bij Philips 1900–1940* (Helmond: 1988), 105–106.

90 Elema, *Opkomst, evolutie en betekenis van research gedurende honderd jaren gistfabriek*, 25, 101; Homburg, Rip, and Small, "Chemici, hun kennis en de industrie," 301.

91 Somsen, *Wetenschappelijk onderzoek en algemeen belang*, 183.

92 Homburg, Rip, and Small, "Chemici, hun kennis en de industrie," 313–315; J. Schweppe, *Research aan het IJ. LBPMA 1914– KSLA 1989: De geschiedenis van het "Lab Amsterdam"* (Amsterdam: Royal Dutch/Shell, 1989), 33–52; Faber, *Kennisverwerving in de Nederlandse industrie*, 207; Somsen, *Wetenschappelijk onderzoek en algemeen belang*, 183–186.

93 J. van den Haar and M. E de Ruiter, *De geschiedenis van de Landbouwuniversiteit Wageningen*, vol. 1: *Van school naar hogeschool, 1873–1945* (Wageningen: Wageningen Agricultural University, 1993), 236–238; H. Buiter, "Koelen en vriezen," in Schot et al., *Techniek in Nederland*, vol. 3, 346; Boersma, *Inventing Structures for Industrial Research*, 165–166; Faber, *Kennisverwerving in de Nederlandse industrie*, 131–134.

94 P. J. Knegtmans, "Onderwijs, wetenschap en particulier initiatief aan de Universiteit van Amsterdam, 1920–1950," in P. J. Knegtmans and A. J. Kox, eds., *Tot nut en eer van de stad: Wetenschappelijk onderzoek aan de Universiteit van Amsterdam* (Amsterdam: Amsterdam University Press, 2000), 83–90.

95 Somsen, *Wetenschappelijk onderzoek en algemeen belang*, 211–212; Heijmans, *Wetenschap tussen universiteit en industrie*, 176–177.

96 Jasper Faber, "C. J. van Nieuwenburg over organisatie van wetenschappelijk technisch werk: Stemmen uit de industrie over toegepast natuurwetenschappelijk onderzoek 1900–1919," *Gewina* 21, no. 1 (1998): 20, 23–24.

97 Boersma, *Inventing Structures for Industrial Research*, 33.

98 H. F. J. M. van den Eerenbeemt, ed., *Van boterkleursel naar kopieersystemen: De ontstaans-geschiedenis van Océ-van der Grinten, 1877–1956* (Leiden: Nijhoff, 1992), 205–207; Olga Amsterdamska, "Beneficent Microbes: The Delft School of Microbiology and Its Industrial Connections," in Pieter Bos and Bert Theunissen, eds., *Beijerinck and the Delft School of Microbiology* (Delft: Delft University Press, 1995), 203.

99 Heijmans, *Wetenschap tussen universiteit en industrie*, 123–125.

100 Hutter, *Toepassingsgericht onderzoek in de industrie*, 35–38.

101 Heijmans, *Wetenschap tussen universiteit en industrie*, 121–123.

102 H. Baudet, *De lange weg naar de Technische Universiteit Delft*, vol. 2: *Verantwoording, re-gisters, tabellen, namenlijsten en bijlagen* (The Hague: SDU Uitgeverij, 1993), 677–681; Boersma, *Inventing Structures for Industrial Research*, 152–157.

103 Boersma, *Inventing Structures for Industrial Research*, 157; G. A. van de Schootbrugge, *50 jaar TPD in beweging: Een halve eeuw natuurkunde voor de praktijk* (Delft: Technical University Delft and Netherlands Organization for Applied Scientific Research, Technical Physics Agency, 1991), 16.

104 N.V. De Bataafsche Petroleum Maatschappij, *De proeffabrieken voor physische en chemi-sche technologie van de Technische Hogeschool te Delft* (The Hague: Royal Dutch/Shell), 1951).

105 De Vries, *Eighty Years of Research at the Philips Natuurkundig Laboratorium*, 123; J. Beyer, "Ingenieursopleiding en textielindustrie," inaugural lecture in the department of Mechanical Engineering, Technische Hogeschool Delft, October 5, 1955.

106 Cf. Lintsen, *Research tussen vetkool en zoetstof*; De Vries, *Eighty Years of Research at the Philips Natuurkundig Laboratorium*, 100–193.

107 Jan Luiten van Zanden, *The Economic History of the Netherlands 1914-1995* (London: Rout-ledge, 1998) 1-6.

108 Harro Maat, *Science Cultivating Practice: A History of Agricultural Science in the Nether-lands and Its Colonies 1863–1986* (Dordrecht: Kluwer Academic Publishers, 2001), 59.

109 J. Eekels, H. H. C. M. Christiaans, and R. H. Kaasschieter, eds., *Ondernemen en vernieuwen: Hommage aan de kleine en middelgrote industrie naar aanleiding van 75 jaar Rijksnijver-heidsdienst* (Delft: Rijksnijverheidsdienst, 1985), 14–15.

110 Jasper Faber, *"Een onafwijsbare plicht": Chemici over het nut van natuurwetenschap en de rol van de overheid, 1900–1940*, report 1996-5, Science and Society Series (Nijmegen: Catholic University, 1996); J. J. Hutter, *Laboratoria in Nederland vóór 1940* (Eindhoven: Eind-hoven Technical University, Technical Sciences Research Center, Innovation and Society, 1986); M. A. W. Gerding, E. H. Karel, and G. E. de Vries, *Van turfstrooisel tot actieve kool: De ontwikkeling van de veenverwerkende industrie* (Zwolle: Waanders, 1997).

111 Technische Rijksvoorlichtingsdiensten ten Behoeve van Handel en Nijverheid, *Verslagen van de werkzaamheden over het jaar 1926* (The Hague: Algemeene Landsdrukker, 1927), 34.

112 Technische Rijksvoorlichtingsdiensten ten behoeve van handel en nijverheid, *Verslagen van de werkzaamheden over het jaar 1925* (The Hague: Algemeene Landsdrukker , 1926), 48–57.

113 G. van Iterson, "De toekomst der rubbercultuur in Nederlandsch-Indië," *De Indische mer-cuur* 39 (1916): 51–56.

114 T. J. van Kasteel et al., *Een kwarteeuw TNO 1932–1957: Gedenkboek bij de voltooiing van de eerste 25 jaar werkzaamheid van de organisatie TNO op 1 mei 1957* (The Hague: Nether-lands Organization for Applied Scientific Research, 1957), 9; *Mededeelingen betreffende de Wetenschappelijke Commissie van Advies en Onderzoek in het belang van Volkswelvaart en Weerbaarheid*, no. 3 (1918); *Mededeelingen betreffende de Wetenschappelijke Commissie van Advies en Onderzoek in het belang van Volkswelvaart en Weerbaarheid*, no. 5 (1919).

115 Jan Al, "Research als overheidstaak," Ph.D. dissertation, Technical University Delft, 1952, 26.

116 C. J. van Nieuwenburg Faber, "De nationale organisatie van wetenschappelijk-technisch werk," *Chemisch Weekblad* 17, no. 7 (1920): 70.

117 Quoted in Al, *Research als overheidstaak*, 30.

118 Went, *Rapport der commissie*, 1.

119 Cf. Wim van der Schoor, "Biologie en landbouw: F.A.F.C. Went en de Indische proefstations," *Gewina* 17, no. 3 (1994): 145–161; Geert Somsen, "Hooge School en maatschappij: H. R. Kruyt en het ideaal van wetenschap voor de samenleving," *Gewina* 17, no. 3 (1994): 162–176; Somsen, "Wetenschappelijk onderzoek en algemeen belang," 201.

120 Went, *Rapport der commissie*, 11–13.

121 Al, *Research als overheidstaak*, 67–69.

122 Paul E. de Hen, *Actieve en re-actieve industriepolitiek in Nederland: De overheid en de ontwikkeling van de Nederlandse industrie in de jaren dertig en tussen 1945 en 1950* (Amsterdam: De Arbeiderspers, 1980).

123 H. C. J. H. Gelissen: "Positieve regionale centraal georiënteerde welvaartspolitiek," in *Werk en streven van Prof. Dr. Ir. H. C. J. H. Gelissen* (Maastricht: Boosten en Stols, 1960), 171–185; Faber: "*Een onafwijsbare plicht*," 25.

124 The Centraal Economisch-Technologisch Instituut was an economic and technological advisory body for industrial investments. The Central Organization for Applied Scientific Research was a cientific research organization.

125 De Hen, *Actieve en re-actieve industriepolitiek in Nederland*, 179, 238–240.

126 Van Kasteel, *Een kwarteeuw TNO*, 21.

127 Ibid., 23–24; Maat, *Science Cultivating Practice*, 82–86.

128 Gerard Alberts: *Jaren van berekening: Toepassingsgerichte initiatieven in de Nederlandse wiskundebeoefening 1945–1960* (Amsterdam: Amsterdam University Press, 1998); Kersten, *Organisatie van en voor onderzoekers*, 8.

129 Netherlands Organization for Pure Scientific Research (ZWO), *Nederlandse Organisatie voor Zuiver-Wetenschappelijk Onderzoek: Voorbereiding en werkzaamheden in de oprichtingsperiode 1945–1949* (The Hague: ZWO, 1950), 6; G. Pascal Zachary, Endless Frontier: Vannevar Bush: Engineer of the American Century (New York: Simon & Schuster, 1997).

130 Netherlands Organization for Pure Scientific Research (ZWO), *Nederlandse Organisatie voor Zuiver-Wetenschappelijk Onderzoek*, 25; Kersten, *Een organisatie van en voor onderzoekers*, 53–164.

131 De Hen, *Actieve en re-actieve industriepolitiek*, 268; Lintsen, *Research tussen vetkool en zoetstof*, 32.

132 H. J. Frietema and G. A. Kohnstamm, "Welke duurzame structuurveranderingen hebben zich laatstelijk voorgedaan . . . ," preliminary advice to the Association for Political Economy and Statistics (Vereeniging voor de Staathuishoudkunde en de Statistiek) (The Hague: Nijhoff, 1947); De Hen, *Actieve en re-actieve industriepolitiek*, 270.

133 Government of the Netherlands, "Nota inzake de industrialisatie in Nederland," Dutch National Budget for 1950 (The Hague: Staatsuitgeverij 1949).

134 Ibid., 36–37; Van Kasteel, *Een kwarteeuw TNO*, 31; Eekels, Christiaans and Kaasschieter, *Ondernemen en vernieuwen*, 17; W. J. Dercksen, *Industrialisatiepolitiek rondom de jaren vijftig: Een sociologisch-economische beleidsstudie* (Assen: Van Gorcum, 1986), 203–205.

135 Ernst Homburg, Aat van Selm, and Piet Vincken, "Industrialisatie en industriecomplexen: De chemische industrie tussen overheid, technologie en markt," in Schot et al., *Techniek in Nederland*, vol. 2, 377–401; Government of the Netherlands, "Vijfde Nota inzake de Industrialisatie van Nederland," Dutch National Budget (The Hague: 1956).

136 F. Henry Brookman, *Het Nederlandse wetenschapsbeleid in Europees perspectief 1945–1975: Een terreinverkenning* (Amsterdam: Free University, Department of History and Social Aspects of Natural Sciences, 1975), 37–64; Dercksen, *Industrialisatiepolitiek rondom de*

jaren vijftig, 199–203; A. Lagaaij and G. Verbong, *Kerntechniek in Nederland, 1945–1974* (Eindhoven: Stichting Historie der Techniek, 1998).

137 A. J. Piekaar, "De organisatie van het wetenschapsbeleid," *Universiteit en hogeschool* 9 (1962–1963): 225.

138 Frits Henry Brookman, "The Making of a Science Policy: A Historical Study of the Institutional and Conceptual Background to Dutch Science Policy in a West-European Perspective," Ph.D. dissertation, Free University, Amsterdam, 1979, 331–332.

139 Ibid., 332.

140 Brookman, *Nederlandse wetenschapsbeleid in Europees perspectief*, 14.

141 Stuart S. Blume, *The Development of Dutch Science Policy in International perspective, 1965–1985: a report to the Raad van advies voor het Wetenschapsbeleid* (Zoetermeer, 's-Gravenhage: Ministerie van Onderwijs en Wetenschappen, Distributiecentrum Overheidspublikaties, 1985); Raad van Advies voor het Wetenschapsbeleid, *Jaaradvies 1990: Een kwart eeuw wetenschapsbeleid* (The Hague: 1990).

142 M. L. de Brauw, *Wetenschapsbudget 1972* (The Hague: Staatsuitgeverij, 1971), 2.

143 Kersten, *Organisatie van en voor onderzoekers*, 186, 319–349.

144 Blume, *Development of a Dutch Science Policy*; Auke van Dijk, Jaap Frankfort, and Tom Horn, *Wetenschaps- en technologiebeleid in Nederland* (Leiden: Adviesraad voor het Wetenschaps- en Technologiebeleid, 1993), 152; Asje van Dijk, "Beleidsinstrumenten in technologiebeleid," in Hans Achterhuis et al., eds., *Technologie en samenleving* (Leuven and Apeldoorn: Garant, 1995), 377–402.

145 Arie Rip, "De gans met de gouden eieren en andere maatschappelijke legitimaties van de moderne wetenschap," *De Gids* 145, no. 5 (1982): 285–297.

146 H. B. G. Casimir, *Het toeval van de werkelijkheid: Een halve eeuw natuurkunde* (Amsterdam: Meulenhoff Informatief, 1983); Kees Boersma and Marc J. de Vries, "De veranderende rol van het Natuurkundig Laboratorium in het Philips-concern gedurende de periode 1914–1994," *NEHA Jaarboek* 66 (2003): 287–313; Homburg, *Speuren op de tast*, 43–51.

147 John Irvine and Ben R. Martin, *Foresight in Science: Picking the Winners* (London: Dover/Frances Pinter, 1984), define strategic research as "basic research carried out with the expectation that it will produce a broad base of knowledge likely to form the background to the solution of recognized current or future practical problems" (4).

148 D. MacKenzie, *Inventing Accuracy: An Historical Sociology of Ballistic Missile Guidance* (Cambridge, Mass.: MIT Press, 1990).

149 For more on the history of the knowledge economy, see Joel Mokyr, *The Gifts of Athena: Historical Origins of the Knowledge Economy* (Princeton: Princeton University Press, 2002).

150 Please see Ernst Homburg and Arie Rip, "Technisch-, industrieel- en landbouwkundig onderzoek in Nederland in de twintigste eeuw," *NEHA Jaarboek voor economische, bedrijfs- en techniekgeschiedenis* 66 (2003): 201–207, for the history of this discussion group.

A ferry carrying the Dutch governor-general, Van Limburg Stirum, across the Lematang River on Sumatra in 1921. Colonial authority was largely based on the technological arsenal of Western powers. The conquest of the East Indian archipelago by the Netherlands was made possible especially by their firearms and larger ships. Effective control over inland regions became possible only after the introduction of smaller and more effective firearms and the deployment of new means of transportation. From the 1850s on, steamboats and steam locomotives provided more speed and flexibility, and after 1900 automobiles soon became important. However, Western technology alone was not enough, as is shown by the use of this ferry.

6 Technology and the Colonial Past

Harro Maat

An overview of twentieth-century technology development in the Netherlands would be incomplete without attention to the Dutch colonies. The Dutch colonial legacy of the first half of the twentieth century has left deep marks in both Dutch society and the former colonies. Two questions are central in this respect: What, exactly, was the position of the Netherlands as a colonial power? And what is the specific legacy of colonialism in the Netherlands and in the former colonies? So far historians have mainly examined these questions only in terms of political-administrative and economic developments.[1] In this chapter I begin to formulate an answer to these questions in terms of colonial technology in the Dutch East Indies. What was the exact nature of technology development in this colony and how did it carry over into the postcolonial period? Because the whole issue of colonization and decolonization in relation to technology development largely involves uncharted terrain in colonial historiography, this chapter can only be exploratory in nature.[2] To avoid getting bogged down in generalities we will focus in more detail on technology developments in the sugarcane industry. Much has been written on this sector already, and this makes it easier to focus more specifically on its technological dimension. Moreover, sugar industrialists invested heavily in technological innovations, and there was also considerable variation in the kind of innovations pursued. Our decision to focus on the sugarcane industry is also motivated by the impression that technology development in this sector offers a representative case of colonial technology development in general. In other words, many of its features can probably be found operating in similar ways in other sectors of the colonial economy.

Before examining technology development in the sugar industry, we discuss several general aspects and developments, notably the influence of major social and institutional changes on colonial technology development. Next, we describe the various actors that left their mark on technology in the Dutch East Indies, its sugar industry in particular. In the remaining part of this chapter we argue that the sugar industry and its technology developed in close interaction with the ethnically and politically diverse colonial society.

Technology and the Colonies

A primary economic motivation for the Dutch to maintain a colonial empire in Asia was the exploitation of its natural resources, which mainly plant products such as coffee, spices, and tobacco. Until 1799, colonial government and colonial entrepreneurship were combined in the United East Indian Company (Vereenigde Oost-Indische Compagnie, VOC) and were mainly geared to organizing trade and transportation; production was primarily in the hands of the indigenous population. For this reason early concerns regarding technical innovation in the Dutch East Indies mainly involved shipping, navigation, and the maintenance of ports. Not that issues related to the production of major crops were entirely neglected. For example, the VOC introduced specific plant species or varieties of species from other regions, so that many crops that are important in the colony's modern export economy are not native to the archipelago.[3] Another, more indirect aspect of VOC promotion of technology development in the colonies related to hiring physicians to protect its staff against tropical diseases and disorders. Various physicians stayed on for longer periods in one of the VOC trading posts and devoted much time to studying the archipelago's rich flora.[4] The VOC also allowed small numbers of other researchers to work in the East Indies. In 1817 the Dutch government, which by then had taken over control of the colonies from the bankrupt VOC, founded the Botanic Garden in Buitenzorg (now Bogor) on the island of Java as a state botanical garden (it is now called the Bogor Botanic Gardens). The purpose of this facility, located in the governmental palace garden, was research efforts geared toward the exploration and exploitation of nature, which also implied innovation-oriented projects. Until the 1860s, however, the botanical garden's budget was too small to do much more than maintain and slowly expand the collection.

Laboratories for the Colonial Economy

In the mid-nineteenth century, in response to developments in Europe, Dutch colonial authorities began to undertake efforts aimed at building laboratories in the Dutch East Indies. One of the scientists involved was G. J. Mulder, a chemist at the University of Utrecht who advised the colonial government on how to set up research facilities.[5] Yet lab research was fairly restricted in scope in the East Indies until the century's final decades. In this half century botanical and chemical research increased considerably. In 1905 the botanical garden and its laboratories became part of the newly created colonial Department of Agriculture (from 1911 the Department of Agriculture, Industry, and Commerce). Earlier the Department of Public Works had

Oil drilling rig number 19, of one of Royal Dutch/Shell's subsidiaries in North-Sumatra, about 1915. Indonesia's oil industry, now a mainstay of its economy, had its origins in the colonial era. Exploration for and exploitation of not only oil but also other resources and minerals such as coal and tin got started in the late nineteenth century. To this end, the Royal Dutch Society for Exploitation of Petroleum in the East Indies, one of the precursors of the Shell corporation, was established in 1890.

been set up to initiate the renewal of infrastructural services. These two departments were unquestionably the main governmental bodies for colonial technology development in the twentieth century.

The Dutch government's increasing attention to research and innovation at the end of the nineteenth century was accompanied by the emergence of several new activities that changed the outlook of both the colonial economy and colonial technology development. Major innovations were associated with the introduction of steam as a power source, the construction of railroads, and the exploitation of minerals. In addition to minerals mining, mainly oil drilling evolved at the start of the twentieth century into one of the most lucrative activities in the colonies.[6] It is closely tied to another characteristic development that started in the nineteenth century and came to full flowering in the twentieth: the strong growth of the private sector, with several businesses evolving into big capital-intensive companies. This growth of business followed major political and administrative reforms, enacted in

1870. These reforms were a response to the "cultivation system," a system of forced cash-crop production by local producers, regulated by the state. The cultivation system, to which I shall return later, was initiated in the 1830s but faced economic decline and political controversy during the 1850s and 1860s. The 1870 reform reduced the economic activities of the government and created legal space for private enterprise. Industrial companies—sugar mills in particular—became interested in doing research as a way to make technological improvements in their operations and thus raise their productivity.

Basically, colonial government and industry relied on the same organizational concept for research and innovation: centralized research institutes (*proefstations,* experiment stations) that served the whole sector. The government stations were funded by public money, the private stations through associations that charged fees to planters in return for supporting them through the experiment stations. This shared approach gave rise to sustained collaboration between public and commercial experiment stations, and also, of course, to disputes about task distribution and mutual coordination. Although the experiment stations clearly left their mark on technology development in the colonies in the twentieth century, they were not the only facilities initiating technological innovation. In all branches of industry innovations were realized without the input of these stations. Moreover, in addition to Dutch technicians, local technicians and employees played a role; in some cases they had received training at one of the schools on Java, or their skills were exclusively based on work experience.

Civilizing Research

The colonial government's large investments in innovation-oriented research was partly a response to similar developments in Europe. A major motivation for setting up research activities was the government's so-called "ethical politics."[7] This was the Dutch version of a "civilizing" mission, similar to what other European colonial powers developed for their colonies.[8] Whereas a liberal approach had characterized Dutch politics in the nineteenth century, around 1900 state intervention became more prominent. In the early twentieth century, the Dutch government's new ethical policies vis-à-vis native subjects were expressed in regulations and initiatives in socioeconomic domains, including investments in science and technology. Their implementation increasingly led to conflicts between the government and colonial entrepreneurs, notably in connection with labor relationships and other social issues. The conflict was most intense after the outbreak of World War I. In the war period companies saw their profits go up because of the increasing demand for all sorts of colonial products, but government saw itself more and more constrained when it came to implementing its ethical politics. Particularly its effort to prevent food shortages through rice imports became a thorny issue because of the limited availability of tonnage and safe shipping routes during the later war years. A factor that further sharpened the contradictions was the rise of nationalist movements on Java. The cautious rapprochement of colonial government toward these pressure groups, combined with the highly unfavorable disposition of these organizations toward the Dutch private sector, was an additional reason for colonial entrepreneurs to engage more actively in political issues. In short, the colonial government's serious effort to defend the interests of the local population affected its involvement with both entrepreneurs and innovation-oriented research, as I will discuss in more detail later in the chapter.

From the mid-1920s on, the relationship between government and business in the area of technology development gradually improved.[9] The economic boom of the second half of this decade was accompanied by major expansion of research and the application of all sorts of technological innovations. This involved not only European companies but also local agriculture and local industries. The economic crisis of the 1930s led to shrinking budgets; hence reduced spending on research and technological innovation caused many expansion plans to be postponed, if not canceled altogether. Because many entrepreneurs who had joined an experiment station association no longer could pay their membership fees, some of these associations felt forced to ask for government support. This led to the establishment of a joint body for coordinating and performing research, which many viewed as a strong improvement of the organization of innovation-oriented research.[10]

During the turbulent years of World War II and the period immediately after the war, and the advent of an independent Indonesian state, the experiment stations were faced with major changes in staffing and financial means. During the Japanese occupation most experiment stations simply continued their activities (albeit on a smaller scale), partly because not all Dutch staff members had been put into internment camps. After the capitulation of Japan research at nearly all experiment stations was started up again. The proclamation of Indonesian independence did not materially affect the stations' operations at first. In 1949, however, all Dutch officials were forced to leave the country, including Dutch employees of government experiment stations. Some ten years later they were followed by their colleagues from the private stations.[11] The expulsion of scientists, technicians, and administrative staff had drastic consequences for staffing and funding, as well as the research effort's continuity. For Indonesia it posed the challenge of finding capable scientists and managers, and especially of locating funding. For the Netherlands it meant a sudden influx of professionals who, given their quite specific training and experience, were hard to employ. In many cases, however, international contacts and the emerging domain of development aid helped to alleviate this problem and personnel could be redeployed. Moreover, several Dutch knowledge institutions that used to focus exclusively on colonial East Indian affairs managed to apply their knowledge and experience in more general or other international contexts, as was true for the Dutch colonial experts.

Sugar Industry Actors

A major feature of technology in the Dutch East Indies is that it developed in a social and organizational context that differed entirely from Dutch society. Because colonial companies and institutions operated in a very specific sociocultural and economic setting, colonial technology cannot simply be understood as Dutch (or Western) technology in an exotic setting. By definition the colonies constituted a meeting place of various cultures and the form and intensity of these cross-cultural interactions were also major variables of technological development. What is more, nearly each and every island of the Dutch East Indian archipelago had a unique social makeup. This was a result from the different geographical conditions and differing ethnic and cultural backgrounds of the inhabitants of the islands but also resulted from the presence and numbers of Europeans, Chinese, and other foreigners. The planters' community on Java and on North Sumatra had very different characters.[12] Likewise, the ethnic division among workers and technicians differed from one branch or company to the next.

In the sugar industry on Java we can roughly distinguish four categories of actors that played a role in sugar technology development. First, the European owners and experts, who performed a leading role in technology development because of their social position and their level of knowledge. Second were the basic technicians active in the fields and factories; in terms of both their training and ethnic background this was a mixed group, comprising Dutch technicians trained at some Dutch or European sugar school, technicians of "Eurasian" descent trained at the sugar school in Surabaya, and others without specific training but who still fulfilled a technical role based on their particular skills and experience. In the third category were the workers, mainly Javanese, who often performed heavy and risky tasks in fields and factories. Finally, the Chinese who worked in the sugar industry constituted a specific group of actors. They differed from the others in sociocultural respects as well as in their specific professional role. Various sugar factories had Chinese owners, but virtually everywhere throughout the sector it was Chinese technicians who were in charge of a crucial aspect of production, namely the sugar's transition from liquid to crystals. As we shall argue, both colonial sugar production and the sector's technological development were determined by cooperation among these various groups of actors.

Technology development in the Dutch East Indies, as our concise overview suggests, was closely linked up with specifically Dutch contexts and colonial pursuits, yet at all levels and in all domains the particular geographic, socioeconomic, and cultural factors linked to the East Indies gave rise to quite divergent processes. Technology development was no exception. How, exactly, this was the case will be elucidated on the basis of technological changes in the sugarcane industry.

Technology and Sugar

Although the European sugar industry on Java was in many respects comparable to the sugar industry in other colonies and countries, two major aspects made it unique: land ownership and the organization of work. In many countries and colonies that cultivated sugarcane, such as Cuba and (until the late nineteenth century) Surinam, the work was done on plantations by slaves imported from Africa. This meant that both land and labor were privately owned. By contrast, on Java manufacturers relied on land and labor that was hired in from the indigenous population. It should be added that during the period of the culitvation system, the local population was left little choice as to whether or not they were willing to supply their land or labor and even after 1870 the companies had several options to secure land and labor by force.[13] In this system the colonial government was a key player, but not the only one. The exploitation of sugar mills, for example, was left to private initiative, while the Netherlands Trading Company (Nederlandsche Handel-Maatschappij, NHM) was in charge of selling the goods on the European market. That the private sector was largely subordinate to the state, however, is suggested by the fact that having good contacts with government was very useful when it came to setting up or taking over sugar mills. Moreover, the NHM was established on the initiative of King Willem I and could also operate as a monopolist thanks to government protection.

For sugar producers the Dutch colonial government's involvement was initially very lucrative, because the most risky aspects of sugar production were covered by the government.[14] In the cultivation system, both sugarcane and sugar as end products were acquired and sold against a fixed price. This is not to say that the producers made money without any effort. Notably in the period from 1840 to 1860, the government put pressure on producers to

The sugar cane industry first expanded under the cultivation system (about 1830–1870), and was later carried on by private companies. The basis of the business was the agricultural concerns that grew the cane. Sugar companies leased land and hired workers, of whom large numbers were needed during the sowing and planting seasons. This photo, from around 1880, shows how the cane was grown: a piece of cane is put into a furrow and covered with soil, and later sprouts grow from it. The furrow facilitated easy irrigation of the plants.

invest in technological improvements in sugar production, to raise both productivity and the quality of the end product, sugar.[15] After 1860 the cultivation system rapidly declined because of mounting political pressure from liberals linked to the call for change by entrepreneurs themselves, who experienced that the NHM, given its monopoly, creamed off profits to their disadvantage.[16] Over two decades government-controlled sugar cultivation was phased out, and the NHM was turned from a trading body into a general bank.

This transformation meant that from then on the sugar producers were in charge of the supply of sugarcane and the sales of their product. However, many producers found out that being responsible for the full process, from planting sugarcane to sales, proved onerous. Furthermore, the rising production of beet sugar was causing sugar prices on the European market to drop sharply, as a result of which sugarcane was sometimes traded below the cost price. Considerable financial investments were indispensable for survival in the last two decades of the nineteenth century, and such support had been supplied by several Dutch institutions, notably the NHM and the Amsterdam Trading Association (Handelsvereeniging Amsterdam, HVA). Essentially these two financiers became the owners of many sugar mills on Java and de facto also the island's largest sugar producers.[17] By investing in technological innovation and tapping new markets in Asia, the sugarcane industry gradually recovered. It realized higher outputs through changes in management and several specific technological innovations.

Organized Innovation

To stimulate technological innovation in sugarcane cultivation and production, toward the end of the nineteenth century the sugar producers on Java began to set up central experiment stations. This organized form of research and innovation was a new phenomenon in the colonial economy. The willingness to collaborate among Java's sugar producers seems remarkable at first sight, mainly because they operated in the same world market and were therefore direct competitors. However, the sector's specific competitive relationships were quite unusual. Sugarcane production strongly depended on factors that applied to a whole region, such as weather variation or the availability of (slave) labor. This meant that mutual competition was not so much a regional affair as a global one, between, for instance, sugar from Java and sugar from Cuba and also between tropical sugarcane and sugar beets from Europe.[18] In short, on Java the producers' joint economic interests outweighed their mutual economic rivalries; there was, as one representative of the sector put it, a "natural solidarity, which has been there from the very start."[19] Soon after the first associations of sugar producers on Java emerged, in the early 1880s, the idea of setting up shared experiment stations presented itself. Although the bad economic situation encouraged collaboration, the founding of associations and experiment stations cannot be simply explained by declining sales or dropping prices. These initiatives were mainly tied to developments in knowledge gathering in European agriculture and the sugar beet industry, which largely originated in Germany.

In the mid-nineteenth century a large number of laboratories were set up in Germany and other Western countries to carry out research on agricultural products, such as manure, feed, milk, and also sugar beets. Recent developments in organic chemistry, led by the German chemist Justus von Liebig, provided a direct incentive in many cases.[20] In addition to providing insights into the nature and effects of all sorts of organic substances, his ideas about experimental science and its application stimulated agricultural organizations to establish experiment stations. These stations were mainly geared toward improving fertilization. The research on beet sugar emphasized its precise composition, the production process, and the chemical additives needed to create a usable product. In 1866 the first laboratory specifically devoted to the sugar industry was set up in Berlin: the Institute for the Sugar Industry (Institut für die Zuckerindustrie).[21] The sugar producers on Java looked to developments in Germany as a model to be followed, and in 1885, 1886, and 1887 they set up three experiment stations for the sugar industry: one in West Java, one in East Java and one in Central Java. The German orientation in this field is further underscored by the large number of German scientists hired to work in these three stations.[22]

The experiment station scientists studied not only chemical substances and processes in sugarcane cultivation and sugar production but also botany. Like organic chemistry, botany had become much more an experimental science during the second half of the nineteenth century, and here, too, German scientists played a leading role.[23] The botanists that were hired by the sugar experiment stations started working right away on a serious plant disease that caused extensive damage to the cane: sereh disease.[24] This is a growth disorder that causes reduced growth, making the cane look similar to sereh, the Indonesian name for lemon grass (*Cymbopogon citratus*). The disaease was effectively controlled initially through agronomic measures and later through resistance breeding. It is thought to be caused by a virus, but the true causes have never been completely understood. Sereh disease only occurred on Java; trained botanists could not only apply their general knowledge about botany and experimental research but also, basically starting from scratch, could develop more knowledge about this particular disease's causes. Thus, the trajectory of work on sereh disease aptly illustrates the relative significance of Europe's (German) laboratories for agriculture and sugar.

Likewise, the sector also had to bring its organization into line with local contexts. Because maintaining three experiment stations appeared too expensive, the one in Central Java was closed prematurely. In 1907 the other two merged into the Association of Experiment Stations for the Java Sugar Industry. The station in East Java was used for agricultural and botanical research, and the one is West Java was used for chemical and technological research. This merger was partly a result of changes in the underlying structure, the sugar producers' associations, which merged in 1894 as a single umbrella organization, the General Syndicate of Sugar Producers in the Dutch East Indies (Algemeen Syndicaat van Suikerfabrikanten in Nederlands-Indië, ASNI).[25] As previously mentioned, most sugar producers on Java did not own the factories they operated; instead, they were employed by Dutch investors to act as general managers. In 1917 a third organization was founded: the Owners' Federation of Dutch East Indian Sugar Entrepreneurs

Sugar is harvested to be studied by the East-Java station, around 1910. Growing sugar cane increasingly involved the application of Western knowledge and cultivation technologies concerning soil preparation, the quality and features of sugar cane, and the timing of the harvest. All of these aspects were extensively studied, principally at experiment stations, of which the first was set up in 1885.

(Bond van Eigenaren van Nederlands-Indische Suikerondernemers, BENISO).[26] In principle, the association of experiment stations was concerned with research and technical issues, ASNI defended the sugar interests on Java, and the owners' federation defended these interests in the Netherlands. In practice, however, the various trade interests, political aspects, and technological concerns were closely intertwined.

The organizational context in which research and technology in the sugar industry on Java developed remained in place until the end of the colonial period.[27] The formula whereby a lab or experiment station for research and innovation worked to the benefit of multiple companies was also applied in other sectors in the Dutch East Indies around 1900, particularly in sectors that, like the sugar industry, relied heavily on tropical crops, such as rubber and tobacco, and that were active on Java and Sumatra exclusively. In contrast to the experiment stations for sugar, however, such stations for other crops were set up in close collaboration with the laboratories of the State Botanical Garden in Buitenzorg. Many of the newly built labs were funded by government money and contributions from planters' associations. After 1905 these semiprivate labs either continued as commercial labs or became part of the colonial Department of Agriculture.

The establishment of experiment stations for sugar and other products had a noticeable effect on the job market for technicians, both directly, through job openings at the experiment stations, and indirectly, as the sugar mills increasingly needed employees who could apply the various innovations and procedures developed by the experiment stations to a factory's specific circumstances. The secondary education system in the Netherlands was a major supplier of technicians: they had to be skilled in machine construction, chemistry, and agriculture. The sugar producers on Java attached great importance to official diplomas and certificates, so as to have some certainty that it was worth paying for a new employee's passage.[28] On Java, in Surabaya, a school for sugar technicians was set up at the start of the twentieth century as well. The experiment stations also hired many academically trained employees. Those who wanted to do research for a short period in the Dutch East Indies could do so in a special lab, built in 1890 on the grounds of the State Botanical Garden, initially called "Visitors' Laboratory" and later on renamed the Treub Laboratory (for Melchior Treub, the director of the botanical garden).[29] The establishment of the Colonial Institute (Koloniaal Insituut) in Amsterdam in 1926, mainly funded by Dutch East Indian industrialists, was a direct counterpart of organized knowledge development in the Dutch East Indian business sector. The preliminary plans for such an institutions had been formulated around 1900. In 1950 it was renamed the Royal Tropical Institute (Koninklijk Instituut voor de Tropen) and gained significance as a major museum, data archive, and research center.[30]

Innovations in Sugarcane Cultivation

In the first years after the establishment of the sugar experiment stations, the research emphasis was on botany, with the aim of finding a speedy and effective solution for sereh disease, which caused large losses. The botanists' approach was basically the same for each crop. Botanists not only studied the plants in detail—the species and varieties, plant anatomy and growth stages—but they also did ecological-agricultural research, focusing on soil, insects, funguses, and other factors that affect the crop. In short, they looked at the total interaction between plant and environment. The German botanist F. Soltwedel, director of the experiment station in Central Java, discovered that some cane varieties were resistant to sereh disease and that also plants at higher elevations did not display the symptoms. The fact that some cane varieties were immune to the disease encouraged botanists on Java to renew their efforts to collect varieties from all corners of the world and test them. A major discovery that was made was a connection between altitude and resistance against sereh disease. At higher elevations the cane did not

The wooden pole in this photo from around 1930 holds a kind of bag of fine-meshed fabric to prevent the cane's flower from being pollinated by random pollen. This protection was necessary to preserve the strain from random pollination, because the flower already had been hand-pollinated using pollen from a selected "father." The 1884 sugar crisis, caused by the effect of competition from beet sugar on the world market, was exacerbated by a serious disease of the cane, sereh, which led to the establishment of a number of experiment stations. These were funded through contributions from the associated sugar companies. Extensive publications reported the results of the many experiments undertaken. The development of new varieties was a major aspect of the stations' activities.

develop the disease and remained resistant when later transplanted to the lower plains. This last discovery turned out to be directly applicable in practice and nearly all sugar producers laid out nurseries in higher areas, the so-called *bibit* gardens.

In the 1880s another major discovery was made on the inflorescence of sugarcane. For a long time it was thought that sugarcane could only be multiplied vegetatively (through cuttings) because it was thought that the plants produce no seeds. This proved to be wrong. Almost simultaneously, Soltwedel and two researchers on Barbados discovered how flowering and seeding occurred in sugarcane. Now breeders were no longer limited by naturally occurring varieties but could develop new varieties through cross-pollination.[31] The direction the fight against sereh disease had to take was now clear and involved combining three features: resistance against sereh disease, a high sugar level, and a high yield in Java's climate. Even though the goal was clear, the task was not simple in a period when the knowledge about breeding processes was still in its infancy. Although biologists were very interested in genetics, they held different views about how qualities are transferred from one generation to the next. Parallel to this theoretical lack of insight was the lack of experience with improving sugarcane. Finding good combinations mainly required a lucky hand, and the length of the growing season meant the results of crossbreeding could be evaluated only many months later. Consequently not until 1921 did botanists and breeders create a sugarcane variety on Java with the needed qualities, a variety named POJ2878 but soon nicknamed "miracle cane."[32] (POJ stands for Proefstation Oost Java.) It took another eight years for this variety to be planted on a large scale, after which sugar producers could do without their *bibit* gardens up in the mountains again, the hitherto most effective strategy against sereh disease. Research on this disease was discontinued, even though its cause was not yet fully identified.

Experiment Stations and Sugar Producers

The research and developments with respect to improving sugarcane suggest that the research effort at experiment stations took place more or less independent of the practice of sugarcane cultivation. In reality, however, there was considerable interaction between scientists at experiment stations and the sugar manufacturers, technicians, and field and factory workers. This knowledge exchange went both ways. Regarding the efforts to improve sugarcane, the stations' botanists benefited significantly from sugar producers' trying out imported varieties of sugarcane and also performing crossings themselves.[33] Likewise, research on sugar field fertilization and irrigation profited much from the experience of and experiments by producers and by the Javanese workers in the fields. The implementation of the experiment stations' research results—such as the introduction of new varieties, cultivation techniques, control of diseases, and knowledge about fertilization and other matters—required more than availability of innovations developed by the experiment stations. One station's successful development a high-yield variety or a certain fertilization method did not mean that these discoveries would work in all areas on Java where sugarcane was cultivated. To allow for such generalization a large number of field experiments were needed. These were performed by experiment stations, and substations were set up in several other locations on Java to further effective implementation of these tests under diverse conditions. Also, innovations were offered to the producers for testing, whereby a positive result could give rise to further and more specific tests. In both cases the sugar producers' cooperation was indispensable, such as their willingness to set aside part of their acreage for experiments. Thus, even though the development of knowledge and innovations in the sugar industry was centralized, it was not altogether cut loose from practice in decentralized areas. This mutual interaction between research, innovation, and the sugar industry also implied that the station scientists became involved in broader issues related to sugar production on Java.

Producers versus Government

Around 1900 the outlook of Dutch colonial politics changed. In the mid-nineteenth century a liberal trend and support of private initiative had led to drastic limitation of the role of government and the abolition of the cultivation system. Around the turn of the century, however, the winds began to blow from the opposite direction. The more active role formulated by the Dutch government in the early twentieth century vis-à-vis its colonies was inspired by "ethical" politics. Enlightened politicians were geared toward what in today's development aid jargon is called "good governance," an administrative system that was not after self-enrichment but held itself accountable to raising the overall population's prosperity. A substantial part of the Dutch government's ethical policy was to be implemented by the Department of Agriculture. In a report on the Department of Agriculture's tasks and responsibilities, the director of the State Botanical Garden, Melchior Treub, underlined the department's role in research.[34] Research and innovation initiated by the department were directed mainly toward improving food agriculture, with rice as the most important food crop by far. Research was also undertaken on cultivating export crops, thus stimulating the native population to get a stake in an economy that was largely run by European

The sugar mills and their extensive fields (ranging from hundreds to several thousands of hectares) were odd Western enclaves in the East Indian landscape and society. They defined the landscape and the immediate environment because of their relatively large spatial presence, which included factory buildings, homes for Western employees, a kampong, and an extensive private railroad network serving the plantations. Another major element was the irrigation system, as shown on this aerial photo from about 1935 of the Meritjan sugar mill, north of Kediri.

companies. The various European planters associations followed these developments with suspicion. Resistance to these initiatives, however, was more limited than that to government involvement in company management regarding labor conditions.[35] Evidently, the effects of ethical politics were felt in research as well.

The sugar companies cultivated sugarcane in fields that were leased from the local population. Most desirable was fertile land with a good water supply, land that that was also very suitable for rice cultivation. This fact constituted the core of a contentious issue that actually came up early, at the very start of the sugar industry's activity on Java and climaxed in the first decades of the twentieth century; it is still debated among historians.[36] The sugar producers felt that the rent paid for the fields and the income from the labor in fields and factories supplied the local population with a net income that exceeded the income earned from the cultivation of food crops. Critics of the sugar industry doubted this and also pointed to the much heavier labor involved in sugar production. Their main argument, however, was that a country in which the food supply was not stable and that had to import rice to avoid famine could not afford to rent out a huge acreage of potential rice fields to entrepreneurs who earned handsome profits through sugar production. This point was especially apposite during World War I, and in 1918 it also became the focus of political struggle, whereby the colonial government tried in vain to reduce the sugar acreage by 20 percent so that these fields would become available for growing rice and other food crops.[37] Whereas the discussion in politics and the media mainly concentrated on the economic and social pros and cons of current land use in Java, in agricultural research the tension between cultivation of sugarcane or food crops was couched in terms of technological issues.

Distribution of Irrigation Water

Water management in the regions where both sugarcane and rice were cultivated was a subject of research for both the sugar experiment stations and researchers from the Department of Agriculture.[38] Experiments focused on the distribution of water and on the effects of irrigation on crop growth. In regions where sugarcane was the dominant crop, irrigation water for sugarcane and for food crops was almost always extracted from the same water source, yet in its distribution priority was given to sugarcane rather than food crops. There was a so-called day-night regulation, whereby sugarcane fields were irrigated during the day and fields with other crops at night. Because it was necessary to be present in the field during irrigation, the Javanese food-crop farmers were, to put it mildly, not very happy with this regulation.[39] To assuage their growing protests against the day-night regulation, in 1915 the department set up several experiments with reservoirs, or *wadoeks*, that filled at night and that farmers could use during the day to irrigate their fields.

The system functioned quite well—there were an estimated two hundred *wadoeks* by 1920—but protests against the water distribution system continued, for two reasons. First, despite the reservoirs it was not possible to irrigate all the fields during the day. A greater concern was that the use of *wadoeks* came with a negative effect on soil fertility. Normally, the silt in irrigation water improved soil fertility, but as a result of the usage of water reservoirs, the silt would sink to the bottom, meaning that it was left behind when during the day water from the *wadoek* entered the fields. In other words, what was seen as a solution to the distribution problem triggered a soil fertility problem. Since there were no precise data on the composition of the silt and its effect on crop growth, opinions differed as to the severity of this problem. The sugar producers of course felt it was grossly exaggerated, but scientists from the Department of Agriculture did take the matter seriously, and this gave rise to another clash between the colonial government and the commercial sector.

Sugar and Rice

Experts had studied the composition of the silt in irrigation water, its effect on soil fertility and its added value to the growth of food crops, since the establishment of the colonial Department of Agriculture, in 1905. This research was continued by P. van der Elst, a botanist who joined the department in 1910.[40] His research focus was on the interaction between soil and the rice plant, mainly concentrating on a rice disease that was common at the time, known under the local name *omo mentek,* or *mentek.* Just as in the case of sereh disease in sugarcane, the exact causes were never discovered, but by making specific changes in cultivation the disease could be curtailed. Already in the first year of his appointment, Van der Elst was convinced that *mentek* was not caused by a nematode, as was then believed, but by specific chemical reactions in the soil involving an imbalance between nitrogen and other minerals, or possibly a poisonous alkaline soil environment. Van der Elst concentrated on studying the effects of tillage and irrigation on soil reactions and the *mentek* symptoms. He identified a correlation between poor tillage, poor irrigation, and the disease, a combination that occurred frequently, especially in sugarcane regions.[41] His research results thus pointed to a negative influence of sugarcane cultivation on rice cultivation. Naturally, this conclusion displeased the sugar producers. Scientists from the experiment station for sugar designed similar tests and, as could have been expected, their results proved the opposite. At first sight this seems a good example of opposite interests that lead to opposite research results. If the Department of Agriculture and the colonial government at large sought to improve the position of the Javanese population, at some point they inevitably would clash with the sugar producers. Indeed, after 1915 their relationship gradually deteriorated.[42]

In everyday practice, however, the situation was more ambiguous. Within the Department of Agriculture not everyone supported Van der Elst's conclusions. Several experts voiced their skepticism, such as J. E. van der Stok, who had first worked at the sugar experiment station but who in 1905 was hired by Melchior Treub to lead the agricultural department's rice research. In 1910 he became director of the sugar experiment station in Pasuruan, where he studied the influence of sugarcane cultivation on food crop cultivation. He published his research results in the 1913 volume of *Archief voor de Java-Suik-*

erindustrie. A year before, however, Van der Stok had already returned again to the Department of Agriculture, on the invitation of Treub's successor, H. J. Lovink. The department's 1912 yearbook, which appeared in 1913, contains a report of the discussion among Van der Elst, Van der Stok, and several other colleagues about the study done by Van der Elst. Some of those present felt the results not to be very convincing and it was proposed to set up a new plan for further research.[43] The sugar experiment station also pursued this issue.

Research and Commercial Interests

Agricultural research for the sugar industry mainly took place at the experiment station in Pasuruan. To gain insight into growth factors under variable conditions, substations were set up in other locations. The substation in Cheribon was led by F. Ledeboer, who advocated undertaking detailed research of the interaction between sugarcane and indigenous food production. Accordingly, during the 1910s he had set up various experiments and his results did not suggest any negative effect of cane on rice. In 1918 Ledeboer gave a lecture before the Technical Association of Sugar Employees in which he made some striking claims. He challenged the sugar producers to take a more serious look at evidence gathered by critics of the sugar industry that showed its negative influence on food agriculture.[44] His challenge was to test the claims on the basis of solid research rather than taking them for granted. Ledeboer complained about the lack of support for his research in this area, despite his evidence that Van der Elst's conclusions could be easily refuted. Ledeboer also urged the sugar industry to do systematic research of both the technological and economical aspects Javanese agriculture, "so that sugar cultivation can smoothly interact with the continuously changing social conditions, rather than having its regular development impeded."[45] Ledeboer's argument, however, came at a time when the sugar industry increasingly began to turn away from the colonial government's ethical politics. Not surprisingly, a proposal for research in an area that overlapped with Department of Agriculture research and that only indirectly and in the long term might benefit the industry received little support. Moreover, the conditions on both sides became more favorable. In the 1920s the food supply gradually improved because of the government's proactive rice purchasing policy, and the sugar sector managed to raise its output based on new varieties without substantial expansion of acreage.

As we have suggested, one should not consider the agricultural research effort for the sugar industry merely as a supplementary service that supplied the sector with innovations. The research's content and direction was partly shaped by the role of the sugar industry in Dutch colonial society. The industry had

The planting out of cane seedlings in pots at the Pasoeroean experiment station, around 1930. The quality of the plant material was crucial to the quality of the cane and the size of the harvest. The breeding material was subjected to extensive testing. The cane's suitability for various kinds of soil, its resistance to diseases and pests, against drought and flooding—all of these aspects were tested, as were those relevant to processing cane in the mill.

to justify the lease of land from the local population for cultivating sugarcane in the face of public opinion and political policy that put the negative sides of sugarcane cultivation in the foreground. The difficult relationship between the sugar sector and colonial government is directly reflected in the sector's agricultural research effort and that of the government, which tried hard to resolve the mutual contradictions. This largely succeeded in the water distribution dispute with the introduction of reservoirs, but it failed in the case of the damaging effects of sugarcane cultivation on food crop cultivation, largely because research failed to settle the issue. This contradiction gave rise to specific experiments and programmatic proposals for agricultural research. As we saw, there was little enthusiasm for those proposals.

The conflict between government and sugar industry lingered in another area of knowledge development as well. The syndicate of sugar industrialists based in the Netherlands (BENISO) felt that the underlying cause of government measures against Dutch entrepreneurs in the colonies lay in the training facility for colonial officials in Leiden. According to the industrialists, this facility exposed future officials to much too progressive views, while painting entrepreneurs as exploiters of the local population. Led by BENISO and the Bataafse Petroleum Maatschappij (BPM), one of the precursors of the Shell Oil Company, the industrialists financed a training facility of their own, which started up operations in 1925 as the University of Utrecht's Faculty of Indology.[46] From this time on, three out of ten colonial officials for the Dutch East Indies were trained in Utrecht.[47] That sugar producers wanted to invest in a Dutch training facility for administrative colonial positions, rather than in research of the interplay between sugar cultivation and the local economy, suggests that they assumed profitable entrepreneurship to depend first of all on a powerful colonial government and much less on a well-developed native economy and society. Despite the many forms of interaction with local population groups, the sugar industrialists considered their sector primarily a Dutch affair—an attitude reflected not only by developments in agriculture-related technology.

Innovations in Sugar Production

The sorts of technological developments in sugar production were very different from those in sugar cultivation. A large part of this difference can be explained directly from the nature of the processes that are relevant in the cultivation of cane and the extraction of sugar. The processing of cane and beet sugar is marked by many similarities. This is why the emergence of the beet sugar industry in Europe and the related attention from scientists and techni-

For a long time the harvested cane continued to be transported by local means of transportation using animals such as water buffalo or sapi as pulling power. Over time railroads and lorries came into use, along with small locomotives. In more remote areas animals still supplied draft power.

cians for sugar extraction had worldwide implications for sugar production. Another important technological factor was the emergence of steam as a source of power and heat. Mutual coordination of the various parts of the process was very important and technological innovations were mostly introduced in mutual interaction as well as in a phased way. The changing political and economic conditions of the late nineteenth century gave rise to a dynamic of ongoing technological innovation. To understand these one must understand the major steps between harvesting sugarcane and bagging dry sugar.

A Step-by-Step Process

After the cane is harvested it is transported to the mill. Traditionally this was done with wooden carts drawn by buffalos or, occasionally, by horses. Around the turn of the century there was a gradual change toward transport by narrow-gage railroad. At first this merely involved laying tracks on which carts would be drawn by oxen or horses.[48] Only later on were draft animals replaced by small steam locomotives. The use of steam traction was only attractive for mills with a large processing capacity that could process cane from a wide area around the factory.

Next, the cane was shredded, before being pressed between large steel rollers. Various pressing mills were used; depending on the type of rollers used, the cane was pressed three, four, or five times in a row.[49] In between the pressings the cane was sprinkled with water so that it swelled, a process called imbibition, which made it possible to extract more sugar from the cane. In time the efficiency of the mills was greatly improved through adaptations in the composition of the metal mill rollers and their groove structure, in combination with optimizing the pressure with the help of hydraulic cylinders.

The juice pressed from the cane underwent various treatments to separate the sugar dissolved in the juice from other ingredients. After a first heating, dissolved calcium oxide was added to bind acids and other substances. Next, the bound particles were filtered out, after which the purified liquid was further concentrated under vacuum. These so-called vacuum pans facilitated low-temperature boiling and saved fuel, and above all it prevented the decomposition of the sugar molecules.[50] Around 1900 the installations consisted of a series of three to five linked pans, called *multiple effect*.[51] Depending on the desired sugar quality, after concentration the liquid could be further purified by adding acids and bases, again followed by filtration.

Next, the purified mass was pumped back into these vacuum pans, where it was boiled to reduce the water level to the point that the liquid saccharose crystallized. The mixture of sugar crystals and molasses or syrup, the pure cane juice, was pumped into centrifuges, to separate the crystals from the molasses. After the crystallized sugar was dried it was put in bags, ready for transport. The molasses still contained up to 20 percent sugar and was used as an ingredient for the production of alcohol such as arrack and rum.

To some extent, technological adjustments of the various parts of the production process were interrelated or mutually influenced each other. For example, new insights based on chemical analyses could lead to equipment adaptations, but a new technology, such as a new filtration press, could also give rise to all sorts of (comparative) experiments. The technological innovations and adaptations generally were more or less random; different companies experimented on their own, achieving occasional success, or tried out new equipment supplied by a specific manufacturer. To a considerable extent, however, this mutual exchange and interaction was orchestrated. These organizational features provide further insight into the character of technology development in sugar extraction.

Experiment Stations and Machine Manufacturers

Shortly after the establishment of the sugar experiment stations, a systematic analysis of the various steps in sugar extraction and processing was initiated in all factories. By the end of the nineteenth century this system of so-called mutual production control was common in the European beet sugar industry.[52] Its basic principle was that in all stages of the sugar extraction process one should repeatedly take samples and determine their composi-

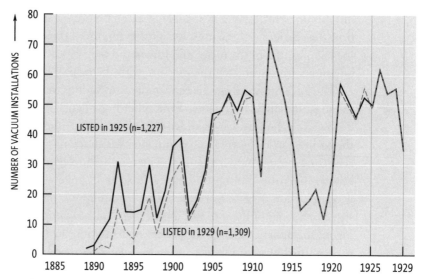

Graph 6-1. Years of first deployment of vacuum installations in Javanese sugar mills, 1884-1929 (For sources see page 603)

Table 6-1. Suppliers of vacuum Installations to the sugar industry on Java in 1929, by number of installations supplied

More than 300	90–130	30–50	20–30	10–20	Less than 10
Stork	Halle Werkspoor Petry Dereux	Piedboeuf Maxwell Borsig Galloway Dürr Mach.Fabr. Breda	Young&Gill Fives Lille Breitfeld Danek Büttner Bousu Fijenoord	Babcock& Wilcox Rombouts Dunkerbeck Hanomag Katendrecht Ruston& Hornsby Wilson	*42 other manu-facturers*

Total 1929 installations: 1,309. (For sources see page 603)

tion. Based on the weight and volume of the cane and its end products (sugar, molasses, and filter residues) one could calculate the saccharose losses. Although at first the sugar producers on Java responded warily, gradually the system gained a foothold, and until about 1910 much attention was paid to the methods of measurement and analysis. This was in great measure because the precise composition of the cane had been hardly studied yet, so that measurements of the first step of the process showed large error margins. Eventually the measurements were standardized and carried out by analysts who worked for the sugar companies. The experiment stations (which mean-

A giant model (scale 1 to 10) of a complete Stork cane-processing factory installation with a capacity of 2,200 tons of cane per twenty-four hours, intended for use in Cuba. The model gave Stork a prominent presence at the Colonial Exposition in Semarang in 1914. Sugar-mill installations and equipment for transport and working the land in colonial cultures were largely based on Western technology and were mostly imported into the East Indies. Over time, the Dutch machine industry managed to acquire a major share of the business, and both Stork, based in Hengelo, and Werkspoor, based in Amsterdam, evolved into leading suppliers, with their own sales and service organizations in the East Indies.

while had merged) could thus restrict their effort to data collecting and processing. The use of fuel was also compared systematically.

This organized knowledge development had a multiple generative effect. For the sugar companies the measurements provided an opportunity for mutual learning. Adaptations and innovations in different facilities could be compared quantitatively and better methods and techniques adopted. Such control allowed experiment stations to compare process stages systematically, which could lead to new recommendations or experiments with other technologies. Periodically, statistics relating to the sugar industry on Java were published. For instance, table 6.1 and graph 6.1 are based on an overview of sugar mill vacuum pans. Graph 6.1 shows a pattern of increasing investments in new installations that lasted about twenty-five years: in 1925 many installations from before 1900 were still in operation, whereas in 1929 there were clearly fewer.

International Contacts

In the first half of the twentieth century, the Dutch East Indies and Hawaii sugar industries were viewed as the most technologically advanced worldwide. In 1898 a handbook written by H.C. Prinsen Geerligs, director of the experiment station in West Java at the time, was published in English on the cultivation and production of sugarcane on Java. Various other sugar experts published in international journals, such as *The International Sugar Journal*, while articles from the *Archief voor de Java-Suikerindustrie* appeared in translation (also in condensed versions). Experiment stations in other countries sometimes even hired a Dutch-speaking person to stay au courant with the latest developments in the sugar industry on Java.[53] It was in the industry's best interests to maintain good international standing and contacts; its good reputation led to invitations from foreign experiment stations and sugar mills, and the exchange of information might subsequently lead to further improvements or new innovations. International recognition also was also a plus with foreign investors and suppliers, and served as a basis for keeping the sector informed about the latest developments in both the economic and technological domain.

Most important, however, were the Dutch East Indian sugar sector's contacts with the Netherlands, best illustrated by developments in machine building for sugar production. All steps of the production process relied heavily on

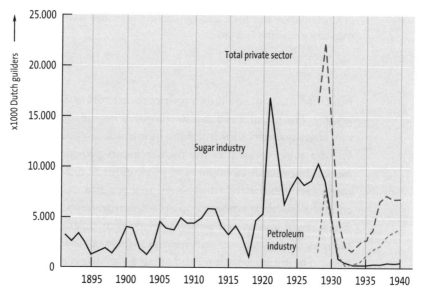

Graph 6–2. Value of machines and machine parts imported to the Dutch East Indies, 1891-1940 (For sources see page 603)

machinery. Graph 6.2 shows that each year millions of guilders were spent on importing new equipment. Most companies supplied machines for specific production processes, not all. An idea of machine building developments can be gained by means of a closer look at the manufacturing of installations, used for boiling the liquid sugar mass. The first installations were produced by European machine manufacturers: Derosne, Cail & Cie. in France; Howard in England; the Amsterdam-based machine builder Paul van Vlissingen; and Dudok van Heel (the precursor of Werkspoor), which also supplied equipment to the Dutch beet sugar industry.[54] These companies generally had offices on Java, either staffed with their own company representatives or run dealers on Java to whom a trade license had been issued. In the late nineteenth century the number of European companies that supplied machines to the sugar industry on Java grew steadily. The two main locations for machine traders and agencies, Semarang and Surabaya, were the main sugar ports. Some established companies in the East Indies built a foundry and developed their own machines for the sugar industry; one of these was Younge & Gill, established in 1858 and later renamed Machine Factory Dapoean.[55] Although no precise figures on multiple years are available, various descriptions suggest that initially Dutch machine manufacturers only had a limited segment of the sugar sector market. It is likely that the East Indian market for machines in the nineteenth century reflected the situation in the European beet sugar industry, in which England, Germany, and France were market leaders. But that changed in the first decades of the twentieth century.

Machine Manufacturer Stork

The largest contributor of Dutch-made machines for the sugar industry in the Dutch East Indies was Stork, from Twente, a region in the eastern Netherlands. This company began to operate in the Dutch East Indian market around the turn of the century and within a few years it had become one of the main suppliers of boiling installations.

Like other companies, Stork established an agency on Java to serve this market. Stork also obtained licenses from foreign companies for the use of certain systems, such as a heating system developed by the well-known English factory Babcock & Wilcox, which was an advantage to Stork.[56] Stork's emergence as a supplier for the sugar industry is partly attributable to the fact that it was a Dutch company and so it profited from nationalist sentiments among sugar industrialists active on Java. Most sugar factories were owned by Netherlands-based companies, so it is likely that decisions on investments were mostly made in the Netherlands. Dutch companies made no secret of their preference for Dutch products. A description of a newly

A 1908 photo of a Stork installation for the Medarie sugar mill, with a production capacity of eighty-six tons of sugar cane every twenty-four hours. In the early 1900s, the Hengelo machine factory Stork would earn a sizable part of its profits from the installation of sugar mills. It could supply nearly all the equipment needed. Prior to being shipped the installation was assembled in the factory for inspection. The buyers, cultivation societies, most of which were based in the Netherlands, thus could get a close-up look at their purchase.

designed sugar mill on Java in the main Dutch engineering journal *De Ingenieur* positively referred to the large number of machines from Dutch companies, affording "the Dutch industry considerable financial advantage vis-à-vis its foreign competitors."[57] Apparently, Dutch machine manufacturers could cash in on nationalist sentiments among sugar producers. Table 5.1 shows that in the late 1920s Stork was by far the leading supplier of installations for the East Indian sugar industry, having a market share of some 30 percent. At this time Dutch equipment manufacturers together supplied about 50 percent of all equipment—up from about 15 percent early in the century. In the wake of their activities on Java the two largest equipment manufacturers, Stork and Werkspoor, also targeted the sugarcane industries of India and various countries in Central and South America, whereby often the two companies closely collaborated.[58] This further underscores the positive effect the Java sugar industry had on Dutch machine industry.

Summarizing we may conclude that technology development in the field of sugar production was characterized by various systematic features. The most far-reaching form of organized technology development was shared production control. Sugar producers' membership the association of experiment stations meant that they actively contributed to innovations through systematic testing and control of their production process. Next in importance is the international exchange of technology, which was organized through publications in English and visits to foreign experiment stations and factories and hosting visitors from abroad. Finally, the growing market share of the Dutch machine industry can be seen as a form of coordination in technology development, even if this was most likely unplanned. By purchasing equipment made by Dutch companies, the sugar industry contributed to this sector's economic and technological development.

Entanglement with a "Eurasian" Industry

From our account of the sugarcane cultivation and sugar refining industries we conclude that the two industries' technologies and processes were different and this is also true of technology development. In sugarcane cultivation, interaction or collaboration with the local population was inescapable and resulted in all sorts of entanglements. Sugar refining, by contrast, seemed to be first and foremost a European affair. The control and improvement of sugar extraction was supported by the work of the experiment stations, and the European machine companies supplied mills, pumps, pans, and other equipment. The relationship with the native population of Java was limited to the factory laborers. Despite this apparent separation, sugar extraction, too, interlocked in various ways with Javanese society.

The sugar industrialists tried hard to create an image of an industry that capitalized on Western rationality combined with a Dutch entrepreneurial spirit, reflected in written materials from the era as well as in visual materials, such as those in a portrait gallery at the colonial exhibition in Semarang in 1914. The photos of the men who had developed the industry mainly depicted white faces.[59] Such presentations masked a much more colorful reality, however. First, the so-called Indian Dutch, persons of mixed ancestry, played a major role in the sugar industry. They generally occupied positions in the technical maintenance of factories and supervised workers in the fields and factories. Because they had both native and Western roots, the Indian Dutch successfully could present themselves as "intermediaries" between Dutch managers and native laborers. A second important social group was the Chinese.

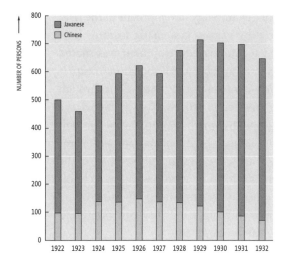

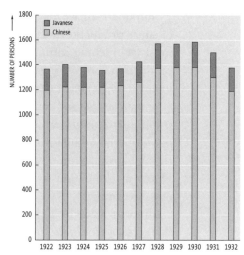

Graph 6-3. Ethnicity of evaporators in the Javanese sugar mills, 1922-1932 (For sources see page 603)

Graph 6-4. Ethnicity of boilers in the Javanese sugar mills, 1922-1932 (For sources see page 603)

Before the Dutch became active in this sector, it was mainly in the hands of Chinese industrialists and despite later domination by Europeans this group never let itself be pushed out of the sector completely. In the early twentieth century, various sugar mills on Java were owned by Chinese entrepreneurs. Chinese employees also were in charge of one crucial phase in the process of sugar extraction: the boiling of the liquid sugar mass in the evaporators, an aspect that required great skill. For good quality sugar the viscosity of the liquid sugar mass had to be brought carefully to a specific level. After most of the water was extracted from the liquid mass, during the boiling process the water level was further reduced until the saccharose crystallized. In this crystallizing, or graining, it was of utmost importance that the sugar crystals attain similar sizes. The water had to be extracted neither too fast nor too slow, and technicians also had to make sure that not all the water was extracted. Boilers kept an eye on the hot mass through an inspection hole. The speed at which splashes ran down the inspection glass was the major viscosity indicator.[60] Graphs 6.3 and 6.4 show that the boiling process was managed almost entirely by Chinese workers. They excelled in handling this boiling technique as no one else and managed to keep this skill largely within their own community. So the Chinese sugar boilers occupied a key position within the sector. Of all sugar mill personnel the Chinese boilers had the highest salary by far, higher than the clerical staff or the Javanese boilers.[61]

In the 1920s and 1930s all sorts of experiments were done to measure the viscosity and crystallization of the liquid sugar mass continuously, thus to automate the boiling process even further. But these tests did not provide applicable results and during the entire colonial period the boiling process remained the almost exclusive domain of Chinese technicians.

The crucial role of Chinese sugar boilers in the sugar industry and the attempts to break this dependence with technological means clarify that also in sugar refining colonial technology development is intertwined with Dutch East Indian colonial society. This entwinement was also recognizable in the organization of the sugar industry. There were many Javanese workers and Chinese owners of sugar mills. In addition, the geographic location of Java was relevant, as it allowed good access to Asian markets. These and other factors account for the fact that until the 1930s the Dutch East Indian sugar industry was rather successful; the sector invested a substantial part of its profits in consolidating and perhaps strengthening its market position. The economic crisis of the 1930s was a huge blow to the industry, but it is hard to determine whether and how the Dutch East Indian sugar industry recuperated because its recovery period lasted only a few years at most, until the outbreak of World War II.

Independence

In the period between 1942 and 1957, two historical events brought the colonial sugarcane industry to an end. Between the time when the Japanese troops occupied the archipelago in 1942 and the liberation in 1945, the sugar industry largely continued to produce, albeit under direct authority of the Japanese army. Although many Dutch experts continued on with their work during the early war years, in the war's finals stages they ended up in Japanese internment camps. During the liberation and subsequent struggle for Indonesian independence sugarcane fields and sugar factories sustained heavy damage. In the period between Japan's capitulation and the transfer of sovereignty, in December 1949, Dutch investors tried to get the sugar industry going again with protection from the Dutch military.[62]

In the tumultuous period of the 1940s and 1950s, Dutch authorities and the sugar industrialists again proved to have divergent policy views. In the 1910s and 1920s the entrepreneurs still argued in favor of rigid colonial rule and opposed the Dutch colonial government's cautious rapprochement with the nationalist movement on Java. Three decades later their positions were reversed. After Indonesia's independence, the Dutch entrepreneurs advocated

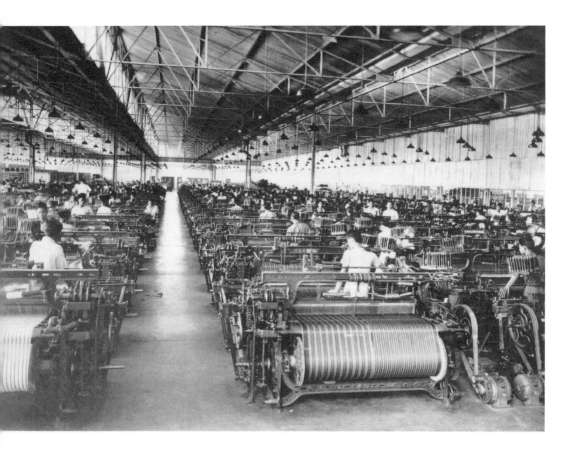

The weaving mill in Preanger, western Java, was a collective undertaking of the Internatio trade firm and four Dutch textile companies from the Twente region of the Netherlands. The mill, which started operations in 1933 as the "the first mechanical [mill] in the Archipelago," was integrated with the traditional local weaving industry. Suzuki looms were selected, partly because these Japanese machines better suited the physique of the Javanese workers. In 1940 the mill employed over three thousand workers. The years of the First World War and the Great Depression saw a dramatic expansion of modern domestic industry in the East Indies, made possible by a relative lack of attention on the part of the colonial power. Various automated weaving mills were founded in the 1930s, as well as a large number of small businesses using modern hand looms.

cooperation and reconciliation with the new Indonesian government, thus hoping to secure their substantial investments and also have input in the industry's reconstruction.

The crisis over New Guinea meant that the Dutch entrepreneurs were forced to leave Indonesia. Their factories were expropriated. The Dutch leadership of private experiment stations was also expelled. They were replaced by Indonesians of Chinese descent who decided to continue only the research aimed at improving sugarcane cultivation.[63] Their exact motivations for stopping research of the chemical processes and technological possibilities of sugar extraction are not known, but are not hard to guess either. In most cases innovation in sugar production coincided with improvements of the equipment and installations marketed—either patented or not—by the internationally operating machine manufacturers. Innovations, then, could be purchased fairly easily. Undertaking research largely had to be motivated by the ambition to keep up a leading international position. Most likely the Indonesian sugar industry lacked both the means and ambition to be a market leader.

The expropriation of the Dutch-owned sugar factories was accompanied by a regulation that Dutch owners had to be compensated somewhat for the loss of capital. Still, Dutch companies were now deprived of an industrial branch in which they had invested for years and that at times had generated large profits.[64] The end of Dutch colonial rule in Asia was not the total end of Dutch activity in the sugarcane industry. For some time Dutch machine factories continued to do business with sugar factories in other countries. Furthermore, the Amsterdam Trading Association (HVA) decided to set up a new sugar mill in Ethiopia. Thus some of the HVA sugar experts who were expelled from Indonesia remained active in the sugar sector. The sugar plantation and factory in Ethiopia were situated in a previously uncultivated and sparsely populated area, which allowed for much more autonomous operations without having to become overly involved with local society. Just like the former possessions in the Dutch East Indies, however, the plantation in Ethiopia had to be given up in the wake of a political revolution.[65] This also made it clear to the HVA that the colonial context in which the Dutch East Indian sugar industry had developed had involved specific historical conditions that were not likely to return.

In the 1950s and 1960s, the technological expertise and knowledge developed in the context of the colonial sugarcane industry and the colonial economy in general lost their almost exclusive focus on the Dutch East Indies. Most Dutch experts expelled from Indonesia found new employment in the Netherlands, a small group moved to Suriname, and others continued their career in colonies ruled by other European nations, notably England.[66] The knowledge institutions in the Netherlands that for decades had provided training and performed research for the colonial economy began to broaden their base. The Colonial Institute in Amsterdam was renamed the Royal Tropical Institute. The training facilities for colonial officials at Leiden and Utrecht were transformed into departments of anthropology and non-Western sociology.[67] Curriculums aimed at colonial agriculture at the agricultural college in Wageningen were converted into general courses on tropical agriculture. Dutch companies no longer had exclusive access to Asia and had to compete in other ways. Since various companies active in the Dutch East Indies had much broader sales and production bases, they managed fairly well to compensate for the loss of the East Indies with activities in other tropical countries. Then, starting in the early 1970s relations between Indonesia and the Netherlands began to improve, giving Dutch companies, engineers, and scientists access to the archipelago once again.

Conclusion

The developments in the Dutch East Indian sugarcane industry during the twentieth century reveal that colonial technology development was largely the work of Dutch experts and Dutch companies. Yet the technology development in the East Indies cannot be characterized as a strictly Dutch affair; rather, it should be understood against the backdrop of colonial society at large. Entwinement with the local economy and local society meant that there were specific inputs from Asian actors. This gave rise to technological solutions and innovation-oriented research that was tied to the local geography, such as the creation of reservoirs for irrigation water and the interrelated research into the effects of silt makeup on soil fertility. There were also specific combinations of Western technology and local expertise, such as the boiling process in sugar factories, whereby European-made pans operated by Chinese technicians led to optimal results. These hybrid forms of technology development were essential for the functioning of the colonial sugar industry and very likely occurred in other sectors of the colonial economy as well. The industry's development also shows that colonial entrepreneurs were often unhappy with their dependence on local society. They tried to present the industry mainly as a European sector that developed thanks to Western science. In their practice Dutch entrepreneurs tried, through technology, to reduce their reliance on local actors and expertise. The introduction of narrow-gage railroad, lorries, and locomotives diminished dependence on carts

This poster announces twice weekly flights from Amsterdam to Batavia by KLM Royal Dutch Airlines in 1935. The entire trip took five and a half days, and involved several stopovers as the map on the poster shows. The quality of the communication lines and connections between the "motherland" and its colony was a subject of great concern, also to the government. Around 1930 it spent large subsidies to keep up a regular connection by air, initially largely for mail. In the 1930s the national airline company, KLM Royal Dutch Airlines, which managed the flights, increased their frequency. The biweekly service in 1930 became weekly in 1931, and twice weekly in 1935. The number of passengers and the volume of the freight were quite limited, however: only 2,816 passengers and thirty tons of freight were transported in 1936. Still, the connection was of great importance for ensuring regular and fast mail delivery.

and buffalos. Although these locomotives were mostly operated by Javanese workers, it is not hard to see that control over both equipment and expertise shifted from the Javanese to the European entrepreneurs. Such appropriation was not successful in all cases, as is suggested by the failed efforts to mechanize the boiling process so as to reduce dependency on the skill of Chinese boilers. These episodes invite us to view colonial technology development as a process whereby Dutch entrepreneurs increasingly succeeded tooling the industry to function on the basis of Western science, European technology, and a Dutch entrepreneurial spirit. But we should not accept this conclusion blindly.

First, Dutch sugar producers rejected certain technological innovations because they were more expensive than the use of local labor and means. The preference for local workers over the introduction of certain machinery was explained not only by their low wage levels but also by their skills and experience with the jobs at hand. Various tasks performed by local workers and technicians could not easily be replaced or improved upon by scientific insights and Western technologies.

Over time Dutch experts and managers developed more awareness of local society and its interests, rather than focusing exclusively on the financial interests of European entrepreneurs. Thus they also implemented technological innovations that were useful to colonial society as a whole. Although the sugar industrialists failed to have much interest in such approach, a concern for local society was, especially among government officials, more or less the starting point for innovation-oriented research.

A final, interrelated, point is that each technological innovation is directly or indirectly linked to specific learning processes. In the operation and application of technical innovations as well as in the development of innovation-oriented research, local personnel received practical training and gained insight into the workings of particular technologies and processes, formally, through training and courses, and, more informally, through on-the-job instructions and learning-by-doing. Such interaction and learning processes took place in all contexts and at all levels. This also explains why, after the Dutch left, the sugarcane infrastructure—its mills and experiment stations—did not completely collapse, but continued to function, albeit in a modified and reduced fashion. Naturally the Dutch sugar experts also underwent a learning process. If, after Indonesian independence, a Dutch company or aid program sent these experts to work in another foreign country, they could draw on the experience they gained from working in tropical circumstances and having collaborated with local technicians and workers in the Dutch East Indies.

Acknowledgements

For their comments on earlier versions of this chapter I am indebted to the editorial board, in particular Nil Disco, Harry Lintsen and Johan Schot, as well as to the authors of this volume and the other contributors to the research project on technology in the Netherlands in the twentieth century. Special thanks are due to Margaret Leidelmeijer and Suzanne Moon for their detailed input and comments.

Notes

1 H. W. Dick, Vincent J. H. Houben, J. Thomas Lindblad, and Thee Kian Wie, *The Emergence of a National Economy: An Economic History of Indonesia, 1800–2000* (Crows Nest, NSW, Australia: Allen & Unwin, 2002); H. W. van den Doel, *De stille macht: Het Europese binnenlands bestuur op Java en Madoera, 1808–1942* (Amsterdam: Bert Bakker, 1994); J. J. P. de Jong, *De waaier van het fortuin: Van handelscompagnie tot koloniaal imperium—De Nederlanders in Azië en de Indonesische archipel 1595–1950* (The Hague: SDUu Uitgevers, 1998).
2 On twentieth-century technology development see Pierre van der Eng, *Agricultural*

Growth in Indonesia: Productivity Change and Policy Impact Since 1880 (Basingstoke: Macmillan, 1996); Margaret Leidelmeijer, *Van suikermolen tot grootbedrijf: Technische vernieuwing in de Java-suikerindustrie in de negentiende eeuw* (Amsterdam: Nederlandsch Economisch-Historisch Archief, 1997); Wim Ravesteijn, *De zegenrijke heeren der wateren: Irrigatie en staat op Java, 1832–1942* (Delft: Delft University Press, 1997).

3 Coffee, tobacco, and cocoa were imported into Java before the nineteenth century, and rubber, cinchona, and palm oil were brought in the nineteenth century. Many food crops were at some point introduced by European or Asian traders.

4 P. J. Florijn, "Geschiedenis van de eerste hortus medicus in Indië," *Gewin* 18 no. 4 (1995): 209–221.

5 Leidelmeijer, *Van suikermolen tot grootbedrijf*, 193–198.

6 J. P. Smits and B. P. A. Gales, "Olie en gas," in Schot et al., *Techniek in Nederland*, vol. 2, 67–89.

7 Elsbeth Locher-Scholten, *Ethiek in fragmenten: Vijf studies over koloniaal denken en doen van Nederlanders in de Indonesische archipel, 1877–1942* (Utrecht: HES Publishers, 1981); Van den Doel, *Stille macht*, 161–214.

8 Alice Conklin, *A Mission to Civilize: The Republican Idea of Empire in France and West Africa, 1895–1930* (Palo Alto: Stanford University Press, 1997).

9 Arjen Taselaar, *De Nederlandse koloniale lobby: Ondernemers en de Indische politiek, 1914–1940* (Leiden: CNWS Publishers, 1998), 311–340.

10 Harro Maat, *Science Cultivating Practice: A History of Agricultural Science in the Netherlands and Its Colonies, 1863–1986* (Dordrecht: Kluwer Academic Publishers, 2001), 66–67.

11 H. Baudet and M. Fennema, *Het Nederlands belang bij Indië* (Utrecht: Spectrum, 1983).

12 Ann Laura Stoler, "Rethinking Colonial Categories. European Communities and the Boundaries of Rule," *Comparative Studies in History and Society* 31 (1989): 134–161.

13 C. Fasseur, *Kultuurstelsel en koloniale baten: De Nederlandse exploitatie van Java 1840–1860* (Leiden: University Press Leiden, 1975), 11.

14 Peter Boomgaard, "Treacherous Cane: The Java Sugar Industry Between 1914 and 1940," in Bill Albert and Adrian Graves, eds., *The World Sugar Economy in War and Depression, 1914–1940* (London: Routledge, 1988), 157–169.

15 Leidelmeijer, *Van suikermolen tot grootbedrijf*, 131–146.

16 De Jong, *Waaier van het fortuin*, 269–280.

17 Taselaar, *Nederlandse koloniale lobby*, 35.

18 Martijn Bakker, *Ondernemerschap en vernieuwing: De Nederlandse bietsuikerindustrie 1858–1919* (Amsterdam: Nederlandsch Economisch-Historisch Archief, 1989), 48–49.

19 H. Ch. G. J. van der Mandere, *De Javasuikerindustrie in heden en verleden, gezien in het bijzonder in hare sociaal-economische betekenis* (Amsterdam: Bureau Industria, 1928), 151.

20 Maat, *Science Cultivating Practice*, 32–33; Margaret W. Rossiter, *The Emergence of Agricultural Science: Justus Liebig and the Americans, 1840–1880* (New Haven: Yale University Press, 1975), 117–171.

21 Bakker, *Ondernemerschap en vernieuwing*, 212–232.

22 Leidelmeijer, *Van suikermolen tot grootbedrijf*, 235–243.

23 Eugene Cittadino, *Nature as the Laboratory: Darwinian Plant Ecology in the German Empire, 1800–1900* (Cambridge: Cambridge University Press, 1990).

24 V. J. Koningsberger, "De Europese rietsuikercultuur en suikerfabricage," in C. J. J. van Hall and C. van de Koppel, *De landbouw in de Indische Archipel*, 3 vols. ('s-Gravenhage: Van Hoeve, 1946–50), vol. 2b: *Genotmiddelen en specerijen*, 353.

25 Leidelmeijer, *Van suikermolen tot* grootbedrijf, 235; Taselaar, *Nederlandse koloniale lobby*, 99.

26 Taselaar, *Nederlandse koloniale lobby*, 102–108.

27 This experiment station continued its operations after Indonesian independence. See

Hermono Budhisantosa, Y. Kurniawan, and H. Effendi., eds., *Indonesian Sugar Research Institute (ISRI)—One Hundred and Ten Years of Service: July 9, 1887–1997* (Pasuruan: ISRI, 1997).

28 For more on this subject, see chapter 5, "The Rise of a Knowledge Society," in this volume.

29 For annual overviews of the staff of experiment stations, see annual editions of Departement van Landbouw, Nijverheid, and Handel, *Verslag van de . . . vergadering van de Vereeniging van Proefstation-Personeel* (Buitenzorg: Archipel Drukkerij, 1912–1938). On the Treub Laboratory, see K. W. Dammerman, "A History of the Visitors' Laboratory ('Treub Laboratorium') of the Botanic Gardens, Buitenzorg, 1884–1934," in P. Honig and F. Verdoorn, eds., *Science and Scientists in the Netherlands Indies* (New York: Board for the Netherlands Indies, 1945), 59–75.

30 Taselaar, *Nederlandse koloniale lobby*, 165–209.

31 Botanically, sugarcane is a grass, and in grasses this is a particularly complicated process. The sugarcane flowers are small and compact. To crossbreed not only must the pollen of plant A be put on the stigmas of plant B but also the pollen of plant B or of any other plant must be prevented from touching the plant B stigmas, by cutting away plant B's stamens prematurely and covering its flower with a fine-mesh bag until one is certain fertilization has occurred—and all this without damaging the flower.

32 All crossings and their many hybrids were systematically numbered. This variety was developed by the botanist Jacob Jeswiet. See Koningsberger, "Europese rietsuikercultuur en suikerfabricage," 350.

33 For example, the variety 247B, much cultivated before the introduction of 2878POJ, was developed by the planter Bouricius. See ibid., 347.

34 M. Treub, *Schematische nota over de oprichting van een agricultuur-department in Nederlandsch-Indië* (Buitenzorg: n.p., 1902).

35 Taselaar, *Nederlandse koloniale lobby*, 261–300.

36 G. R. Knight, "Did 'Dependency' Really Get It Wrong? The Indonesian Sugar Industry, 1880–1942," in J. Th. Lindblad, ed., *Historical Foundations of a National Economy in Indonesia, 1890s-1990s* (Amsterdam: Royal Netherlands Academy of Arts and Sciences, 1996), 155–173.

37 Taselaar, *Nederlandse koloniale lobby*, 303–306.

38 Control of irrigation was divided among the Department of Agriculture, the Department of Civil Public Works, and the "Binnenlands Bestuur," the Interior Administration (see Ravesteijn, *Zegenrijke heeren der wateren*, 221–226).

39 Suzanne M. Moon, *Constructing "Native Development": Technological Change and the Politics of Colonization in the Netherlands East Indies, 1905–1930* (Ann Arbor: University of Michigan Press, 2000), 102–107.

40 *Jaarboek van het Departement van Landbouw in Nederlandsch-Indië*, 1910 (1911); Moon, *Constructing "Native Development*,*"* 107–112.

41 Moon, *Constructing "Native Development*,*"* 108.

42 Taselaar, *Nederlandse koloniale lobby*, 306–311.

43 *Jaarboek van het Departement van Landbouw, Nijverheid en Handel in Nederlandsch-Indië* (1912 (1913), 108–111.

44 F. Ledeboer, "De suikercultuur en de inlandsche landbouw," *Archief voor de Suikerindustrie in Nederlandsch-Indië* 26 (1918), stated: "Increasingly those facts are seen to be so convincing, that people feel justified in using them as weapons against the sugar industry without further research and evidence" (2128).

45 Ibid., 2135.

46 Taselaar, *Nederlandse koloniale lobby*, 315–320; C. Fasseur, *De indologen: Ambtenaren voor de Oost 1825–1950* (Amsterdam: Prometheus, 1993), 412–425.

47 Fasseur, *Indologen*, 426–428.

48 J. Mulder, "De voor- en nadeelen, zoowel financieël als in de praktijk, van riettransport per

rail," speech to the fourth conference of the Algemeen Syndicaat van Suikerfabrikanten on Java, Soerabaia, 1900.

49 H. C. Prinsen Geerligs, *Korte handleiding tot de fabrikatie van suiker uit suikerriet op Java en in Suriname*, 5th ed. (Amsterdam: De Bussy, 1930); A. J. van der Linden, "De inrichting van een suikerfabriek op Java: Voordracht gehouden in de vergadering van het Koninklijk Instituut van Ingenieurs van 29 Juni 1918," *De Ingenieur* 33, no. 37 (1918): 698–725.

50 At high temperatures, saccharose decomposes into noncrystallizable glucose and dextrose.

51 Prinsen Geerligs, "Korte handleiding tot de fabrikatie van suiker," in Noell Deerr, ed., *Cane Sugar: A Textbook on the Agriculture of the Sugar Cane, the Manufacture of Cane Sugar, and the Analysis of Sugar-House Products*, 2nd ed. (Altrincham: Norman Rodger, 1921).

52 Cf. Leidelmeijer, *Van suikermolen tot grootbedrijf*, 249–256.

53 Roger Knight, "Sugar, Technology, and Colonial Encounters: Refashioning the Industry in the Netherlands Indies, 1800–1942," *Journal of Historical Sociology* 12, no. 3 (1999): 218–250.

54 Bakker, *Ondernemerschap en vernieuwing*, 28–30.

55 Arnold Wright, ed., *Twentieth-Century Impressions of Netherlands India: Its History, People, Commerce, Industries and Resources* (London: Lloyd's Greater Britain Publishing Company, 1909), 540.

56 Stork, *Stork: 120 jaar industriële dynamiek* (Utrecht: Stichting Matrijs, 1989), 22. C. Beets, "De machinefabriek van gebr. Stork te Hengelo" *De Ingenieur* 29, no. 25 (1914): 462–471.

57 Van der Linden, "Inrichting van een suikerfabriek op Java," 698.

58 Werkspoor, *Werkspoor 1827–1952: Gedenkboek uitgegeven ter gelegenheid van het honderd vijfentwintig jarig bestaan op 9 februari 1952* (Amsterdam: 1952), 156–157.

59 Knight, "Sugar, Technology, and Colonial Encounters," 236.

60 Prinsen Geerligs, "Korte Handleiding tot de Fabrikatie van Suiker," 105.

61 Philip Levert, *Inheemsche arbeid in de Java-suikerindustrie* (Wageningen: Veenman, 1934), 137.

62 J. A. M. Goedkoop, "Handelsvereeniging 'Amsterdam'" 1945–1958, herstel en heroriëntatie," *Jaarboek voor de Geschiedenis van Bedrijf en Techniek* 7 (1990): 219–240.

63 Budhisantosa, *Indonesian sugar research institute*.

64 Van der Eng, *Agricultural Growth in Indonesia*.

65 Goedkoop, "Handelsvereeniging 'Amsterdam,'" 1945–1958."

66 Maat, *Science Cultivating Practice*, 105–108.

67 Peter Kloos, "Het ontstaan van een discipline: De sociologie der Niet-Westerse volken," *Antropologische Verkenningen* 7, no. 1–2 (1988): 123–146.

J. B. Bakema (1914–1981) and J. H. van den Broek (1898–1978) with a scale model of their design for the Civil Technology Building of the Technical University Delft, in the 1960s. Their practice became world-famous with their design of the Lijnbaan in Rotterdam (1949–1953). As professors at Delft they trained many architects. Their control over spatial design is rendered quite concrete in this model. Technicians such as urban architects and civil engineers claimed that their professional expertise equipped them to make fundamental decisions as to the appropriation of portions of the public domain.

7 Technology as Politics: Engineers and the Design of Dutch Society

Dick van Lente and Johan Schot

The second half of the nineteenth century is known as the heyday of Dutch liberalism.[1] The European crisis of the years 1848–1850 had led in the Netherlands to a liberal constitution, and in the ensuing years the new liberal order was consolidated. For instance, the king's power was further restricted; the cultivation system in the Dutch East Indies, by which the government had extracted tremendous wealth from the colony, was abolished in favor of private enterprise; the newspaper tax was abolished; and workers obtained the right to strike.[2] In the absence of a powerful conservative countercurrent, liberal views prevailed in Dutch politics for the time being. Other oppositional movements such as the confessional parties (Roman Catholic, Protestant) and the socialists would grow more influential toward the end of the century.

The older generation of liberals adhered to the classic liberal tenet that government should leave society's organization as much as possible to private initiative, but from the 1870s on a new, left-wing liberalism gained ground whose adherents had different views on the relationship between the state and its citizens. They no longer conceived of the state as a community of taxpayers, but rather as a productive collaborative community, to which all active citizens from all ranks belonged. Furthermore, at the end of the nineteenth century left-wing liberals adopted the socialist view that those who for reasons beyond their control were unable to provide for themselves also counted as full citizens, and therefore had the right to vote and were entitled to support from the government. Unfettered development of the individual continued to be a basic premise for the new liberals, but they felt the state had to create the proper conditions for it to take place. Partly in coalition with the new confessional parties, they realized a large number of new laws, such as the first act prohibiting child labor (1874), a law limiting working hours (1889), the Housing Act (1901), which stipulated basic quality standards for homes, and the Public Health Act (1901).[3] Laws enacted in 1859 and 1863 provided a foundation for the improvement of primary and secondary education, such as a new type of advanced high school with a modern curriculum and the new Technical University Delft. Starting in 1860, the state created a national rail network to boost the country's economy, linking the older private railroads and building new ones; in the 1870s the connection of the ports of both Rotterdam and Amsterdam to the North Sea was greatly improved; and from 1890 agricultural research stations were set up where farming cooperatives could have their feed and fertilizer tested. Small wonder, then, that government expenditures significantly increased in this period: from a little over 69 million guilders in 1853 to 194 million in 1913. However, this spending decreased as a percentage of the gross domestic product, from 12 percent to 8 percent, which shows that the new government activities were made possible by the Dutch economy's rapid growth.[4]

New efforts at reform were even more conspicuous at the local level, due to the fact that the new constitution made municipalities responsible for the provision of major services such as health care and education. Per capita local spending grew from 7.57 guilders in 1862 to 25.64 guilders in 1907, with by far the largest growth in expenditures for public works.[5] In the 1890s the City of Amsterdam took over the water system, gas service, urban transport system, and telephone system from private companies. The city only granted contracts to companies that paid decent wages and had their employees work no more than eleven hours per day.[6]

The emergence of new forms of state intervention and investments in technological development rested in a firm belief in science and progress, not only among left-wing liberals but also among socialist and confessional politicians.[7] H. Goeman Borgesius, a left-wing liberal leader, wrote in 1875, "The human spirit mocks nature's resistance." Steam, he felt, was the driving force behind the stormy social and economic changes in his day and age. Steam engines made work less burdensome, and railroads and steamships brought different peoples into proximity. "It seems as if merely the sound of the steam whistle is putting coarseness and barbarism to flight."[8]

The Rise of the Engineers

The new activities undertaken by the Dutch national and local governments represented the beginnings of what after World War II would evolve into a welfare state. Its rise implied that many domains became subject to government intervention, including technological development. In this chapter we discuss the specific role of engineers in these social developments, focusing on their contribution to and autonomy from state intervention in infrastructure, polder development and public housing.

Several sociologists and social historians have argued that the influence of the state's bureaucracy, to which more and more engineers belonged, has been increasing in the Netherlands since the late nineteenth century. This development is attributed to a variety of causes. Industrialization and the growth of towns gave rise to substantial problems with respect to public health and public order. These concerns prompted new interventions, supported by new views about the legitimacy of state intervention. Specific professional groups, such as physicians and engineers, organized to improve their social position and advanced scientific and technological solutions to social problems.[9] The combination of these two developments—a state that took responsibility for more social tasks and the professionalization of groups whose expertise was aimed at carrying out those tasks—caused Dutch politicians to be more and more dependent on the input of state officials who possessed engineering expertise. They helped to identify and frame issues for collective debate and they proposed specific policies and identified salient points for negotiation. To the extent to which engineers consolidated bureaucratic power within national administrations, they institutionalized and routinized their influence. Since they were directly involved in specific issues and policies much longer, they tended to be much better informed than a minister or a member of Parliament. Although the role of engineers and other experts was basically service-oriented, they wielded substantial political influence. This might result in the "displacement of politics": the transfer of agenda setting and problem solving from the political sphere to the bureaucracy, so that government policies were increasingly shaped by the new class of (technological) experts, rather than by ideologically charged debates in Parliament.[10]

Followed to its extreme, the involvement of experts could lead to a form of technocracy in which technical experts would take over the government from the politicians.[11] As we will see, several engineers, in the Netherlands and elsewhere, advocated this type of government, arguing that modern society was too complex to be governed by politicians, who after all were amateurs in the fields they were responsible for. This was, however, not what happened. Many engineers preferred to work in the shadow of elected government power and gain as much power and autonomy as possible within the bureaucracy. This chapter traces the extent to which engineers were successful in creating a large mandate for themselves in various periods during of the twentieth century. It also explores how engineers and other experts framed the various issues at hand. We look at specific areas—the building of polders, cities, and other infrastructures—that transformed not only the physical appearance of the Netherlands but also its self-image and the lives of many citizens. As we will show, engineers often disagreed about the proper

The new Moerdijk Bridge was one of the most talked-about bridges of the time when this photo was taken, in 1871. It would span the Hollands Diep, a broad estuary of the Rhine and Meuse rivers, southeast of The Hague. Two bridge components near Moerdijk are being readied to become part of what was to be the longest bridge in Europe at the time. In the second half of the nineteenth century, liberalism prevailed in Dutch politics and society. Liberal governments did not refrain from state interference, however. Government explicitly supported or implemented major projects, such as the national railroad network and the bridge over the Hollands Diep.

solutions to specific social problems, and in the course of the century specialists from other disciplines, such as biologists, economists, and sociologists, joined the debates, proposing other types of solutions. Expert opinions were thus heavily divided, and experts pursued different political agendas. In the Netherlands, as elsewhere in Europe, the political system became dependent on expert knowledge, and experts acquired a robust mandate to pursue their political ambitions. In this sense technocracy became an important building block of the political system. However, the rise of technocracy does not imply the disappearance of political debate and the role of politicians, as is often wrongly assumed—not only by the experts themselves but also by contemporary commentators and political historians.

Technocratic Ambitions in a Liberal Climate, 1850–1914

The role of engineers in the Dutch government's national and local policies was a major political theme in the second half of the nineteenth century.[12] We will first examine the way engineers tried to gain influence through the Rijkswaterstaat, the state agency responsible for public works, especially water management and, increasingly, railroad construction. Subsequently we discuss how engineers operated at the city level.

From the 1840s on, leading engineers employed by the Rijkswaterstaat, in particular L. J. A. van der Kun and F. W. Conrad, aimed to expand their influence on infrastructure policies. They pleaded with the government not only to appoint a chief inspector to head this agency but also to set up a special council staffed by experienced engineers who would advise the minister. These plans met with great resistance from members of the Dutch Parliament and from a number of subsequent ministers of interior affairs, who exercised formal authority over the Rijkswaterstaat. A high official wrote that a chief inspector would acquire too much power, which would reduce the minister to "a mere tool" and make him dependent on advice from experts in telegraphy, railroads, and road and canal construction.[13] The new constitution of 1848 had placed the state budget under parliamentary control and the members of Parliament were not prepared to hand it over to the state engineers.

Rijkswaterstaat and the Railroads

The issues of appointing a chief inspector and the influence of engineers were brought into focus in the context of Dutch railroad construction.[14] From the first company-owned railroad line in the Netherlands, completed in 1839, the building effort had been in the hands of private companies, and the state played only a supervisory role. The engineer L. J. A. van der Kun, who in the early 1840s had been director of the Rhine Railroad Company and who after his return to the Rijkswaterstaat in 1845 became the minister's main adviser on railroads, was one of the outspoken proponents of railroad construction by the state. Yet in the early 1850s the government hoped that the Dutch railroad network would be further developed through private initiative. When this did not happen, Minister G. C. J. van Reenen appointed Van der Kun chief inspector and head of the Rijkswaterstaat agency in the Ministry of Interior Affairs, while also inviting him to report on the issue of the need for railroads and the desirability of either private or state operation of them. In his 1857 report Van der Kun argued that in terms of infrastructure the Netherlands was lagging far behind other countries, as evidenced, for instance, by the poor connection of its ports to the hinterland. He advised the state to take charge of constructing the major lines and set up a special agency for managing the work.[15] This proposal was rejected by a parliamentary commission, and shortly thereafter the Dutch Parliament decided that private companies should take charge of construction, possibly with the help of state subsidies. One of the arguments against railroad construction by the state was that a government agency would always work more expensively and produce less quality than private companies.

The Rijkswaterstaat continued to do studies of the technical problems involved in railroad construction, such as bridging the country's major rivers. When a bill that would have granted to private companies the right to carry out construction work was voted down in Parliament in 1859, Minister of Interior Affairs F. A. van Hall seized the opportunity and managed to get a bill passed that was based on the idea that the state itself would undertake railroad construction. He succeeded because he promised members of Parliament to link major towns in their electoral districts to the railroad network. The construction effort was handed over to a special agency, staffed by hydraulic engineers. Although the influence of Rijkswaterstaat engineers on the decision of state-run construction was mainly indirect, it is clear that in the 1840s and 1850s the influence of Van der Kun in particular should not be underestimated. He developed, documented, and communicated all arguments in favor of their influence.[16]

The establishment of the new agency for state railroads and Van der Kun's successful march to the top of that agency did not as yet automatically imply that henceforth leading engineers would have a greater influence on policy. Although Van der Kun served as a kind of unofficial minister of Dutch water and transport infrastructures, upon his death in 1864 Minister J. R. Thorbecke refused to appoint a new chief inspector, preferring to receive advice on major policy issues from multiple sources. Two years later the decision of his successor, H. Geertsema, to appoint F. W. Conrad as chief inspector of the agency met with protests from Parliament, but appointment went through. Upon Conrad's death it was again decided not to appoint a new chief inspector.

In the 1870s it became obvious that managing public works as well as other areas of trade and industry policy was developing into too big a job for the Ministry of Interior Affairs, and when in 1877 a new cabinet was formed it was decided to create a separate Ministry of Water Management, Trade and Industry (Ministerie van Waterstaat, Handel en Nijverheid). In nearly every subsequent cabinet an engineer was appointed to take charge of this ministry. Members of Parliament repeatedly warned that he would function

Construction of the walls of the new sea lock in the Noordzeekanaal, in mid-1893. Despite its declining political power and administrative clout, the Rijkswaterstaat still could boast of striking technical achievements at the close of the nineteenth century. A sea lock in the Noordzeekanaal was needed to accommodate the increasing size and draft of ocean-going vessels. Construction of the heretofore largest lock in the world was a serious and impressive undertaking. Electric power, still a novelty, would be used to run the lock: move the gates and fill the compartments with water.

too much at the behest of his fellow engineers and cause this ministry's expenditures to rise inexorably. For this reason, one-man leadership of the Rijkswaterstaat, too, continued to be a contentious issue. A chief inspector as head of the Rijkswaterstaat was even likely "to outshine the minister [of water management, trade and industry] in charge," stated a parliamentary commission in 1881.[17] Still, the minister got things his way and that same year he appointed a new chief inspector.

After 1890 the prestige and power of the Rijkswaterstaat declined. Most national-level large-scale projects such as railroads and canals were completed, which explains why the agency increasingly concentrated on maintenance and less on the development of new projects. Furthermore, the agency had trouble finding talented engineers because much more interesting work was available in construction companies, engineering firms, and the Dutch colonies. The 1908–1909 discussion about the construction of the Afsluit-dijk, a thirty-kilometer-long dyke topped by a causeway closing off the Zuiderzee, the shallow inlet covering the heart of the country, was revealing. Many top engineers in the agency doubted the technical feasibility of this huge project. The minister therefore decided to entrust the project to a new agency in which he could appoint engineers of his own choice.[18] Conse-quently, after World War I, carrying out the new major national project of which the Afsluitdijk was to be the first part, the so-called Zuiderzee Works, was entrusted to different organizations.

> In the early twentieth century, then, the status of Rijkswaterstaat engi-neers was ambiguous in Dutch society. On the one hand their num-ber and political presence had significantly gone up in the second half of the century. They had played an important role in initiating Dutch railroad and water management infrastructures. In nearly every cabi-net an engineer was put in charge of the newly established Ministry of Water Management, Trade and Industry, which also had quite a wide-ranging focus. At the same time, however, the mandate of the Rijkswaterstaat was still largely restricted by the discussions in the Par-liament and the cabinet. In particular, the Dutch Parliament followed Rijkswaterstaat engineers' exploits critically, even with suspicion, and their mandate was often discussed and disputed.

Public Housing and Town Planning

Engineers also sought to influence the shape of municipal public housing projects and town planning. In the 1870s Dutch towns began to see rapid growth and the demand for inexpensive housing increased accordingly. Until then housing construction had almost exclusively been a matter of private enterprise.[19] Contractors, motivated by quick profits, rapidly turned out housing blocks. Relying on poorly educated workers and inferior materials, they built as many homes as possible on lots as tiny as possible. Because of fierce competition between these contractors and because renters were in no position to demand minimal standards and enjoyed no legal protection, most of these houses were of poor quality. Another form of private initiative involved housing corporations that offered working-class families a simple yet decent home for a low rent. The rents collected had to cover the con-

An almshouses' project developed in 1864 by the Association for Improving Working-Class Housing in Dordrecht. In the second half of the nineteenth century, particularly in large cities, there were initiatives to build suitable rental homes for lower-class tenants.

struction cost. These corporations provided only a limited number of homes as a percentage of all housing needs—for example, in Amsterdam, 7 to 8 percent of the houses built between 1852 and 1902—and generally they catered to the more prosperous workers.[20]

The Local Government Act of 1851 gave Dutch municipalities the authority to formulate building ordinances and design the layout of streets, yet only rarely did towns intervene to the advantage of their poorer residents. Increasingly, however, lower-class housing began to be a source of concern to the social elites. Politicians, legal experts, architects, physicians, and also engineers joined a public debate that in 1901 would lead to the adoption of the Housing Act and the Public Health Act. This social legislation reflected medical-hygienic considerations motivated by moral and political concerns: a well-built and nicely decorated home would keep workers from drinking in cafés and would encourage decent behavior, and thus contribute to political stability and a healthy and productive working class. Already in 1853 the Royal Institute of Engineers had reported on the issue of how to build inexpensive homes that met the "doctrines of public health."

From the 1880s onward various engineers became preoccupied with the so-called social question.[21] A major example is Cornelis Lely, who as head of the Ministry of Water Management, Trade, and Industry was one of the driving forces behind social legislation, notably the 1901 Industrial Injuries Act. Lely was irritated by engineers in public service who strictly focused on their technical tasks and refused to consider the social meaning of their work. In 1904, for example, he wrote to his son:

> If he [the engineer] limits himself to designing works while leaving the assessment of their usefulness and so on to others, such as aldermen, council members, ministers and so on, this engineer will merely remain the technical servant of those who are in charge as aldermen or council members in municipalities, as representatives in provincial councils, in firms as directors and so on. But where possible, I feel, we should strive to gain control of this management ourselves, and this implies we should not remain technicians exclusively.[22]

Here Lely clearly formulated his technocratic ambition, one that was shared by other Dutch engineers.

In the 1890s a commission of the Association of Civil Engineers (Vereeniging van Burgerlijke Ingenieurs) argued for a larger role for engineers and chemists in public health; they had more knowledge than physicians of technologies that helped prevent diseases, such as sewers and garbage incineration.[23] Moreover, subjects such as economics and factory hygiene ought to be incorporated into the training of engineers. In 1894 J. C. van Marken, an engineer and the owner of a chemical factory in Delft, also advocated training "social engineers" to be well informed about economics and social issues. Hired by companies or the government, they could develop practical plans for improving the living and labor conditions for workers and their families.[24] The Sociotechnical Association of Democratic Engineers and Architects (Sociaal-Technische Vereeniging van Democratische Ingenieurs en Architekten), a forum of young socialist-oriented engineers set up in 1904, stimulated public discussion about social problems among engineers. Many of its members attained prominent positions in politics and public service. While Rijkswaterstaat engineers increasingly concentrated on technical management tasks, a new generation of engineers was emerging that had great confidence in their power to solve social problems through technological means, and they were also prepared to act. To implement their ideas they zoomed in on public housing and town planning in particular.

In this area they joined forces with a handful of architects and a new group of professionals, female housing supervisors who collected rents for housing corporations, which gave them an opportunity to find out more about the needs and wishes of renters, to whom they gave advice on issues such as education and household management. Because of their familiarity with the workers' everyday lives, these supervisors also offered advice to their housing corporations on the design and layout of new public housing, and occasionally this advice was taken.

The 1901 Housing Act seemed to provide opportunities to new professionals such as these engineers, architects, and housing supervisors. This legislation obliged municipalities to map the local housing situation and formulate building standards for new homes, while also providing for state loans to purchase land and finance construction and management of housing. The law was meant to encourage private companies or housing corporations to start building homes, but municipalities were encouraged to do so as well. Local governments had to take stock of the number of dwellings and the number of occupants per dwelling, check the quality of the houses, declare houses unfit for human habitation, and take measures aimed at improving and expanding the total supply of houses. Obviously, this was quite a big job. Young Sociotechnical Association engineers such as A. Keppler, P. Bakker Schut, and A. Plate became directors of city housing agencies in Amsterdam, The Hague, and Rotterdam, respectively. Until World War I, however, the Housing Act had little practical effect. Cities' and housing corporations' contributions to the yearly housing output did not exceed 5 percent until 1912, partly because of uncertainties about rents and the financial responsibilities of national and local governments, difficult procedures pertaining to land purchases, and resistance from business and taxpayers, who were represented on local councils.[25] The market continued to prevail, and businessmen had much influence on politics.

The discussion on town plans developed for areas of The Hague and Tilburg provide good examples of the balance of power between engineers and local elites around the turn of the century.[26] In 1908 the architect H. P. Berlage presented a design for the inner city of The Hague to its city council.[27] Up to this time, town planning basically came down to sketching a planned network of streets, which was in accordance with the Local Government Act, but Berlage's plans comprised much more. He was one of the first in the Netherlands who felt that urban interventions should be based on viewing the city as an integrated whole: one should aim not only for a hygienic city with good traffic circulation, but also for a beautiful city. This view followed the groundbreaking 1890s work of the Viennese architect Camillo Sitte, but the influence of Baron Haussmann, the Parisian prefect who in the 1850s and 1860s, with the help of two engineers and a surveyor, had supervised the complete redesign of the Parisian inner city, was equally visible. Haussmann's partly technocratic project included plans for wide boulevards and new sewers, water works and street lighting. The aim was to create a technically advanced and disciplined city where potential epidemics

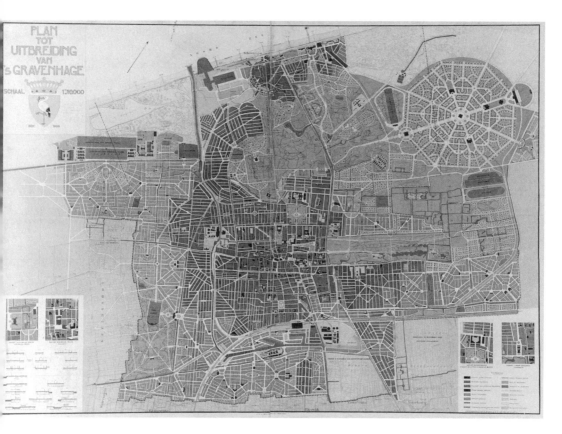

An appealing and very detailed expansion plan prepared by the architect H. P. Berlage for the city of The Hague in 1908. Around 1900, urban design and urban facilities became a major domain for technicians and civil servants. A separate public works agency was set up, and gas, water, and electricity services were managed by municipal agencies, and architects such as Berlage became more involved in urban expansion and planning.

or public insurgency could be easily suppressed. Just like Haussmann, in his design for The Hague Berlage planned central areas in the shape of squares, surrounded by public buildings and interconnected by main avenues lined with monumental buildings. Berlage was a socialist who as a member of the Amsterdam Health Commission had visited the homes of the poor. Tomorrow's socialist architecture, he felt, should be geared toward making a better city for everyone. But he received no mandate from the city government to develop such views. His plan required large-scale expropriation, but the elites of The Hague did not support this route and so only parts of the plan were realized, leaving many properties unaffected.[28]

Tilburg was a textile mill town where in 1901 the local government decided to have the measures called for by the Housing Act performed by the local Public Works Department because local taxpayers were not prepared to see the town's bureaucracy grow any further.[29] This department, however, proved unable to cope with all the work involved. The town decided to appoint J. H. E. Rückert, trained as engineer at the Royal Military Academy (Koninklijke Militaire Academie) in Breda, to take charge of the effort after it had repeatedly been urged by the national government to do so. He took up his new job energetically in January 1913. First he traveled to Germany to take a course in urban development, a field that was virtually nonexistent in the Netherlands. After his return he published an article in a national journal for engineers (*De Ingenieur)* in which he claimed that urban development was the rightful domain of engineers rather than architects. Then he developed a plan for the expansion of the city that was based on a socioeconomic analysis and inspired by the English garden city movement and on Sitte's and others' work. He designed six standard homes for various income categories and urged local government to start funding housing construction instead of waiting for private initiative to do so.[30] Quite soon he clashed with the alderman in charge of public works, who was also the owner of a cigar factory and despised Rückert's activist stance. The mayor and the aldermen refused to set aside money for purchasing land. Moreover, local contractors felt that his standards for housing were too high.[31] Rückert had no weapons at his disposal to fight the opposition he encountered. His mandate, in other words, proved too limited and in January 1917 he resigned.

Thus, as these two cases show, the Housing Act had only marginal effects in most Dutch towns and cities until World War I. Although engineers managed to gain major posts in several cities—for example, on health commissions (in part replacing physicians on these commissions)—their intervention scope remained limited. Town boards were generally not willing to carry the extra cost of public housing. Slum clearance was slow because often there was no alternative housing for the occupants, and contractors argued that building regulations made it unprofitable to build homes for the lower classes.[32]

The period 1850–1914 marks the emergence of a group of engineers in local and national government who had technocratic ambitions, meaning they were convinced their technological expertise especially qualified them to be in supervisory positions and tackle social problems. The political elites responded ambivalently to these implied claims. On the one hand, the need for government and expert guidance of specific social processes was now recognized. It was felt, for example, that the design and building of working-class homes could not be left to contractors anymore without technical super-vision from government. On the other hand, politicians wanted to maintain political control and provide as much scope as possible to private business. This position was very nicely formulated in 1905 by Pieter Lodewijk Tak (1848–1907), who held a seat in Parliament as a member of the Social Democratic Workers Party (Sociaal Democratische Arbeiders Partij) and who was a well-known publicist. He argued that government increasingly had responsibility to carry out tasks that required technical expertise; he cited port construction and public hygiene as examples. Although this need for expertise was hard to reconcile with democratic decision making, decisions could not be left up to technicians because their opinions varied as often as those of politicians. Tak therefore advocated a decision process in which a variety of specialized commissions participated, including technicians, but with the politicians having final decision-making power.[33]

Engineers Plan Polders, Architects Experiment, 1914–1940

During World War I the Netherlands remained neutral and the government warded off the threat of economic chaos caused by the drastic reduction of trade by drastically regulating the economy, through ad hoc agencies in close consultation with business leaders and largely without involving Parliament. After the war these regulatory agencies were disbanded, but more than pre-viously the country's economic policies were formulated as a result of formal deliberations between government, employers' organizations, and labor unions. There was much talk about "managing" the economy and society in general, implying that the system of democracy (general suffrage had been introduced only recently, in 1918), was forced on the defensive again. Many commentators argued that there were too many parties and that politicians spent too much time deliberating and failed to show resolve.[34]

This criticism goes to the essence of the issue of the role of expertise and the mandate of technicians. Many commentators who felt that the "igno-rant masses," now in a position to vote, were incapable of making sensible choices, a view that even gave rise to doubts as to the credibility of members of Parliament as well. Some prominent Dutch politicians and intellectuals argued in favor of putting the management of the economy into the hands of experts. One of them was M. W. F. Treub (1858–1931), who in the 1890s as city board member of Amsterdam led the takeover of utility companies and during World War I supervised the drastic reorganization of the economy. In 1918 he took part in the elections as leader of the Economic Union

(Economische Bond) a party that favored entrusting economic and foreign affairs to top-level experts, thus shielding these policy areas from party politics. This party won only three seats, yet the view that the "common interest" was better served by expert governance than by parliamentary decision making gained ground.[35] The Social Democratic party perspective on a plan-based economy, as formulated in the party's white papers "Report on Socialization" ("Socialisatierapport," 1921) and "Labor Plan" ("Plan van de arbeid," 1935), also gave priority to expert guidance of the economy as a way to end the waste resulting from competition, advertising, and the production of luxury goods. It was not a coincidence that several prominent engineers were among the plan's authors, such as Th. van der Waerden, J. Goudriaan, and J. W. Albarda, who were all members of the Association of Democratic Engineers and Architects. Their views on steering the economy flowed from their criticism of capitalist economy rather than criticism of democracy, but there was a clear resemblance with other proposals to reorganize the economy, based on expert advice.

Some argued for a prominent role of engineers in particular rather than of experts in general. For instance, right after World War I the influential engineer I. P. de Vooys, a professor of mechanical technology at the Technical University Delft (in the 1930s he became director of the synthetic fiber company AKU), claimed that high productivity and rational handling of limited natural resources, essential to a well-oiled economy, could only be realized with the help of engineers. Similarly, in public governance engineers could strengthen the state's efficiency by using the systematic logic that was part of their field.[36] In 1934 the Dutch Technocratic Union (Nederlandsch Technocratisch Verbond) was established, following the example of the American organization Technocracy Inc., founded in 1933 by Howard Scott.[37] One of this group's leaders claimed the goal of technocracy was "prosperity for all, based on a scientific organization of our society, which allows for deploying man's scientific powers in expert ways to solve both physical and psychological social problems." He understood "expertise" more broadly than the art of engineering, even though most members of his organization were engineers.[38]

To what extent did such views fall on fertile ground with the wider public? And did they lead to a broader mandate for engineers? The level of support among the population for delegating more power to the experts is hard to assess. It is clear, however, that there was a widespread endorsement of new technologies, which suggests a degree of confidence in technological experts. For example, The popular press closely followed the latest developments in radio, television, and aviation. Inventors were honored as heroes.

A souvenir game, given to customers who bought a tin of Verkade cookies, produced by the well-known cookie company to commemorate KLM's victory in the 1934 air race from Mildenhall, Suffolk to Melbourne, Australia. KLM's plane, *De Uiver* (*The Stork*), won the race on corrected time. In those days countless mementos and souvenirs were issued to commemorate aviation feats—the record-setting flights of KLM pilots gave the country a morale boost in the midst of the Depression. One year earlier some 15,000 people had flocked to Schiphol Airport to greet the returning *Pelikaan*, another KLM plane, which in December 1933 had made a record-setting flight to the East Indies, a feat that occasioned countless mementos and aviation souvenirs and garnered the crew a royal medal.

When in 1927 the national newspaper *Algemeen Handelsblad* (*General Trade Times*) asked its readers to mention six individuals whom they considered to be "the greatest men alive," hundreds of respondents mentioned a total of 114 names. Edison topped the list with 70 percent of the votes; Mussolini, as the incarnation of energetic innovation, came in second with 66 percent, followed by Marconi (50 percent), Einstein (42 percent), Ford (25 percent), and Lindbergh (25 percent). Lindbergh's historic transatlantic flight had occurred five months earlier.[39] The first flights of Royal Dutch Airlines to the Dutch Indies elicited waves of enthusiasm. After one such flight, in 1933, the pilot A. Viruly wrote that it was as if the Netherlands had turned into "one big aviation family."[40] This hardly implies a political mandate, but it is perhaps indicative of a social climate in which the notion of having engineers further design the Netherlands would meet with little resistance. To what extent were engineers able to exploit this social climate? We will assess the nature of the mandate for engineers for two cases: the development of new polders and building of public houses.

Polders

Between 1931 and 1967, the Netherlands reclaimed 165,000 hectares of land from what had been the Zuiderzee, a shallow two-thousand-square-mile inlet of the North Sea. The design and planning of this new land was almost completely in the hands of various groups of technological experts.[41] Because experts were given such a broad mandate, this case allows us to explore what happens when technocracy becomes a dominant governance mode. One important conclusion we reached is that struggles over the aims and operation of the project did not stop simply because decision making was largely in the hands of technocrats. The displacement of the project from the public realm of political rivalry and party politics in Parliament to the technical realm resulted in a struggle among a variety of experts and bureaucrats that largely took place outside the realm of parliamentary politics. All sorts of experts—on the basis of newly discovered interests and related new types of expertise—laid claim to the right to influence the design of the new land. Civil engineers, agricultural engineers, urban developers, and social demographers all had something to say about the creation of the new polders. In many ways the polders' resulting physical and social structures are a reflection of the work and political-social ideals of these technocrats and the balance of power among them.

The wish to close off the Zuiderzee to create new agricultural land, create a freshwater reservoir, and create a protective buffer against floods had been discussed several times after 1850. Many plans were developed but successive cabinets seemed unable to make up their minds about going ahead with the vast project.[42] In 1886 several prominent individuals—mayors, members of national and provincial parliaments, water management officials, and executive directors of local chambers of commerce—decided to form a lobby group to promote the project, the Zuiderzee Society (Zuiderzeevereeniging), and hired the young engineer Cornelis Lely to develop a new plan. Lely intended to close the Zuiderzee by means of a barrier dam and create four polders, which would cover more than half of the area of the former inlet. In 1891 he became minister of water management, and he put together a commission to study his plan's feasibility. In 1894 the commission reported positively and recommended that the state take charge of the project, but the government's fall from power prevented Lely from proposing a bill. In the ensuing years the Zuiderzee Society unsuccessfully pushed a new bill. Even Lely, who in 1897 became minister again, appeared to have given up on pursuing the issue. Other political priorities prevailed, such as social legislation and the Achin War in the Dutch East Indies. Moreover, the agricultural crisis of the 1880s had led to a diminishing demand for land and decreasing fees for leases on land, which would make it hard to recover the cost of new polder projects.

The World War I era marked a turnaround for a number of reasons: financial, economic, and political. In 1913 Lely again took charge of the Water Management Ministry. From the very beginning he negotiated draining of the Zuiderzee as part of the government's program, claiming that land lease rates had increased significantly, that his predecessors had exaggerated the cost of the project, and that since 1901 the state's income had doubled and so it could afford to undertake the project. Moreover, after war broke out, the government position changed and it felt called upon to intervene in the economy in major ways. In 1915 Lely wrote to his son: "We are ready to do something big and because the war has changed our notion of money so radically people are no longer as afraid of taking up [public] works."[43] In the early years of the war it became clear that Dutch agricultural land was not sufficient to feed the entire population, meaning that during the war the country would have to rely on importing food. Then, in January 1916, there was a storm surge that flooded large areas around the Zuiderzee. This event effectively buried all doubts about the project's future. In 1918 Parliament adopted the Zuiderzee Act by acclamation. It involved a skeleton law granting the government a broad mandate: the inland sea would be closed off and partly drained according to Lely's plans.

Many observers viewed the Zuiderzee project in terms of the country's national greatness. That in the interwar period technological feats could take on major symbolic significance is shown by the following passage from

The larger-than-life engineer Cornelis Lely, the minister of water management, armed with a shovel, drafting tools, and a toy army of building equipment and laborers, in preparation for the definitive reckoning with the god of the Zuiderzee. The cartoonist, Louis de Leeuw, saw it as an unequal fight, yet the god caused another big flood in 1916.

Nieuwe Rotterdamsche Courant of May 28, 1932, when the barrier dam had just been completed and the first polder had been pumped dry:

> Regardless of one's view of the Zuiderzee drainage issue, and even if together with the Zuiderzee fishermen one feels saddened and down-hearted, the work also has a technological side to it for which we can only have admiration. It appeals to our imagination and commands our deep respect. It is therefore appropriate to congratulate the engineers who supervised the project with their accomplishment, as well as all those who through their labor have contributed to it. They have shown the Netherlands at its best.

The project was not put into the hands of the Rijkswaterstaat. Many of its engineers considered it an impossible undertaking and Lely himself felt that this agency was a sluggish bureaucratic organism that made insufficient use of the latest technologies.[44] A new entity, the Zuiderzee Works Agency (Dienst Zuiderzee Werken), was established in 1919 to oversee the project. Lely entrusted its management to two hydraulic engineers, H. Wortman and V. J. P. de Blocq van Kuffeler, who had been involved in the development of his plan. In the ensuing years this new agency mostly employed civil engineers from the TU Delft who developed a strong esprit de corps, as well as contempt for outsiders, notably engineers from the agricultural college in

Wageningen, whom they dubbed "potato engineers."[45] The new agency was supervised by an advisory commission, the Zuiderzee Council, chaired by Lely. The brief economic crisis following World War I caused a delay of several years, but in 1925 work was finally started on building the Afsluitdijk, the main barrier between the Zuiderzee and the North Sea, and draining the Wieringermeer in order to make the first of four planned polders; it was completed in 1931. Next, the Noordoostpolder, Oostelijk Flevoland, and Zuidelijk Flevoland were turned into polders, their draining being completed in 1942, 1957, and 1967, respectively.

The planned layout of the polders was hotly disputed from the start.[46] The plans developed by the Zuiderzee Works Agency for the Wieringermeer were based on water management considerations exclusively. Because the elevation of the new polder varied, it was divided into four sections, each with its own ground water level. The waterways for drainage and shipping were largely planned along these sections. Roads ran parallel to them and where they intersected it was foreseen that villages would spontaneously develop (as had happened with earlier polders). The division of the farmland in lots was determined by the drainage technology. This approach to new polder design had been criticized by the well-known nature preservation activist Jac. P. Thijsse in 1915. He hoped "the new polders will not be parceled according to the block system and the land will not be exploited up to the last square meter merely for the sake of its monetary value." Thijsse argued that the people who went to live there would also need something more: "The polders should look nice and also, let's say, be pleasurable."[47] In 1922, the chairman of the Dutch Architects' Union (Bond van Nederlandsche Architecten), S. de Clerq, who was also on the Zuiderzee Council, raised the issue of the future polders' "aesthetic development." Without opting for a specific building style, he felt there should be coherence. He also felt that Zuiderzee Works engineers were likely to have no expertise on this issue. In 1925 yet another voice joined in when the sociologist H. N. ter Veen published a critical study on the Haarlemmermeerpolder's settlement in the 1850s that made an impression on policymakers involved in the Zuiderzee Works. His study revealed that government had restricted itself to draining the Haarlemmermeer, after which the land had been sold to the highest bidder. The building of roads and farms and other structures had been left entirely up to private initiative, causing the polder to develop quite slowly, in both social and economic terms. To speed up the process, the government should take charge of the new polders' socioeconomic design. Even before the first new polder was drained, then, other experts unambiguously challenged the expertise and authority of the Zuiderzee Works to design it.

In 1926 the government, at the urging of Parliament, set up a commission (the sociologist Ter Veen was secretary) to explore how the design of all new polders could best be managed. This commission proposed a new type of organization: a "public body" that would be better protected against "harmful political meddling by Parliament" than the Zuiderzee Works Agency, an agency controlled by the ministry.[48] The result was the establishment, in 1930, of what later on would be named the Wieringermeer Directorate. Led by agricultural engineers, it was in charge of building polder villages and selecting farmers to lease the polders. These two executive bodies, the directorate and the Zuiderzee Works Agency, were charged with carrying out and overseeing the work of creating, laying out, and populating the polders—a collaboration that proved a trying affair.

This arrangement did not quite deal with all the issues that had led to criticism of the design of the first polder, the Wieringermeer. A new professional group, urban developers, criticized its one-sided agrarian focus. Most urban developers were civil engineers, who after the introduction of the 1901 Housing Act had been hired by local governments to give shape to public housing policies and urban expansion. In 1918 they organized themselves into the Netherlands Institute for Public Housing, led by a Social Democratic lawyer, D. Hudig. Several nature conservation professionals also joined this institute. It published a journal of public housing (*Tijdschrift voor volkshuisvesting*), organized an international conference in 1924 with renowned professionals such as H. P. Berlage and Ebenezer Howard, and set up an Urban Development Council chaired by the Delft-trained construction engineer and architect M. J. Granpré Molière (1883–1972).[49]

One of the issues of concern to the Institute for Public Housing was regional planning.[50] In the 1920s the rapid growth of towns and road infrastructures and concern for the increasing loss of unspoiled nature led to plans to control the purchase of land for residential and commercial development. The discipline of regional planning was applicable not only to urban expansion but also to the spatial development of a region and, ultimately, the whole country. In 1938 the Institute for Public Housing organized a conference to discuss a national plan. Dutch and other urban developers used the work of Patrick Geddes, a Scottish biologist and botanist who was also known as an innovative thinker in the fields of urban planning and education, to develop a regional planning method. Its starting point was the slogan "Survey before plan": first carefully describe the physical and social structure of an area ("place, work, folk," as Geddes put it), analyze its "driving forces," and subsequently integrate them into your plan.

The new polders gave Dutch town planners a unique opportunity to apply Geddes's insights, and after its 1924 conference the Institute for Public Housing offered its assistance to the Zuiderzee Works Agency. The head of the Zuiderzee Works, De Blocq van Kuffeler, refused to commit himself, however, which prompted the Institute for Public Housing to form a Zuiderzee Commission, chaired by Hudig, whose membership included Granpré Molière's colleague the architect P. Verhagen, and W. G. Witteveen, director of the Urban Development Agency of Rotterdam. It reported that the polders' design had to rely on town planning expertise. Shortly thereafter the Institute for Public Housing urged the minister to make a regional plan for the new Wieringermeer polder. In an article in the 1926 volume of *Tijdschrift voor Volkshuisvesting*, Hudig released precisely such a plan. It argued that the polder's planning should be based on a balanced integration of four functions: living, working, leisure, and traffic. An "organic road network" would breathe life into the polder's residential areas. Not surprisingly, the approach they advanced differed completely from that of the agricultural and hydraulic engineers.

The lobbying of the town planners resulted in the appointment of Granpré Molière the as an adviser at the Zuiderzee Works Agency in 1926. Although he could do little regarding the polder's general design, with his students he did have major influence on the planning of villages, which was based on very specific views of society and social life. During his study at Delft, Granpré Molière had been a member of the Sociotechnical Association of Democratic Engineers and Architects and had made a name for himself in 1916 with the design of Vreewijk (with Verhagen), a working-class neighborhood in Rotterdam that looked like a village, inspired by the British garden city idea and the factory villages. With three thousand houses built by 1930, Vreewijk was the largest "garden village" in Europe.[51] Granpré Molière called it a remedy against urban desolateness and degeneration and the mental disintegration caused by World War I.[52] After 1924, when he became a professor at Delft, lecturing in urban development and other fields, he won such a following that people began to refer to the "Delft school of urban design."

In his polder work Granpré Molière sought to counterbalance the unsettling effects of industrialization and urbanization, processes that he considered to be at odds with people's fundamental social and spiritual needs. He liked the idea of polder communities without the urban degradation of large buildings, pretentious shops, and vulgar advertising. Village design had to foster harmonious forms of social life. He designed each village with a green zone in the middle, surrounded by public buildings. The Catholics, Protestants, and Dutch Reformed each had their own church building on the village's outskirts, at the end of one its main roads. If the local population grew, it would be easy to expand the village.

The 1930 reclamation of the Wieringermeerpolder added a significant amount of new land: 20,000 hectares. Its development was carefully prepared: the plan was completed ten years after the initial drafts. Not only the spatial aspects but also its socioeconomic development were planned in detail, including open space, agricultural businesses, and three villages. Most homes came with substantial backyards, to be used as vegetable gardens, as can be seen in this 1934 photo of the Torenstraat in the village of Middenmeer.

Thus, the design of the new Dutch polders in the 1920s and 1930s involved a host of social, geographical, and technological concerns, hence it was a politically charged affair. Although the government and Parliament had delegated the polder design to the experts, this generous mandate and technocratic mode of governance did not imply the absence of political discussion. On the contrary: the debates among the experts were in many ways quite political, centering on how society would develop on the new land and what level of management was needed. Would settlements automatically come into being in a spatial structure determined by water management concerns? Or should settlements be planned in advance, and if so, how? Should the new land be designed exclusively to maximize agrarian output, or were aesthetic aspects relevant as well? The groups of experts substantially differed on these points, not just because of their different expertise but also because of divergent views of society.

Public Housing

As discussed earlier, the Housing Act of 1901 had few practical effects during the first fifteen years after its passage, but as in the case of the Zuiderzee Works, World War I caused a turnaround. During the war, housing construction stalled because of rising prices of construction materials; rents went up and a housing shortage developed.[53] A 1919 survey revealed a nationwide shortage of 57,550 homes.[54] Starting in 1916 the Dutch government began to increase financial support for building corporations and stepped up the pressure upon town governments to start building cheap homes. Municipalities were obliged to build temporary housing and housing corporations could get direct financial support from the government, with no interference by local government. After 1918 the pace of home building increased significantly, undertaken mainly by private companies and housing corporations, and less by local governments. In 1921 the Housing Act was amended in order to make it possible to force municipalities to support private housing construction.[55] For the first time there were clear nationwide standards for the design of houses for subsidized housing that echoed the discussions from the previous decades: no "alcoves"; separate bedrooms for parents, boys, and girls (so at least three bedrooms); and standards regarding water supply, waste discharge, humidity control, and "tightness" to exclude drafts.

This period of sustained government intervention lasted about five years. From 1921 on state funding was gradually reduced, construction costs went down again, and in 1925 P. J. M. Aalberse, the minister of trade and industry, judged that the housing shortage had been solved and that financial support for housing corporations could be discontinued. From that moment, housing construction once again became largely a market-regulated affair. Housing corporations were geared toward basic, inexpensive, and efficient designs, as well as better management of the available housing stock. Public housing mainly became dependent on local incentive policies and therefore on local political relations.

In the immediate postwar era working-class homes were usually built in small neighborhoods that functioned more or less like a village: two-story homes at the edge of town whose front facades faced the sun and came with much "green."[56] The English garden cities, which had become more widely known in the Netherlands as a result of a contest sponsored by the Sociotechnical Association in 1913, served as a source of inspiration. Vreewijk, a district reflecting the aversion to the unhealthy and immoral aspects of urban life, embodied the ideals of the late-nineteenth-century housing reformers. Where space was scarce, as in the inner cities, architects designed closed blocks common in the nineteenth century, but this time based on designs that met the new standards and came with aesthetic attention to facades (most spectacularly in those by Michiel de Klerk and the Amsterdam school). Back-to-back homes were no longer built. Moreover, the new homes were more spacious and no longer had alcoves; though many lacked a shower or bathtub, a bathhouse or a shop providing hot water could be found nearby.

Furthermore, some neighborhoods were built according to radically new principles. This occurred mostly in towns and cities where the Social Democrats had gained positions as aldermen after the 1918 elections (the total number of Social Democratic aldermen went up from ten to eighty-seven, in seventy-three municipalities). Many of them took charge of public housing in cities, together with like-minded directors of local public housing agencies. They tended to give commissions to architects in the modern school, such as those associated with Theo van Doesburg's journal *De Stijl*. Some of them belonged to the international functionalism movement, known in the Netherlands as New Building (Nieuwe Bouwen).[57] They met in two societies: "De 8," based in Amsterdam, and "Opbouw" (Construction), based in Rotterdam (in 1932 the two merged). Several members had a civil engineering background. The working-class neighborhoods they designed differed in two respects from the new public housing neighborhoods mentioned previously. First, they developed a Cubist formal language, based on pursuing beauty through the composition of surfaces and straight lines rather than by means of wood or brick ornamentation. Second, they actively sought to use new materials and production methods such as concrete and prefabricated elements.

At a conference of the national organization of housing corporations in

1918, when the housing shortage was peaking, a fierce debate erupted on mass production of dwellings. The director of Amsterdam's Building and Housing Inspection Agency (Bouw- en Woningtoezicht) proposed to solve housing problems by standardizing types of homes and using standard building materials, a view supported by Modern architects such as J. J. P. Oud and H. P. Berlage. "Housing production should become mass production," Berlage claimed.[58] However, most architects and public officials involved disagreed and warned against dull uniformity. The architects probably feared that the spread of standard designs and the elimination of ornamentation would undermine their position in the decision-making process.

Still, various Modern housing construction projects were carried out. In Rotterdam, Arie Heykoop, a member of the Social Democratic Workers Party, became alderman in charge of public housing in 1918, and A. Plate, an engineer who was a member of the Sociotechnical Association, was appointed as the first director of the municipal housing agency.[59] Within a year, Plate asked the Modern architect Michiel Brinkman to design a working-class housing project for the Spangen district. The design was revolutionary: 273 dwellings shared communal laundry space, drying lofts, and a bathhouse. The front doors opened onto an inner courtyard and the second-floor apartments were connected through a kind of gallery that could be reached by elevator. There was a flat roof for sunbathing—a feature that elicited fierce resistance from city council members who felt that this would encourage immoral behavior. Although the collective facilities and costs also met with criticism, Heykoop managed to get the plan passed.

In this same period J. J. P. Oud designed several housing blocks with very innovative standard homes that gave streets an entirely new look. One of the projects, the famous Witte Dorp in Oud-Mathenesse, was built in 1922. Although the government was already cutting down on funding for housing, the Temporary Housing Act made it possible to build the homes after all. The project involved a social housing experiment, with problem families placed in between functioning families, so that the latter could provide support to the former.[60] The problem families first received specific training in the basics of living, in a smaller housing project called "Arbeid adelt" ("Work ennobles"), supervised by a female housing inspector.[61] Oud, like many housing reformers since the late nineteenth century, was convinced that homes and neighborhoods could be designed in ways stimulating "good conduct." The neighborhood was closed off from the outside world and had a central playground, with the services building in the middle. Kitchens were small: they were intended not as living space but merely as workspace for housewives. Parents and children slept in separate bedrooms.[62]

Modern architects tried to influence the behavior of the occupants through their home layouts. Kitchens were situated at the side of the street or gallery, so that mothers could watch their children and also so that family life in the living space would be shielded from the eyes of the outside world. The houses designed by the Amsterdam school, like those by Oud, had small kitchens, so that they could not be used as living space. Some rooms came with built-in beds.[63]

After 1925 these experiments in public housing were harder to fund and thus became much scarcer. The most famous example was the Bergpolderflat, built in 1934 to a design by W. van Tijen, J. A. Brinkman, and L. C. van der Vlugt for a private company that concentrated on working-class housing construction. These men were convinced that affordable and healthy homes could be built without subsidies by applying the most advanced technology and production methods: a steel frame and prefab parts.

Meanwhile Modern architects continued to study more rational construction methods and more efficient home design. For example, in 1933 and 1934 the Dutch Architects' Union organized a design contest for "inexpensive working-class homes," which resulted in designs that on hindsight can be said to anticipate postwar housing construction. Many designs featured open lots with parallel, straight-lined housing blocks and stories with almost identical units that could be accessed via a gallery. Incidentally, the jury's conclusion was that sound construction and inexpensive construction did not really go together.

More than previously, future residents were involved in such activities. In particular the Netherlands Association of Housewives (Nederlandse Vereniging van Huisvrouwen) was active in this area. It encouraged translation of foreign handbooks on the "Taylorization of households" and it regularly consulted its constituency as to their wishes for home design. The Bruynzeel kitchen, *Bruynzeelkeuken*, installed in many public housing units after World War II, was inspired by earlier foreign kitchen models promoted in the Netherlands by the Housewives Association and by the Holland kitchen, *Hollandkeuken*, whose design had been commissioned by the association.[64]

In this respect a major source of inspiration was the International Congress of Modern Architecture (Congrès Internationaux d'Architecture Moderne), established in 1928 with the Swiss architect Le Corbusier (1887–1965) as its driving force.[65] This organization, chaired by Cornelis van Eesteren from 1930 to 1947, favored technical solutions for social problems and frequently assumed a slight contempt for politics. Le Corbusier felt that housing construction should follow the example of Ford's series production of

Housing block G. Callenburgstraat/Tjerk Hiddes de Vriesstraat, in Amsterdam, built in 1933, with its communal pool in the inner courtyard. The ideals of light, air, and healthy physical culture were given expression in new social housing built after World War I.

automobiles because a home was "une machine à habiter," "a machine for living," a view echoed by Oud when he called his design for a "minimal home' for the working class district De Kiefhoek a "Ford for living."[66] Concrete and elevators made it possible to build high-rises, allowing many people to live on a fairly small footprint of land. Because enough green space was left in between apartment buildings, residents enjoyed ample sunlight and fresh air.[67]

In International Congress of Modern Architecture's opening manifesto urban development was described as "the organization of collective living in towns and countryside."[68] Public housing was high on the organization's list of priorities. In 1932 the Dutch architect J. B. van Loghem argued that the problem of public housing construction could be solved by getting "big industry" involved—though in many cases this large-actor role would end up being filled by national or local government.[69] Van Loghem also stated: "If the art world is not to get the short end of the stick in Rotterdam, eventually we must put a stop to this damned fanaticism among politicians, while artists and engineers—meaning those who possess the skills, have the competence to do something that has substance—should be allowed to be in charge as well."[70] This aversion to politics was characteristic of technocratic thinkers (as we saw already in the case of Treub and Lely).[71]

A final example of the role of engineers in public housing is a study on the importance of sunlight for people's health that was commissioned in 1931 by the Amsterdam Society of Architects "De 8." Officials from public housing and preventive medicine, architects, and the Royal Institute of Engineers (Koninklijk Instituut van Ingenieurs) participated. B. Merkelbach and W. van Tijen wrote:

> Is it only about this sun? Not at all, for essentially a much deeper issue is involved. Within the urban development domain it concerns the expression of an attitude toward life, one that is also expressed in clothing, sports, education, way of life, etc. This changed attitude can best be described as a growing need for pleasure in life, a joy we believe to be attainable for everyone, if only our physical and spiritual needs are brought into the right balance, and if only we rationally capitalize on and take into account all the forces and opportunities that are present in humans and society.[72]

This statement encapsulates the technological optimism of the New Builders: it suggests confidence in their power to create a better society through good spatial design. The designer of the Wieringermeer villages, Granpré Molière, expressed the same confidence, albeit at the service of a more traditionalist social ideal. Shortly after World War I, several Dutch cities afforded modern architects a broad mandate for realizing their technological solutions to social problems. However, these experiments and a small number of projects, such as the Bergpolderflat in the 1930s, represented only a minor share. Most Dutch housing construction during the interwar period was fairly traditional and resulted from open-market competition. Although the work of the New Builders was influential and attracted international attention, both government and the market afforded them only marginal space.

After World War I, then, engineers, architects, and urban developers gained a greater foothold in Dutch society, partly because of prevailing skepticism about political democracy, criticism of entrepreneurship (on the part of the Social Democrats), and increased trust in experts and technology in general. It became an accepted and proven point of view that a technical approach offered a good option for tackling social problems. The activities of the groups mentioned contributed to the further growth of the intervention state. They prompted government to action by explicitly calling on it to intervene and by showing that the state could realize social objectives through large-scale technical projects, which would instill a sense of national pride and unity to boot. In the case of public housing construction, the actual mandate still lagged far behind the ambitions of architects and urban developers. Although engineers were given great liberties in designing the new polders, this did not mean they implemented their views without any further debate. Within their mandate there was rivalry among the various experts: politics was increasingly displaced from the realm of politicians to the realm of technological experts and their projects.

Architects, Urban Developers, and Social Demographers in a Changing Political Climate, 1940–1970

In many ways the effects of World War II on Dutch society were similar to those of World War I, but they were more intense and lasted longer. Just as in the period after the first war, all sorts of technological experts gained unprecedented opportunities for realizing ideas they had developed earlier. This time their mandate seemed even more widely accepted because destruction was more substantial, prosperity created more room for big projects, and trust in experts and the leading role of government had perhaps grown larger than before the war. Just like in the prewar period, however, the power

The hydraulic engineer Johan A. Ringers (center, with glasses), during a working visit in his role as a board member of the Oranje-Nassau Mines, around 1947. Ringers (1885–1965) played a leading role in the systematic development and postwar reconstruction of the Netherlands. Among other things he was director-general of Rijkswaterstaat, government commissioner for reconstruction, and, from 1945 to 1946, minister of public works and reconstruction. Moreover, he was chair of the Association of Delft Engineers and vice-chair of the Royal Institute of Engineers.

relations among the different groups of experts were strongly determined by political priorities set by others.

The most tangible challenge after World War II was the country's physical reconstruction; countless buildings and facilities had been destroyed and land had been flooded on purpose. It was only natural that engineers were given a leading role in cleaning up. During and immediately after World War II the reconstruction effort was led by J. A. Ringers, an engineer with the Rijkswaterstaat, who set up the Reconstruction Agency (Wederopbouwdienst) which was led by civil engineers. After the war its effort was continued by the Water Management and Reconstruction Commissioners Board (College van Algemene Commissarissen voor Waterstaat en Wederopbouw). In each province a chief engineer and director of public housing supervised the reconstruction effort and checked on everything: urban development plans, housing designs, use of materials.[73]

As for economic reconstruction, leading experts proposed a plan-based approach, which initially seemed to have wide social and political backing.[74] Best known among these experts were the Social Democratic Party member Hein Vos, an engineer, and Jan Tinbergen, an econometrist who had contributed significantly to the Social Democrats' 1935 "Labor Plan." Tinbergen referred to his field as an "engineering" science. They felt that government should direct the economy on the basis of statistical data and models. The Central Planning Bureau (Centraal Planbureau), established in September 1945 to provide the scientific basis for this policy, never obtained such a full mandate and it soon fulfilled the more modest role of advisory body. The model of governance that ultimately prevailed was corporatist: policies were not simply imposed from above, but resulted from ongoing deliberations between government and representatives from employers and employees in the Labor Foundation (Stichting van de Arbeid), set up in May 1945. But since the negotiators were experts who relied on econometric analyses, this corporatist model had unmistakable technocratic overtones.[75]

In postwar cabinets and in Parliament there was solid consensus on the need to invest in technology development.[76] It was clear that the Netherlands, especially after losing the East Indies in 1949 and with declining employment in agriculture, would have to capitalize on exporting industrial wares. This called for investments aimed at overcoming the country's assumed technological backwardness, notably vis-à-vis the United States, and expenditures for R&D rose dramatically.[77] In 1946 two new research institutes were set up: the Foundation for Fundamental Research of Matter (Stichting Fundamenteel Onderzoek der Materie) and the Mathematical Center (Mathematisch Centrum). Both later became part of Zuiver-Wetenschappelijk Onderzoek, the Dutch Organization for Pure Scientific Research, set up in 1950. Minister Rutten approvingly quoted the founders of the American organization that served as model, the National Science Foundation: "Today no nation is stronger than its scientific resources."[78] Two new technical universities were established—at Eindhoven, in 1956, and Twente, in 1961—to ensure a sufficient number of engineers, and the number of technical students in higher education rose rapidly. The number of full professorships at the Technical University of Delft grew from 57 in 1950 to 167 in 1962, and in this same period the number of students rose from 5,125 to 7,615.[79]

Substantial monies were tied up in science and technology policies.[80] Still this policy area, like the economic policy domain, was hardly subject to parliamentary control,[81] mainly because policies were designed by foundations such as the Foundation for Fundamental Research of Matter, the Institute for Nuclear Research, Reactor Centrum Nederland, and the Organization for Pure Scientific Research. These organizations relied on input from government, business, and scientists and largely developed through co-optation as members of the organizations appointed new members of the same organizations.[82]

This overall approach was supported by Dutch society at large. Based on an international comparative study, the political scientist A. Lijphart has characterized the attitude of the Dutch regarding political authorities in the 1950s as remarkably docile: voters tended to give politicians a broad mandate.[83] More specifically, there was support for a dirigiste economic policy, so as to avoid a 1930s-like crisis. Large technological projects invoked admiration. A 1958 survey asked Dutch youngsters "what the Dutch can be genuinely proud of." About one third of the respondents named hydraulic engineering projects (these were the early years of the Delta Works, of which more below).[84] The 1957 exhibition "The Atom" showed how inexpensive power feeding all sorts of electric appliances would make life more enjoyable. Among the 750,000 visitors was the popular historian Geert Mak, who forty years later still had fond memories of the exhibition. It seemed to be a successful attempt to make the country "nuclear power-minded," as the chairman of the association of Dutch power companies put it.[85]

Around 1960 some researchers concluded that the Netherlands had in fact become some sort of technocracy. The sociologist P. Thoenes wrote that the country was governed by "new regents," distinguished by their "instrumental competency," which implied a rather ambivalent judgment. On the one hand, their leadership might allow "the Dutch people to achieve a sense of unity, which in previous periods could only be found in coping with shared suffering or danger or in military victories or defeats. Drainage and diking projects, development plans, industrialization projects, statutory industrial organizations, the emphasizing of the idea of national economy—all imbue the notion of 'the Netherlands' with more content and reality." On the other hand, Thoenes argued that regardless of "the modesty with which this new elite presents itself, the integrity of its operations and the democracy-mindedness of its individual members, as a group it basically has anti-democratic views. It is a group to which our country owes a lot, since we have entrusted [to] it major and difficult tasks as to our future. But the task of having this elite function in a genuine democratic frame is one we cannot leave up to the elite itself."[86]

In the next section we examine what the mandate afforded these new "regents" amounted to in practice.

"Well-designed land" was what the planners of the Zuiderzee reclamation project had in mind for new polder land. The rational parceling of the Noordoostpolder is still easily visible in the current polder landscape. Here the Espelerweg, looking west from Emmeloord in 1992.

The Noordoostpolder and the Oostelijke Flevopolder

The economic crisis of the 1930s delayed development of the Noordoost-polder (Northeast Polder), but in 1936 the Dutch Parliament voted in favor of getting started with its empoldering. A major reason was the employment this would provide. In many respects the new polder's design involved a process that was analogous to that of the Wieringermeer. First the Zuiderzee Works Agency engineers decided on a global plan centering on the water-ways and roads, then a land division plan was developed. The top priority was to create efficient agricultural acreage, to allow for swift recovery of the large investments. As in the Wieringmeer, the Water Management Ministry deter-mined the layout of the polder, after which Granpré Molière and his col-leagues began work on designing villages. Again architects and the Institute for Public Housing protested and again experts disputed each other's expert-ise, sometimes fiercely so. For example, the Zuiderzee Works' Van Kuffeler held that unlike the Zuiderzee Works and the Building Department of the Wieringermeer Directorate, the urban developers had insufficient insight "into the countless data and the nature of the prospective land," whereas A. D. van Eck, from the directorate's Building Department, sneered that architects "will even design a building with evident flaws, as long as its exterior will look nice."[87]

During the war the German occupation worked to the advantage of the interests of planners and urban developers. After all, the Germans wanted to design plans for Europe as a whole. In 1941 the National Plan Agency (Rijksdienst voor het Nationale Plan) was set up as part of the Reconstruction Agency, led by J. A. Ringers, followed one year later by the establishment of provincial planning agencies. Because the polders became part of a national plan, the urban planners obtained a powerful new ally in the struggle to get themselves involved in the polders' construction. Although it was too late to exercise much influence on the design of the Noordoostpolder, these urban planners were not going to allow the planning of the two southern polders to be guided purely by hydraulic and agricultural considerations.

Despite Van Kuffeler's and Van Eck's unfriendly words about architects and urban planners, Granpré Molière was again asked to design villages and provide advice on urban development. This is why most villages in the Noordoostpolder have the same Delft school character as those in the Wieringermeer. Granpré Molière recommended engaging private architectural firms as well, because the Building Department could not handle the job alone. Van Eck grew unhappy with the Delft school's monopoly on polder design, and when modern architects from "De 8" and "Opbouw" approached him with the request to design a polder village to be presented at the International Congress of Modern Architecture conference on human settlements, held in 1949 in Bergamo, he readily accepted. The result was a village that strongly differed from the others: Nagele. Prominent New Builders such as G. Rietveld, A. van Eyck, J. B. Bakema and C. van Eesteren contributed to it.

It was unprecedented for modern architects to contribute to polder design. The greater input of sociologists was another novelty.[88] The sociologist H. N. ter Veen was an energetic scientific entrepreneur. He became professor in Amsterdam in 1933, and he founded the Sociographic Work Fellowship (Sociografische Werkgemeenschap), which conducted a series of empirical studies, such as of entrepreneurs in the Amsterdam Jordaan district and on migration from Amsterdam to the nearby village of Badhoevedorp. In 1936 he set up a scientific foundation that up to 1943 issued twenty-six publications. Ter Veen explained to his students at the university in Amsterdam that they ought to become "social engineers," seeking to "calculate the forces in society in order to build on the knowledge thus gained."[89]

This formulation closely tied in with the urban development program of "Survey before plan." Ter Veen put in substantial efforts to find employment for his students, notably with the economic-technological institutes and town-planning departments of major cities.

Urged in part by the National Plan Agency, both polder agencies hired sociologists during the war. The Wieringermeer Directorate hired two of Ter Veen's students, E. W. Hofstee and S. Groenman. They criticized their mentor's view that socially the new land would develop more or less automatically because only the most entrepreneurial farmers would be ready to establish themselves in the polder. They felt that the government should select settlers by paying attention not only to skill and financial ability but also to qualities relevant to community formation, such as membership of agricultural organizations and cooperatives. Moreover, the two sociologists predicted that the thus selected "progressive farmers" would take on a lifestyle that would be more akin to that of urban culture: the relationship between farmers and workers would become more pragmatic and the farmers would be less thrifty and buy more durable luxury goods. Hofstee and Groenman saw this trend as inevitable, and argued that it had to be taken into account in designing the polder.

This brought Hofstee and Groenman into sharp conflict with the Delft school designers of the villages as to where the agricultural workers were supposed to live on the farm, with other workers in small hamlets, or in villages. The sociologists felt that the workers could no longer be considered a community with the tenants, they should be allowed to live away from the farms, whereas Granpré Molière and the architects saw this as a case of giving in to the zeitgeist, allowing individualism to proceed. The outcome was a compromise: small groups of agricultural workers' homes were built not too far from farms. As it turned out, the agricultural workers increasingly decided to live in the polder's villages.

The sociologists had had doubts about the possibility of guiding community formation through selection of farmers all along, and then other developments further complicated their social planning effort. S. L. Mansholt, the Dutch minister of agriculture from 1945 to 1958, introduced a new policy that emphasized large-scale, efficient production.[90] This forced small farmers in other parts of the country out of business, and the new polders were seen as a perfect solution: they could build new lives as modern farmers there. The same applied to farmers from Zeeland, the province stricken hardest by the 1953 flood. Furthermore, there was also the rule that workers who had been involved in preparing the soil for cultivation were given preferential treatment if they wanted to start a farm on the new land. All this implied that only a limited portion of the land (about 20 percent) could be assigned to farmers selected by the directorate.[91] But this did not need to be a problem, the sociologists reasoned, as long as there were a few leading figures among the new farmers. As it turned out, the community formation process in the Noordoostpolder followed a familiar Dutch pattern of pillarization: they were strongly inclined to organize themselves according to confession or political creed, just like the rest of Dutch society, rather than to pursue the formation of a new community.[92]

The planning and design of the Oostelijke Flevopolder, fully drained by 1957, evolved in a much different way. From the start sociologists and urban developers played a prominent role, at the expense of the role of agricultural and hydraulic engineers. The two southern polders were closer to the Randstad, the densely populated western part of the country, which increased the importance of national planning concerns and coordination with the specific needs of the Amsterdam region. In the late 1930s the lobby against a polder exclusively designed by hydraulic and agricultural engineers was joined by nature conservation organizations. In 1943 a working group was set up, which represented over fifty organizations, for the Landscape of the Zuiderzee Polders, chaired by P. G. van Tienhoven (also the chairman of the Society for the Preservation of Nature). This working group, which included the famous architects C. van Eesteren and W. Dudok as well as a representative from aquatic sports clubs, tried to convince the Zuiderzee Works Agency and the Wieringermeer Directorate of the need to focus the new polder's design not only on creating agricultural land but also on providing woodlands, areas for aquatic sports, and a more varied landscape. Dudok referred to "a synthesis between a cerebral engineering effort and the wider human concern in which the aesthetic urge presents itself."[93] The organization of urban planners also voiced its views again, but at first the two polder agencies hardly seemed to welcome these various interventions.

This changed in 1946 when J. F. R. van de Wall succeeded De Blocq van Kuffeler as head of Zuiderzee Works Agency. In the previous decade this agency had lost power because all emphasis had been on the development of the Noordoostpolder, a project supervised by the Wieringermeer Directorate. Van de Wall responded to the new trends by setting up a Town Planning Department, headed by Mrs. E. F. van den Ban, an engineer, and with the urban developers L. S. P. Scheffer and Van Eesteren (chair of the International Congress of Modern Architecture and professor of urban development at the TU Delft) serving as advisers. To enhance the quality of the negotiations involved, in 1949 Van de Wall established an Advisory Commission for the Design of the Southern IJsselmeer Polders, staffed not only by the

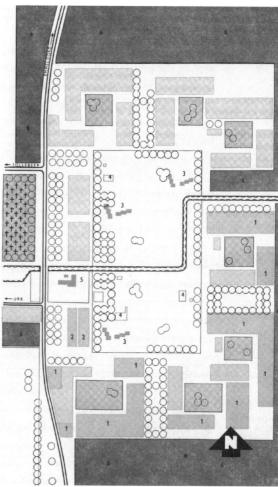

"Village under construction." The first designs for the towns and villages in the
IJsselmeerpolders all started on the drawing table. The Noordoostpolder was planned
to have one central town, surrounded by ten villages. The plan for Nagele was
designed between 1948 and 1953 by a group of private architects, who also designed
the actual buildings. The main idea was a large central communal space surrounded
by residential courtyards. Each compound was planned to have three similarly
designed schools and four churches. A stand of large trees, offering shelter and also
creating a fence, completed the general design. The plan's key: (1) housing, (2) shops, (3)
school, (4) church, (5) plantings. The photo at right was taken in Nagele in 1956.

Zuiderzee Works Agency's three urban developers but also by experts from the Wieringermeer Directorate and the National Plan Agency, including many Institute for Public Housing members. In the 1950s this last organization published several reports on the threat of overpopulation in North and South Holland: the postwar baby boom and a large-scale influx from other provinces would cause major growth in these western provinces: the population was projected to increase from 4.3 million residents at the end of World War II to 5.4 million in 1960.[94] It was clear that the new polders would play a part in tackling this problem.

The Zuiderzee Works Agency had also hired a sociologist, Ch. A. P. Takes, who had earlier worked with the National Plan Agency. He undertook an extensive study of the influence of services nuclei in North Holland and Friesland and other nearby regions on the new land and then, in collaboration with Groenman and Hofstee, designed a plan for establishing polder villages. The sociologists stressed the need for larger villages because farmers gradually were embracing a more urban lifestyle and would demand a higher level of service facilities than had been made available in the Noordoostpolder. The Town Planning Department, too, designed a parceling plan that strongly deviated from the plans for the earlier polders: the dike north of Lelystad would be curved to create a bay; slightly curved roads would cause the polder to be less rigidly divided and created agricultural lots that were not exclusively rectangular; and much space was set aside for small groves of trees. In short, the town planning lobby and organized nature and sports lovers got things their way.

Zuiderzee Works presented the plan before the Wieringermeer Directorate was consulted, which therefore responded very negatively to it. The head of the directorate, Van Eck, dismissed it as "romantic," rejecting the wooded areas and oblique parcels as uneconomic, and presented a plan of his own that was fully in line with the earlier polder designs. He had support from Hofstee and the urban developers E. van Embden and W. Bruin, both students of Granpré Molière; they all felt the polder should primarily be used for agriculture. Ultimately, in 1956, the minister opted for the Zuiderzee Works plan. As the collaboration between both polder agencies hardly improved over the years, in 1963 the IJsselmeer Polders Agency (Rijksdienst voor de IJsselmeerpolders) was set up as a successor to the Wieringermeer Directorate, with a number of new leading figures who were better trained to apply the new planning ideas.

The planning of Lelystad opened a new chapter in the conflict between the Zuiderzee Works Agency and the Wieringermeer Directorate.[95] In 1956 the minister commissioned Van Eesteren, at that time the best-known Dutch Modern town planner, to make a design for the town, supervised by a planning commission that included Hofstee, Scheffer, and Van Embden. Van Eesteren had already acted as adviser for the Zuiderzee Works and in the design of Nagele he had productively collaborated with Van Eck, from the directorate. Although this seemed a solid basis for productive collaboration, a good survey—generally considered a necessary basis for any major plan—was missing. Van Eck, who still resented his defeat by the Zuiderzee Works, refused to have his agency, the Wieringermeer Directorate, do such a preliminary study. Van Eesteren approached the new town as the overall project's crowning achievement and—unlike Granpré Molière and his students, who favored designing villages that gradually expanded—wanted to put in a prestigious town in one fell swoop, in the style of the New Building movement. His plan assumed a population of 100,000, but in the absence of preliminary studies this was a highly uncertain figure. Some thought that no more than 25,000 people would ever live in Lelystad. The Wieringermeer Directorate and most members of the Planning Commission argued for starting construction in the center, so that from the very beginning the town would have a heart around which it could expand. Van Eesteren, however, wanted to start with the planned housing in the southwest section and move from there to the center, an approach that more or less required the plan's full realization at one time. This effectively killed the plan, and in 1964 he gave up and returned the assignment to the government. In the ensuing years Lelystad was developed without a clear underlying plan. A series of architects and planners, with the IJsselmeer Polders Agency serving as coordinator, designed neighborhoods that were subsequently realized, starting with the downtown area and the neighborhoods to the north and south of it. At any point it was possible to take into account new wishes and needs.

One of the reasons why Van Eesteren's plan failed was the introduction of "process-planning" in the course of the 1950s. Instead of using "blueprint plans," as had occurred with the first two polders, builders started working with "open plans," which could be adapted whenever new developments and changing insights made it advisable to do so. For example, agriculture's increasing mechanization significantly reduced the number of farmhands needed, resulting in the building of fewer villages than initially planned; simultaneously migration from the Amsterdam region became more central, requiring construction of more homes for commuters and more recreational areas.

The experience of polder construction showed that after World War II technological experts from various disciplines enjoyed a broad mandate. This meant not so much the end of political discussion but rather its displacement. Because there were two polder agencies that worked with experts from various disciplines and often with strongly competing insights, the mandate remained quite vulnerable to interference from government or Parliament. New priorities—such as those linked to population growth, migration to the Randstad, and decline of the agrarian population—became part of political debates between experts and influenced both the debates' outcome and the groups' power relations. For instance, the planners gained influence at the expense of the agricultural experts. Accordingly, the polder agencies' decision to hire new experts and to give them such a prominent role was decidedly political in that it implied a choice for another physical-social design of the polder.

This is not to say that the society that came into being in the polders fulfilled what the planners had had in mind. For one thing, the farmers developed a more urban style of living than the Delft school designers had envisioned, and it also became clear that "pillarized" subcultures prevailed in the new polders rather than integrated communities. Likewise, the government's policy of using the polders to solve the problems of small farmers and to help victims of the 1953 flood certainly did not help to realize Ter Veen's dream of "a new society on new land."[96]

Public Housing Construction

In the 1980s, Dutch historians conducted a debate as to whether World War II had truly reflected a break in Dutch history. The debate's outcome was that continuity prevailed: despite efforts aimed at fundamental political renewal, the old pillars and parties returned, at most slightly changed in a few cases. The pressing problems of reconstruction and decolonization pushed aside political renewal and the old structures provided some sort of security, for which, apparently, there was a deeply felt need in this period.[97]

Other historians, however, argued for a silent breakthrough for which mainly architects and town planners were responsible.[98] Regarding the government's role in these areas, World War II did mean a real break: after fifteen years of private initiative playing first fiddle, from 1945 onward public housing was managed by the government until the late 1960s. Government intervention was thus sustained much longer than after World War I. The difference can be explained by three factors: the destruction was much greater this time around; the population's growth accelerated, going from 8 million at the start of the war to almost 11.5 million in 1960 (in the 1950s it was expected that by 2000 a population of some 20 million would be reached); and, finally, not just democracy but also central planning was more taken for granted.

During World War II, when construction virtually came to a standstill, several architects intensively studied town planning methods and quality public housing.[99] The Reconstruction Agency, led by Ringers, initiated the first efforts to put these ideas into practice. When, during the war years, the agency had no work because of a shortage of materials, Ringers had his people study efficient building technologies and make an elaborate inventory of the housing stock; they also studied the anticipated population growth and family size, so as to be ready to meet the projected need for housing after liberation as swiftly as possible. Ringers and several leading architects also set up study groups. Van Tijen and Merkelbach were leading figures in the Study Group on Housing Architecture, which took stock of the views of many architects. Their report showed great confidence in the possibility of contributing to a better world through home construction. It required a plan-based approach, government coordination, and standardization. The study group developed an elaborate program as to what homes should be like, whereby it distanced itself from the thirties notion of the minimum home. Homes had to have *surplus*, extra space in which people could pursue their hobbies. Another study group, of which Van Tijen was also a member and which was led by A. Bos, director of Public Housing in Rotterdam, concentrated on neighborhoods as living communities that offered a counterbalance to the drawbacks of urban settings (such as too much stimulus and noise, loneliness amid the crowd). In his explanation of a plan for The Hague, the architect Dudok formulated this "neighborhood idea" as follows:

> The community feeling has to be expressed in the neighborhood by a community center with a neighborhood hall, which not only serves shared physical interests—neighborhood medical care, contact with local government—but which also allows people to guide and give shape to spiritual and cultural life (clubs, societies, lectures, etc.).[100]

Such ideas had great influence on postwar housing construction in the Netherlands.[101] The neighborhood concept constituted the basis for the design of new housing developments in the southern part of Rotterdam, where all basic facilities had to be "at stroller distance" from homes.[102] Similarly, views on home layout and standardization of the construction process were influential. The Dutch government had a strong hold on new public housing construction projects via its Central Directorate of Public Housing. This body, set up in 1946 as part of the Ministry of Reconstruction and Public Housing, carefully inspected each housing construction plan. When many proposals failed to meet quality standards, the Central Directorate of Public Housing had several "standard homes" designed to serve as models. Housing corporations could also use them as more or less ready-made designs. They

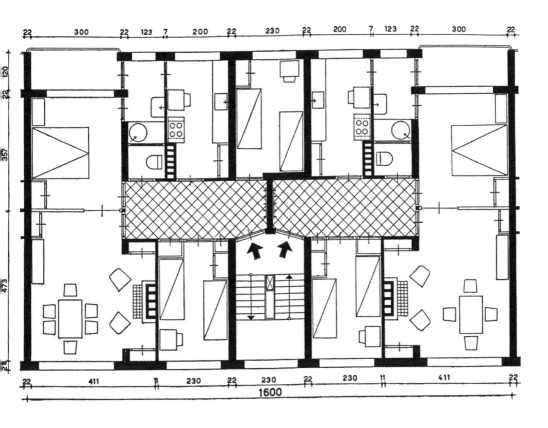

A "Series N, Type A" plan for a housing unit comprising apartments for two families, one couple with two children and one with four (note the number of chairs at the dining tables). This unit of seventy-six square meters of surface area (*oppervlak*) provides six *slaapplaatsen* ("sleeping places") in the lefthand apartment and four in the smaller apartment. The wet areas—kitchens and bathrooms—are placed next to each other. After World War II, standardized housing was constructed on a large scale, according to maps that had been developed collaboratively as part of a comprehensive housing construction program by a working group of architects who were members of the Study Group on Efficient Housing. The chairman of the group was the Delft professor J. H. van den Broek, who had already studied housing issues in depth in the years before the group's founding in 1948. There were seven housing variants, with sometimes multiple types, depending on family size.

were produced by the Dutch Architects' Union's Housing Architecture Group. Of course many architects had mixed feelings about this project as it might cause them to be relegated to the sidelines.

Government direction was also expressed in the "Prescriptions and Guidelines" ("Voorschriften en Wenken") on quality standards for subsidized housing construction, which first appeared in 1946. Elements of the old hygienist agenda, such as exposure to sufficient sunlight, were included, as well as directions concerning the layout of homes, the size and form of rooms, and construction guidelines aimed at ensuring good quality for minimal costs. These ideas were closely related to many of the ideas developed by Van Tijen and his colleagues during the war, yet now lacking the "surplus" they had deemed so crucial. The architects vehemently protested against these guidelines. In subsequent versions (1951, 1965) the standards were slightly raised, but they continued to lag behind the program formulated by the Study Group on Housing Architecture during the war.[103]

"These guidelines," its authors wrote in 1946, "are not meant to be applied rigidly…. One should view them mainly as an illustration of the spirit in which the intended homes should be designed, not as an attempt at unhealthy leveling." Still, rather uniform housing was the result, because the government set quite detailed minimum standards and did not fund anything beyond, and also promoted standardization.[104] In 1944 the Rational Building Foundation (Stichting Ratiobouw) had been established to select labor-saving construction methods based on assembling prefabricated elements. Only construction with approved systems was subsidized. The Study Group on Efficient Housing, set up in 1953 under the auspices of Bouwcentrum—a collaborative network of architects, contractors, housing corporations, the government, and unions—developed "optional plans," creating housing designs that met the "Prescriptions and Guidelines" and that elsewhere could serve as models. The government subsidized the construction of these homes—40,000 by 1970. The same applied with respect to a standard apartment building developed by top architects in collaboration with Bouwcentrum, the Central Organization for Applied Scientific Research, and others, of which many were built as well. As a result, in many locations a repetitive series of neighborhoods emerged consisting of moderate high-rise apartment buildings, alternating with low-rise buildings, with broad stretches of green in between. Later on this model became a symbol of all that was wrong with postwar public housing construction.

From the outset there was criticism of the role played by engineers in charge of public housing construction and employed by the Ministry of Reconstruction and Public Housing. Members of Parliament worried about far-reaching centralization of policies and the one-sided technical definition of the building task. One representative argued in 1946 that as a result of this approach, "cultural values, religious interests, social wants, ownership relations, and even financial transactions [were] seen as engineering problems and thus couched in rationalist and materialist terms."[105] The head of the Interior Ministry's Public Housing Department, P. A. van der Drift, also protested the far-reaching influence of the Reconstruction and Public Housing Agency on housing design, subdividing lots, and the positioning of housing blocks. In June 1945 he wrote to Minister Beel:

> Public housing ... should be led by a minister and a ministry that put social solidarity first. This solidarity is the engine; the building etc. is the wheel—is secondary. The wheel is an instrument, has *qua talis* no heart. The country will always wish to have a responsible minister of whose social intentions one can be assured. This type of intent will never be sought in the head of a technical department—most certainly not with your colleague at Public Works and Water Management [Ringers], to whom technology prevails of course.[106]

The direct necessity to step up the building effort silenced such protests, but the same objection was raised later on. Modern architects who favored putting technology at the service of social aims actually shared this view. In a posthumously published report (1975), Van Tijen, the pioneer of gallery apartment buildings, called high-rise building in public housing an international failure.[107]

To what extent have Dutch architects and urban developers shaped the postwar built environment on the basis of their ideas about a better society? The prominent role of Modern architects in shaping the built environment stands out, especially when compared to the dominant position of the more traditionalist architects who designed the polders. The prominence of architectural traditionalism in the polders can be explained by the prestige of Granpré Molière in the 1910s and 1920s. His work on the garden city Vreewijk perfectly qualified him for building villages in polders. His belief in the harmonious rural community agreed well with the agrarian focus of the first polders. Although the Wieringermeer Directorate's Construction Department, led by Van Eck, was open to other ideas, as the village of Nagele exemplifies, it gave priority to efficiency and was hardly inclined to foster discussion among architects. This is why Van Eck gladly made use of the experienced Granpré Molière, who also had several students who could be deployed easily.

Ideas regarding public housing, by contrast, did not apply primarily to the countryside but to cities, which explains the major role of modern architects in the cities, and the later polders, which were more oriented toward the western, urban part of the country. Public housing advocacy mainly attracted Social Democrats and architects who adhered to the views of New Building; many of them were also inspired by socialist ideas. With the help of new technologies and design principles, these architects wanted to build a more healthy, playful, and just world. Because public housing had to be made widely available inexpensively, they developed designs that relied on new materials and standard parts. After World War II, when the country was faced with an enormous housing shortage, such ideas were particularly opportune.

However, many Modern architects were eventually disappointed by actual practice, in which their influence turned out to be more limited than they expected. There are basically three reasons for this. First, a persistent housing shortage and a limited government budget meant that inexpensive mass production came first and housing quality took a backseat. Second, the government issued long-term contracts to contractors using approved systems for very large housing projects, which minimized the role of architects.[108] Finally, the subsidies were tied to such strict conditions that in bilateral negotiations the Central Directorate of Public Housing officials, according to one historian of public housing, could "basically bend the design practices of individual architects ... to their will."[109]

Not only was the influence of modern architects, planners, and designers restricted by government actions, but also they had to share influence with the representatives of the users. A host of organizations—including housing corporations and Women's Advisory Commissions (*Vrouwen Advies Commissies*)—tried to improve the quality of homes, despite the prevailing trend of stinginess. The Women's Advisory Commissions managed to have countless small changes made in home designs regarding such features as the direction in which doors opened, the layout of rooms, the exact location of sockets and power points, an extra sink, and so on. At the national level, the Netherlands Household Council (Nederlandse Huishoudraad) was established right after the war and brought several women's organizations under one umbrella. [110]

The influence of planners and designers was further increasingly restricted by people's choices. For example, modern interior designers associated with the Good Living Foundation (Stichting Goed Wonen), founded in 1946, tried to promote a light, Modern style of home design characterized by simple lines and lots of space, but had little success. A 1956 study by the Bouwcentrum revealed that most people, especially those from the working classes, preferred heavy furniture and rooms crowded with furnishings because they

Youths in Amsterdam's Goudriaanstraat seem to be enjoying the playground in the stretch of green and recreational space in between the housing blocks in this 1960 photo. Urban expansion of the fifties and sixties was marked by the more or less massive character of repetitious housing blocks that amounted to a planned ensemble.

found this more cozy and stylish.[111] With rising prosperity, the available options increased, more houses were built by the private sector, and the influence of government decreased accordingly. Apartment living became unpopular and most people wanted a "family home with a garden." This led to almost out-of-control suburbanization, which signaled the failure of the 1960s government effort to concentrate expansion in certain designated regions.[112] In other words, it was the market that began to prevail, rather than planning.

In light of the architects' mandate, the development of public housing in the three decades after 1940 provides a paradoxical picture. At first sight the New Builders had huge influence: housing developments based on their designs and ideas developed in the thirties and during the war mushroomed in the postwar years. Nevertheless it is understandable that Modern architects complained about their limited mandate. After all, the government successfully pushed its policy of making public housing as affordable as possible. It gained wide support because the housing shortage was the biggest problem facing the postwar government. This undermined the designers' mandate, something that, paradoxically, had already been compromised by the principle of standardization favored by Modern architects themselves. As with the polders, government interventions ultimately determined actual decisions, much to the architects' frustration. In the 1960s, when design—and designers—gained ground again, the tide began to turn.

Politicization and the Building of a New Mandate, 1970–1990

The wide public support enjoyed by the Dutch government during reconstruction quickly eroded in the course of the 1960s. The leaders of pillarized organizations all faced resistance and demands for participation. This new spirit largely emerged among the generation born shortly after World War II. On average these young people were much better educated than their parents. At universities and colleges, which saw steady growth, they had learned to think critically about society and this pitted them against the prevailing patriarchal power relationships. The ruling elite did not try to push back, however; they gave in to the new demands for participation without much resistance. Many politicians, intellectuals, and high officials, in the Netherlands and elsewhere, believed that changes were under way that could not be stopped.[113] From the start, therefore, major Dutch protest movements enjoyed support from established, scientific and technically trained experts. This can be seen in the history of the rise of the environmental movement, which clearly built upon older movements that had voiced concerns about the industrialization and modernization of Dutch society.

The concern about the demise of nature as a result of industrial growth and urban sprawl was not new in the 1960s.[114] It had surfaced at the end of the nineteenth century, leading in 1905 to the founding of a powerful organization, the Society for the Preservation of Nature (Vereeniging tot Behoud van Natuurmonumenten). Conservationists influenced the thinking of regional and urban planners. In the late 1940s some professors from the Agricultural College at Wageningen warned against unbridled exploitation of nature and the risks of chemical insecticides. In 1963 the Dutch Central Planning Bureau pointed to the substantial funds needed to fight air and water

Highway designers came up with technical solutions such as wildlife crossings over highways. Nature became integrated into spatial planning. To allow for the mobility and territorial requirements of larger wildlife, so-called "ecoducts" were designed and special facilities, like badger tunnels, were built for smaller animals.

pollution. Starting in 1963, grassroots groups began to protest openly against air pollution in the port district of Rotterdam and the establishment of a petrochemical company, Progil, near Amsterdam; and they began to pursue protection of fragile ecological systems. Older, established conservationists were among the initiators. The Provo movement made young people familiar with the environmental issue and spurred them to activism. After 1970, and especially after the appearance of the report by the Club of Rome, *The Limits to Growth* (1972), the movement grew rapidly. In the course of the 1970s six hundred to seven hundred new environmental grassroots groups came into being.[115] The involvement of prominent people understandably attracted the attention of the Dutch media, and legislative measures taken to protect the environment. Various environmental organizations received government funding, enabling them to hire staff and develop their own expertise. Cities, provinces, and the state began to employ environment officials, and advisory councils began to welcome representatives of the environmental movement. The Dutch political parties—first the small ones on the left, followed in the late seventies by the Labor Party and, to a lesser degree the Christian Democrats—paid increasing attention because polls revealed that many voters were worried about the environment.

Another good example is the urban renewal movement, which aimed at creating more agreeable living conditions in cities, and which evolved along similar lines as the environmental movement, using ideas developed by an earlier generation. It was hardly surprising that architects themselves began to resist the rigid functionalist building of the 1950s and 1960s. Although Modern design had been based on their own conceptions and ideas, standardization resulting from government regulations had left cites strewn with buildings that did little credit to architects. In 1959 the young architects J. B. Bakema, H. Hertzberger, and A. van Eyck took over the editorship of *Forum*, a journal for architects. In their first issue they published "the story of another concept" (*Het verhaal van een andere gedachte*), which spelled out an alternative to what they saw as cold housing and dull urban design: a small-scale, hospitable, humanist, individualist, colorful, varied architecture, inspired by old towns and primitive cultures. They felt that cities should be "veritable interiors of the community at large, so that all residents know who and where they are."[116] The implementation of such a program would once again give architects a major role, after standardization had marginalized them and government regulations had shackled them.

In 1962, *Bouwkundig Weekblad* (*Construction Weekly*) organized a contest on the issue of rigid regulations in housing construction. The jury identified as a trend in the responses: a sense of "the city as an exciting interplay of differentiated forms of housing." An influential 1961 study by a TU Delft professor, N. J. Habraken, "The Carriers and the People" ("De dragers en de mensen"), criticized "mass housing construction" that failed to take into account specific wishes by individual residents.[117] The author nicely captured the ethos of individualism, which now was given more consideration because of rising prosperity and which conflicted with standardized housing production. He felt there should be room for "personal involvement and decisions," the formulation of individual desires, and people's "need to show off and shape their own environment."[118] This way of thinking seamlessly fitted the striving for democratization and participation that grew prevalent in many areas. It also is of a piece with resistance to large infrastructure projects by the Rijkswaterstaat and the standardized approach to public housing. Each of these areas came under criticism in the 1960s.

The Rijkswaterstaat

In the 1950s and 1960s the Rijkswaterstaat reached the height of its prestige with the Delta Works project.[119] Its plans were developed by the Delta Commission, established by the government right after the 1953 flood. It was manned by several leading engineers in water management, including J. A. Ringers, De Blocq van Kuffeler, and Johan van Veen, the head of the Research Section on Estuaries, Tidal Rivers, and Coasts, which in the 1930s had extensively studied tidal movements and riverbeds in the area. Another member was the famous econometrist Jan Tinbergen, who did economic calculations. In 1954, three years before the government and Parliament adopted the commission's plans, the Rijkswaterstaat started up the effort. Until the early 1970s this agency had plenty of autonomy and it was admired for a series of hydraulic innovations. Thousands of people visited the Delta Works, which were also widely covered by the media. The project evolved into a symbol of the Netherlands. Other large infrastructural projects made a big impression, such as the traffic cloverleaf at Oudenrijn (1968) and the highway "diamond" around Rotterdam, including the Van Brienenoord Bridge (1965) and the Benelux Tunnel (1967). In the 1960s the budget of the Rijkswaterstaat quadrupled. Clearly the agency had a broad mandate from both the political system and the public, and it developed extensive autonomy in the 1950s and 1960s. This autonomy was mainly challenged, as earlier in the case of the polders, by other experts, notably planners from the State Agency for the National Plan. For example, in the case of highway construction they asked to take into account not only economic aspects, but also aspects tied to recreation, scenic beauty, and population spread. The Rijkswaterstaat's leading officials managed to limit the influence of the State Agency for the National Plan into the 1960s by dismissing the planners' naïveté while touting their own expertise.[120]

In the late 1960s the public mood began to change vis-à-vis technocrats and planners. This was first reflected in the appearance of new environmental organizations, such as the Waddenzee Preservation Society, founded in 1965, which resisted plans to turn the Waddenzee (a tidal marsh in the northern Netherlands, between the mainland and the offshore islands) into a polder, and the Study Group Oosterschelde, founded in 1968, which wanted to prevent the closure of this estuary, the final stage of the Delta Plan. In 1973 the predominantly Social Democratic Den Uyl government took office. It pursued environmental policies proactively, and it appointed a committee to reconsider the closure of the Oosterschelde. The changing power relations were clearly reflected in the makeup of this committee: a biologist, a planner, an environmental expert, a fisheries expert, an economist, and just two civil engineers. They decided to install an adjustable flood barrier, which would protect the land but also preserve the existing ecological balance and allow for local fisheries to stay in business. The design of this barrier required detailed insight into the Oosterschelde's ecosystem. Consequently, the ecologists and biologists of the Environmental Section of the Rijkswaterstaat, set up in 1969, played a much more prominent role. From the 1970s, ecological concerns constituted a permanent element of water management designs, attributable also to the increasingly knowledgeable response by the environmental movement. In order to steal its thunder and become the dominant institution in Dutch water management again, the Rijkswaterstaat implemented a "much broader reprofessionalization" that incorporated new ecological goals and related expertise.[121]

In road construction a similar development occurred.[122] In 1971 the Ministry of Public Works and Water Management wanted to start construction of the A27 from Breda to Hilversum, taking into account local zoning plans and after consulting the National Forest Service and the State Highways "Aesthetic Care" Bureau. A TU Delft student of planning, Jaap Pontier, studied the plan and concluded it meant the end of Amelisweerd, a wooded area near Utrecht. He wrote a report in which he put forward his objections and proposed two alternative routes. He founded the Working Group Amelisweerd, which counted local politicians and many students among its members and was supported by, among others, the Federation for Preservation (Bond Heemschut), the World Wildlife Fund, and the Dutch Mycology Association. Also in 1971 the city of Utrecht and the minister of public works and water management agreed on one of the alternative routes proposed by the Working Group Amelisweerd. This success lent impetus to renewed activism. The Roads Commission, which advised the minister on road construction, began to include representatives from environmental organizations, and citizens began to grasp new legal strategies to buttress their participation in decisions on road construction. This made the decision process slower and more complex.

In 1973, after the first oil crisis, the Dutch government was forced to make serious budget cuts, and the Ministry of Public Works and Water Management was an easy target for the Den Uyl government. The ministry's expenditures as a percentage of the national budget dropped from 7.9 percent in 1971 to 2.8 percent in 1981 and 1.6 percent in 1993.[123] A large portion of the budget was spent on the costly Oosterschelde Dam, and road construction came to a halt. In the 1980s, the era of the Christian Democratic governments led by Ruud Lubbers, cuts were also achieved by privatizing divisions of the Rijkswaterstaat and by contracting out to private contractors projects

The controversy concerning the planned road section of the A27 highway near Amelisweerd received continued local support after the occupation of the wood by protesters was ended roughly. The objections against the bulldozer mentality of the planners are well illustrated by this mural produced by the anarchist collective KARKAS, painted in 1985 in the Ganssteeg in Utrecht. The action group was canny enough to add its own bank account number with the slogan "Give for Amelisweerd." Shortly after it was painted the mural was vandalized—another expression of urban culture.

that always had been part of this agency's core business, such as the flood barrier of the New Waterway between the port of Rotterdam and the sea, completed in 1997.

From the 1970s to the 1990s competing political agendas played a role in the construction of flood barriers in the Oosterschelde and the New Waterway, integral water management, public enquiry procedures in highway construction, and the declining budget of the Rijkswaterstaat. This agency lost much of the relative autonomy it had enjoyed in the 1950s and 1960s. The decision-making process surrounding water management and infrastructures became highly politicized, meaning that citizens—whether organized or not—gained influence, and politicians felt forced to cater to the public's new desire for participation. Moreover, the agency lost some of its core activities to private companies. In the 1990s, reflecting its reduced power, the Rijkswaterstaat pursued a new type of mandate: serving as expert and supervisor in "open and interactive planning procedures," which would allow all potentially interested parties to be involved in major hydraulic engineering and infrastructural projects—a challenging mission because there were so many conflicting interests.[124] The development to a more open and interactive planning process was visible not only in the construction of wet and dry infrastructures but also in the planning of new polders.

Zuidelijk Flevoland (Southern Flevoland)

The southern Flevoland polder, drained in the years from 1959 to 1967, became the most varied of the IJsselmeer polders in terms of its land use. Its design was in the hands of the IJsselmeer Polders Agency, and reflected the sociopolitical themes of the era. This time around major conflicts between various agencies and experts, as in the case of Lelystad, were avoided. The IJsselmeer Polders Agency mobilized all possible sorts of expertise and tried to serve divergent interests. In the 1950s and 1960s it became evident that the polder would be used to alleviate the Randstad's population growth. In 1958 a major commission issued a report, "The Development of the Western Part of the Country ("De ontwikkeling van het westen des lands") for which agency planners had done a great deal of background work. The report argued that the cities of the western part of country should not develop into a single mega-city and that the area of lakes and agricultural land in the central part of the country should continue to be preserved as a "green heart." This meant that new space had to be found for urban development. The southern IJsselmeer polders were designated as one such area, an idea also embraced in the first "Memorandum on Land Use Planning" ("Nota ruimtelijke ordening"), which was issued in 1960.

Ultimately land uses in the new southern polder were apportioned as follows: less than 50 percent for agriculture, 12 percent as nature preserve, a little more than 10 percent for urban structures, and almost 25 percent for "woodlands and recreational facilities."[125] Initially the area between Lelystad and Almere, the polder's lowest-lying area and not yet entirely drained, was set aside for industry, but eventually it was decided to let it evolve into a nature preserve, called De Oostvaardersplassen ("the Eastern Boating Lake"). In 1981, when the railroad was put in, its planned route was changed quite a bit without provoking much criticism.

The experience with Lelystad, designed by Van Eesteren from scratch as a whole new town, had taught experts that this wasn't really sensible. This is why the town of Almere was designed to have several nuclei ("poly-nuclear"): a new subdivision could be added as needed and at any point one could adapt the existing plans to deal with new problems or insights.[126] Critics dismissed the agency plans for Almere, which involved pleasant low-rise suburbs without any urban appeal, as anti-urban. This character was attributed to the fact that the agency, just like its precursor the Wieringermeer Directorate, was led by agricultural engineers from Wageningen. This character fitted in with the earlier polders' planning—from the villages by Granpré Molière to the failure of Van Eesteren's large-scale urban plan. R. H. A. van Duin, the agency director, said his agency had opted for low-rise and ample green zones because this is how most Dutch wanted to live. He did not object to the criticism leveled at his agency because he and his staff had been responsive to popular sentiment. The planning of Almere was in the hands of a project bureau that involved a collaboration among architects, agricultural architects, engineers, urban developers, and others from all relevant traditions and disciplines represented in the Netherlands, from the Delft school to students of Van Eesteren and Bakema. One of the architects later wrote in his dissertation that the bureau "wanted Almere to be all things to all people." It should help solve overpopulation elsewhere, be a stimulating environment for a quite varied population (including minorities), and have a balanced ecology but also an urban culture.[127] Environmental concerns—the Club of Rome's *Limits to Growth* report was published during the city's planning stage—led to a vision of urban mobility that minimized automobile traffic and promoted public transport, bicycles, and strolling.

In short, the design of the southern polder was much less a product of engineers and other experts behind their drawing tables. Van Duin's approach totally differed from that of the previous generation of designers. Granpré Molière and Van Eesteren, despite being opposites in some respects, shared a design approach that was based on a vision of tomorrow's society. Van Duin and his staff, it seemed, wanted to put themselves at the service of the current population's wishes and of the political priorities of the time, such as concern for the environment and need for a new nature preserve—priorities about which apparently, notwithstanding the political noisiness of the early 1970s, broad consensus existed after all. Put in the terms we introduced at the beginning of this chapter: they legitimized their mandate not with a visionary argument about a technological road toward a better future but with an argument for having divergent social aspirations fuse harmonically in the present.

Housing Construction and Spatial Planning

In the 1960s, criticism of monotonous, standardized housing grew stronger.[128] Professor N. J. Habraken of the TU Delft proposed a new approach to construction, starting from the distinction between the basic building and all the built-in elements. Prospective residents would be allowed to participate

in the design of the built-in elements, which constituted their immediate living environment. In 1964, to develop these ideas further, Habraken and a
number of architectural firms founded the Architects Research Foundation
(Stichting Architecten Research). Many architects followed the example of the
foundation, even though the flexibility they aimed at clashed with strict government regulation. Contractors began to offer various models of built-in
elements. Gradually this approach grew common and it became part of government regulations. The 1992 Building Code, which replaced local regulations, even stipulated that the layout of homes should be left flexible.

This development was also a response to growing diversity in the demand
for housing since the 1960s. For example, the growth of the number of students, the trend to postpone starting a family, and the arrival of "guest workers" from other countries led to more single-person households, with quite
diverse demands as to what a home should offer.[129] It was only natural, then,
to start designing homes that could be used by various types of residents
whose specific wishes designers did not know in advance. The most extreme
form of residents' participation in the design of homes and neighborhoods
was the 1970s' "participation architecture," whereby prospective residents
were given far-reaching opportunities to influence home designs.[130] The
architects involved, according to a public housing historian, Noud de Vreeze,

> … viewed the input of residents in plan development processes as a
> major opportunity for innovative architecture and urban develop-
> ment. They wanted to get rid of their traditional position as techni-
> cal and aesthetic experts in the building process, and saw the architect
> of the future functioning as organizer of public participation proce-
> dures that would "automatically" lead to architectonic and town-plan-
> ning concepts. In a process of "de-professionalization" the architect
> would only have to make his knowledge understandable and com-
> municate it to the group for which he worked in such way that the
> group could apply it.[131]

This brief phase in public housing was marked by democratization and sur-
rendering the notion of town planning as the realization of utopian projects.
Opponents have often dismissed the building style of this period as "new
frumpiness" (*nieuwe truttigheid*) because houses were built in an old-fashioned style featuring a lot of ornamentation. This may have been an understandable reaction to the sober functionalism of the 1950s and 1960s but it
also underscored the distance we had come since the 1920s and 1930s, when
designers and housing supervisors still tried to prescribe home design as

much as possible because they saw it as a way of civilizing its residents.

A major effort of the 1970s involved urban renewal—revitalizing inner cities. Because of rapid postwar suburbanization, many Dutch inner cities had become dilapidated. Those who could afford it moved to the suburbs, leaving behind those of little means, including students and elderly people. In the 1960s, city boards began to tear down old neighborhoods, which were replaced by shopping malls, office high-rises, and new traffic infrastructures (Hoog Catharijne in Utrecht is a good example). But these efforts generated protest in neighborhoods such as Nieuwmarkt in Amsterdam and Oude Westen in Rotterdam in which students and social workers often played a major role. They found a willing ear among democratization-minded architects, young left-wing politicians, and government officials. As a result, similar procedures as the ones used in large infrastructural projects were introduced: residents were involved in the plans, and often local government even helped them by funding the hiring of experts. In other words, a new alliance was forged between architects, young politicians, and local officials who wanted to enhance their visibility, and the residents of "disadvantaged neighborhoods." When the Den Uyl government was in office, politically this alliance had the prevailing wind. Yet, paradoxically, architects who wanted "to build for neighborhoods" were as dependent on various government funds as their colleagues had been during the 1950s, although the rules were now set on the basis of other political preferences. In the 1980s, when government began to cut back drastically, public housing construction dropped from 54 percent of all new home construction in 1982 to 35 percent in 1988 and 24 percent in 1992.[132]

In the 1970s and 1980s, then, the role of architects and urban developers changed in similar ways as that of the Rijkswaterstaat engineers and the designers of Zuidelijk Flevoland. The old paternalist mandate embodied in the visionary designer of a better society quickly lost legitimacy because residents' organizations, the environmental movement, and other interest groups demanded a say in design processes. In response these technical experts began to present themselves as a new type of expert, namely as designers who were able to combine and integrate all the various wishes of all stakeholders in specific designs.

Technocracy in the Netherlands

In this chapter we have concentrated on the role of engineers, architects, and urban developers in the design and planning of Dutch society in the twentieth century, arguing that its development has been codetermined by all sorts of technical experts. By the late nineteenth century, industrialization and urbanization had led to an array of social problems. Increasingly, a variety of experts began to offer help in solving these problems. This involved more than providing strictly technical services, as most of these experts articulated sustained views on creating a better society and their own indispensable role in it. In the period from 1850 to 1914 a number of these groups, notably physicians and engineers, acquired a more prominent role in several policy areas, but politicians sought to curtail these professionals' mandate, fearing what was later called "technocracy" and trying to retain maximum control of policy decision-making processes themselves, while simultaneously imposing as few limitations as possible on market activity.

World War I marked a turnaround. Various urgent problems led to more intervention by the central government, while a political climate emerged in which there was more willingness to allow experts substantial policy space. In the new polders, water management engineers and, to a lesser extent, agricultural engineers and architects even obtained a broad mandate regarding the design of the newly reclaimed land. Other experts, such as urban developers and sociologists, also tried to gain influence on polder design, with limited success, at least for the time being. In several major cities, Modern architects and urban developers briefly enjoyed a broad mandate, notably regarding public housing, but after government withdrew its subsidies they were forced to confine themselves to their drawing tables. Their highly utopian and technocratic plans were resurrected after the war and deeply influenced postwar building policies.

The period between 1940 and 1970 can be considered the heyday of the influence of technological experts, a situation caused by World War II, which had various major direct and indirect effects. The German occupiers had stimulated planning efforts, and total social collapse at the end of the war gave rise to strong public support for centrally led reconstruction. Huge population growth and the widely held view that the Netherlands had to industrialize further stimulated new planning. In this climate technical experts managed to acquire a broad mandate. Society's postwar physical infrastructure—homes, polders, and road infrastructures—was increasingly designed by social demographers, engineers, town planners, and architects. Competing concerns and approaches were discussed and resolved less in Parliament or the press and more in the offices of planning firms and agencies. Accordingly, the discussion's prevailing ideas had more to do with international views and perspectives, as developed, for example, in the International Congress of Modern Architecture, than with party programs and the pillarized structure of Dutch society.

"Measuring is knowing" is the motto of the technical researcher who aims to control living and dead matter. Water flows didn't escape the engineers' eagerness to measure things. Measurements took place in situ but also in scale models, such as here, in the Water Course Laboratory in Delft around 1940. The effects of interventions in the water system were mapped and represented as numbers.

Conversely, in this period the limitations of the mandate also became quite clear: Modern architects saw their plans being cut to the bone and mainly deployed for building large quantities of standardized inexpensive homes. The polder planners had to accept that the new land was merely used to solve problems of the "old land," notably to accommodate the Randstad's overpopulation and the victims of the 1953 flood in Zeeland. In the course of the 1960s, politics fell under the spell of democratization and the public's striving to participate in all sorts of decision-making processes. This undermined the autonomy of engineers. Democratization and technocratic planning did not go together well. By 1960 some planners, architects, and urban developers had already given up on the notion of "blueprint planning" and had switched to process planning and allowing prospective residents a say in the actual design of new homes. In a number of areas, such as urban renewal and environmental policies, new movements were given ample space in the 1970s. Alliances emerged between technical experts, the new movements, and young new-left politicians, both at the national level and in major cities. Each of these groups tried to increase its influence by acquiring a new mandate in the new political constellation. Established institutions such as the Rijkswaterstaat responded by incorporating politically relevant expertise, such as environmental technology. This was precisely the same tactic the Zuiderzee Works Agency had applied in the 1950s, when it hired modern urban developers and social demographers to help in the planning of the southern polders.

Exploiting their growing mandate and using an array of technical interventions, planners and designers contributed crucially to shaping Dutch society in the twentieth century.

The history told in this chapter makes clear that the rise of technocracy should not be confused with the idea that experts fully took control. We propose to define the rise of technocracy as a process in which political issues were pursued and resolved by experts who were deeply divided and represented specific political agenda's.[133] Technological experts designed their plans on the basis of various views and ideals about society. The design of polders, cities, and homes was the result of debates among those experts and, of course, their struggle for recognition and power. The outcome of this change was strongly determined by what we rather loosely could call "the political climate." Changing political priorities often determined which group of experts could play a dominant role in a specific period. These priorities in turn were determined by major concerns among the public and by the dominance of specific political groups. Examples of such public concerns include employment, a major factor in the decision to start construction of the Noordoostpolder; the housing shortage, which was seen as "public enemy number one" and which determined the choice for sober and standardized housing and the striving for citizens' participation, which generated "participation architecture" in the cities.

These interactions between citizens, politicians, and businesses, and the influence of various sorts of experts have been too little studied in the Netherlands and other Western European countries and the United States. This chapter is an attempt to initiate the exploration of this complex subject. Further research in this area is likely to be productive and rewarding.

Acknowledgements

For their comments on earlier versions of this text, the authors would like to thank Bert Altena, Ed Taverne, and the editors of the series on the history of technology in the Netherlands in the twentieth century, notably Liesbeth Bervoets, Adri Albert de la Bruhèze, Ernst Homburg, Mila Davids, Ruth Oldenziel, Arie Rip, and Harry Lintsen.

Notes

1 For an insightful account of Dutch liberalism, see Siep Stuurman, *Wacht op onze daden: Het liberalisme en de vernieuwing van de Nederlandse staat* (Amsterdam: Bert Bakker, 1992), in particular 357–377. See also his *Les Libéralismes: La théorie politique et l'histoire* (Amsterdam: Amsterdam University Press, 1994).

2 On the cultivation system see chapter 6, this volume, by Harro Maat, "Technology and the Colonial Past," in particular note 12.

3 See also I. de Haan, "Treub en de sociaal-democratisering van de maakbaarheid," in J. W. Duyvendak and I. de Haan, *Maakbaarheid: Liberale wortels en hedendaagse kritiek van de maakbare samenleving* (Amsterdam: Amsterdam University Press, 1997), 74.

4 J. L. Van Zanden and A. Van Riel, *The Structures of Inheritance: Dutch Economy in the Nineteenth Century* (Princeton: Princeton University Press, 2004). On the basis of this low percentage and the low levies on imports and social taxes, these authors refer to the Netherlands as the most liberal country of Europe. The increasing government expenditures show it involved an ever more activist liberalism.

5 Hans Knippenberg and Ben de Pater, *De eenwording van Nederland: Schaalvergroting en integratie sinds 1800* (Nijmegen: SUN, 1988), 162.

6 See also D. Damsma and P. de Rooy, "'Morele politiek': De Radicalen in de Amsterdamse gemeentepolitiek 1888–1897," *Tijdschrift voor sociale geschiedenis* 19, no. 1 (1993): 115–128.

7 Dick van Lente, "Ideology and Technology: Reactions to Modern Technology in the Netherlands 1850–1920," *European History Quarterly* 22 (1992): 383–414.

8 H. Goeman Borgesius, "Stoommachines en vokswelvaart," *Vragen des Tijds* (1876): 27–29.

9 E. S. Houwaart, "Medische statistiek," in H. W. Lintsen et al., eds., *Geschiedenis van de techniek in Nederland: De wording van een moderne samenleving, 1800–1890*, 6 vols. (Zutphen: Walburg Pers, 1992–1995), vol. 2, 19-45; H. van Zon, "Openbare hygiëne," in Lintsen et al., *Geschiedenis van de techniek in Nederland*, vol. 2, 47–79; E. S. Houwaart, "Professionalisering en staatsvorming," in Lintsen, *Geschiedenis van de techniek in Nederland*), vol. 2, 88–92; E. S. Houwaart, "Medical Statistics and Sanitary Provisions: A New World of Social Relations and Threats to Health," *Tractrix: Yearbook for the History of Science, Medicine, Technology, and Mathematics* 4 (1992): 81–119.

10 On the rise of new experts see Harold Perkin, *The Third Revolution: Professional Elites in the Modern World* (London: Routledge, 1996); Frank Fischer, *Technocracy and the Politics of Expertise* (Newbury Park, Calif.: Sage, 1990), and on their role in twentieth-century political and economic development see Charles S. Maier, *In Search of Stability: Explorations in Historical Political Economy* (Cambridge: Cambridge University Press, 1987). On engineers in the Netherlands, see H. W. Lintsen, *Ingenieurs in Nederland in de negentiende eeuw: Een streven naar erkenning en macht* (The Hague: Martinus Nijhoff, 1980); on physicians see E. S. Houwaart, *De hygiënisten: Artsen, staat en volksgezondheid in Nederland, 1840–1890* (Groningen: Historische Uitgeverij, 1991); on civil servants in the Netherlands, see N. Randeraad, "Ambtenaren in Nederland, 1815–1915," *Bijdragen en mededelingen betreffende de geschiedenis de Nederlanden* 109, no. 2 (1994): 209–236, and V. Veldheer,

Kantelend bestuur: Onderzoek naar de ontwikkeling van taken van het lokale bestuur in de periode 1851–1985 (Rijswijk: Sociaal en Cultureel Planbureau, 1994).

11 There is a rich literature on technocracy, but the concept is also often used in a rather loose way to refer to the influence of technology or engineers. What we mean by technocracy is the situation when engineers get a strong mandate from the state to set the political agenda and implement it. For a discussion on the concept see F. Fischer, "Technocracy and the Politics of Expertise," in Claudio M. Radaelli, ed., *Technocracy in the European Union* (London: Longman, 1999); P. Weingart, *Die Stunde der Wahrheit? Vom Verhältnis der Wissenschaft zu Politik, Wirtschaft und Medien in der Wissensgeschellschaft* (Weilerwist: Velbrück, 2001), in particular chapter 4, "Wissenschaftliche Expertise und Politische Entscheidung." On the Netherlands see J. A. A. van Doorn, "Corporatisme en technocratie—een verwaarloosde polariteit in de Nederlandse politiek," *Beleid en Maatschappij* 8, no. 5 (1981): 134–149; J. A. A. van Doorn, *De laatste eeuw van Indië: Ontwikkeling en ondergang van een koloniaal project* (Amsterdam: Bert Bakker, 1994), chapters 3 and 4. We have borrowed the concept of mandate from H. van Lente, *Promising Technology: The Dynamics of Expectations in Technological Developments* (Delft: Eburon, 1993). For developments in the United States see William E. Akin, *Technocracy and the American Dream: The Technocrat Movement 1900–1941* (Berkeley: University of California Press, 1977), and Edwin T. Layton Jr., *The Revolt of the Engineers: Social Responsibility and the American Engineering Profession* (1971; reprint, Baltimore: Johns Hopkins University Press, 1986).

12 Our account is based on, among others, Eric Berkers, *Technocraten en bureaucraten: Ontwikkeling van organisatie en personeel van de Rijkswaterstaat, 1848–1930* (Zaltbommel: Europese Bibliotheek 2002); Lintsen, *Ingenieurs in Nederland in de negentiende eeuw*, 172–187, 253–260, 267–274; M. S. C. Bakker, "Overheid en techniek," in Lintsen, *Geschiedenis van de techniek in Nederland*, vol. 6, 121–129.

13 See Berkers, *Technocraten en bureaucraten*, 41, 171.

14 Lintsen, *Ingenieurs in Nederland in de negentiende eeuw*, 172–186; Ruud Filarski and Gijs Mom, *Van transport naar mobiliteit: De transport revolutie* (Zutphen: Walburg Pers, 2008).

15 Lintsen, *Ingenieurs in Nederland in de negentiende eeuw*, 177.

16 Ibid., 186.

17 Berkers, *Technocraten en bureaucraten*, 80.

18 E. Berkers, "Kustlijnverkorting en afsluittechniek," in Schot et al., *Techniek in Nederland*, vol. 1, 73–74.

19 Our account is based in particular on M. J. J. G. Rossen, *Het gemeentelijk volkshuisvestingsbeleid in Nederland: Een comparatief onderzoek in Tilburg en Enschede, 1900–1925* (Tilburg: Stichting Zuidelijk Historisch Contact, 1988); E. M. L. Bervoets, "Betwiste deskundigheid: De volkswoning, 1870–1930," Schot et al., *Techniek in Nederland*, vol. 6, 119–141, and E. M. L. Bervoets, "Woningbouwverenigingen als tussenschakel in de modernisering van de woningbouw 1900–1940," Schot et al., *Techniek in Nederland*, vol. 6, 143–159; A. J. de Regt, *Arbeidersgezinnen en beschavingsarbeid: Ontwikkelingen in Nederland, 1870–1940—Een historisch-sociologische studie* (Meppel: Boom, 1974), chapter 7; Noud de Vreeze, *Woningbouw, inspiratie en ambities: Kwalitatieve grondslagen van de sociale woningbouw in Nederland* (Almere: Nationale Woningraad, 1993). See also P. M. M. Klep et al., eds., *Wonen in het verleden, 17e – 20e eeuw: Economie, politiek, volkshuisvesting, cultuur en bibliografie* (NEHA, Amsterdam: Nederlandsch Economisch-Historisch Archief, 1987); J. Nycolaas, *Volkshuisvesting: Een bijdrage tot de geschiedenis van woningbouw en woningbouwbeleid in Nederland, met name sedert 1945* (Nijmegen: SUN, 1974); H. Buiter, "Werken aan sanitaire en bereikbare steden, 1880–1914," in Schot et al., *Techniek in Nederland*, vol. 6, 25–49; A. van den Bogaard, "De geplande stad, 1914–1945," in Schot et al., *Techniek in Nederland*, vol. 6, 51–73.

20 De Regt, *Arbeidersgezinnen en beschavingsarbeid*, 177–178.

21 Lintsen, *Ingenieurs in Nederland in de negentiende eeuw*, 298–325.

22 Bakker, "Overheid en techniek," 127.

23 Lintsen, *Ingenieurs in Nederland in de negentiende eeuw*, 306, 311; Bervoets, "Betwiste deskundigheid."

24 Ibid., 314–316.

25 De Vreeze, *Woningbouw, inspiratie en ambities*, 134–137.

26 On Rotterdam, see P. Th. van de Laar, *Stad van formaat: Geschiedenis van Rotterdam in de negentiende en twintigste eeuw* (Zwolle: Waanders, 2000), 265–273.

27 V. van Rossem, "Berlage: Beschouwingen over stedebouw," in Sergio Polano et al., *Hendrik Petrus Berlage: Het complete werk* (Alphen aan de Rijn: Atrium 1988), 46–66; V. van Rossem, "Een keerpunt in de Nederlandse stedebouw: Plan Zuid," in Karin Gaillard and Betsy Dokter, eds., *Berlage en Amsterdam Zuid*, catalogue of the exhibition "Berlage en Amsterdam Zuid" (Amsterdam: Municipal Archive, 1992), 9–25.

28 Jan Bank and Maarten van Buuren, *1900: Hoogtij van burgerlijke cultuur* (The Hague: Sdu uitgevers 2000), 142–146; Schuursma, *Jaren van opgang*, 412–419; E. H. Kossmann, *The Low Countries 1780–1940* (Oxford: Oxford University Press, 1978), 450.

29 The following discussion draws on Rossen, *Gemeentelijk volkshuisvestingsbeleid*, chapter 5, and Kees Doevendans, *Stadsvorm Tilburg, historische ontwikkeling: Een methodologisch morfologisch onderzoek* (Eindhoven: City of Tilburg and Technical University Eindhoven, Faculty of Architecture, 1993), 105–150.

30 Tilburg was the first city with an urban expansion plan. See H. van der Cammen and L. A. de Klerk, *Ruimtelijke ordening: Van plannen komen plannen—De ontwikkelingsgang van de ruimtelijke ordening in Nederland* (Utrecht: Het Spectrum, 1999) 46.

31 For an elegant extract on technical standards versus right of ownership, see Rossen, *Gemeentelijk volkshuisvestingsbeleid*, 67.

32 Bervoets, "Betwiste deskundigheid"; De Vreeze, *Woningbouw, inspiratie en ambities*, 137.

33 P. L. Tak, "De gemeente," in J. W. Albarda and H. E. van Gelder, eds., *P. L. Tak: Herdrukken uit de Kroniek* (Rotterdam: Uitgeversmaatschappij v/h Wakker, 1908), 249–354. We are grateful to Adrienne van den Bogaard for mentioning this article to us.

34 P. de Rooy, "Een zoekende tijd: De ongemakkelijke democratie, 1913–1949," in Aerts, *Land van kleine gebaren*, 227–229.

35 De Rooij, "Een zoekende tijd," 192. On M. F. W. Treub, see De Haan, "Treub en de sociaal-democratisering van de maakbaarheid," 73–88.

36 I. P. de Vooys, *Techniek en maatschappij: De beteekenis der techniek voor de maatschappelijke evolutie in verleden en toekomst* (Amsterdam: Van Kampen, 1921), especially 123, 125, 128–130, 137, 141, 142.

37 For the history of this movement see Akin, *Technocracy and the American Dream*, chapter 2.

38 A. Luwel, *De technocratie: Theorie en beweging* (Kampen: Kok, 1980), 67.

39 Eric Smulders, "Het wonder van morgen: De televisierage in Nederland, 1928–1931," master's thesis, Erasmus University, Rotterdam, Faculty of History and Arts, 1993, 71–72.

40 A. Viruly, "Onze Pelikaan," in *De wereld van boven* (Douwe Egberts, n.p. [1935]), 34–35. Consumers who bought Douwe Egberts coffee could send in coupons that came with a packet of coffee and receive a copy of this book, which candidly promoted aviation. Book in 'Van Lente's collection. M. L. J. Dierikx, J. W. Schot, and A. Vlot, "Van uithoek tot knooppunt: Schiphol," in Schot et al., *Techniek in Nederland*, vol. 5, 122.

41 This section draws especially on Zef Hemel, *Het landschap van de IJsselmeerpolders: Plannning, inrichting en vormgeving* (Rotterdam: NAi, 1994); Coen van der Wal, *In Praise of Common Sense—Planning the Ordinary: A Physical Planning History of the New Towns in the IJsselmeerpolders* (Rotterdam: 010 publishers, 1997); D. J. Wolffram, *70 jaar inge-*

nieurskunst: Dienst der Zuiderzeewerken, 1919–1989 (Lelystad: Stichting Uitgeverij de Twaalfde Provincie, 1997); D. J. Wolffram, *Zeeuwse pachters in de Noordoostpolder: Selectie en bijdrage aan de sociale opbouw, 1945–1962* (Lelystad: Stichting Uitgeverij de Twaalfde Provincie, 1995); A. J. Geurts, *De "groene" IJsselmeerpolders: Inrichting van het landschap in Wieringermeer, Noordoostpolder, Oostelijk en Zuidelijk Flevoland* (Lelystad: Stichting Uitgeverij de Twaalfde Provincie, 1997); and J. T. W. H. van Woensel, *Nieuwe dorpen op nieuw land: Inrichting van de dorpen in Wieringermeer, Noordoostpolder, Oostelijk en Zuidelijk Flevoland* (Lelystad: Stichting Uitgeverij de Twaalfde Provincie, 1999).

42 Berkers, "Kustlijnverkorting en afsluittechniek"; Willem van der Ham, *Heersen en beheersen: Rijkswaterstaat in de twintigste eeuw* (Zaltbommel: Europese Bibliotheek 1999); G. P. van der Ven, ed., *Leefbaar laagland, geschiedenis van de waterbeheersing en landaanwinning in Nederland* (Utrecht: Matrijs, 1993), 237–244. For more detailed information, see G. L. Cleintuar, *Wisselend getij: Geschiedenis van de Zuiderzeevereniging, 1886–1949* (Zutphen: Walburg Pers, 1982), especially 77, 151, 169, 184, and 251–285.

43 Cleintuar, *Wisselend getij*, 283.

44 Van der Ham, *Heersen en beheersen*, 42–44, 53–54; Wolffram, *70 jaar ingenieurskunst*, 41, 97.

45 Wolffram, *75 jaar ingenieurskunst*, 46; Van der Ham, *Heersen en beheersen*, 291.

46 See the titles mentioned in n. 45 and also Gerrie Andela and Koos Bosma, "Het landschap van de IJsselmeerpolders," in Koos Bosma and Liesbeth Crommelin, eds., *Het Nieuwe Bouwen: Amsterdam, 1920–1960*, catalogue of the exhibition at the Stedelijk Museum Amsterdam (Delft: Delft University Press, 1983), 142–174.

47 Geurts, "*Groene*" *IJsselmeerpolders*, 23.

48 Wolffram, *Zeeuwse pachters in de Noordoorstpolder*, 15.

49 Koos Bosma, *Ruimte voor een nieuwe tijd: Vormgeving van de Nederlandse regio 1900–1945* (Rotterdam: NAi, 1993), 148–179.

50 Van der Cammen and De Klerk, *Ruimtelijke ordening*, 59–66, 72.

51 Bank and Van Buuren, *1900*, 148; Schuursma, *Jaren van opgang*, 24.

52 Van der Cammen and De Klerk, *Ruimtelijke ordening*, 58–59.

53 For statistics on Dutch housing construction between 1902 and 1991, see De Vreeze, *Woningbouw, inspiratie en ambities*, 130–131.

54 Rossen, *Gemeentelijk volkshuisvestingsbeleid in Nederland*, 85.

55 Ibid., 88 (graph).

56 The following summary largely relies on De Vreeze, *Woningbouw, inspiratie en ambities*.

57 Ben Rebel, *Het Nieuwe Bouwen: Het functionalisme in Nederland 1918–1945* (Assen: Van Gorcum, 1983); Auke van der Woud, *Het Nieuwe Bouwen internationaal: CIAM, volkshuisvesting, stedebouw* (Delft: Delft University Press, 1983).

58 Bervoets, "Betwiste deskundigheid."

59 Van de Laar, *Stad van formaat*, 359–363; Hans Ibelings, *Nederlandse architectuur van de 20e eeuw* (Rotterdam: NAi, 1995), 40–41. On A. Plate see also Len de Klerk, *Particuliere plannen: Denkbeelden en initiatieven van de stedelijke elite inzake de volkswoningbouw en de stedebouw in Rotterdam, 1860–1950* (Rotterdam: NAi, 1998).

60 Rossen, *Gemeentelijk volkshuisvestingsbeleid in Nederland*, 223–224. Another example of engineers' work is the garden city of Pathmos in Enschede.

61 Rebel, *Nieuwe Bouwen*, 191.

62 R. Dettingmeyer, "De strijd om een goed gebouwde stad," in Bosma and Crommelin, *Nieuwe Bouwen*, 34.

63 De Vreeze, *Woningbouw, inspiratie en ambities*, 481; Rebel, *Nieuwe Bouwen*, 189–192.

64 R. Oldenziel and M. Berendsen, "De uitbouw van technische systemen en het huishouden: Een kwestie van onderhandelen, 1919–1940," in Schot et al., *Techniek in Nederland*, vol. 4, 57–61; E. M. L. Bervoets, M. Th. Wilmink, and F. C. A. Veraart, "Coproductie: Emancipatie van

de gebruiker? 1920–1970," in Schot et al., *Techniek in Nederland*, vol. 6, 161–195.

65 Van der Woud, *Nieuwe Bouwen internationaal*; Rebel, *Nieuwe Bouwen*, 61–64.

66 Van der Cammen and De Klerk, *Ruimtelijke ordening*, 86.

67 On Le Corbusier and Modern architecture in general, see Richard Weston, *Modernism* (London: Phaidon, 1996).

68 Van der Cammen and De Klerk, *Ruimtelijke ordening*, 88.

69 Ibid., 86.

70 Van de Laar, *Stad van formaat*, 361.

71 Cf. Fischer *Technocracy and the politics of expertise*, 22, 24.

72 De Vreeze, *Woningbouw, inspiratie en ambities*, 199–200.

73 E. M. L. Bervoets and F. C. A. Veraart, "Bezinning, ordening en afstemming, 1940–1970," in Schot et al., *Techniek in Nederland*, vol. 6, 215–239; Van der Ham, *Heersen en beheersen*, chapter 6; H. T. Siraa, *Een miljoen nieuwe woningen: De rol van de rijksoverheid bij wederopbouw, volkshuisvesting, bouwnijverheid en ruimtelijke ordening, 1940–1963* (The Hague: Sdu Uitgeverij, 1989); Koos Bosma and Cor Wagenaar, eds., *Een geruisloze doorbraak: De geschiedenis van architectuur en stedebouw tijdens de bezetting en wederopbouw van Nederland* (Rotterdam: NAi, 1995).

74 Kees Schuyt and Ed Taverne, *1950: Prosperity and Welfare* (Basingstoke, U.K.: Palgrave Macmillan, 2004) 77–83, 106–110.

75 W. J. Dercksen, *Industrialisatiepolitiek rond de jaren vijftig: Een sociologisch-economische beleidsstudie* (Assen: Van Gorcum, 1986); Th. L. M. Thurlings, "Overheid en bedrijfsleven," in G. A. Kooy, J. H. de Ru, and H. J. Scheffer, eds., *Nederland na 1945: Beschouwingen over ontwikkeling en beleid* (Deventer: Van Loghum Slaterus, 1980), 207; J. W. van Deth and J. C. P. M. Vis, *Regeren in Nederland: Het politieke en bestuurlijke bestel in vergelijkend perspectief*, 2nd ed. (Assen: Van Gorcum, 2000).

76 Dercksen, *Industrialisatiepolitiek rond de jaren vijftig*, 192–221. On the high expectations of science and universities in those years, see Kossmann, *De lage landen 1780-1989: Twee eeuwen Nederland en België*, part 2: "1914-1980" (Amsterdam: Agon, 1986) 319–320. Unfortunately, this part of Kossmann's book appeared in a later Dutch edition and has not been translated.

77 It was 470 million in 1959, 3.2 billion in 1972, and 4.8 billion in 1978. See F. Messing, "Het economische leven in Nederland 1945–1980," in D. P. Blok et al., eds., *Algemene Geschiedenis der Nederlanden*, part 15 (Haarlem: Unieboek, 1982), vol. 15, 182.

78 G. Alberts, F. van de Blij, and J. Nuis, eds., *Zij mogen uiteraard daarbij de zuivere wiskunde niet verwaarlozen* (Amsterdam: Centrum voor Wiskunde en Informatica, 1987), 22; see also Schuyt and Taverne, *1950*, 120–126, 333.

79 H. Baudet, *De lange weg naar de Technische Universiteit Delft*, part 2: "Verantwoording, registers, tabellen, namenlijsten en bijlagen" (The Hague: Sdu Uitgeverij, 1993), 761.

80 Dercksen, *Industrialisatiepolitiek rond de jaren vijftig*, 213, 219.

81 Fennema, "Politikologisch onderzoek naar het kernenergiebeleid," 65–85; Dercksen, *Industrialisatiepolitiek rond de jaren vijftig*.

82 For an overview of these developments, see chapter 5 in this volume.

83 A. Lijphart, *Verzuiling, pacificatie en kentering in de Nederlandse politiek*, 3rd ed. (Amsterdam: De Bussy 1979), chapter 9.

84 J. Goudsblom, *De nieuwe volwassenen: Een enquete onder jongeren van 18 tot 30 jaar* (Amsterdam: Querido, 1959), 180; "welfare provisions" came in second, with a quarter of all respondents and far ahead of all other issues mentioned. The Delta Works project was accompanied by much publicity and propaganda; see Schuyt and Taverne, *1950*, 135–136.

85 G. Mak, *De eeuw van mijn vader* (Amsterdam: Atlas, 1999), 437; G. P. J. Verbong and J. A. C. Lagaaij, "De belofte van kernenergie," in Schot et al., *Techniek in Nederland*, vol. 2, 239.

86 P. Thoenes, "De nieuwe elite," in A. N. J. den Hollander et al., eds., *Drift en koers: Een halve*

    eeuw sociale verandering in Nederland (Assen: Van Gorcum, 1968, 3rd printing), 330–331.
87 Geurts, *"Groene" IJsselmeerpolders*, 53.
88 On the sociologists, see E. Jonker, *De sociologische verleiding: Sociologie, soicaal-demo-*
    cratie en de welvaartsstaat (Groningen: Wolters-Noordhoff 1988), 90–94; M. Gastelaars,
    Een geregeld leven: Sociologie en sociale politiek in Nederland 1925–1968 (Amsterdam: SUA,
    1985), 81–87; Van Woensel, *Nieuwe dorpen op nieuw land*, 71–77.
89 Schuyt and Taverne, *1950*, 81.
90 See chapter 8 of this volume.
91 Wolffram, *Zeeuwse pachters in de Noordoostpolder*, 32.
92 Ibid., chapters 3–6. Pillarization (*verzuiling*) refers to the separation, in Dutch and Belgian
    society, between different social groups. These societies were (and in some areas, still are)
    "vertically" divided in columns, or "pillars" (*zuilen*), according to different religions or ide-
    ologies.
93 Geurts, *"Groene" IJsselmeerpolders*, 102
94 Van der Cammen and De Klerk, *Ruimtelijke ordening*, 116.
95 Petra Brouwer, *Van stad naar stedelijkheid: Planning en planconceptie van Lelystad, en*
    Almere 1959–1974 (Rotterdam: NAi 1997), chapters 1 and 2; Van der Wal, *In praise of Com-*
    mon Sense, 151 passim.
96 H. N. ter Veen, "Op nieuw land een nieuwe maatschappij: Het Zuiderzeeprobleem," in
    Mensch en Maatschappij 6, no. 4 (1930): 313–329.
97 For a concise account of this period, see J. C. H. Blom, "Nederland sinds 1830," in J. C. H.
    Blom and E. Lamberts, eds., *Geschiedenis van de Nederlanden* (Rijswijk: Nijgh en Van Dit-
    mar, [1993]), 347–349.
98 See Bosma and Wagenaar, *Geruisloze doorbraak*.
99 Bervoets and Veraart, "Bezinning, ordening en afstemming"; De Vreeze, *Woningbouw,*
    inspiratie en ambities, 233–249; Bosma and Wagenaar, *Geruisloze doorbraak*, part 3.
100 De Vreeze, *Woningbouw, inspiratie en ambities*, 245.
101 Ibid., 234, 252, 255, 270 passim, 274, 287.
102 R. Daalder, *Als de dag van gisteren: Honderd jaar Rotterdam en de Rotterdammers*, part 8:
    "En de bouw van hun stad" (Zwolle: Waanders, 1990), 195.
103 De Vreeze, *Woningbouw, inspiratie en ambities*, 274.
104 Bervoets and Veraart, "Bezinning, ordening en afstemming."
105 Van der Ham, *Heersen en beheersen*, 171–172.
106 Siraa, *Een miljoen nieuwe woningen*, 62.
107 De Vreeze, *Woningbouw, inspiratie en ambities*, 329.
108 Ibid., 456.
109 Ibid., 473.
110 For a discussion on the influences of women's advisory commissions, see Wiebe E. Bijker
    and Karin Bijsterveld, "Women Walking Through Plans: Technology, Democracy and Gen-
    der Identity," *Technology and Culture* 41 (2000): 3, 485–515.
111 I. Cieraad, "De gestoffeerde illusie: De ontwikkeling van het twintigste-eeuwse woning-
    interieur," in Jaap Huisman, Irene Cieraad, Karin Gaillard, and Rob van Engelsdorp Gaste-
    laars., eds., *Honderd jaar wonen in Nederland, 1900–2000* (Rotterdam: Uitgeverij 010,
    2000), 74–75; Wies van Moorsel, *Contact en controle: Het vrouwbeeld van de Stichting*
    Goed Wonen (Amsterdam: SUA, 1992), chapter 2.
112 Schuyt and Taverne, *1950*, 217.
113 James C. Kennedy, *Nieuw Babylon in aanbouw: Nederland in de jaren zestig* (Amsterdam:
    Boom, 1995), incorrectly identifies this as a typically Dutch phenomenon. Cf. Arthur Mar-
    wick, *The Sixties: Cultural Revolution in Britain, France, Italy, and the United States, c. 1958–*
    c. 1974 (Oxford: Oxford University Press, 1998). See also Andrew Jamison and Ron Eyer-
    man, *Seeds of the Sixties* (Berkeley: Universty of California Press, 1994).

114 On the environmental movement, we rely on Jacqueline Cramer, *De groene golf: Geschiedenis en toekomst van de Nederlandse milieubeweging* (Utrecht: Jan van Arkel, 1989). See also Andrew Jamison, Ron Eyerman, Jacqueline Cramer, *The Making of the New Environmental Consciousness: A Comparative Study of the Environmental Movements in Sweden, Denmark and the Netherlands* (Edinburgh: Edinburgh University Press, 1990), in particular chapter 4, on the Netherlands.

115 Cramer, *Groene golf*, 42–43.

116 Van der Cammen and De Klerk, *Ruimtelijke ordening*, 176; Van der Wal, *In Praise of Common Sense*, 60.

117 Koos Bosma, Dorine van Hoogstraten, and Martijn Vos, *Housing for the Millions: John Habraken and the SAR (1960–2000)* (Rotterdam: NAi, 2000).

118 De Vreeze, *Woningbouw, inspiratie en ambities*, 412.

119 Van der Ham, *Heersen en beheersen*, chapter 7; Schuyt and Taverne, *1950*, 146–150.

120 Van der Ham, *Heersen en beheersen*, 288–289.

121 An account of these events can be found in C. Disco and M. L. ten Horn–van Nispen, "Op weg naar een integraal waterbeheer," in Schot et al., *Techniek in Nederland*, vol. 1, 191–197; see also Wiebe Bijker, "The Oosterschelde Storm Surge Barrier: A Test Case for Dutch Water Technology, Management and Politics," *Technology and Culture* 43 (2002): 3, 569–584.; Cornelis Disco, "Remaking "Nature": The Ecological Turn in Dutch Water Management," *Science, Technology and Human Values* 27 (2002): 2, 206–235.

122 Van der Ham, *Heersen en beheersen*, 324–328.

123 Ibid., 331.

124 Bosch and Van der Ham, *Tweehonderd jaar Rijkswaterstaat*, 302–303.

125 Geurts, *"Groene" IJsselmeerpolders*, 176.

126 Brouwer, *Van stad naar stedelijkheid*, chapters 3 and 4.

127 Van der Wal, *In Praise of Common Sense*, 201.

128 The following discussion draws on De Vreeze, *Woningbouw, inspiratie en ambities*, 336–451, Van der Cammen and De Klerk, *Ruimtelijke ordening*, 115, 131, 143–244, 165–267, 171, 179, 186–287, 259–361. For the 1950s and 1960s, see also H. Ibelings, *De moderne jaren vijftig en zestig: De verspreiding van een eigentijdse architectuur over Nederland* (Rotterdam: NAi, 1996), 30.

129 F. van Poppel and H. van Solinge, "Ontwikkelingen in de levensloop," in Corrie van Eijl, Lex Heerma van Voss, and Piet de Rooy, eds., *Sociaal Nederland: Contouren van de twintigste eeuw* (Amsterdam: Aksant, 2001), 57–60.

130 De Vreeze, *Woningbouw, inspiratie en ambities*, 422–425, 474, 484–489.

131 Ibid., 474.

132 Ibid., 346–347.

133 See, for example, Fischer, *Technocracy and the Politics of Expertise*, 15, 22.

The first cigar-making machine of the 1920s, shown here on a pastel by Herman Heyenbrock, was a "bundling" machine that supplied a semifinished product, the cigar's inner core. This machine's further development was soon cut short by the economic crisis of the thirties: to maintain employment, a law was passed prohibiting further mechanization in this sector. The cigar industry had been a major Dutch industry since the 1850s, accounting for a large portion of in exports. It owed this success partly to low wages, which also slowed down innovative mechanization. On top of that it proved hard to mechanize the processing of tobacco.

8 Technology, Productivity, and Welfare

Jan Pieter Smits

Dutch twentieth-century economic development has been characterized by a lively dynamic; the per capita growth of the gross national product (GNP) was at least twice as large as that in the nineteenth century. This figure would have been even more spectacular if the country's population increase had not been such a major factor during much of the twentieth century.[1] The ongoing GNP growth, the implied rise of the per capita real income, and the substantial decrease of income inequality have contributed to a solid and much broader economic basis of prosperity. Almost all Dutch households were given ample opportunity to participate in consuming durable goods, which used to be an exclusive privilege of the elites.[2] In this chapter I focus on the specific role of technological developments in this economic growth.

The American economic historian Joel Mokyr has developed a model that allows him to identify how processes of technological and economic progress are interrelated, in which he distinguishes between macro and micro innovations. Macro innovations he characterizes as exogenous; they cannot be seen as responses to price changes or other economic incentives, but involve the development of radically new technologies, such as steam engines or electromotors. By contrast, micro innovations are small improvements of technological achievements that result not only from learning processes in factories where these new technologies are produced (*learning by doing*), but also from learning processes in the market through the interactions between users and producers (*learning by using*).[3]

Referring to this distinction between macro and micro innovations, one can analyze the relationship between technological development and economic growth in a dynamic framework. From time to time new macro innovations are introduced in society, followed by a series of small improvements, which leads to cost reduction and raised output. Likewise, increasing investments in human capital (such as educating the working population) allow for the maximally effective application of new technologies. This in turn makes it possible to raise labor productivity even further. At one point, however, the limit of this growth is reached. Sooner or later it is no longer possible to add cost-effective improvements to a specific technology, meaning that its growth potential is exhausted. Labor productivity will go up again only if the introduction of a macro innovation (a new key technology such as the electromotor) gives rise to a new cycle of micro innovations.

Economic historians have devoted much attention to cycles of macro innovations tied to a series of micro innovations. They argue that these cycles are responsible for sustained economic "waves" lasting some forty-five to sixty years (the so-called Kondratiev cycle).[4] Rather than trying to reconstruct such a cycle, in this chapter I seek to link the rise of new macro innovations to the development of the Dutch economy's output, measured in terms of GNP growth, competitiveness, and the development of *total factor productivity* (TFP), an indicator reflecting the combined effect of labor and capital productivity, or, put differently, the increase of efficiency through labor and capital deployment in production processes.[5]

But our analysis of the output of the Dutch economy in the twentieth century is not limited to these strictly economic factors. I will also trace the extent to which the Dutch population has benefited from the Dutch economy's increased labor productivity and competitiveness. Whether the population's *welfare* has increased in the same measure is, after all, a relevant concern. A comprehensive analysis of the economic effects of technological results will also take into account the potential downsides of growth, such as more income inequality or environmental damage.

Technological Development and Labor Productivity

To explain the long-term development of the Dutch economy from the vantage point of technology, Jan Pieter Smits, Bart van Ark, and Herman de Jong have made an econometric analysis of a series of data with respect to the country's economic output and productivity.[6] Their aim was to find out whether structural shifts or turning points could be identified in the Dutch economy's growth pattern. Their analysis revealed two such moments, in 1916 and 1975.[7] Some analyses have established that the year 1860 was such a year, and that after that year the Dutch economy began to show more dynamism.[8] Strikingly, however, they found that 1860 was not significant. Their data indicate that the Dutch GNP's growth rate fundamentally increased only after World War I and that 1916 marked a structural turning point. From a technological perspective, looking at total factor productivity (TFP), World War I also represented a structural break. After World War I, TFP growth in the Netherlands was significantly higher than in the previous period. The second turning point, around 1975, reflected a clear slowing of the growth of GNP. Thus, on the basis of recent economic-historical research

Table 8.1. Dutch percentage of the total number of patents applied for by foreigners in the American market, 1880 to 1993, by decade

1880 – 1889	0.3
1890 – 1899	0.5
1900 – 1909	0.6
1910 – 1919	0.8
1920 – 1929	1.4
1930 – 1939	2.7
1940 – 1949	3.7
1950 – 1959	6.1
1960 – 1969	3.9
1970 – 1979	2.8
1980 – 1989	2.4
1990 – 1993	2.1

(For sources see page 603)

Table 8.2. Average annual percentage growth of labor productivity in the Netherlands and Northwestern Europe, 1870–1994[a]

	The Netherlands	Northwestern Europe
1870 – 1890	1.3	1.5
1890 – 1913	1.3	1.6
1913 – 1929	3.2	2.1
1929 – 1950	0.4	1.5
1950 – 1973	4.3	4.4
1973 – 1979	3.5	2.8
1979 – 1987	2.6	2.2
1987 – 1994	1.5	2.1

a. Labor productivity is measured in terms of the GNP per hour worked. Northwestern European countries are Austria, Belgium, Denmark, Finland, France, Germany, the Netherlands, Sweden, Norway, Switzerland, and Great Britain.

we can identify three stages in the development of the Dutch economy: a first stage of growth that started in the 1860s and ended around 1916, a second stage that ended around 1975, and a third stage that started in 1975 and for which no end point can be established as yet.

To what extent is it possible to link up these stages of growth with stages of technological development? Can we identify the same shifts or turning points in both processes? A first, preliminary, exploration of this issue involves a consideration of the historical development of Dutch patents.

A good indicator, for example, is the Dutch share in the total number of patent applications in the American market (see table 8.1). These data suggest that the level of technological development in the Netherlands was still fairly low prior to World War I.[9] In the period from 1920 to 1960, however, the share of Dutch patents increased by more than five percentage points, after which Dutch technological development showed a clear proportional decline. The contribution of Dutch inventions dropped from 6.1 percent in the 1950s to 2.1 percent in the early 1990s. In the development of patent applications we can thus identify shifts that more or less coincide with shifts in the economic development of the Netherlands. The drop in patent applications started earlier than the slowing of productivity growth, but this can be explained by assuming that a leveling off of technological innovativeness has a delayed effect on labor productivity.[10]

To find out more about the extent to which technological developments have influenced the process of economic growth it is useful to map Dutch labor productivity growth. Table 8.2 provides information about this growth compared to that in a number of other northwestern European countries. The figures in table 8.2 seem to confirm the pattern of growth in economic and technological development just outlined. Until World War I (the first stage), Dutch labor productivity growth is below the northwestern European average. The result of this fairly slow productivity growth was that in around 1913 the Netherlands—which at the start of the nineteenth century had been the second richest country in the world, trailing only Great Britain—had been surpassed by many other countries.[11] This table also shows that the years around World War I mark a turning point. Especially in the period from 1913 to 1929 the Dutch productivity growth rate is very high, and in the period from 1950 to 1979 its average growth rate tops that of all the countries in northwestern Europe. From the late 1970s on, Dutch productivity growth starts to level off and falls significantly below the northwestern European average by the mid-1980s. Although the data on Dutch labor productivity growth when compared internationally seem to agree quite well with the three stages of economic growth identified, their correlation is not perfect. Whereas the period from about 1916 to the mid-1970s can be viewed as a single stage of growth, the period from 1929 to 1950 had a markedly different development in terms of labor productivity. This decelerated growth seems at odds with there being no break in the pattern of economic and technological growth (patents and TFP) in the 1916–1975 period. Importantly, the developments during the Great Depression of the 1930s and World War II clearly show that technological development does not automatically lead to labor productivity growth. In the remainder of this chapter I analyze why the Netherlands in comparison to other countries in the period between World War I and the 1970s (the years from 1929 to 1950 excepted) managed to show strong growth, whereas growth rates up to 1914 and economic growth in the most recent period are below the level of growth realized in other countries in Europe. To what extent can these differences be attributed to technological developments? Which specific forms of technology were of importance in the various stages of economic growth and why did the Netherlands manage to reap the benefits of some of these technologies optimally in the period since World War I until the mid-1970s, but was less successful in this respect during the two other stages of growth?

Dutch Economic Specialization Pattern

As a fairly small open economy the Netherlands has been quite dependent on developing its position in the world market. To gain insight into the

In the 1930s and 1940s the output of truck farming was almost as important as that of arable farming, and the former played a larger role in exports. Small businesses prevailed in this period: nearly half of the businesses were smaller than one hectare. The truck-farming sector was also labor-intensive. Vegetables were partly grown in cold frames, and flowers and bulbs were grown outdoors, as seen here in Haarlemmerliede, in 1941.

Dutch economy's accomplishments in combination with the role of technological development, it is therefore useful to identify the sectors in which the Dutch have specialized and consider the significance of technological development for this specific pattern of specialization. Table 8.3 provides a basic overview of the competitiveness of a number of Dutch products in the world market and hence the pattern of specialization of the Dutch economy in the mid-1980s.

This list clearly reveals the importance of agriculture for the Dutch economy. It is remarkable in fact that a small economy such as the Dutch manages to produce more than half of the global trade for a number of truck farming and livestock products. In addition to agrarian products and foodstuffs, chemical products and fuel make a major contribution. From the electrical products sector—put on the map mainly because Philips significantly contributed to the economy into the 1970s—only light bulbs appear on this

Table 8.3: Competitiveness of the Dutch economy as percentage of total world trade accounted for by Dutch products, by product, in 1986

Cut flowers	63.9	Pork	31.8
Eggs	61.1	Fertilizer	31.8
Hogs	56.6	Raw alkyds	31.3
Plants, flower bulbs	56.4	Light bulbs	30.6
Milk	53.1	Acyclic hydrocarbons	30.3
Cocoa powder	48.6	Eggs from birds	29.9
Tomatoes	43.4	Beer	29.6
Natural gas	40.1	Tractors	28.8
Potatoes	35.5	Tobacco products	27.8
Cocoa butter	32.4	Gas oil	26.2

(For sources see page 604)

Table 8.4: Relative advantage of Dutch industry, 1906–1950[a]

	1906	1928	1938	1950
Food	107	121	243	147
Raw materials	51	51	31	66
Metal	16	52	41	95
Machinery	4	55	36	73
Transportation	[b]	87	75	54
Chemicals	616	385	201	194
Textiles	48	106	89	151

a. Data for the period 1928 to 1950 are derived from Ingvar Svennilson, Growth and Stagnation in the European Economy (Geneva and The Hague: United Nations, Economic Commission for Europe, 1954), 293.
b. 1906 transportation figure subsumed under "machinery."

(For sources see page 604)

Table 8.5: Relative advantage of Dutch industry, 1970, 1980, and 1998

	1970	1980	1998		1970	1980	1998
Unprocessed foodstuff	89	111	136	Non-metal products	131	81	76
Alcohol and tobacco	69	134	120	Iron and steel	99	73	100
Raw materials	106	121	106	Metal products	122	108	109
				Machinery and			
Fuel	89	160	163	transportation	83	70	86
Natural oils / fats	141	159	157	of which:			
Chemicals	117	116	117	telecommunications	123	90	91
Leather	100	62	42	electric machinery	127	99	62
Rubber	91	79	102	light bulbs	183	137	135
Cork and wood products	115	149	134	Clothing	199	211	112
Paper	100	115	127				
Textiles	127	103	86				

(For sources see page 604)

list. Overall, the list shows that the Dutch economy focuses on primary products, a pattern of specialization that has deep historical roots. And the Netherlands has always had a competitive edge in agrarian products.

The historical development of the comparative advantages of Dutch business is quantified in table 8.4 for various types of economic activity. Export data constitute the basis for this calculation. In establishing a sector's competitiveness the share of a specific Dutch branch of industry (such as textiles) in world trade is compared to the share of the total Dutch exports. For example, when the share of Dutch textiles in this sector's world trade is 10 percent, while all Dutch products together contribute 5 percent of world trade, the proportional advantage of the Dutch textiles sector is (10/5) x 100 = 200. Table 8.4 and 8.5 provide an overview of the proportional competitiveness of the various sectors.

The figures higher than 100 suggest a clear relative advantage for the sector involved, and those below 100 suggest a relative disadvantage vis-à-vis other sectors in the Netherlands.

Unfortunately, only quite general trade data are available on the first half of the twentieth century, which is why we can only identify broad categories. For example, the strong export growth of electric machines and light bulbs in the 1930s remains hidden in table 8.4, these product groups being included in the more general category of machine building. Still, the figures offer some indication of the Dutch economy's pattern of specialization. First, even before World War I the Netherlands had a clear competitive edge in the area of chemical products. This preeminent position is mostly based on fertilizer exports; other Dutch chemical sectors did not have a strong position in the global market. At this stage the Dutch chemical industry focused on large-scale production of fertilizer and almost all of it was exported.[12] Similarly, food and luxury foodstuffs did fairly well, partly because of the export of fruits, vegetables, meat, dairy products, and margarine.[13] The Dutch metal, machine, and textile industries were poorly developed in this period. In the subsequent, interwar, period these differences among sectors shrank somewhat. Dutch steel and machine building, which came to the forefront during the so-called "second industrial revolution,"[14] displayed major growth after World War II (although the country did not yet enjoy a large comparative advantage), and during the entire first half of the twentieth century Dutch food and luxury foodstuffs managed to gain an increasingly stronger position in the international market. Since the 1960s the available data on international trade have been far more extensive, and this makes it possible to analyze the Dutch economic specialization pattern in much greater detail. Table 8.5 shows that at the start of the 1970s, when the country's economic and commercial expansion peaked, its international economic success was mainly linked to the agriculture and food cluster, the chemical industry, and the electronics sector. These sectors, as well as a number of fairly small labor-intensive business sectors such as clothing, had the highest comparative advantages.

Starting in the late 1970s, Dutch competitiveness began to show a strong decline. The electronics sector showed a dramatic decline. In particular the diminished comparative strength of light bulbs, electric machines, and telecom products is striking (see table 8.5). There was also a marked decline in labor-intensive industries such as textiles and leather. The rise of new low-wage producers in southern and eastern Europe, as well as in Southeast Asia and Latin America, played a major role in the proportional decline of these industries in the Netherlands.

Our analysis of Dutch export development reveals that the increase in the Dutch share of world trade in the period from 1916 to 1970 is mainly accounted for by these three clusters—agriculture and food, chemicals, and electronics—and its relative decline in the more recent period is largely an effect of the electro-technological industry's decline and of the lowering of the competitiveness of a number of labor-intensive business sectors. From the 1970s on, the Dutch economy concentrated more and more on primary products. Above all, this involved agrarian products and, by extension, food and luxury foodstuffs. Moreover, it is possible to note a comparative advantage in heavy chemicals, with strong ties to oil refining.

How can we explain this specific pattern of specialization? And what role do technological factors have in this specialization process? Traditional trade theory emphasizes the presence and cost of production factors.[15] It is assumed that a particular country specializes in the production and export of goods depending on production factors that are widely available and hence fairly cheap. One may expect, for instance, that an economy with large quantities of (cheap) agricultural land at its disposal will specialize in arable farming, whereas a country with a large and well-trained labor force will instead specialize in high-tech products. Technology plays almost no role in this model. The available quantities of labor, capital, and land determine what will be produced and how. Moreover, this model leaves no room for innovation—making efficient use of factor inputs that allow a country to specialize in sectors in which it has no obvious advantages.

A single glance at Dutch production and export statistics tells us that the Dutch pattern of specialization cannot be explained by means of traditional trade theory. Despite the relative scarcity of agricultural acreage and the late discovery of oil and natural gas reserves, the Netherlands gained a comparative advantage in agrarian and chemical production at an early stage. This specific pattern of trade specialization can be explained with reference to technology-related variables. The high scores for agricultural products, food-stuffs, and chemicals align with high productivity levels in these product groups, notably in the period after 1916.[16]

Prior to 1950 high output in agriculture mainly applies to soil productivity, the yields per hectare or per animal, rather than labor.[17] In the period after World War II Dutch agriculture diversified substantially, and labor productivity in greenhouse growing and the livestock sector in particular reached record heights. A study by Herman de Jong on the Dutch industry's comparative labor output relies on international output data for various sectors and shows that Dutch labor productivity in the food industry was high by international standards. In the mid-1930s the production per worker in the Netherlands was almost 7 percent above the British level and even almost 40 percent above the German level.[18] In the chemical industry the pattern is less clear. Although in this sector labor productivity is higher in the Netherlands than in Belgium (+6 percent) and Great-Britain (+7 percent), it is *lower* than that in Germany (-5 percent). De Jong also shows that in a major segment of the food industry—particularly in bread and flour production—high labor productivity can largely be attributed to large companies and a high level of mechanization via electromotor use. By contrast, the electrification level of Germany's bread and flour factories was only 58 percent that of Dutch factories.[19]

This issue of product specialization can also be studied from other perspectives. It is reasonable to assume, for instance, that the Dutch industry's strong emphasis on food and chemical products is related to the position of the Netherlands in international trade networks. In particular, the port of Rotterdam plays a major role in international transit of crude oil and all sorts of raw agrarian materials. These international distributive services led to the development of major processing industries such as the petrochemical industry near the Rotterdam harbor in Pernis. Still, the strong position of the port of Rotterdam cannot be explained with by its favorable location alone. This port realized huge growth in the twentieth century because it successfully capitalized on technological developments.[20]

1890–1916: Slow Development in the Steam Era

In the decades prior to World War I the Dutch economy's competitiveness eroded. The Dutch share of world trade was no longer growing and in some years even showed a decline, and its productivity growth lagged behind that

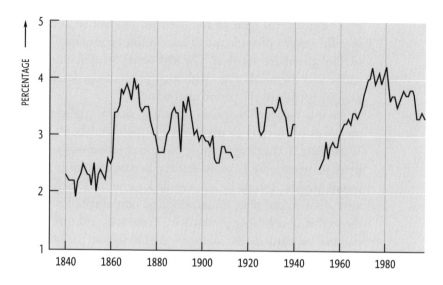

Graph 8.1. Dutch percentage of world trade. (For sources see page 605)

of other countries (see graph 8.1).[21] This is quite surprising because historians often describe the 1890–1916 period as one of major economic development. Some authors go as far as to characterize this period as a "takeoff" period.[22]

Actually, though, it is not difficult to account for this paradox. Although the Dutch GNP grew strongly and the country industrialized, this growth and industrialization were mainly realized by companies that operated in the domestic market, rather than leading to more exports (with some exceptions such as margarine and fertilizer). Why did Dutch exports grow so slowly? An analysis of the cost development in the Dutch industry shows that there was in fact a problem with competitiveness.[23] Real production costs strongly increased after 1890. As a result only the largest companies were able to operate adequately in the international market. It is no coincidence that precisely in the 1890–1916 period the structure of industrial business markedly changed in the sense that very many small and medium-sized businesses closed down. The emphasis within Dutch industry increasingly shifted toward large companies. This "restructuring" of industrial business, including the loss of production capacity in the early phase of this process, caused a temporary deceleration of export growth. After World War I exports picked up steam again, mainly because small and medium-sized companies also began to display a new growth dynamic as a result of the introduction of the electromotor.

In the nineteenth century the Netherlands became important internationally as a major center of cocoa processing, partly thanks to specific knowledge of cocoa-powder processing developed there. Dutch companies exported throughout the world. In their ads companies used the services of famous artists from abroad, such as the designer of this 1893 poster, the French Art Nouveau illustrator Adolphe-Léon Willette. In their visual message many companies relied on traditional Dutch imagery and symbols. Another major export company, Philips, used images of Dutch folkloristic dress and wooden shoes in its ads for light bulbs.

In respect to this period it is hard to link the lack of export-guided growth and stagnating productivity development (compared to other European countries) unequivocally to specific technological developments. Dutch technological development has been marked by the interweaving of various kinds of technologies that, taken together, have generated favorable economic effects, such as in the agricultural sector.[24] Still, it is possible to identify several key technologies that were applied in many sectors and therefore could produce major economic effects.[25] In the 1850–1890 period the incidence of steam engines in the country's total machinery grew from 5 percent to over 60 percent.[26] The Netherlands, however, was quite late in introducing steam engines, a disadvantage that was never entirely recouped. It should be noted that there were good institutional, economic, and technical reasons for not introducing the steam engine sooner.[27] Basically, levels of demand were too low for modern technologies such as steam engines to be used on a profitable basis. As a result of changes in real income as well as a number of institutional changes in the second half of the nineteenth century these scale constraints were removed and the application of steam took place in a number of specific sectors, such as the textile and paper industries. These were also the sectors that showed the strongest productivity growth.

Table 8.6 shows that this growth was relatively low in branches marked by minimal input of steam engines, for example, in the chemical and food-processing industries, segments of the economy in which the use of steam engines was not widespread.

In the second half of the nineteenth century the diffusion of steam power was very successful in shipbuilding and the exploitation of minerals (by 1850 these sectors relied exclusively on steam power), as well as in the paper industry, where steam engines had replaced all existing machines by 1860. In the food

Table 8.6. Average annual productivity growth in several industrial sectors from 1865 to 1913 and the use of steam power, expressed as number of steam engines as percentage of the total number of machines in use, in 1890

	Productivity growth	Percentage of steam engines
Total industry	+ 1.8	61
textiles	+ 1.8	97
metal and shipbuilding	+ 3.1	95
paper	+ 4.4	100
chemicals	+ 1.1	45
food	+ 1.1	43

(For sources see page 604)

and chemical industries, however, machines driven by water mills and wind mills still had the upper hand. In these sectors the introduction of steam technology was either not possible on technical grounds or not desirable from an economic point of view. The food industry used some steam engines, but they were mostly fairly small and did not lead to a significant rise in labor productivity.

The interrelation between stagnating productivity development and growth of exports on the one hand and the introduction of the steam engine on the other can also be formulated in a different way. The Dutch economic structure—with a strong emphasis on agriculture and service industries and a preponderance of fairly small companies—proved less susceptible to productive application of steam power than, for instance, the economies of Great Britain and Belgium. This caused productivity development to come to a standstill, and as a consequence Dutch companies were less capable of profiting from the growth of world trade.[28]

1916–1970: A Period of Strong Growth

The 1916–1970 period saw unprecedented economic growth in the Netherlands. The Dutch GNP grew prodigiously, the TFP increased, and the competitiveness of the Dutch economy showed strong improvement. If we look at the process of electrification we can conclude that this specific technological advance caused the powerful economic development. This process occurred very rapidly in the Netherlands, and it appears to account for as much as 50 percent of the growth of factor productivity. This strong economic growth can further be explained by the significant investments made in the knowledge infrastructure.

The Significance of the Electromotor

The electromotor can be seen as one of the key technologies of the "second industrial revolution," which sparked off all sorts of technological and organizational processes of change.[29] The introduction of the electromotor triggered the rationalization of production processes and had a substantial influence on increases in productivity. Herman de Jong has calculated what share of industrial growth can be attributed to the electrification of production processes, performing his calculations with the help of growth accounting analysis. The basic principle of this method is fairly simple. Starting from production growth De Jong tried to figure out how much of the growth was attributable to an enlargement of the factor inputs of labor and capital and

The interwar period saw significant improvements in labor productivity in the
Netherlands, partly because of the deployment of more modern means of production.
The production of textiles on automated looms continued to expand. The increasing
availability of electricity also made it possible in more rural areas to set up a
mechanized loom or, with a minor investment, transform a manual production facility
into a mechanical one. In the southeastern part of the province of Northern Brabant
this gave rise to a regional textile industry made up of a network of mechanized
companies, including the Swinkels weaving mill in Deurne, shown here, which was
established in 1920. It consisted of simple looms in a simple building, basically a
wooden barrack over one hundred meters long.

how much to more efficient use of these inputs. This increased efficiency is referred to as the already introduced total factor productivity, or TFP. De Jong performed this analysis first by using common data on the available stock of capital goods and then by using an alternative capital indicator: the aggregate value of electric machinery value.

If electrical machines are used as an indicator for capital instead of the more usual figures for the entire stock of capital goods, TFP growth proves to be only half as large. It can be concluded, then, that a substantial part of TFP growth can be attributed to the diffusion of electric machinery (see table 8.7).

This large productivity effect is a direct effect of the labor, capital, and raw materials saved by applying this key technology. Electric machinery was less expensive to purchase, lighter weight, and smaller (while delivering the same power). This machinery was easier to operate, required less maintenance, was safer and quieter, and produced less stench than steam engines. It could also be switched on and off easily and thus could meet short-term energy needs, unlike other types of machinery such as steam engines, which were much less flexible in use. Moreover, companies had to reorganize major parts of their production process if they wanted to realize the potential of using electric machinery in a decentralized way. Many companies began with a central position for their machinery, but switched to a decentralized arrangement. Occasionally steam and electric power were used side by side.[30] Calculations indicate that what economists call "embodied," or physical, technological development has influenced industrial growth in great measure.

The Netherlands was very successful in the period from 1916 to 1970 at diffusing technology throughout most sectors of the economy. The electro-motor, one of the principal general-purpose technologies of the "second industrial revolution," was introduced early in the Netherlands; the diffusion of this new key technology occurred almost as quickly as in the United States.[31] First, steam technology—elsewhere the main driver of the nineteenth-century industrial revolution—in the Netherlands was never applied

Table 8.7: Average annual percentage growth of real output, labor, capital, and Total Factor Productivity (TFP) in Dutch industry, 1921–1960

	Production	Labor	Capital	TFP
1921-1938	4.14	1.34	3.74 (normal)	1.8
			6.10 (electrical)	0.9
1938-1960	5.06	3.00	3.24 (normal)	1.9
			5.46 (electrical)	0.8

(For sources see page 604)

on a large scale. This meant that technological development was less subject to aspects of path dependency than was the case in countries that had extensive steam infrastructure. Second, electromotors could be easily deployed in sectors that were strongly developed in the Netherlands but had relied on steam power only marginally, such as the food and chemistry sectors. These sectors, where the productivity growth before the war had been comparatively low, now realized a strong increase in labor productivity. Third, the electromotor had major influence in particular on small and medium-sized companies, which in the 1920 and 1930s constituted the core of Dutch industry. Precisely these sorts of companies were well placed to profit from the new electricity technology because the cost was affordable, and actions that previously had been performed with the help of human power could now be taken over by machines, for less money. Thus, electricity provided a solution to the problems facing entrepreneurs in the 1930s: "To industries that were mechanizing their activities the electromotor served as an aid that allowed them to modernize their business without drastically changing its character" and "With respect to the many manual tools, electricity was the perfect aid that helped raise a worker's output instantly."[32]

Finally, the Netherlands, because of its neutral status during World War I, profited much from the postwar boom in aggregate demand.[33] The new technology also ultimately played to workers' advantage: Labor unions successfully bargained a substantial reduction of working hours while wages did not change.[34] In practice this meant a strong rise of the labor cost per product unit. To counter an overall rise of the production cost, companies felt forced to invest in labor-saving technologies.[35] The diffusion of the electromotor was stimulated, then, by an increase of the demand and a rise of the (relative) labor cost.

Table 8.8: Investments in education and R&D, 1921-1992, as percentage of their contribution to GDP

	Education	R&D	Total
1921	2.8	–	2.8
1929	3.5	–	3.5
1938	3.3	0.2	3.5
1947	2.6	0.5	3.1
1960	4.5	1.5	6.0
1973	6.6	1.9	8.5
1979	6.5	1.9	8.4
1987	5.2	2.3	7.5
1992	4.6	1.9	6.5

(For sources see page 604)

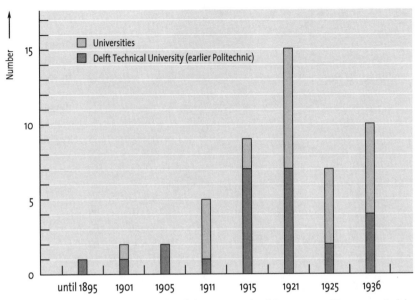

Graph 8.2. Appointments to boards of directors and advisory committees accepted by chemistry professors at Delft technical university and other universities
(For sources see page 605)

Investments in the Knowledge Infrastructure

Dutch economic growth in the twentieth century is marked particularly by an increase of the knowledge intensity of production. Table 8.8 shows how investments in education and research and development evolved over the course of the century. Around 1900 a knowledge infrastructure had come into being in agriculture and various industrial sectors.[36] In agriculture, experiment stations and information services had been established. Delft's strongly improved polytechnic became more important to Dutch industry, while various associations were established where industrial and university engineers could meet, such as the Netherlands Society of Mechanical and Marine Engineers (Nederland Vereniging van Werktuig- en Scheepsbouwkundigen), founded in 1889, and the Netherlands Society of Electro-technology (Nederlandsche Vereeniging voor Elektrotechniek), founded in 1895. Four years later the two merged into the Royal Institute of Engineers (Koninklijk Instituut van Ingenieurs, KIvI), with the periodical *De Ingenieur* as its official journal. In this period, however, Dutch companies continued to be dependent on importing knowledge. Prior to World War I only a few companies, such as Philips and Royal Dutch, had established research labs, but during the war a strong demand emerged for products not produced in the Netherlands. This led to the organization of a process of intensive knowledge acqui-

Students of the State Agricultural Winter School in Schoonedijke, Zeeland, are immersed in the fine details of engine technology (around 1954). The success of Dutch agriculture rested on the effective dissemination of knowledge and widespread basic education. Education of various types was widely available at various levels. For instance, there were special winter courses for farmers, when their workload was at its lowest level.

sition, partly because importing knowledge had become a more complicated matter. It is telling that the number of engineers and scientists employed by industry increased by 50 percent between 1913 and 1918 (see graph 8.2).[37]
In the interwar period the Dutch knowledge infrastructure rapidly expanded. Medium-sized companies in various industrial sectors began to set up their own laboratories. Their research effort was always geared to concrete products and processes rather than fundamental research. Typically, these labs were small, employing only a handful of people.[38] Yet the total expenditures for R&D went up significantly in the interwar period, while collaboration between companies and between companies and universities was further strengthened and the Dutch daytime technical education system expanded, from 5,247 enrolled students in 1900 to 31,143 students in 1938.[39]

Accelerated growth and expansion of the knowledge infrastructure also took place after World War II. This war granted science and technology major social status because the victory was in part attributed to the scientific and technological achievements of the United States, and many scientists and engineers were quite willing again to put their knowledge at the service of goals they considered socially relevant. Still, the debate on how to realize these goals persisted: Could social goals be realized by encouraging basic scientific research or by setting specific socially relevant goals and funding

research to meet them? Between 1947 and 1979 knowledge-related investments as percentage of the Dutch GNP increased from 3.1 percent to 8.4 percent, but by the 1979–1992 period it had gone back to 6.5 percent, its decline
mainly dating from after 1973.[40] Investments in R&D in the Netherlands
were mainly made by the country's five largest multinationals. Between 1959
and 1964 they were responsible for two thirds of the industrial R&D expenditures, but after 1970 that figure decreased to 55 percent in 1992.[41]

It is difficult to reliably determine the effect of investments in the knowledge infrastructure on economic development. One such indicator may be
the strong correlation between investments in R&D and the growth of
granted patents, which are assumed to be reliable indicators of innovation
and resultant productivity. It also appears that sectors that invested more in
R&D were also granted more patents.[42] Furthermore, proportionally the
greatest investments were made in R&D in agriculture and food, chemicals,
and electro-technology—sectors that could also operate competitively in the
global market. Of course these correlations are difficult to specify in great
detail. Moreover, what mattered was not only pure investments in R&D but
also and even more so the increasing cooperation and network formation
among industries and between universities and industries. An analysis of
studies of individual companies and sectors shows, for instance, that investments in R&D allowed companies to diversify and open new markets so as
to realize further growth. This applied to radio production by Philips and
the chemicals division of Shell.

The Central Sugar Corporation (Centrale Suiker Maatschappij) diversified into sugar-based chemicals and in 1949 it started production of synthetic
resins, followed one year later by production of vitamin B. The State Mining corporation (De Staatsmijnen) began to produce raw materials for synthetic fibers partly using knowledge it had developed on its own.[43]

Continuity or Discontinuity Between 1916 and 1970?

The figures on investments in R&D and granted patents do not show any
major discontinuities in the 1916–1970 period, but in the 1930s these data do
diverge widely from data on changes in labor productivity. In the years from
1929 to 1938 Dutch labor productivity showed a strong decline, yet all the
relevant data on technological knowledge indicate a rather steady increase.
How can this be explained?

Contrary to the prevailing picture that emerges from the relevant literature,
the 1930s was not exclusively an era of economic downturn. Dutch industry had

an even pattern of growth during the period from 1921 to 1960 (see also table 8.7). The TFP increase from 1921 to 1938 matched that of the years from 1938 to 1960, and the growth figures of capital input were also quite similar in both periods. De Jong explains the processes of rationalization and efficiency increase during the crisis years as resulting from an active effort on the part of entrepreneurs to compensate for the high (nominal) wages, which put much pressure on the profit margins of Dutch industrial companies.[44] Because of several specific circumstances, however, these investments did not lead to higher labor output.[45] During the 1930s crisis, the Dutch government made sharp cutbacks, thus strongly reducing its own spending. In addition, the public's purchasing power developed more slowly than in the 1920s. Finally, there was pressure on exports because the government conducted a monetary policy that resulted in an overrated guilder. The effect of these developments was a market contraction (despite increased technological knowledge and increased efficiency of production), which made it necessary to reduce production.[46]

Similarly, World War II did not immediately cause a major disruption of Dutch economic development. The study by Hein Klemann has shown that May 10, 1940, when the Germans occupied the Netherlands, was by no means the end of economic activity in the Netherlands. In fact, the Dutch economy experienced growth in the first years of war as German orders kept pouring in. As a consequence, we have no need of an economic miracle to account for the quick postwar recovery.[47] After the war, just as after World War I, the economy once again simply entered a new phase of growth. As we saw earlier, this growth can be partly explained by ongoing investments in capital goods, education, and R&D that resulted in a high TFP. Moreover, in this phase it was shown that investments made in the 1930s could turn out to be profitable after all. The significance of the import of American technology was limited, even though large groups of Dutch industrialists visited American companies in the context of the Marshall Plan to learn more about their new technologies. A careful reading of reports by productivity commissions tells us that Dutch entrepreneurs were highly impressed by the technological level of American industry but that they generally realized it could not be automatically transferred to Dutch industrial settings.[48]

At the same time, however, Dutch industry developed an underlying structure that would make it vulnerable. Whereas in many Western countries the emphasis was on large-scale and capital-intensive industries, the Netherlands actively tried to secure rather traditional, labor-intensive sectors such as textiles as well.[49] This focus of the country's industrialization policy was a direct outcome of its controlled wage policy and the wish to secure industrial employment in the country's less urbanized provinces. The Dutch

A view of the already automated weaving room of the largest independent company in the textile sector, Van Heek & Co., in Enschede around 1955. The Dutch textile industry flourished after 1945. At its peak, around 1960, it employed some 100,000 laborers, largely in big automated plants such as Van Heek's. Especially after 1960 wages in the Netherlands rose quickly. For industrial companies it became increasingly difficult to compensate for the rising cost of wages through further automation.

government, by keeping wages and prices low, tried to turn the Netherlands into an island within Europe where things were affordable, thus enlarging the country's competitiveness. This policy was successful. When, at the end of the 1950s, the policy of wage restraint had to be abandoned because of increasing labor-market shortages, nominal wages began to go up strongly. This put a lot of pressure on the competitiveness of labor-intensive sectors such as textiles and shipbuilding. In the 1960s these sectors invested again in expansion and further mechanization, but eventually they could no longer compete internationally because of the high wage level.

After World War II the Netherlands also invested in further developing its primary industries. This tendency was reinforced by the Dutch government's energy rate policy. The government tried to stimulate industrialization by keeping the cost of energy low, and the discovery of natural gas deposits of Slochteren made such a policy possible.[50] In the early 1960s, Dutch energy rates were among the lowest in the world. Although the average energy price in the Netherlands was linked to the price of oil in the world market, the large industrial users could purchase energy for very low rates. The price differential (namely, the difference between the world market rates and the price paid by the industrial sector) was paid for by the consumers. This differential price policy has become one of the main underpinnings of Dutch energy rate policies and has positively influenced the competitiveness of energy-intensive sectors. In 1992 the cost of energy in the Netherlands was two thirds of the Belgian level and as low as 45 to 50 percent of the level in Denmark and Germany.[51] This low rate encouraged Dutch industry to focus on forms of production that relied heavily on raw materials. Particularly the chemical industry profited from this policy. In the first postwar years, oil refining in Pernis grew rapidly. The refineries mainly concentrated on processing crude oil entering the port of Rotterdam in bulk shipments. After the early 1960s, the Dutch petrochemical industry received an extra boost from the low energy rates. In addition, greenhouse agriculture, road transport, and segments of the metal industry saw strong growth because of the low energy rates. From the early 1960s, the country increasingly gained a competitive edge in energy-intensive forms of production.[52] The shift in industrial emphasis also caused an acceleration of productivity growth. The energy-intensive branches generally were also highly capital-intensive and were therefore characterized by relatively high levels of labor productivity.

This analysis shows that World War I clearly marked the beginning of a phase of growth acceleration, and that World War II did not mark a break in economic development. It seems safe to conclude that the technological basis for Dutch economic growth in the period from 1916 to 1975 was quite stable. If much technological potential remained untapped in the 1929–1936 crisis, after the war this potential was turned into productivity growth because of the improved institutional situation, which also contributed to unprecedented economic growth in the years from 1950 to 1973.

Decelerated Technological Development and Productivity Growth after the 1970s

From the early 1970s, the Western world, the Netherlands included, was faced with a sharp deceleration of economic growth. This can be explained in part by the energy crises. After World War II the economic structure of many Western countries increasingly became energy-intensive, and this was especially true of the Netherlands, so it is no wonder that the Dutch economy was hit hard by the oil crises.[53] Especially the 1979 oil crisis, which resulted in excessively inflated energy prices that negatively influenced the earning power of many sectors, led to a major recession. Data from the Central Statistics Bureau (Centraal Bureau Statistiek) reveal that this deceleration was strongest in energy-intensive sectors such as the chemical industry.[54] Earlier, with artificially low energy rates, energy-intensive forms of produc-

Table 8.9: Average annual percentage growth of GDP, by component (labor, capital, human capital), investments in R&D, and Total Factor Productivity (TFP), 1947-1994

	GDP	Labor	Capital	Human capital	R&D	TFP
1947–1973	5.1	0.9	5.9	1.5	11.8	1.7
1973–1979	2.7	−1.4	3.9	0.9	4.8	1.1
1979–1987	1.2	−1.9	2.3	0.0	2.5	0.7
1987–1994	2.5	−0.7	2.8	2.6	2.2	0.2

(For sources see page 604)

Table 8.10: Comparison of percentage growth, by technological and economic indicators, between 1971–1975 and 1986–1990 periods

	GDP/employee	R&D/employee	Investments/employee
OECD[a]	36.2	59.0	12.7
Japan	24.0	76.1	34.7
United States	− 10.3	− 27.7	- 1.4
Great-Britain	− 0.5	− 26.6	12.2
The Netherlands	− 14.0	− 47.7	- 25.8

a. OECD = Organization for Economic Cooperation and Development
(For sources see page 605)

tion had functioned as engines of economic growth; now they contributed to the slowing of economic growth, a slowdown that lasted until the mid-1980s. When energy prices returned to their old level, sectors such as the chemical industry could strongly grow again. Both oil crises, however, revealed the particular vulnerability of the Dutch economy to exogenous price fluctuations in the world market for petroleum products.

Other factors as well fed into the economic crisis. By means of growth accounting one can map the development of economic growth before and after 1973. Table 8.9 clearly shows that from the early 1970s onward, Dutch economic growth began to slow down. The 1973–1994 period showed less than half the growth realized in the 1947–1973 period. This drop occurred in particular in R&D investments, TFP growth, and investments in physical capital. An international comparison of these growth rates may reveal even more clearly the loci of the problems (see table 8.10).

The slowing of economic growth in the Netherlands was very substantial compared to that in other countries. This negative productivity development can be attributed to a reduction in the level of investment in both physical plant and knowledge. The drop in investments in R&D per employee is particularly high. In the Organization for Economic Cooperative Development generally, between the early 1970s and the late 1980s investments in R&D

per employee increased by almost 60 percent, but in the Netherlands they fell by over 45 percent. Plainly, the 1970s mark a break in the trend of long-term technological development. In 1971 the Netherlands ranked second on the OECD list of industrial R&D intensity (the amount of money invested in industrial R&D per unit of industrial production). From the early 1970s on, the Netherlands' rank started to fall, down to eighth in 1990.[55] Another factor in the low productivity development was that the growth potential of the technologies linked to the economic growth in the period after 1916 was largely exhausted by around 1970. Apparently the main technologies of the second wave of growth, such as the electromotor, were abundantly disseminated across the economy by the early 1960s. The productivity slowdown of after 1973 can be partly explained by this exhausting of existing technological opportunities as well. It should be added that the slow productivity growth in the Netherlands cannot simply be explained by a shift of the economic emphasis from a high-output industrial sector to a relatively low-output service sector, because in the Netherlands this shift was not significantly larger than in other countries.[56]

The marginal growth of Dutch labor productivity in the most recent period raises questions about the application of new information and communication technologies in the Netherlands; although this country played a minor role in the development of these technologies, Dutch entrepreneurs were perfectly capable of successfully integrating technologies in their production processes that were developed elsewhere. In the late twentieth century, a rapid and successful diffusion of new (information and communication) technologies took place in the industrial sector, which saw substantially enhanced growth of labor productivity, but there were problems in the service sector.[57] Despite its serious investments in information and communication technologies, its productivity growth continued to be low, judged by international standards. The limited influence of these investments appears related to rigid organizational structures that complicate successful implementation of the new technologies.[58] Time will tell whether the new technologies will cause economic acceleration of the service sector as well, which is crucial to the Dutch economy. If the Netherlands does not profiting sufficiently from the new information and communication technologies—much as it failed to cash in on steam technology at the end of the nineteenth century—the country's economy will see marginal growth at best in the years ahead.

Welfare and Income Inequality

So far I have discussed primarily productivity and competitiveness. Briefly, up to the 1970s Dutch production grew ever more efficient, but from the end of the 1970s the country began to lose ground in terms of competitiveness. It can also be asked whether Dutch society at large has capitalized on the strong pre-1970s growth or whether its reduced competitiveness over the last few decades has also led to a decline in welfare. To analyze the welfare effects of economic growth, I will attempt to answer several questions. First, has economic growth led to more even income distribution? Second, what development activities unfolded in households—which are ignored by many economic analyses as a dimension of production. Third, to what extent did economic growth harm the natural environment? All these aspects, once analyzed, will be combined in a new indicator representing welfare development.

In recent years, economic historians have studied the development of income inequality more extensively. Much of this work was done with Kuznets's theory of modern economic growth in mind.[59] This theory posits that in the long run economic growth is accompanied by a decrease of income inequality. In the short term, however, other trends can be observed. When a new technology is introduced, the demand for highly skilled labor can increase so strongly that temporarily this labor earns a premium. This so-called *skill premium* causes the income gap between educated and unskilled labor to grow wider. This effect gradually vanishes again after training or educational systems are adapted to accommodate the demands linked to the new production techniques, doing away with the specific labor scarcity involved and thus the need for a skill premium.

Quantitative research demonstrates that throughout the twentieth century, income inequality has significantly decreased in the Netherlands.[60] This leveling can largely be accounted for by the fact that the share of wages as part of the national income grew strongly. This growth can be attributed not only to a rise of the nominal wages, but also to expansion. The average number of workers per company strongly increased in the period from 1890 to 1970. Because the wages of workers are more equally distributed than the profits of capital owners, it is possible to explain the overall decrease of income inequality in the twentieth century partly by the structural shift of the economic emphasis from capital to labor.[61]

A trend of income leveling can also be observed within the group of wage earners. Because large groups of workers have gained access to the education system, in particular in the period after World War I, and the supply of skilled labor went up, wage differences between skilled and unskilled labor continued to be fairly limited. In this period the rapidly growing level of organization among workers probably played a role as well, as it allowed labor unions to prevent large pay differences from being implemented.[62] In two periods, however, there were (short-lived) increases in income inequality: in

An office, in 1965, of the insurance company Nillmij in The Hague. The decline of the number of jobs in industry since the 1960s has been offset, more or less silently, by the growth of the service sector. Despite ongoing mechanization, as illustrated by the various office machines seen in the photo, this sector provided a lot of employment opportunities for the time being.

the 1920s and the 1990s. This increased income inequality occurred in periods in which new key technologies were introduced in the Netherlands: the rapid diffusion of the electromotor in the twenties and of the information and communication technologies in the nineties. Both cases probably led to a skill premium for educated labor. International sources suggest that highly skilled workers in sectors with relatively much use of electromotors were paid higher wages.[63] It seems very likely, then, that new technologies caused a new demand for new forms of highly skilled and expensive labor.

In the 1990s an additional factor was responsible for raising income differences in the Netherlands. The income position of those with few skills was negatively affected not only by the information technology revolution but also because of the increasing competition from low-wage countries, in Southeast Asia in particular.[64] This last factor is evident in many Western countries, but it is felt more deeply in the Netherlands, with its labor-intensive industrial production structure. Moreover, this country has one of the most open economies worldwide. In 1987 exports were 44 percent of GDP, whereas the average rate within the OECD was 24 percent, just over half the Dutch rate![65]

Considering the full spectrum of factors involved, it can be concluded that on balance the kind of economic growth experienced by the Netherlands in the period after 1890 has led to a reduction of income differences and to increased welfare. The technology shock of the 1920s (the rapid electrification of Dutch industry) led to a momentary increase of the reward of educated versus uneducated labor. As would have been expected by Kuznets, after this increase, income inequality later quickly dropped again.

Household Production

A crucial adjustment of basic national income involves the registration of labor that takes place in households, and thus outside the market mechanism. There is no reason to consider household-related activities such as food preparation, cleaning, and care as nonproductive. By mapping these alternative forms of production and relating them to the leisure time people have available it is possible to obtain a more complete picture of prosperity and welfare development. With the help of time and motion studies and by assigning financial value to time spent on an activity it becomes possible to take household activities and welfare effects into account in economic analyses.

One obstacle to doing this is that there are no good data available on how time was spent in the Netherlands in the first seven decades of the twentieth century, but an approximation can be made of the development of the eco-

A cooking course for men, held in the late 1950s in the office building of the IJssel power plant in Zwolle. Men have gradually started performing more household tasks, but most housework is still done by women, even though efforts have been made to try to teach household skills to men.

nomic value of household production using experimental estimates and scarce data. This estimate shows that household production rose sharply after World War I.[66] Unknown is an whether increased production in households was tied to fewer hours of work. For example, one may assume that the introduction of washers has led to a gain of time for each load of laundry, but not for total time spent on laundry because women began to do the laundry multiple times per week. The potential gain of this new technology was thus immediately canceled by the higher- quality outcome that could be produced. Of course, this caused an increase of household productivity and thus, presumably, of welfare. The mechanism of increased utilization of available time for raising household productivity has been demonstrated for various countries.[67] It can be concluded that the household sector significantly contributed to the development of welfare.

This contribution has mainly come from women (see table 8.11). This table shows that in 1975 women devoted a substantial share of their time to

Table 8.11: Percentage of men's and women's time spent in formal and informal labor, 1955–1997

	1955-56	1962	1975	1988	1997
Men					
Formal labor	10	9	14.1	15.8	16.6
Household and family	6	5	5.0	6.7	7.8
Total	16	14	19.1	22.5	24.4
Women					
Formal labor	3	4	3.6	6.2	8.8
Household and family	28	26	17.0	18.7	12.9
Total	31	30	20.6	24.9	21.7

Note: Statistics Netherlands and the Social and Cultural Planning Office of the Netherlands have conducted surveys on how much time individuals spend on a variety of activities: productive activities (both formal labor and household work), leisure, and time spent sleeping. Taken together, these activities add up to the full twenty-four hours.
(For sources see page 605)

household labor, whereas men spent much more time on labor outside the household. Men's share in household tasks was limited, but it did grow. Strikingly, the workload of women was higher than that of men. In the years 1975 and 1988 men devoted 19.1 percent and 22.5 percent, respectively, of their time to productive activities, whereas women in these same years spent 20.6 percent and 24.9 percent, respectively, of their time productively. In 1997, however, a change began to set in, one that concerned leisure time.

Development of Leisure Time

In order to attempt to quantify the increase of welfare it is also interesting to consider whether job pressure went up or down in the twentieth century. A comparison of the average number of hours worked per country in the European Union shows that in the Netherlands people worked the least (household activities are not calculated).[68] This can be partly explained by the high labor productivity of Dutch employees: when they work, they apparently generate more output and income in less time. Still it is strange to realize that Dutch employees work fewer hours than people do anywhere else in the EU. Notably in the past decades time as a factor in society is increasingly under pressure. The solution seems to be tied to the difference in formal working hours between men and women. Dutch women, who entered the formal labor market only relatively late in the twentieth century, have generally worked part-time. Dutch men, by contrast, have put in more hours of formal work over the past decades. Because men have also started to do more household chores, the amount of leisure time must have significantly dropped.[69] The trend of more leisure that started at the beginning of the twentieth century has thus come to a standstill. In the late nineteenth century the struggle for leisure time evolved into one of the key concerns of the labor movement. In 1955 and 1956 leisure amounted to twenty-four hours per week, but by 1962 it had gone up to 29 hours.

Around 1960 the five-day workweek was implemented.[70] Accordingly, the Central Statistics Bureau concluded, "It is not impossible that this development has to be interpreted as a typical expression of 'living with prosperity' in which 'leisure time' is considered one of the most valued goods."[71] Next it shows that at the start of the 1960s Dutch women had 25.4 hours of leisure per week, which is significantly less than men, who had 32.4 hours of leisure per week.[72] These data suggest another side of the growth process: although productivity in the market sector and in households strongly increased in the twentieth century, at the start of the twenty-first century the working Dutch population has much less leisure time than earlier generations.

Economic Growth and Environmental Damage

Environmental damage has occurred in all historical eras, and every human activity has some effect on the natural environment. Still, there are reasons to believe that the contemporary environmental problem differs fundamentally from that in earlier periods and that this era's economic growth has led to excessive pressure on nature and the environment. The main reasons for this are the following[73]:

- *Spatial integration.* In the premodern era economic development occurred in a limited number of locations, near deposits of natural resources or sales markets. Because of improved traffic networks, in the course of nineteenth and twentieth centuries economic activities have gradually spread across the entire country. This has strongly increased the pressure on nature.
- *Natural resources usage.* In the premodern economy, growth was mainly based on human power and horse power, as well as on wind and water energy. Because the energy capacity was limited, economic growth could never be substantial. This changed in the course of the nineteenth and twentieth centuries, when the discovery of large quantities of coal, oil, and natural gas facilitated virtually unlimited growth of the economy. The burning of these fossil fuels, however, has caused severe environmental problems such as acid rain and additional global warming.
- *Technological development.* Modern economic growth is mainly based on technologies and production processes that are highly polluting and generate a great deal of waste. For instance, the strong growth of productivity

The beach near Scheveningen on a sunny day in the spring of 1935. In the course of the twentieth century, people got increasing amounts of leisure time, so that they could plan more vacation trips. The seashore became a major destination.

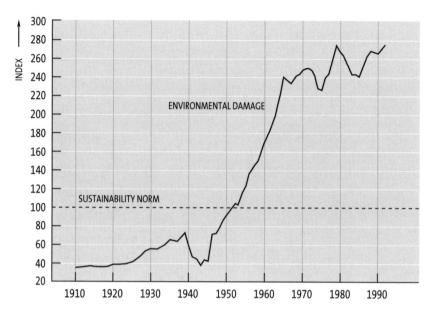

Graph 8.3. Volume of environmental damage in the Netherlands, 1910–1995, indexed to a sustainability standard of 100.
(For sources see page 605)

in agriculture is largely based on large-scale application of fertilizer and chemical insecticides. The petrochemical industry's growth has also had adverse effects on the environment.

In order to analyze the relationship between economic growth and environmental damage in the long run an environmental damage index has been put together that covers nearly the entire twentieth century (see graph 8.3).[74] Even though environmental damage estimates are complicated and the margins of error substantial, the general trend identified can be viewed as reliable. The graph's environmental damage involves the emission of harmful gases such as carbon dioxide, sulfur dioxide, and nitrogen oxides, which are in large measure responsible for global warming and acidification of the natural environment. The graph also shows the increase in the volume of waste: manure surplus produced by the dairy sector, polluted soil near industrial companies, and the growing mountain of garbage generated by households. Finally, estimates are made of the coal, petroleum and natural gas quantities that are used up in production.

The horizontal line (the sustainability standard) in graph 8.3 indicates that up to this level pollution does not exceed the sustainability standard as

Staff of the Inspectorate of Electro-Technical Materials (KEMA) study a power plant's air pollution in 1977. Environmental noise and damage predate the modern industrial era, but the large-scale concentration of modern industrial activity and the use of new substances did mark a new phase in the volume of pollution. Initially concerns about air pollution were largely limited to problems it could cause within company premises, while outside pollution was only noted if, like soot or smoke, it was visible. Not until the 1960s there was systematic attention to this kind of pollution. Mapping it proved easier than eliminating it, though.

Table 8.12: Individual sectors' environmental damage as percentage of total damage, 1950–1995

	1950	1960	1970	1980	1990	1995
Passenger transport	1.1	1.2	2.0	4.2	3.2	2.8
Goods transport	3.4	3.8	4.5	3.9	2.9	5.4
Utility companies	7.2	7.2	7.3	8.0	4.3	3.3
Other companies	72.6	72.9	69.8	68.0	70.7	66.7
Families	15,7	14,9	16,3	15,9	18.9	21.7
Total	100	100	100	100	100	100
Consumers' share	16.8	16.1	18.3	20.1	22.1	24.5

(For sources see page 605)

formulated in Dutch national environmental policy plans. In other words, since the early 1950s the damage to the environment has reached such a level that the standard of sustainable economic growth can no longer be met; our natural resources have been used up recklessly. Although there was a stabilization of the damage to the environment in the course of the 1980s and 1990s, the level still continues to be unsustainable. The plateauing of the growth of environmental damage can be explained by two factors. First, a number of environmental problems are actually diminishing in size, particularly the problem of acidification caused by the emission of sulfur dioxide. The introduction of catalytic converters on vehicle engines has substantially reduced the emission of this poisonous gas. Second, since the 1960s the economic emphasis has increasingly shifted toward the services sector, one that on average is less polluting than the industrial sector.

Table 8.12 gives an overview of the degree in which the various sectors are responsible for the environmental problem. First considering the role of business and industry, as previously observed, in the twentieth century the Dutch economy used up ever more fossil fuels such as coal, oil, and natural gas. Notably the chemical industry, the metal industry, greenhouse agriculture, and transport and hauling saw strong growth as a result of the differential energy price policy pursued by the government in these years.[75] This policy led to high productivity growth in these sectors, but the burning of fossil fuels also resulted in substantial air pollution. This increased problems such as acidification and global warming.

The growing environmental problem was not caused only by industry. Consumers contributed their share as well. Table 8.12 shows that their contribution to total environmental damage went up strongly in the period from

1950 to 1995. This development can be explained in part by economic factors. Per capita consumption exploded after World War II. Christian Pfister has analyzed the postwar environmental problem in terms of what he calls the 1950s syndrome.[76] He, too, views the fifties as a watershed in the relationship of economic growth to environmental damage. His analysis strongly emphasizes postwar sociocultural changes in the Western world, pointing in particular to changing consumption patterns such as increased automobile driving by households and the rise of single-person households, which use more energy per capita than multiperson households.

When we consider long-term environmental damage we see that it broke the sustainability limit only after World War II, but economic growth had a negative influence on welfare before this. The problems that became visible at that time had emerged earlier; resulting from the process of industrialization, they had surfaced in all Western countries. In the fifties and sixties, the energy-intensive and polluting sectors in the Netherlands were given a strong boost by specific government policies. The discovery of the natural gas deposits of Slochteren allowed the government to supply the Dutch industrial sector with very cheap energy. Similarly, almost all households were hooked up to the natural gas system, which further raised energy consumption. The ecological effects of this policy were far-reaching, because quite polluting sectors such as the chemical industry, the metal industry, and utilities production saw strong development.

Economic Growth and Welfare in the Netherlands: A Long-Term Perspective

In the previous sections I discussed various welfare aspects of technological development and economic growth. Here I try to combine these various developments in a single indicator, a welfare index, which can be tracked with the growth of GDP. This will make it possible to trace to what extent processes of technological development and economic growth did or did not lead to more welfare. The calculations involved are quite experimental and have a large margin of error.[77]

By comparing the historical development of this index with the conventional national income data, we can see in which periods economic growth has led to either an increase or lowering of welfare. The calculation of welfare starts from the national income data, but they are corrected for factors that either raise welfare or lower it and that are not incorporated into the system of national accounts or this is done incorrectly. The welfare index takes into account not only change in income inequality, household pro-

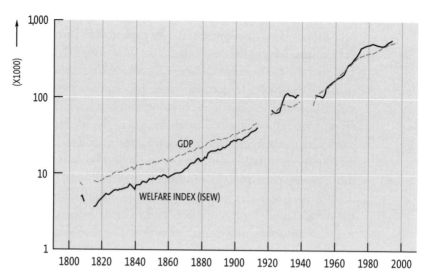

Graph 8-4. Comparison of per capita GDP with Index of Sustainable Economic Welfare (ISEW), 1800–2000, using a semi-logarithmic scale (in constant 1990 guilders)
(For sources see page 606)

duction, and environmental damage but also, for example, the stock of capital per worker. The reason is simple: when economic growth is to be sustainable, the economic growth potential of future generations should not be affected, while growth of the capital goods stock should not be lagging behind that of the labor volume. Nor should investments be based on borrowed foreign currency because of the interest to be paid later on. This is a reason for earning penalty points as well.

Graph 8.4 compares the development of the common GDP with the newly constructed welfare index, known as the ISEW (Index of Sustainable Economic Welfare). Whenever the welfare index is lower than the GDP, it means that economic growth has had negative effects on the population's welfare. Not surprisingly, at the start of the nineteenth century welfare in the Netherlands was way below the GDP, but major progress was made in the course of the nineteenth century. In particular in the period after 1870, the population's welfare increased substantially. This trend persisted in the early twentieth century and even became stronger. Especially during World War I and the 1920s, welfare grew much more rapidly than the GDP. At the start of the 1930s the welfare level of the Dutch population was even higher than could be expected from the GDP data. In the early 1930s, however, the tide turned. During the crisis of the 1930s and World War II, the population's welfare sharply declined, even more than could be expected from the economic growth rates. Yet quite soon Dutch society managed to cope with the

Table 8.13. Gap between GDP and the Index of Sustainable Economic Welfare (ISEW) as a percentage

	Income Inequality	Productive Household Labor	Environmental Damage	Net Capital Growth	Total Gap
1815	− 23	13	0	− 19	− 51
1870	− 21	13	0	− 6	− 36
1913	− 10	15	0	− 8	− 14
1931	14	24	0	5	48
1950	0	22	− 2	− 17	− 4
1982	4	40	− 9	23	40
1992	− 2	28	− 7	8	9

(For sources see page 605)

blows of war and economic crisis. From the 1960s the welfare level was structurally higher than could be expected based on the GDP figures. In the early 1980s this situation came to an end again: the quality of life began to decline.

As pointed out, the major advantage of the welfare index is that the various aspects of welfare, such as the degree of income inequality and the size of environmental damage, can be compared directly with each other simply because they are all expressed in monetary terms. It is possible, then, to trace in detail which factors have been primarily responsible for the index's deviant development. Table 8.13 analyzes the gap between the GDP and the welfare index for a sample of years and which welfare components account for this development. This table needs some clarification. The figures provide an idea of the degree in which prosperity (GDP) and welfare (ISEW) diverge. In 1815, for example, we see that the gap between GDP and the welfare index is 51 percent, GDP being higher than the welfare index. From this we can deduce that in this year welfare was substantially below the GDP level.[78] We can also trace which components account for a major part of the difference between GDP and welfare index. For instance, in 1815 it turns out that high income inequality and negative net capital growth has a strong welfare-lowering effect; together they are responsible for a "reduction" of the GDP of 42 percent. Household labor, by contrast, provides a positive contribution of 13 percent of the regular GDP. Table 8.13 shows, then, the specific contribution of the major welfare components to either raising or sinking economic welfare.[79]

The table shows that the increase of welfare in the nineteenth century can be mainly attributed to the decrease of income inequality and the reduced using up of the stock of net capital goods. These two developments fit in well with the observed transition from premodern to modern economic structure that occurred in this period.[80] The income distribution became less uneven in particular through the revision of the tax system in the 1850s and a reduction of the gap between the reward for skilled and unskilled labor.[81] The stronger capital growth especially in the second half of the nineteenth century can be attributed to various factors, such as the development of relative factor prices and an improvement of the climate for investment, but mainly because in this period mechanized production, partly based on steam technology, became profitable from an economic perspective. As a result, after 1850 the stock of net capital goods began to grow markedly faster than the labor volume, thus meeting the demand for sustainable growth. Accordingly, the largest component of the strong increase of welfare in the 1920s was a further reduction of income inequality.

The second most important factor was the production by family households. Notably the increasing use of durable consumer products led to a substantial increase of household production. Whereas the use of durable consumer products in the late nineteenth century was still very limited, this variable accounts for approximately one quarter of the (positive) difference between welfare and GDP indicators at the start of the 1930s. The decline of the welfare index between 1931 and 1950 can chiefly be attributed to the destruction of capital that occurred during the war and seriously affected the economy's productive power. Furthermore, particularly during the crisis of the 1930s the income distribution became a lot more uneven, which also reduced ownership of durable consumer products.

The strong growth of welfare in the 1945–1982 period mainly occurred in family household production, as well as in net capital growth. In this period the sharp rise of the expenditures for reducing environmental pollution began to have a large impact on welfare. From the early 1980s, GDP growth and increases in welfare no longer went hand in hand. Even though the economy was growing, the extent of environmental damage increased, and income inequality showed a dramatic increase. Both environmental damage and rising levels of income inequality resulted in a considerable slowing down of the growth of welfare. Most recently, income inequality has increased significantly. The gap between the rewards of employers and employees has become wider; at the same time the income difference between skilled and unskilled labor has increased substantially (the *skill premium*). Moreover, spending on the environment is still a major expenditure. Finally it can be observed that there is pressure on both the growth of production within family households and the increase of the net capital goods stock. This last development in particular is worrisome, because this may undermine the economy's long-term capacity to grow.

A photo of a Unilever frozen-foods plant. In the 1960s Unilever entered the market for frozen foods and ready-to-eat-meals. After World War I large companies tried to widen their base through diversification, at first concentrating on exploiting the full potential of already available raw materials or already acquired technical competencies. After World War II this trend gained momentum, as changes in the market and in consumers' preferences were better anticipated and accommodated. For example, manufacturers in the food industry catered to the desire for increased variation in the diet and for convenience foods.

Conclusion

In this chapter I have shown that three stages of economic growth in the Netherlands can be identified in the period between 1890 and the present. Until World War I, the period in which steam technology prevailed, the Dutch economy saw relatively slow growth. Next, until the mid-1970s, Dutch labor productivity rose sharply when compared to international standards, and there was also major interrelated growth of the competitiveness of Dutch business. In recent decades, however, this growth has been subject to erosion. The new information and communication technologies have not yet raised productivity growth in the Netherlands, as has occurred in many other Western countries.

The powerful development of the Dutch economy in the twentieth century can be partly attributed to the rapid diffusion of the electromotor. Also in the area of "non-embodied" technological progress, meaning investments in the knowledge infrastructure, the Dutch economy showed strong development in the period after World War I. The intensity of industrial R&D was very great when compared to that of other countries until the late 1960s. Major investments in creating new knowledge were made principally by the Dutch multinationals.

Unmistakably these multinationals have left their mark on economic and technological developments during much of the twentieth century. In few other countries have large corporations contributed such a significant share to the national economy as in the Netherlands. It is no coincidence that the three sectors in which these multinationals have been particularly active—agriculture and food, the chemical industry, and electrical engineering—were also exactly the sectors in which the Netherlands had a relative edge in world trade. Small and medium-sized companies also cooperated intensively in powerful innovation networks, especially important in agriculture.

The Dutch economic specialization pattern has deep roots in the country's history. A major source of Dutch industrial might has been industrial companies that concentrate on processing raw materials. These inputs partly came from the Dutch agrarian sector itself and were partly imported from the colonies. Thus, in the Netherlands there has been a close link between agriculture, colonial trade, and industrial development. One could make the argument that many of the economic activities in which the Dutch excelled during the twentieth century were already core economic activities in the seventeenth and eighteenth centuries. This prompts the question as to whether Dutch entrepreneurs continue to focus on a revitalization of their tradition of trade capitalism, in which the figure of the businessman-

The former cattle market in Utrecht, in the late 1960s. The new "sacred cow"—the auto—has driven out the original users. Until the 1960s, car ownership was just a dream for the working classes, but this changed drastically as economic progress was made. Growing car ownership and high prices of new cars led to the emergence of a lively market in secondhand vehicles.

entrepreneur prevails and innovation is mainly a matter of good business sense, not ideology—knowing where to buy the technologies you need and only engaging in innovation when no other alternatives are available.[82]

Economic historians may be enthusiastic about the Dutch economy's strong growth until the 1970s and be somber about its more recent sluggish development. However, what is the relevance of indicators such as productivity and competitiveness if we want to learn more about the welfare of the population at large? Is growth the only objective worth pursuing and measuring? This question is quite legitimate because in the 1960s thinking in terms of growth was increasingly criticized. When considering the entire period it may be concluded that economic growth in the long term has indeed led to more welfare. Remarkably, perhaps, the biggest increase of welfare occurred in periods of strong economic growth (1913 to 1929 and 1950 to 1973), whereas there was a noticeable decline of welfare during the economic crises of the 1930s and 1980s. If we focus on the role of technological development in this process of growth, it is striking that Dutch households, partly because of the diffusion of all sorts of technological aids throughout the twentieth century, have increasingly contributed to welfare. Moreover, the introduction of new technologies did not cause larger income inequalities except in the 1920s and the 1990s. These positive effects, however, were in part canceled by the strong growth of energy-intensive technologies, which caused substantial harm to the environment and called for defensive expenditures in order to contain this damage at least somewhat. At the start of the nineteenth century the Netherlands' economic development largely had negative effects on the country's welfare development, but by the end of the twentieth century, the economic growth of the Netherlands had resulted in a reasonably high level of welfare.

Acknowledgements

Many thanks are due to my fellow authors for our pleasant discussions on this project over the years. In particular I would like to mention Johan Schot and Harry Lintsen, with whom I have had many fruitful discussions about the issues addressed in this chapter. As editor of this volume Johan Schot has also greatly contributed to realizing the actual text of this chapter. In addition, I would like to thank my colleagues at Groningen for attentively seeing to it that the final text would have sufficient economics and economic history in it.

Appendix 1: Calculation of Household Production

Despite the fact that many useful and productive activities take place within households, they are rarely accounted for in the calculation of national income. This is because household activities take place outside a market mechanism, implying that no formal monetary value can be attached to them. Of course it is unsatisfactory that activities that are so basic to our social well-being and economic functioning remain invisible on these formal grounds, and efforts have been made to correct this failure to account for household production by linking it to a so-called shadow price, as has also been done regarding environmental damage.

Before assigning a shadow price, first the volume of household production has to be measured. This is fairly simple. Essentially the same method of calculation is followed as for professional household services such as those performed by servants hired by families to perform household tasks and thus are part of the market: it is established how many people perform household production activities and how much time they need for them. This labor input is multiplied by an average wage per hour of work, to arrive at a total wage for household production.

Basically the same calculation is done for the part of household production that does *not* involve the market. The Central Statistics Bureau has data on the number of households and the number of persons per household. For the 1800-1913 period we have to rely on reconstructed national accounts.[83] Next, an estimate is made of the number of hours worked per household, broken down by men and women. From the 1950s onward, we have reliable information on the use of time within households; for the 1930s we have a few varied materials. For the period preceding the 1930s we assumed a fixed number of hours worked. This is reasonably in line with the information we have about technological development in households. Before 1930 most household activities were not yet highly mechanized.

The number of hours worked in family households is multiplied by an average hourly wage, roughly based on the wages paid to servants or other professions that perform activities that are similar to the labor involved in household work. In this way the wage sum of the household sector can be calculated. As is the case with the output of servants as reflected in national accounts, it is assumed that the wage income adequately reflects production. To do justice to technological developments and the ensuing labor productivity increases from a time series perspective, an estimate of these technological developments is made. As is done more often in national accounts, the development of realized technological growth is calculated on the basis of the growth of the amount of "capital" per worker, in this case defined as the number of sustainable consumer goods produced per unit of labor. These data can be derived from the national accounts. Goods that do not raise household productivity, such as a car, bicycle, or radio, are included, which implies that the growth of productivity is slightly overestimated.

Appendix 2: Calculation of Costs of Environmental Damage

The calculation of the costs of environmental damage, and in particular the macroeconomic costs as a percentage of the GNP, is still in its infancy. Calculations that are made commonly concern the most recent period, but the data presented in this chapter are for a much longer time frame. Internationally there is as yet no agreement on how environmental damage should be quantified. The calculations presented here have been presented to colleagues both at home and abroad who specialize in this area. Still, the calculation's results are only a rough indication of the extent of environmental damage.

The calculation of environmental damage involved several steps. First, volume data for different categories of environmental pollution were established, following the approach of the Central Statistics Bureau, which isolates several environmental themes: global warming, acidification of surface water, manure surpluses, and the waste problem. Acidification was quantified by measuring the emission of poisonous gases that would be responsible for these environmental problems, primarily carbon dioxide (CO_2) for global warming and sulfur dioxide (SO_2) and sodium oxides (NO_x) for acidification. In line with common practice, series of data on the emission of these poisonous gases (so-called emission series) are calculated on the basis of energy usage. Data regarding energy consumption are itemized for each type of energy (coal, petroleum, natural gas, and electricity) and for each sector of use (agriculture, industry, utilities, transport, households, and others). These data could be derived from energy statistics collected by the Central Statistics Bureau. The emission series were multiplied by emission factors, or conversion factors, which allow calculation of how much poisonous gas (CO_2, SO_2 or NO_x) is released in burning fossil fuels, for each gas. This procedure is followed in order to do justice to the differences in pollution intensity between the various forms of burning energy. Based on conversion factors derived from Dutch national accounts, the series concerning the emission of various gases are weighed in terms of total series regarding global warming and acidification. The overfertilization problem has been calculated on the basis of agricultural statistics collected by the Central Statistics Bureau; the series on overfertilization, like those for environmental problems, was calculated according to the National Accounting Matrix for Environmental Accounting) of the Central Statistics Bureau. Historical series regarding both household and industrial waste are only available for the most recent period. For periods before 1970 it is assumed that the growth of waste went hand in hand with the development of industrial production and household consumption. Official Central Statistic Bureau figures were used whenever possible. For the period from 1800 to 1913, the historical national accounts were used.[84]

These partial series are weighed and combined, resulting in a single total index for environmental damage, whereby assigning a relative value to each environmental problem is a challenge. It is assumed that the problems seen as most serious by society should be given most weight. Weighting factors were derived from the various versions of the *National Environmental Policy Plan*, which indicates costs that are committed to tackling the environmental problems involved, on the assumption that the government's prioritizing of efforts to reduce various forms of environmental pollution is a valid reflection of society's "environmental demand."

Finally, the total index of environmental damage thus calculated is set alongside a sustainability standard, a boundary that should not be crossed because otherwise the sustainability of economic growth will be at issue and the natural living environment will be affected in irresponsible ways. Of course, the definition of a sustainability standard is also a difficult matter given this concept's divergent definitions. Here too it was decided to start from the standards articulated in the *National Environmental Policy Plan*, which again were assumed to be a reasonable reflection of ideas in society concerning sustainability. By relating the historical series of the total environmental damage to the sustainability standard (see graph 8.3) it can be determined at which moment the environmental damage surpassed the sustainability standard. From that moment—in the early 1950s—economic growth went hand in hand with declining economic welfare, resulting from increasing environmental damage.

Finally a *price* must be assigned to the environmental damage, so that we can express this damage as a percentage of the GNP. This part of the calculation is the hardest because the environment is not a product that is traded in the market and for which a market price is paid. Yet it is possible to determine a shadow price by means of a detour. On the basis of the *National Environmental Policy Plan* we know which costs have to be made to bring back the current level of environmental pollution to the sustainability standard. In addition we know in which period the current environmental problem emerged: after the early 1950s. The total amount of environmental damage in the final year may then be reversed to the preceding years in which the damage accumulated. Because the estimates of the volume of environmental damage are itemized for the various economic sectors, it can be traced which sectors of economic life are responsible for the environmental problems that emerged. This will make it possible to do a good cost-benefit analysis of the economic growth process.

Notes

1 The gross domestic product (GDP) is used as a standard for measuring a country's economic achievements and often also as a standard for comparing welfare among countries. The calculation is based on the outputs of both the private sector and the government. The GDP can be seen as the sum total of the added value produced by agriculture, industry, and services (including government). Data on the Dutch GDP are derived from R. J. van der Bie and J. P. Smits, *Tweehonderd jaar statistiek in tijdreeksen, 1800–1999* (The Hague: Centraal Bureau Statistiek and University of Groningen, 2001). Between 1820 and 1913 the highest per capita annual growth of production was 0.9 percent, and the average growth rate in the period from 1913 to 1995 is 1.8 percent. Historical GDP data are downloadable at http://www.ggdc.net/ databases/hna.htm

2 On the development of wages and income distribution, see Jan Luiten van Zanden, "De egalitaire revolutie van de twintigste eeuw," in Corrie van Eijl, Lex Heerma van Voss, and Piet de Rooy, eds., *Sociaal Nederland: Contouren van de twintigste eeuw* (Amsterdam: Aksant, 2001), 187–200. See also Hettie Pott-Buter, "Arbeid en inkomen," in Hettie Pott-Buter and Kea Tijdens, eds., *Vrouwen: Leven en werk in de twintigste eeuw* (Amsterdam: Amsterdam University Press, 1998), 33–56.

3 Joel Mokyr, *The Lever of Richness: Technological Creativity and Economic Progress* (Oxford: Oxford University Press, 1990), 12–16.

4 Chris Freeman and Luc Soete, *The Economics of Industrial Innovation* (London: Routledge, 1997), 17–22. See also the most recent articulation of this vision in Chris Freeman and Francesco Louçã, *As Time Goes By: From the Industrial Revolutions to the Information Revolution* (Oxford: Oxford University Press, 2001).

5 The TFP is calculated by means of so-called "growth accounting." Essentially it is based on factorizing the growth of the GDP. First it is traced to which extent growth can simply be explained through an increase of the input of labor and capital. The remaining part that is not accounted for and that therefore can be seen as an increase of the efficiency with which labor and capital are deployed in the production process is called TFP. For a further explanation of this issue as well as a consideration of the analytical meaning of the TFP concept, see Angus Maddison, *Dynamic Forces in Capitalist Development: A Long-Run Comparative View* (Oxford: Oxford University Press, 1991), 157–160.

6 Jan Pieter Smits, Herman de Jong, and Bart van Ark, *Three Phases of Dutch Economic*

Growth and Technological Change, 1815–1997 (Groningen: University of Groningen, Groningen Growth and Development Center, 1999).

7 The analysis was performed with the help of the Chow test; see G. C. Chow, "Test of Equality Between Sets of Coefficients in Two Linear Regressions," *Econometrica* 28 no. 3 (1960): 591–605.

8 See Jan Pieter Smits, Edwin Horlings, and Jan Luiten van Zanden, *Dutch GNP and Its Components, 1800–1913* (Groningen: University of Groningen, Groningen Growth and Development Center, 2000); see also Jan Luiten van Zanden and Arthur van Riel, *The Strictures of Inheritance: The Dutch Economy in the Nineteenth Century* (Princeton: Princeton University Press, 2005) 263 passim.

9 The Netherlands' limited share of the total number of patents is shown in particular in the light of the fact that in the twentieth century the Netherlands contributed 2 to 3 percent of global production. This estimate is based on Maddison, *Dynamic Forces in Capitalist Development*.

10 This observation agrees well with observations in the international literature. Robert Gordon, "Interpreting the 'One Big Wave' in U.S. Long-Term Productivity Growth," in Bart van Ark, Simon K. Kuipers, and Gerard H. Kuper, eds., *Productivity, Technology and Economic Growth* (Boston: Kluwer Academic Publishers, 2000), claims that the strong growth of labor productivity until the early 1970s has been a unique phenomenon in history. He feels that economic growth was stimulated through such spectacular breakthroughs that productivity growth reached record highs, thus also having long-term after-effects. Drawing on innovation research by Alfred Kleinknecht, Gordon shows that the number of product and process innovations realized in the 1920s and 1930s was unprecedented. See Alfred Kleinknecht, *Innovation Patterns in Crisis and Prosperity: Schumpeter's Long Cycle Reconsidered* (London: Macmillan, 1987).

11 Maddison, *Dynamic Forces in Capitalist Development*, 6–7.

12 E. Homburg and H. van Zon, "Grootschalig produceren: Superfosfaat en zwavelzuur, 1890–1940," in J. W. Schot et al., *Techniek in Nederland*, vol. 2, 281–284.

13 For an overview, see A. H. van Otterloo, "Nieuw producten, schakels en regimes, 1890–1920," in Schot et al., *Techniek in Nederland*, vol. 3, 249–261. See also Marlou Schrover, *Voedings- en genotmiddelenindustrie: Een geschiedenis en bronnenoverzicht* (Amsterdam: Nederlandsch Economisch-Historisch Archief, 1993).

14 On the "second industrial revolution," see chapter 4 of this volume, in which the origin and meaning of the term is explained: "The American business historian Alfred Chandler has outlined an influential model of historical development in which the tendency toward scale increases plays an important role. In his view, the years between 1890 and 1930 are the crucial period, one that some even refer to as the 'second industrial revolution.' During this time, all sorts of factors—such as better communications, improved technologies and the rise of mass markets—created the preconditions for large-scale production."

15 E. Heckscher, "The Effect of Foreign Trade on the Distribution of income," *Ekonomisk Tidskrift* 21 (1919): 497-512; Bertil Ohlin, *Inter-regional and International Trade* (Cambridge, Mass.: Harvard Economic Studies, 1933).

16 This agrees well with findings of evolutionary economists such as Giovanni Dosi,

Keith Pavitt, and Luc Soete, who believe that trade specialization can be largely explained by technology factors. They observe a major interrelationship between market shares of product groups and levels of relative labor productivity; see Giovanni Dosi, Keith Pavitt, and Luc Soete, *The Economics of Technical Change and International Trade* (New York: Harvester Wheatsheaf, 1990).

17 J. Bieleman, "De landbouw en het sociaal-economisch krachtenveld," in Schot et al., *Techniek in Nederland*, vol. 3, 19. See also J. de Vries, "The Decline and Rise of the Dutch Economy, 1675–1900," in Gary Saxonhouse and Gavin Wright, eds., *Technique, Spirit and Form in the Making of Modern Economies: Essays in Honor of William N. Parker* (Greenwich, Conn.: JAI Press, 1984), 149–189; Jan Luiten van Zanden, *The Transformation of European Agriculture: The Case of the Netherlands* (Amsterdam: Nederlandsch Economisch-Historisch Archief, 1994); Merijn Knibbe, *Agriculture in the Netherlands, 1851–1950: Production and Institutional Change* (Amsterdam: Nederlandsch Economisch-Historisch Archief, 1993), 193–195.

18 H. J. de Jong, *De Nederlandse industrie 1913–1965: Een vergelijkende analyse op basis van de productiestatistieken* (Amsterdam: 1999), chapter 2. See also H. J. de Jong, *Catching Up Twice: The Nature of Dutch Industrial Growth During the 20th Century in a Comparative Perspective* (Berlin: Akademie Verlag, 2003).

19 De Jong, *Nederlandse industrie*, 89–90.

20 See also H. van Driel and J. W. Schot, "Het onstaan van een gemechaniseerde massagoedhaven in Rotterdam," in Schot et al., *Techniek in Nederland*, vol. 5, 75–95.

21 Trade data derived from Maddison, *Dynamic Forces in Capitalist Development*.

22 J. A. de Jonge, *De industrialisatie in Nederland tussen 1850 en 1914* (reprint, Nijmegen: SUN 1976).

23 J. P. Smits, "The Determinants of Productivity Growth in Dutch Manufacturing, 1815–1913," *European Review of Economic History* 4, no. 2 (2000): 223–246.

24 J. Bieleman, "Boeren werd *agri-business*: Een synthese," in Schot et al., *Techniek in Nederland*, vol. 3, 232.

25 On the concept "key technology," see J. W. Schot, H. W. Lintsen, and A. Rip, "Betwiste modernisering," in Schot et al., *Techniek in Nederland*, vol. 2, 22–25.

26 Ronald M. Albers, *Machinery Investment and Economic Growth: The Dynamics of Dutch Development, 1800–1913* (Amsterdam: Nederlandsch Economisch-Historisch Archief, 2002). Total machinery also includes electrical machines as well as more traditional forms such as water mills and windmills. The increase of steam in total machinery is calculated as the number of horsepower of steam engines as a percentage of total horsepower of all types of machinery.

27 H. W. Lintsen, "Een land zonder stoom," in H. W. Lintsen, ed., *Geschiedenis van de techniek in Nederland: De wording van een moderne samenleving, 1800–1890*, 6 vols. (Zutphen: Walburg Pers, 1995), vol. 6, 51–63, and H. W. Lintsen, "Een land met stoom," in Lintsen, *Geschiedenis van de techniek in Nederland*, vol. 6, 191–205.

28 For an analysis of the limitations of the use of steam engines for small companies, see M. Davids, "Van Stoom naar stroom: De veranderingen in aandrijfkracht in de industrie," in Schot et al., *Techniek in Nederland*, vol. 6, 271–283.

29 See, for example, the international standard work in this area: S. Schurr and B. C. Netschert, *Energy in the American Economy, 1850–1975: An Economic Study of Its His-*

tory and Prospects (Baltimore: John Hopkins University Press, 1960). For the definition of the concept of "second industrial revolution," see Schot, Lintsen, and Rip,
"Betwiste modernisering," 21.

30 Davids, "Van stoom naar stroom," 276; De Jong, *Nederlandse industrie*, 184–187.

31 De Jong, *Nederlandse industrie*, 190.

32 A. de Graaf, "De industrie," in P. B. Kreukniet, ed., *De Nederlandse volkshuishouding
tussen de twee wereldoorlogen: Bijdragen tot de sociaal-economische vernieuwing*,
vol. 8 (Utrecht: Spectrum, 1952), 18.

33 Ronald Jan van der Bie, *"Een doorlopende grote roes": De economische ontwikkeling
van Nederland, 1913–1921* (Amsterdam: Tinbergen Institute, 1995).

34 Ivo Kuijpers, *In de schaduw van de grote oorlog: De Nederlandse arbeidersbeweging
en de overheid, 1914–1920* (Amsterdam: Aksant, 2002).

35 R. M. Albers and H. J. de Jong, "Industriële groei in Nederland, 1913–1929: Een verkenning," *NEHA-Jaarboek voor economische, bedrijfs- en techniekgeschiedenis* 57 (1994):
444–490.

36 For an overview of these developments, see chapter 5 in this volume. See also Jasper
Faber, *Kennisverwerving in de Nederlandse industrie, 1870–1970* (Amsterdam: Aksant,
2001), especially chapter 2.

37 Faber, *Kennisverwerving in de Nederlandse industrie*, 33.

38 Ibid., 262.

39 For these figures and other data on technical daytime education, see chapter 5 in
this volume and Bart van Ark and Herman de Jong, "Accounting for Economic
Growth in the Netherlands Since 1913," *Economic and Social History in the Netherlands*
7 (1996): 223.

40 Van Ark and De Jong, "Accounting for Economic Growth in the Netherlands," 237–
239.

41 Jan Luiten van Zanden, *The Economic History of the Netherlands 1914–1995* (London:
Routledge, 1997) 40; Centraal Bureau Statistiek, *Kennis en economie 1999: Onderzoek
en innovatie in Nederland* (Voorburg: Centraal Bureau Statistiek, 1999), 89–90.

42 Faber, *Kennisverwerving in de Nederlandse industrie*, 39, table 2.1.

43 For a development of some of these examples, see chapter 9 in this volume. For
more details on network formation see chapter 5 in this volume.

44 De Jong, *Nederlandse industrie*, 178–181.

45 J. P. Smits, "Economische ontwikkeling, 1800–1995," in Ronald van der Bie and Pit
Dehing, eds., *Nationaal goed: Feiten en cijfers over onze samenleving (ca.), 1800–1999*,
published to celebrate the centenary of the Centraal Bureau Statistiek (Voorburg:
CBS, 1999).

46 See also J. L. van Zanden, *De dans om de gouden standaard: Economisch beleid in de
depressie van de jaren dertig* (Amsterdam: Vrije Universiteit, 1988).

47 Hein A. M. Klemann, *Nederland 1938–1948: Economie en samenleving in jaren van
oorlog en bezetting* (Amsterdam: Boom, 2002).

48 Economic Cooperation Administration, Technical Assistance Program, "The Way
America Works," report of the tour of the first Dutch labor team to the United States
of America (November–December, 1949; n.p. [1950]). See also Frank Inklaar, *Van
Amerika geleerd: Marshall-hulp en kennisimport in Nederland* (The Hague: SDU, 1997),

especially 327–345.

49 De Jong, *Nederlandse industrie*, 291–300.

50 J. P. Smits and B. P. A. Gales, "Olie en gas," in Schot et al., *Techniek in Nederland*, vol. 2, 87.

51 Dutch Ministry of Economic Affairs, *Toets op het concurrentievermogen: Achter-grondrapportage* (The Hague: 1995), 45.

52 Centraal Bureau Statistiek, *De Nederlandse energiehuishouding: Uitkomsten van maand- en kwartaaltellingen* (Voorburg: 1947–1971).

53 J. P. Smits, "Omvang en oorsprong van de milieuschade, 1910–1995," in Van der Bie and Dehing, *Nationaal goed*, 235–254.

54 Centraal Bureau Statistiek, *Nationale rekeningen* (Voorburg: 1980-1986)

55 P. Patel and K. Pavitt, "Uneven (and Divergent) Technological Accumulation Among Advanced Countries: Evidence and a Framework of Explanation," in Daniele Archibugi and Jonathan Michie, eds., *Trade, Growth and Technical Change* (Cambridge: Cambridge University Press, 1998), 55–82.

56 Bart van Ark, *Sectoral Growth Accounting and Structural Change in Postwar Europe* (Groningen: University of Groningen, Groningen Growth and Development Center, 1995).

57 Centraal Plan Bureau, "Groei arbeidsproductiviteit op sectorniveau en bedrijfs-niveau," in *Centraal Economisch Plan 1998* (The Hague: CPB, 1998), 125–137; Bart van Ark, Lourens Broersma, and Gjalt de Jong, *Innovation in Services: Overview of Data Sources and Analytical Structures* (Groningen: University of Groningen, Groningen Growth and Development Center, 1999).

58 Van Ark, Broersma, and De Jong, *Innovation in Services*.

59 Simon Kuznets, *Modern Economic Growth: Rate, Structure and Spread* (New Haven: Yale University Press, 1966).

60 A. van den Berg and J. Hartog, "Honderd jaar ongelijkheid: Inkomensverschillen sinds het einde van de negentiende eeuw," in Van der Bie and Dehing, *Nationaal goed*, 109–126.

61 Figures about the categorical income distribution—the distribution of the national income across wage and capital income—are derived from Van der Bie and Smits, *Tweehonderd jaar statistiek in tijdreeksen*.

62 Van Zanden, "Egalitaire evolutie van de twintigste eeuw," 195.

63 Claudia Goldin and Laurens F. Katz, "Origins of Technology-Skill Complementarity," *Quarterly Journal of Economics* 113, no. 3 (1998): 683–732.

64 Adrian Wood, *North-South Trade, Employment and Inequality: Changing Fortunes in a Skill-Driven World* (Oxford: Oxford University Press, 1994).

65 See Maddison, *Dynamic Forces in Capitalist Development*, 326.

66 For the calculation of household production see appendix 1.

67 The classic study on this subject is Ruth Schwarz Cowan, *More Work for Mother: The Ironies of Household Technology from the Open Hearth to the Microwave* (New York: Basic Books, 1983). For an excellent overview from an economic-historical perspective, see Joel Mokyr, *The Gifts of Athena: Historical Origins of the Knowledge Economy* (Princeton: Princeton University Press, 2002), 198–215. His thesis is that women opted for household work because it would improve the entire family's health.

Women made this choice partly as a result of new insights about the role of dirt, nutrition, and childcare in the etiology of disease. Mokyr also discusses extensively the extant studies of the ways women spend their leisure time.

68 See, for instance, international statistics cited by Maddison, *Dynamic Forces in Capitalist Development*, 270–271.

69 Pascale Peters, "Vrije tijd als sluitpost," *Index* 7, no. 9 (2000): 2–3.

70 Centraal Bureau Statistiek, *Vrije-tijdsbesteding in Nederland, 1962–1963* (Zeist: 1964 to 1966).

71 Ibid.

72 Ibid.

73 J. P. Smits, "Economische groei en de aantasting van natuurlijke hulpbronnen: Theoretische beschouwingen met een toespitsing op de Nederlandse situatie," *NEHA-Bulletin* 11, no. 1 (1997): 3–33.

74 Smits, "Omvang en oorsprong van de milieuschade," 235–254.

75 For an analysis of growing production and use of energy-consuming devices in the Netherlands, see also G. P. J. Verbong et al., "Ter introductie," in Schot et al. *Techniek in Nederland*, vol. 2, 11–15.

76 Christian Pfister, *Das 1950er Syndrom: Der Weg in die Konsumgesellschaft* (Bern: Haupt, 1996).

77 For more explanation of the calculation and the welfare index, see E. Horlings and J. P. Smits, "De welzijnseffecten van economische groei in Nederland, 1800–2000," *Tijdschrift voor Sociale Geschiedenis* 27, no. 3 (2001): 266–280. The welfare index, called the index of sustainable economic welfare, was developed by Herman E. Daly and John B. Cobb Jr., in *For the Common Good: Redirecting the Economy Towards Community, the Environment and a Sustainable Future* (London: Beacon Press, 1989).

78 This gap is calculated as follows: gap = ([ISEW - GDP] / GDP) X 100.

79 It should be pointed out that the components from the first four columns do not add up to the last column with sum totals. Among other things this is because a number of components are not reflected in this table. See also Horlings and Smits, "Welzijnseffecten van economische groei in Nederland," 277.

80 E. Horlings and J. P. Smits, "A Comparison of the Pattern of Growth and Structural Change in the Netherlands and Belgium, 1800–1913," *Jahrbuch für Wirtschaftsgeschichte* 38, no. 2 (1997): 83–106.

81 A. Vermaas, S. W. Verstegen, and J. L. van Zanden, "Income Inequality in the 19th Century: Evidence for a Kuznets Curve?," in Lee Soltow and Jan Luiten van Zanden, eds., *Income and Wealth Inequality in the Netherlands 16th-20th Century* (Amsterdam: Spinhuis, 1998), 145–174.

82 This suggestion was made earlier by J. W. Schot, "Innoveren in Nederland," in Lintsen, *Geschiedenis van de techniek in Nederland*, vol. 6, 236, partly based on J. L. van Zanden, *Arbeid tijdens het handelskapitalisme: Opkomst en neergang van de Hollandse economie, 1350–1850* (Bergen: Verloren, 1991), 16.

83 See Smits, Horlings, and van Zanden, *Dutch GNP and Its Components*.

84 Ibid.

A DSM production location, near the A76 highway, in 1968. In 1900, few inhabitants of this area in the south of Limburg could have imagined that two generations later the landscape would be dominated by busy highways and large industrial complexes. By the sixties the Netherlands had become an industrial nation, but from the mid-1960s the importance of industry to employment declined again. Such large-scale chemical complexes generally offered employment only to a limited number of people.

9 Technology, Industrialization, and the Contested Modernization of the Netherlands

Johan Schot and Dick van Lente

In the twentieth century the Netherlands developed into a genuine industrial nation again, just as it had been in the seventeenth century, during its Golden Age. Many Dutch commentators and politicians had expressed a longing for modernization through industrialization since the late eighteenth century,[1] but between 1820 and 1860 the industry's contribution to the national income in fact decreased, while that of agriculture increased and that of the services sector continued to be at a high level.[2] From the 1860s onward, the importance of industry began to go up, and in a process spanning a full century industrial activity even turned into the engine of Dutch economic and social development. As a result of this long wave of industrialization, the country's industrial growth outpaced its overall economic growth.[3] Particularly after World War II, the renewed emphasis on industrialization also caused a major shift in employment from agriculture to industry. In the late nineteenth century 30 percent of the Dutch working population was active in the agrarian sector, but this figure dropped to slightly over 10 percent by 1960, partly because of the rapid mechanization of Dutch agriculture. In the postwar years sustained industrialization was also accompanied by unprecedented mass consumption, even though, we argue below, consumption patterns did not merely follow changes on the supply side, as many economic historians incorrectly assume.

To characterize the significance of technological development in Dutch history, this chapter concentrates on changes in the nation's economic structure during the twentieth century, notably its striking industrialization and the strong growth of industry, agriculture, and consumption.[4] Although we will discuss these areas separately, our concern is with the more general picture that thus emerges. This strategy allows us to gain more insight into specific shifts or breaks and to identify what is most characteristic in the Dutch economic modernization process. Industrialization serves as a particularly useful focus, not just because of obvious differences in this respect between the twentieth and the nineteenth centuries in the Dutch case, but also because the country's fairly late industrialization trajectory did not always follow a straight line. There has been much debate on the nature of Dutch

Specialization in agriculture took off starting around 1850, but only after 1950 did this process pick up steam. The meat and dairy sectors saw the emergence of specialized businesses such as hog, dairy, livestock, and poultry farms. In 1988 meat and dairy farms were responsible for 61 percent of Dutch agricultural production.

industrialization, its role in the country's modernization, and the various alternative paths explored. We proceed as follows. For both agriculture and industry we first outline a standardized account of the main developments as pictured in the current socioeconomic scholarship. Subsequently we specific analyse how a history-of-technology perspective enriches (and sometime contradicts) this picture. When we discuss industry, we also include consumption, since we seek to show the relationships between industrial development and what happened in the market. This chapter has a double aim: It seeks to present a history of Dutch industrialization from a history-of-technology perspective and also to explore the value added by use of this perspective compared to earlier interpretations. Thus, it should be read as a historiographical exercise that shows what a history-of-technology perspective contributed to a general economic history

Industrializing Agriculture

The Netherlands has always had a major agricultural sector, and from 1850 this sector became strongly geared toward innovation because the market grew rapidly as a result of demographic developments and rising wage levels at home and abroad (notably in Great Britain). This growth mainly occurred in cattle breeding and truck farming, which caused Dutch farmers to focus more strongly on these sectors. The agricultural crisis of the 1880s struck Dutch arable farming with particular force,[5] as farmers had to contend with huge price drops resulting from very inexpensive grain imports from the United States and other countries, made possible in part by the emergence of fast transoceanic steamer connections. But cattle breeding also came under pressure for other reasons: overproduction and low prices. The agrarian sector used to react to such sales crises by cutting back production which would help to make prizes to go up again. In this period, however, a new response emerged. Farmers embraced further specialization in truck farming and cattle breeding, a response made possible by the availability of cheaper fertilizer and feed (grain and corn), which allowed for more intensive land use.

The first major structural change in Dutch agriculture thus involved a stronger emphasis on exports of dairy and truck farming products and imports of large quantities of raw materials (fertilizer, feed). Although this focus on exports reflected a well-established tradition for Dutch farmers from the coastal regions, its enhancement was made possible by new developments allowing for specialization in livestock, poultry, hogs, cheese, and butter, as well as vegetables, fruits, and flower bulbs. And on the sandy soils of the inland regions farmers began to shift their activities also toward cattle breeding at the expense of arable farming. They had always produced crops such as rye, buckwheat, and potatoes for the market, but now they used these crops increasingly as feed. Many farmers did not specialize completely in one type of farming. Instead they continued to combine arable farming with cattle breeding, which changed only after 1950, when they began to opt for dairy, poultry, hog farming, or cattle breeding exclusively. In 1988 cattle breeding (for dairy and meat production) accounted for 61 percent of Dutch agricultural production, while truck farming, including greenhouse agriculture, contributed 30 percent and arable farming only 9 percent.

A second major structural development in Dutch agriculture is that, surprisingly, after 1880, small farm businesses became a more dominant factor while increases in production and productivity were occurring—a development that lasted until after World War II. Particularly in the coastal provinces farmers began to rely less on agricultural workers after 1880; farming became a family business, even though wage labor continued to play a role. Elsewhere, notably in the inland regions with sandy soils, wage labor was minimal.[6] These regions mainly had small farms in terms of acreage, and in this period their number increased and the number of large farms declined.[7] Consequently, between 1900 and 1950 the number of individuals employed in agriculture increased as well, from 552,000 to 747,000.[8] This increase did not keep pace with the country's population growth, so the percentage of those

This photo, from around 1950, shows mowing being done with a mechanical mower, with the help of horse power; two farmhands are needed for gathering the hay into sheaves. Until the 1950s most agriculture was labor-intensive, even though in arable farming machines were increasingly introduced for the most time-consuming activities such as sowing and harvesting. After 1950 horses and farmhands would soon disappear from the farm.

active in agriculture as part of the total employment dropped, but only slowly until 1950.

After 1950 the size of Dutch farm businesses gradually increased and many small farms disappeared, which strongly reduced agriculture's share in total employment. Although the family farm continued to be dominant,[9] it became integrated much more strongly in a complex network of businesses: food industries, machine builders, producers of input materials such as feed and fertilizers, and providers of services. Concepts such as "agribusiness" and the "agriculture-food cluster" grew common. Dutch businesses competed very well in the world market, notably with cut flowers, flower bulbs, hogs, pork, milk, cream, and fresh tomatoes. So the period around 1950 marked a break in Dutch agriculture: mixed farming disappeared and the sector as a whole embraced "produce, specialize, and rationalize" as its new motto, accompanied by scale increases and labor reductions.

Socioeconomic historians attribute the absence of major scale increases prior to 1950 to the rise of cooperatives and active government intervention.[10] By forming purchasing cooperatives, especially for feed and fertilizer, fairly small farming businesses could still profit from scale advantages. In addition, farmers set up cooperatives for processing and selling dairy products, sugar, and potatoes, as well as cooperative auctions for trade in truck farming products. In response to the crisis of the 1880s, the Dutch government had started to devote more money and attention to agricultural education, research, and information, which gave rise to a well-developed R&D system, the so-called OVO-triptych: research (*onderzoek*), information (*voorlichting*), and education (*onderwijs*). As part of this effort, experiment stations were set up and traveling instructors disseminated agricultural know-how on farms throughout the country. The combination of specialization and institutional development made it possible for Dutch farmers to compete in the world market again.

Because of the generally favorable agrarian economic situation in the wake of the 1880s crisis, the income development of Dutch farmers showed a positive trend. In the course of the 1920s, however, as prices began to go down again, it was once more a challenge for farmers to make a living. Initially these problems could be solved by improvements in productivity. In 1928, however, grain prices began to drop rapidly, showing a decline within three years of about 60 percent. By the end of 1929, the prices of dairy products followed suit. The effect on farmers' income was dramatic.[11] To tackle the crisis, the government adopted a whole series of measures, in close collaboration with the sector. Farmers received subsidies that allowed them to sell their products competitively on the world market, and measures were taken to protect the home market as well.

Why did the Dutch government provide farm businesses—rather than industrial firms—such major support and protection in the 1930s? First, government had gained much experience in intervening in agriculture during the crisis of the 1880s. Second, World War I had led to a guided economy, marked by production regulations, price setting, and regulations pertaining to consumption. The measures taken during the 1930s crisis merely represented a next step, albeit a substantial one, because now the government also intervened in market prices. At the same time, the principle of the government's protecting agriculture was not contested. Finally, the intense electoral

"Eat more Dutch eggs. They are nutritious and tasty." A postcard from the 1920s, part of a government-supported campaign encouraging Dutch people to consume Dutch agricultural products. Such campaigns were a response to a serious agricultural crisis in the late 1920s caused by falling prices. The national government took steps— production limitations and fixed pricing—that deeply affected the agricultural sector.

struggle in the 1930s between confessional and liberal parties for the farmers' vote also contributed to price intervention. It was easy for Dutch farmers, who since the turn of the century had been well organized via cooperatives and associations, to find a willing ear among politicians for their agenda—a phenomenon that after World War II became known as the "green front." Because of subsidies and protection, Dutch farmers earned a fairly good income in the mid-1930s, which kept farmers on their land. The number of businesses, notably very small ones, even increased, and more family members worked on family farms.

Given this situation it was hard to introduce labor-saving technologies, except in arable farming, which was mechanized to a limited extent in order to deal with expensive peak demand for labor during the harvest.

After World War II, overall prosperity began to increase. Of course, farmers wanted to share in this, but earning significantly more income was only possible through substantial productivity growth, which the size of many Dutch farms prevented. Sons of farmers had little prospect of running a profitable farm business of their own, so they increasingly looked for opportunities outside the agricultural sector. Consequently, the number of those employed in agriculture rapidly declined; simultaneously the Dutch government's policies concentrated increasingly on restructuring small farm businesses, lowering cost prices, and raising labor productivity. In the 1950s the government geared this restructuring to increasing the size of small farms, rather than buying farmers out. Buying out small farmers became common practice in the sixties, after the establishment of the agricultural Development and Restructuring Fund (Ontwikkelings- en Saneringsfonds) in 1963. At the same time, the formation of a protected European market improved sales opportunities. This market was in fact a continuation of the subsidy system with which the Dutch had already gained experience in the 1930s. It was no coincidence, then, that precisely the Dutch greatly contributed to developing a European common agricultural policy. Dutch farmers operated successfully not only in this European market but also outside it, in the unprotected world markets, as shown by the comparatively large share of Dutch agricultural products traded on a global scale. This was accompanied by rapid mechanization and specialization geared to grain, dairy cattle, hogs, poultry, and truck farming. Although many small farms disappeared as part of this transformation, most farms were still family run; establishing very large farms was difficult due to the limited availability of land and the high land prices.

In explaining structural developments of Dutch agriculture, most historians have presented the role of technology mainly as an exogenous factor. The introduction of the steamship, for instance, was instrumental in creating

"Fordson: Costs half. Does double." An ad from the 1930s, designed to appeal to the
Dutch market, presents American Fordson tractors as cheap and powerful. The number
of small farms grew during the years of crisis, and mechanization was not a major
focus, resulting in low sales of American Ford tractors in the Netherlands.

larger transatlantic markets and thus in lowering prices of grain by making cheap imports feasible. This provided a major stimulus for specialization in cattle breeding and truck farming, facilitated in part by the introduction of cheaper fertilizer and feed (also made possible by better transport connections). Prior to 1950, cheap labor and high land prices led to an emphasis on use of technologies that helped to raise crop productivity, whereas afterward the rapidly rising wages made mechanization profitable and thus led to an emphasis on the boosting of labor productivity. Many socioeconomic historians supplement this neoclassic economic explanation with an institutional one by pointing to the importance of the OVO-triptych and cooperatives, which not only allowed for effective application of crop technologies and, later, mechanization, but also enabled farmers to profit from scale benefits. They point to two major breaks: the agricultural crisis of the 1880s and the turnaround after 1950, marked by accelerated mechanization and replacement of small-scale mixed farming by large-scale specialized agribusiness.

Inventive Farmers and the Cooperative Alternative

Jan Bieleman and Peter Priester in their history of agricultural technology in the Netherlands, as published in *Techniek in Nederland in de twintigste eeuw* (*Technology in the Netherlands in the Twentieth Century*), confirm that the 1880s and 1950 are major turning points, but they emphasize that many changes had already started in the 1850s.[12] They specify also the role of technology. Technological development was initially aimed at raising the output of small farms by optimizing land use—for example, through the introduction of feed and fertilizer to boost production, drainage of new land, and improved cultivation and breeding techniques, which raised crop yields and milk production. Prior to 1950 the agrarian sector and the Dutch government had little use for mechanization, as suggested by the decreasing imports of farming equipment in the 1930s.[13] Labor-saving technologies were not implemented, even if they were available and known to Dutch farmers. After 1950 however, technological development became geared to reducing labor needs through mechanization. As a result, the total volume of human and animal labor needed in agriculture decreased between 1950 and 1980 by two thirds. While horses almost completely disappeared from the farm, the total assets (machines, buildings, and livestock) saw an 80 percent increase.[14]

The introduction of the tractor became the cornerstone of mechanization. A tractor was only profitable on farms of sufficient size, but its purchase also implied the purchase of a host of other interrelated equipment. In arable farming, the threshing machine, the self-binder, and the mowing machine became integrated in harvesters (combines). As a result, the number of man-hours devoted to grain cultivation dropped from about 195 man-hours per hectare in 1950 to 15 or 20 in 1975.[15] In dairy farming, mechanized milking was introduced, followed by cubicle stalls, milk tanks, milk piping, and tank trucks, resulting in a nearly completely automated chain from stall to milk-processing plant. In 1940, 3 percent of Dutch livestock was milked by machines; in 1970 this figure was 90 percent.[16]

Between 1960 and 1980 the number of hours required annually to milk, tend, and feed one dairy cow declined from 330 to 80.[17] Detailed examples of mechanization and reduced needs for manual and animal labor provided by Bieleman and Priester suggest that frequently these processes involved difficult, altogether new challenges.[18]

Priester and Bieleman also examine the effect of the development of crop and cultivation techniques that were deployed in the period after 1950 to allow for maximum mechanization. Livestock, crops, and the landscape itself were adapted to machines. For example, drainage and land consolidation as farmers exchanged small plots of land to create larger or more continuous parcels improved the land's bearing power and the ability of machines to maneuver. In the context of the OVO effort, breeding farms selected cows that were best adapted to mechanical milking, such as animals that needed no aftermilking or, later on, had udders suitable for milking machines).[19] Experiment stations adapted grain varieties to combine-harvesting and beet seed to precision sowing machines.[20] The increased interweaving of technologies was partly a targeted and coordinated process. Technologies were often selected because of their fit in the ensemble of technologies. For instance, mow-threshing was favored over swath mowing because it tied in better with the development of new wheat varieties.[21]

Finally, Bieleman and Priester discuss the largely underrated role of inventive farmers in the current literature: They picture Dutch farmers as innovators and problem solvers.[22] For a number of motivated and inventive farmers, designing machines may have been even more important than farming itself. A 1961 handbook on agriculture stated that "many improvements are invented by users, rather than, as one perhaps would expect, just by engineers in factories behind their drawing tables.... In our country, too, we can point to countless examples of such improvements."[23] Bieleman and Priester extensively document such innovations by farmers and their diffusion among farmers—small improvements involving tractors, mechanical drainage, plowing techniques, and harvesting of hay rotary mowers and tedders (which spread hay for drying).[24]

These insights from the history of technology enrich and qualify insights in the history of Dutch agriculture generated in the current socioeconomic literature. The role of specific technological developments proves to be more complex and differentiated than assumed in this literature, and all of these developments are also mutually interconnected. However, these additions do not seem to challenge the established account of Dutch agricultural development in the twentieth century. There is one major difference, however. A history of technology approach interprets the nature of the change process differently.

To understand this, we zoom in on 1950 as a major turning point. Historians generally present the development of small farming in the interwar period as a hardly efficient outcome of the 1930s economic crisis.[25] From a history of technology perspective as developed in this volume the notion of efficiency itself is problematized. It is perceived as part of a struggle between various technological alternatives and the social and cultural preferences they embodied. From this vantage point, technological development is conceived as a path in which visions and social choices are articulated and tested. Eventually such a process results in a technological regime whereby social preferences harden and become anchored in specific technologies (products and processes) that are subsequently optimized. In the history of Dutch agriculture we see a struggle between two alternatives: small-scale versus large-scale farming. The first model was that of the small, autonomous farm that could survive very well in the global market with the help of purchasing and processing on a cooperative basis and active government support. This alternative cannot be considered a relic of an old farming system that is bound to vanish, but should be regarded as the outcome of a quite productive strategy developed by Dutch farmers toward the end of the nineteenth century. As a direct response to the agricultural crisis of the 1880s, they had shifted to cooperative purchasing and cooperative selling and processing of agricultural products. This enabled them to stay in charge of processing agricultural products for the international market and keep industrialists from completely dominating the sector through large-scale factory-processing of agricultural products. The success of these farmers' initiatives can be illustrated with reference to the rapid increase of the quantity of milk processed in cooperative dairy factories: 19 percent of the milk in 1895, 48 percent in 1903, and 66 percent in 1910.[26] The cooperative alternative was successful also because the government supported small farmers. It gave rise to a new regime, with technological developments aimed at raising the productivity of land, crops, and livestock.

During the 1930s no one disputed either the support of agriculture in

The Sint Lambertus dairy factory in Veghel in 1926, which was closed down in 1985. In the interwar period the cooperative farm movement was a visible presence in country and city alike. Central purchasing, meat processing and storage, and especially the omnipresent dairy factories testified to the importance and size of the movement. Cooperative purchasing, processing and selling of dairy allowed small businesses to realize benefits of scale.

general or of small farming in particular. In 1934 the Dutch government set up a commission to study the difficult position of small farming businesses. It concluded that under normal conditions small farms should certainly be able to function well, but that temporary financial support was required to deal with the crisis. As a result of this study, in 1936 a separate Small Farms Businesses Service (Dienst Kleine Boerenbedrijven) was set up to provide financial support and information tailored to this type of farming.[27] Dutch agricultural organizations often did not view mechanization as a preferable route toward improvement. When in 1937 the progressive agricultural information officer A. P. van den Ban argued that there was room for only three harvesters in the Netherlands, this should be viewed as a claim about a preferable future rather than as a realistic prediction.[28]

After World War II a new situation gradually emerged in which small farming was increasingly defined as a problem. Government and agricultural organizations put their stakes on raising production and lowering costs via

This 1968 photo of the Bommelerwaard shows a survey being done as part of the process of land consolidation, which often involved careful measuring. Especially after World War II, land consolidation became a major tool in the government's effort to encourage the creation of larger farm businesses. Land consolidation made it possible to improve drainage, bearing power, and the tilling of agricultural land with machines. This alignment of cultivation techniques and mechanization raised labor productivity, even if here and there it led to the destruction of picturesque rural landscapes.

mechanization. They did so from the viewpoint that modernization was inevitable in the face of future international competition, if farmers' income was to be improved. An additional factor was that the government wanted to deploy agriculture's labor potential for industrialization.[29] Initially this policy was not aimed at actually eliminating small farming (businesses of fewer than ten milk cows or five hectares of land), even though the goal was to increase the size of farms.

At the same time, both government and agricultural organizations were working to strengthen cooperative alternatives. For example, in the period from 1947 to 1953 the Dutch government subsidized farming equipment cooperatives, which facilitated mechanization in regions with sandy soil. Small farmers also experimented with shared use of milking machines, to allow dairy farmers with fewer than ten cows to reap the benefits of mechanized milking. The milking machine was mounted on a truck that could serve several farms.[30] In the 1950s the policy aimed at farm expansion in particular concentrated on consolidation of holdings. With improved access to specific plots by means of draining or reparceling the farm fields, machines could be deployed more efficiently.

In the course of the 1950s it became clear that this policy would lead to the demise of small mixed farms. The government, together with agricultural organizations, introduced the model of the specialized large-scale family farm as a new alternative, even if these organizations did not yet openly say that the policy would be to restructure farming in this direction. From the early 1960s on, however, the restructuring of small mixed farms was explicitly stimulated by all parties. For many farmers it became a matter of getting bigger to stay in business or quitting farming altogether. The Development and Restructuring Fund was set up to help increase the size of businesses and restructure small businesses, now openly supported by organized agriculture. The OVO activities were no longer geared toward helping small farmers to succeed. Even cooperative farms, once set up by small farmers, were now geared to producing as inexpensively as possible for mass markets. Although a quite powerful movement of free farmers (*vrije boeren*) denounced this new agricultural policy, while also sharply criticizing the agricultural organizations that no longer served their interests, in the course of the 1960s their resistance lost terrain.[31] Gradually small mixed farming businesses were simply pushed out of the market. In 1950, 31 percent of the farms were smaller than five hectares, but by 1970 only 11 percent were.[32] The new agricultural regime was designed to support large-scale, mechanized and specialized farming only.

Thus we argue that a history-of-technology perspective as developed in this volume allows us to lay bare the stakes for the various actors in this development. Small-scale farming ultimately became untenable not only because labor became more expensive but also because it was an outcome of a struggle between various coalitions of actors who held very different visions of the desired future for Dutch agriculture. This outcome became apparent in the early 1960s. In the interwar period small-scale farming still flourished. Yet we also would like to emphasize that large-scale farming incorporated an important feature of small-scale farming: the emphasis on making farming into a family business.

In the next section we look at industry, following the same narrative structure as that used in the precious section. First we present an overall picture of the principal developments, then we discuss the most prominent interpretations available in the socioeconomic literature, and we finish by reinterpreting the narrative from a history-of-technology perspective.

An Industrial Nation Again, at Last

The most significant increases in Dutch industrial production occurred between 1921 and 1938 and between 1950 and 1965. This growth surpassed that of Germany, Belgium, and the United Kingdom. Until the 1890s the Dutch industrial structure was dominated by the consumer goods industry, which since 1860 had seen rapid growth. It applied to sectors such as printing, diamonds, paper, textiles, breweries, and tobacco. In addition, from 1870 the metal industry and the construction industry were growing, and, after 1890, shipbuilding, machine building, and the chemical industry as well. In 1913 three quarters of the added value in Dutch industry was produced in food and luxury foodstuffs, metal, clothing, and textiles.[33] The first two continued to be strong in the ensuing decades, while over time the clothing and textiles industry lost its number one position to the rapidly growing petroleum and chemical industry. Especially after World War II major expansion occurred in oil refining, bulk chemicals, plastics, and pharmaceuticals.

After 1890, overall industrial activity increased and also large businesses saw strong growth. Between 1870 and 1920 several major companies came into being that were to dominate Dutch industry throughout the twentieth century, including Royal Dutch/Shell, Unilever, and Philips, which became the largest companies and were never challenged in their rankings, followed in size by DSM, Akzo, and Hoogovens.[34] In the 1920s the three leading companies evolved into multinationals with various divisions active in many countries. There was a significant size gap between the six leading companies and all the other large businesses on the list of the one hundred largest companies (in terms of assets and number of employees). Such a wide gap was not present in the United States, England, and Germany;[35] This situation where six large companies set the tone for many other substantially less influential ones defined Dutch industry's characteristic dual structure in the twentieth-century.[36]

Although specific data for a proper analysis of developments in the sector of smaller and medium-sized businesses are largely absent, several general trends can be discerned.[37] For one thing, the relative share of small business employment declined. Between 1889 and 1930 the share of total employment accounted for by companies with fewer than fifty employees went from 84.6 percent to 56.3 percent.[38] By 1950 the percentage of those working in companies with fewer than one hundred employees was 17 percent.[39] In the ensuing years this share would decline further, to 13 percent.[40] The contribution of medium-sized businesses to the economy clearly increased. At the end of the nineteenth century this trend was seen as a threat and a movement to

The Amsterdam diamond-cutting firm Zeldenrust in 1911. Dutch industrial development did not take place equally rapidly in all sectors. The first spurt of industrialization, which took off in 1860, was based on the production of consumer goods. The main sectors affected were textiles and paper, breweries, printing companies, tobacco, and diamonds. Although companies in this sector mechanized in the twentieth century, like Zeldenrust, they would continue to operate on a small scale.

preserve the crafts emerged in the Netherlands (as in other European countries), supported by government.[41] Soon, however, government policy became increasingly geared to supporting innovation and expansion in small and medium-sized companies, rather than to preserving the crafts tradition. Although the success of this new movement geared to medium-sized players is hard to assess on the basis of the available data, it is clear that medium-sized businesses in dynamic branches such as the chemical industry and machine building played a major role.[42]

In the course of the 1960s the share of employees in Dutch industry began to decrease, even though many companies were still flourishing. By the 1970s, however, more and more companies met with setbacks. This caused the demise of many companies in textiles, clothing, and shoes, as well as shipbuilding; the mines in Limburg were shut down and the six leading companies cut many jobs. Still, the long wave of industrialization—which in the Netherlands took a full century—had generated strong energy-intensive and capital-intensive process industries (paper, metal, bulk chemicals, and oil

Coal or ore is loaded onto barges in the Waalhaven in 1928. A loading bridge made it possible to transfer the load from ocean ships directly onto river barges. Traditionally the Netherlands had a strong international services sector in addition to a major agricultural sector. This was in part determined by the country's river delta geography and its colonial ventures, but also by rapid mechanization of port activities, which allowed Rotterdam to develop into a major transit port for Europe.

refining) and agro-industries linked to thriving Dutch agriculture, while in the 1960s energy-intensive industries received an additional boost from the recent availability of cheap natural gas. These industries continued to play an important role in the last decades of the twentieth century.

A major structural feature of Dutch industry was its even distribution throughout the country. Labor-intensive industries (textiles, clothing, leather, electro-technology) were mainly concentrated in Twente and North Brabant, while capital-intensive industries were mainly found in Holland. Modest industrial development took place in river regions (brickworks), mining developed in Limburg, and farming-related industries thrived in North Brabant and East Groningen. In the early 1960s it was still easy to identify the nation's industrial regions on a map, but after the subsequent decline of industrial employment this became increasingly difficult. The restructuring of several major sectors undermined the pattern of regional specialization. In the closing decades of the twentieth century, Dutch regions were no longer as strongly influenced by specific industries.[43]

Socioeconomic historians have identified several factors that account for the comeback of industry and big business in the Netherlands, after its early glory days in the seventeenth century, and the subsequent (partial) deindustrialization. The first major phase of modern industrial growth, from 1860, is mainly related to four developments. First, the domestic market grew because of population growth and rising wages. Second, transport infrastructures expanded significantly (railroad and tram tracks, canals, ports), largely through government investments, which led to market integration. Third, the price of coal, the main industrial fuel, went down. Fourth, industry profited from economic liberalization by a lowering of taxes and duties and a reduction of the country's debt, which stimulated the food and luxury foodstuff industry. The metal industry saw strong growth because of the demand for steam engines, railroad equipment, and ships. The Netherlands also profited from its colonial possessions, which made it possible for a company such as Royal Dutch/Shell to emerge. Not only specifically Dutch developments facilitated growth; the substantial growth of world trade in this period also helped, and Dutch companies with a strong international commercial network were especially well positioned to profit from this.

The ongoing development after 1890 is linked in particular to the so-called "second industrial revolution." The country's late industrialization began to coincide with a new international wave of innovation in oil refining, the chemical industries, steel, electro-technology and machine building (electromotors and internal-combustion engines). The largest, most successful companies in the Netherlands are found precisely in these sectors, as well as in the food industry. This wave of innovation gave a strong impetus to the development of big industry, as socioeconomic historians such as Jan Luiten van Zanden have argued, because the new technologies could only be deployed in a profitable manner in large-scale production, partly because of the high investment costs (*economies of scale*). In this reasoning they follow the views of the oft-cited American business historian A. Chandler.[44] This expansion, Van Zanden argues, subsequently led to the need to serve sizable markets and this was only possible by further enlarging business size and by investing in distribution, marketing, and management. Because various companies were active in the same markets, however, this gave rise to overproduction and falling prices in the 1920s and 1930s. This was a strong incentive to make various new products from the same raw materials (*economies of scope*)—a policy geared to diversification. Consequently, large diversified and multinational companies with numerous divisions began to emerge. This development, according to Van Zanden, is also visible in the Netherlands, where large diversified companies such as Unilever, Royal

Dutch/Shell, Philips, DSM, and AKU (later part of Akzo, which later became Akzo Nobel) came into being. For expansion one needed innovation in process technologies and for diversification one needed product innovation. It is no coincidence, then, that precisely this new type of business invested heavily in R&D. During the 1930s the six leading multinationals in the Netherlands already paid for 50 percent of the total industrial R&D and after World War II this share went up to 70 percent. During the twentieth century this intensified R&D also led to Dutch companies registering an increasing number of patents in various countries.[45]

Van Zanden explains this pattern of expansion and diversification with reference to a combination of newly emerged opportunities (innovations and new markets) and the development of a business strategy aimed at exploiting them, marked by investments in expansion followed by diversification and investments in R&D and development of professional management. The companies that successfully deployed this strategy became so-called first movers: it was hard for others to catch on because these leaders continued to have specific advantages over their challengers. Only few companies that entered the market later on managed to join the first movers. The Netherlands was blessed with both a number of first movers and several successful challengers, which resulted in its half a dozen leading companies that successfully competed internationally.

Van Zanden's analysis focuses on the leading large companies, but many Dutch companies successfully followed a different strategy, as Keetie E. Sluyterman and others have demonstrated.[46] Their strategy was based on collaboration and on producing several niche products for export. Collaboration implied that companies agreed on their shares for the Dutch market for specific products, and thus avoided competition. In 1930 one third of the companies on the list of one hundred largest Dutch companies engaged in such arrangements, and the government supported this type of cartel formation. In addition, some medium-sized Dutch companies' products competed successfully on export markets with those of large foreign companies.[47] Examples of such companies and products are Nederlandsche Gist- en Spiritusfabriek (a yeast and spirits company); the glue and gelatin factory Delft; the candle factory Gouda; Noury & van der Lande, a flour and oil company that developed bleaching agents for flour; Organon, which produced insulin and hormones; Norit, which made a sugar bleaching agent; Van Berkel's Patent, which was quite successful at manufacturing cutting machines and automatic scales; and Stork, which specialized in, among other things, machines for the sugar industry. This type of business was commonly run by family members instead of hired professional management, and personal ties were important.

A former weaving mill of the Van Heek company in Enschede, in 1971. This photo is indicative of the decline of the textile industry and is an image of the deindustrialization that marked the Netherlands at the close of the twentieth century. Until recently, some Dutch regions and cities had quite specific industrial identities, which young children learned about in geography lessons at school. For example, Enschede, on the country's eastern border with Germany, was closely identified with the textile industry. In 1950 the textile sector, along with clothing production, was responsible for almost two thirds of local employment. This proportion began to decrease after 1960, but there were still almost 22,000 jobs in this sector. Ten years later more than half of these were gone.

Chandler has claimed that this type of company was much less successful than the managerial firm, as illustrated by British industry, but Sluyterman and Winkelman arrive at the opposite conclusion in the case of the Netherlands.[48] They view family capitalism as a successful strategy for a country such as the Netherlands, with many fairly small businesses.[49]

To explain the vitality of medium-sized Dutch business, it is important to point not only to strategies of collaboration, specialization in a limited number of products, and the role of family ties but also to the significance of newly available technological opportunities. For instance, in the late nineteenth century a major advantage of the new generation of motors was that they could be deployed in production processes on a small scale and in a flexible manner. After 1895 electromotors spread in the Netherlands at a pace comparable to that in the United States, the leader in this measure. Electromotors fitted in well with the comparatively small average size of Dutch companies, and it helped them to compete in the world market.[50]

Other factors besides economic ones, however, must be considered in an examination of the trajectory of modern Dutch industrialization. Historians have pointed to the large role of the "pillarized" social system in which the labor class and their unions were incorporated in three different ideological blocs: catholic, protestant and socialist, and leaders of these blocs negotiated on various political issues. After World War II the socialist leaders agreed to pacify labor and keep wages low for quite a long time, in order to allow for rapid industrialization. Another consequence was that few strikes occurred in the Netherlands. Historians also mention the stimulating effect of the country's neutral position in World War I, which enabled it to engage in several industrial activities, leading to the establishment of Hoogovens, a major iron and steel company, and the salt industry. Moreover, several Dutch manufacturers temporarily had no major competitors in domestic and international (German) markets. The profits thus made were invested in machinery, particularly in the immediate postwar years but also later.

Finally, for the period after World War II historians underscore the major role of Dutch industrialization policies and also aid under the Marshall Plan. Although the importance of Marshall Plan aid was mainly psychological, its direct economic effect, for instance on financing capital goods, should not be underestimated. This effort created a climate in which industrial interests were seen to be self-evident and provided forceful backing to the Dutch government's industrialization policies. Moreover, in the wake of the Marshall Plan, Dutch businesses took over the "American way of life" approach, marked by productivity development, mass production, and mass consumption. Without much debate, medium-sized Dutch companies began to pursue standardization and type restriction.[51] In the 1950s the family business lost ground: coerced decisions to expand and compete in the European market signaled the need for more capital, while many businesses felt obliged to attract external management expertise.[52]

The industrial society that around 1960 had come into being in the Netherlands soon met with new challenges, however, as shown, for instance, by the rapid collapse of the textile industry. After 1960 there was a new wave of investment, triggered—as in the 1920s—by rising wages, but eventually these investments hardly helped labor-intensive sectors such as shipbuilding and clothing and textiles, which could no longer compete with low-wage countries. The options for raising labor productivity in these industrial branches were exhausted. Although this applied less to process industries, such as chemical and food industries, in these industries, too, increased costs would cause problems in the 1970s and 1980s. Three waves of cost increases followed each other: wages, energy prices, and costs tied to environmental pollution. Combined with the effects of a strong guilder this meant that in the 1970s Dutch companies increasingly had trouble competing in the world market, causing a weakening of the nation's industry.

The general pattern of Dutch industrialization and its explanation by socioeconomic historians can be summed up as follows. In the twentieth century the country went through a period of major industrialization, a process that took off after 1860, when a series of new companies came into being while older companies recuperated. During and after World War I industrial activity noticeably accelerated, while the newly emerged group of six leading companies consolidated its position. The crisis of the 1930s gave rise to further scale increases, diversification, and new investments in R&D, strategies that were taken up again after World War II, which itself had little restraining influence.[53] Overall, Dutch twentieth-century industrialization was marked by a slow yet steady process of growth followed by a "big bang" in the postwar period,[54] after which in the 1960s industrial efforts halted because it no longer proved possible to deal with rising costs through scale increases and increasing productivity. After 1965 industrial employment began to drop again. Socioeconomic historians explain Dutch industrialization and deindustrialization by referring to a combination of economic factors (such as demand fluctuation and factor costs), institutional developments (such as the abolition of restrictive government regulation after 1860), and the development of a stable and pillarized political system and industrialization policy. In addition, incidental factors—such as two world wars, geography, colonial possessions, and the discovery of natural gas—were influential. Socioeconomic analyses do mention technology, but its role is never elaborated. As we show below technology's role in Dutch twentieth-century industrialization does warrant a closer look.

Technology and Industrialization

Based on history of technology research as published in the volumes of *Techniek in Nederland in de twintigste eeuw* (*Technology in the Netherlands in the twentieth century*) we focus on four issues which help to gain a deeper understanding of the Dutch industrialization path. First of all, in the twentieth century the Netherlands evolved into a nation that began to explore and also cash in on its mineral resources. Rather than being an accidental or externally motivated development, this sustained effort aimed at exploiting the nation's natural resources by means of technological interventions, was a direct result of decisions by businessmen, engineers, and cabinet leaders, which were partially rooted in a new kind of techno-nationalism.[55] This had major ramifications for the Dutch industrialization path.

Second, industrial expansion cannot simply be linked to capital intensity. In the period before 1950 raising capital intensity was certainly not viewed by all companies, cabinets, and engineers as a goal to be pursued. Van Zanden and socioeconomic historians too easily assume a continuous process of scale increases, whereas in fact there was a clear break after World War II, notably in the chemical and petrochemical industries, sectors where capital intensity and expansion have been quite dominant factors.

Third, although many socioeconomic historians have referred to the strategy of diversification, they have hardly studied its technological basis. Instead they focused primarily on economic rationales. For example, diversification in the chemical industries was not only propelled by the need to compensate for overcapacity caused by scale increases, but also involved strategic investment, geared to exploiting networks of materials. Insight into these networks and the related research strategy is of great importance to explain the emergence of large companies as well as the particular direction of industrial diversification in the Netherlands.

Finally, industrialization was not only a matter of businesses, governments, and unions, but also of concern to consumers and their organizations. Some socioeconomic historians may have addressed the economy's demand side, but they did so mainly quantitatively, in terms of market growth. Consumers' choices and the formation of mass markets are rarely covered in their analyses, if at all, perhaps because the assumption is too easily made that mass consumption merely follows mass production. We will argue that Dutch consumers and consumer organizations made major contributions of their own to industrialization and industrial expansion.

The Discovery of Minerals

Until the early twentieth century, coal mining and the exploitation of other minerals in the Netherlands took place on a modest scale, especially when considering the extensive available reserves. In the course of the century, however, the Dutch energetically embarked on efforts to find and exploit natural resources such as coal, salt, oil, and natural gas. This endeavor was partly related to the outbreak of World War I, which brought home the importance of having and being in control of one's own natural resources. Yet another major factor was the reinvigorated sense of nationalism that became prominent at the end of the nineteenth century in Europe. Such nationalist senti-

In the twentieth century the Netherlands was discovered to possess a rich supply of
mineral resources, which led to a variety of new economic activities. This characteristic
wooden derrick from the mid-1950s was used to extract salt from the earth in
Hengelo. To extract the salt, lukewarm water is pumped through pipes underneath
the derrick inside the underground salt layers to dissolve the salt. The salty mixture is
pumped up and transported to a plant for further processing. An extensive chemical
complex came into being there, one of the precursors of the current Akzo Nobel
corporation.

ments were geared toward the need to explore domestic mineral resources as a driving force behind establishing a national industry. Engineers served as major spokespersons for this mode of nationalism. As Professor I. P. de Vooys put it in 1920, in an address to the Social-Technical Society of Democratic Engineers and Architects, minerals were "the necessary foundation of a national community."[56] Engineers introduced the notion that exploiting and winning mineral resources required a systematic, scientific, and large-scale national approach.

Exploratory drillings by private companies in the second half of the nineteenth century had established that the southern half of the province of Limburg had much coal in its soil, but initially those drillings did not lead to major extraction efforts. This is why in 1901 the Dutch Parliament accepted Minister Cornelis Lely's proposal to have a state company exploit the coal fields not yet granted in concession. In the light of this plan a quite sizable region was set aside for developing a model mine, in which the latest technologies could be tested and where making profits in the short run was not a first priority.[57] This was the beginning of a large state company, De Staatsmijnen (State Mines), which proved to be of major significance for the Dutch energy supply and ultimately became the major chemical company DSM.

Two years after the establishment of De Staatsmijnen, the State Minerals Exploration Agency was established on the initiative of Minister Lely. Random exploratory drillings undertaken by private companies bent on making fast profits would be replaced by a systematic exploration of the entire Dutch terrain, and this was defined as a government task. As the new director of this agency, W. A. J. M. van Waterschoot van der Gracht, put it, "Together we must insist on retaining the benefits of our coal fields for our own country."[58]

In 1909 new geologic insights led to what would prove to be the agency's most spectacular find: enormous salt deposits near Winterswijk, close to the German border.[59] By the end of World War I it became the headquarters of the Royal Dutch Salt Company (Koninklijke Nederlandse Zoutindustrie). Salt was used in households and the food industry, but also by water treatment companies, in the textile and in fertilizer industries. Electrolysis made it possible to extract chlorine from salt. From the 1930s, and especially after World War II, chlorine developed into one of the chemical industry's major substances (the "hammer in the toolbox" of industrial chemists) and its increased use in chemical industries is seen as a "technical regime change."[60] From 1950 this substance was transported by train to Pernis, the headquarters of the country's largest chlorine user, Bataafsche Petroleum Company (the Dutch subsidiary of Royal Dutch/Shell), which produced, among other things, pesticides, epoxy resins (solvents), and glycerin (a base material for many products). In the 1960s the Royal Dutch Salt Company, via a series of mergers, eventually became part of Akzo Nobel, which is one of the two largest chemical companies in the Netherlands. After World War II the chemical industry and the public utilities companies were the most rapidly expanding sectors in Dutch industry.[61] Without the government's systematic exploration effort initiated in 1903, however, this sector would not have achieved such prominence in the Dutch economy.

Likewise, the discovery of petroleum and natural gas resulted from a sustained exploration effort that involved a technological driven expertise.[62] As early as 1850 the Dutch government had dispatched junior engineers to the Dutch East Indies to do exploratory drillings. When in 1890 the Royal Dutch Petroleum Company took a concession there, it marked the beginning of a large series of new drillings and much geologic soil research, not only in the East Indies but also in other parts of the world. Soon these efforts led to this company's spectacular growth, which received a further boost through its merger with the Shell Transport and Trading Company in 1907. From the 1930s a subsidiary of this new multinational, Bataafsche Petroleum Company, began systematic oil drillings in Dutch soil as well, with support from the government. Because Royal Dutch/Shell was seen as a national industry, the State Minerals Exploration Agency played no role in this effort. After World War II, this oil drilling was formally arranged in a joint venture with the American company Esso (Standard Oil—later Exxon).

Although the subsequent discovery of huge natural gas reserves, first near Coevorden (1948) and later near Slochteren (1959), was more or less a byproduct of this effort, it would greatly influence the postwar economic structure of the Netherlands. In the 1960s the Dutch government placed immense trust in nuclear energy as a cheap source of electricity for the near future. This is why it not only supported nuclear energy through large subsidies but also decided to use up the newly found natural gas reserves quickly, first, to stimulate the country's further industrialization, and also to fund the growing system of social services, especially after the coal mines in Limburg were closed down again and many miners became unemployed. As a result, Dutch industry could buy natural gas for very low rates, which gave rise to the creation or expansion of highly energy-intensive industries, such as greenhouse farming, aluminum, and chemicals.

As our argument underscores, the extraction of large quantities of minerals from Dutch terrain has had an enormous influence on Dutch twentieth-century industrialization and the country's economic structure. Unfortu-

A chemical plant of the DSM company. After centuries in which wood, peat, and wind were the main sources of energy, in the twentieth century the Netherlands switched to coal and, after the 1950s, natural gas and petroleum, the result of a sustained search for mineral resources. Gas and petroleum became also important as raw materials for all kind of industries. In 1968 DSM, too, switched from coal to natural gas or petroleum as a basis of its chemical products and to this end began to use its first naphtha cracker, an installation to break down oil chemically into substances needed to produce many plastics.

nately, the pursuit of more industrial activity has caused major pollution and contributed to the rapid deterioration of the natural environment.[63] To put the contemporary Dutch economy's strong energy dependence in a long-term perspective, the Netherlands was already an energy-intensive economy in the early seventeenth century, when the energy sources were wind and peat. This period was followed by a long decline in relative energy use, and around 1900, when wind and horsepower still supplied 30 percent of the demand for energy, the Dutch economy was marked by quite low energy intensity. In the ensuing decades private and industrial energy use significantly increased and shifted toward fossil fuels, especially coal.[64] The energy intensity of the Dutch economy was quite similar to that of other Western European countries. In this respect, however, the 1950s mark a turning point, when coal was almost completely pushed out of the energy market by natural gas and oil, and, more important, the energy intensity per product strongly went up in the Netherlands, which was not the case in its neighboring countries. Moreover, this increase proved structural: energy-intensive industrial sectors, notably chemicals and metal, grew at a much faster pace than industrial sectors with lower energy use.

The focus on exploiting Dutch natural resources embraced at the end of the nineteenth century as an element in the building of the nation involved lengthy and at times difficult technology-based search processes. It led to both failures and successes, but whatever the outcome, insight into these exploratory efforts and their results is essential to understanding Dutch industrialization, as well as the emergence and success of large companies such as Royal Dutch/Shell, DSM, and Akzo Nobel, which have partly defined the modern Dutch industrial landscape.

Expansion as a Form of Speculation on the Future

Expansion and diversification have not been the only successful responses to growing markets, competition, and new technological possibilities. Small scale is often incorrectly associated with inefficiency. Sluyterman and Winkelman convincingly demonstrated that medium-sized companies may operate very effectively in the world market,[65] while Rienk Vermij in this volume has shown that notably before World War II expansion was not pursued automatically and that business consultants frequently pointed to the risks implied in anticipating larger prospective sales.[66] The phrase used in this context was "speculation on the future." The Dutch government considered excessive scale increases undesirable because they would lead to unemployment. The Industrialization Bureau (Bureau Industrialisatie), set up by the Dutch government in 1936, was assigned the task of formulating a national industrialization policy by starting from the notion of high-quality, labor-intensive, and knowledge-intensive products. The motto was "Specialization rather than mass production." Companies such as Philips, which explicitly pursued expansion through mechanization, repeatedly ran up against boundaries regarding the quality of its mass-produced light bulbs, for example. Some handwork could not yet be mechanized, as in radio sets, where many types and small series made standardized assemblage complicated.[67] In the final product industry, scale increase was just one option among others.

One alternative option, for instance, involves the relevance of technological flexibility. As Mila Davids and others have argued, the ability to produce products in small series and adapt them to customers' specific wishes has guided technological development in the final product industry. Expansion linked to series production might lead to a reduction of precisely that valuable flexibility.[68] This explains, for instance, why in the 1930s Dutch dockyards did not opt for riveting machines and the use of high-speed steel, which allowed very fast cutting. These technologies were only advantageous in large series production. Another example is electrification of the production process, which occurred more rapidly in the Netherlands than in other European countries. The rapid diffusion of the electromotor has to be viewed as part of a strategy aimed at flexible production. Electromotors, unlike steam engines whose steam boilers had to be fired up, could be linked efficiently to individual machines and could be turned on and off whenever necessary, increasing flexibility. The very significance of this flexibility and the prevalence of medium-sized businesses explain why before World War II, assembly lines were rarely adopted in the Netherlands. It is ironic that the assembly line, which economic historians regard as the technology of large-scale production par excellence, was used in the Netherlands in small companies producing varying series. About half of all Dutch companies that relied on assembly line production in 1940 were clothing workshops. They did not use assembly lines to facilitate mass production but mostly to discipline workers.[69]

Another alternative growth option concerns the load factor as it played a role in, for instance, electricity production. Because of the use in the twentieth century of alternating current, which cannot be stored, supply and demand had to be constantly balanced and peak load had to be avoided. Consequently, electricity plants pursued not a larger scale per se but optimal use of their capacity—they spread the load factor—so that their machines did not have to be turned on and off too often. In the history of public utilities—electricity and natural gas—local and regional versus national and international scale is also a core issue. Geert Verbong has shown that in the Netherlands, after 1910 the provincial companies and networks became the dominant scale of electricity supply.[70] The option of a national grid, as realized in other countries, was in fact discussed, but the Netherlands developed its own system. Natural gas was initially supplied locally, even though in the interwar period there were also examples of regional gas systems maintained by DSM and Hoogovens (so-called "distance gas"). The Dutch energy supply was turned into a nationwide system after World War II, whereby in the case of electricity the necessary changes were already realized under German pressure during the war. The huge natural gas reserves discovered near Slochteren in 1959 were exploited by the national government in order to push through a national gas distribution network as advocated in the so-called "Esso master plan," strongly inspired by the American model: "Put in a natural gas transport network that reaches into all corners," was the new slogan, and this is exactly what was done.[71] In 1967, 1320 kilometers of main transport network was finished, over 200 kilometers more than the original plan.[72]

After World War II, expansion and scale increase (of companies) was viewed positively. For instance, the government advocated the establishment of a capital-intensive chemical base industry. With the government's support, various chemical companies experienced strong growth.[73] Even so, initially the scale increases of the Dutch chemical industry were modest, at least when compared to those in other countries in the same era. A 1951 internal "Memorandum to the Board" of Royal Dutch/Shell concluded that the Dutch chemical industry was poorly integrated and quite heterogeneous, small-scale, and individualist, and there were no large chemical plant complexes.[74] Although Dutch companies had grown in previous decades, they continued to be smaller than some of their foreign competitors. The memo's conclusion was still valid by the late 1950s, when there were 120 chemical companies with more than one hundred employees.[75]

In the 1960s, however, the situation changed completely. Between 1963 and 1973 bulk chemicals saw spectacular growth, and Dutch companies became leading suppliers to chemical industries throughout northwestern Europe. How to account for this spectacular growth? Ernst Homburg has offered three explanations.[76] First, the Dutch companies' customers, many of them German chemical companies, switched from coal to petroleum and natural gas, which were supplied by Dutch companies. Second, the government decided on a policy of making the natural gas discovered near Slochteren available to domestic companies for low rates so as to stimulate the nation's industrialization. This worked very well: low energy rates served as a magnet and drew in many foreign manufacturers that opted for large-scale production. Third, several pivotal technological breakthroughs occurred. Until around 1950 the production capacity of synthesis-units for ammonia was

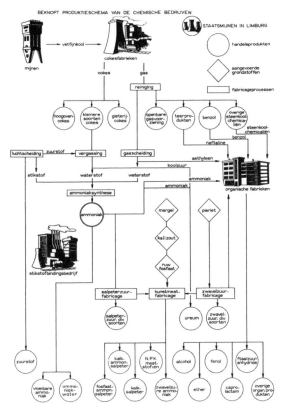

A diagram from 1950 depicting the stages of processing coal into various intermediary and end products. The fat coals mined by DSM in the Limburg mines were not suitable as fuel for stoves, but they could be processed into different kinds of cokes, which are suitable for use in blast furnaces. The processing of residual products gave rise to a diversification based on an analysis of substance networks. Cokes and gas coming from the cokes factory formed the basis for ammonia, nitric acid, urea, sulfuric acid, alcohol, and all kinds of artificial fertilizers. Several plants were needed to process these substances.

about 20,000 tons per year. In the 1960s this capacity increased to 350,000 tons as a result of the development of *single train* ammonia factories. Expansion was not a continuous process, but occurred in spurts, and by the 1960s many Dutch companies managed to realize their 1950s dream of expansion.

Diversification, Networks of Materials and Expansion

According to Van Zanden, the process of scale increase (economies of scale) is closely linked up with the notion of diversification (economies of scope).[77] He argues that the major investments associated with new technologies have led to the emergence of very large companies. Subsequently, however, the market became saturated, and these companies ended up in a murderous competition. Their response was diversification: developing new products with the same materials, which allowed them to serve new markets. In discussing diversification, an essential aspect is often neglected: the existing competencies, technologies, and opportunities for exploiting materials. To understand the direction of this diversification, one must consider the options and choices made and tried out by companies and engineers. Precisely on this point technology history can provide interesting new insights that lead to a better understanding of the timing and specific direction of diversification, which in turn is of great importance for gaining insight into the development of the overall economic structure.

The chemical industry well illustrates the dynamic of diversification prompted by expansion because of huge competitive pressures and also shows how sound analysis of this diversification requires insight into networks and technical qualities of raw materials.[78] During World War I, Royal Dutch/Shell considered entering the dyestuffs industry. The company converted its oil into gasoline, one byproduct of which was toluene, a major raw material for the production of dyestuffs and explosives. The company next did analyses of networks of raw materials and intermediary products needed for the production of end products. These analyses indicated that the company could never realize the same diversification as the German dyestuffs industry because of the inadequate quality of its raw materials and of possible synthesis routes. This partly explains why the company decided to focus on breaking down and hydrogenating heavy residue fractions for conversion into car gasoline, a product for which the demand was growing, and to identify other products that might be produced, such as ammonia. For years the hydrogenating trajectory caused many problems and hardly produced results. In the late 1920s, eventually, the company decided to build ammonia plants. Thus Royal Dutch/Shell again entered the chemical sector. In 1928 a new research program was started, whereby the articulation of a network of materials, as in 1918, played a crucial strategic role. Starting from propane and butane as base materials, the company outlined in detail which products could be made from them. The options were systematically tried out, but not all succeeded. Ultimately, however, this exploratory approach resulted in production of pesticides, plastics, and an array of other products.

Thus for the nature and direction of Royal Dutch/Shell's diversification, technological issues leading to diversification have been crucial, quite apart from economic expansion mechanisms leading to overproduction and diversification. Specific choices were tested on the basis of qualities of raw materials and their analysis, and some of them failed. This delayed the planned diversification effort or guided it into another direction, as happened right after World War I.

Other companies, such as DSM, engaged in similar processes.[79] The coal extracted from the state mines was not suitable for stoves in homes, but it could be processed into coke, which was used in blast furnaces. Accordingly, DSM decided to build a coke plant, the Emma, which started production in

1919. It generated coal tar and coke gas as byproducts. The tar, used to make engine fuel and tar for road construction, was partly sold at home or in the colonies and partly exported to Germany, where large chemical companies used it to make dyestuffs, explosives, and pharmaceuticals.[80] The gas was used as fuel for the coke furnaces and was sold to nearby towns for local gas supply. DSM's new coke plant, Maurits, which started production in 1929, generated even more gas. To make the plant profitable, other applications of coke gas were found in the production of fertilizer according to a new procedure, on the market since 1925. This new invention soon led to the establishment of numerous nitrogen plants. Consequently, when DSM's Nitrogen Fixation Company (Stikstofbindingsbedrijf) started operations in 1930 there was much overproduction and prices fell. The response of DSM was further diversification: it also began to produce other nitrogen fertilizers that in some respects were better than the earlier fertilizer, yet the company capitalized on already available raw materials and intermediary products. For example, it began to produce nitric acid and, subsequently, nitrates based on its ammonia production. It also managed to win ethylene from the available coke furnace gas, which was later used to produce alcohol. After World War II, DSM explored further diversification into new products such as nylon, PVC, polystyrene, detergents, and pharmaceuticals, and proved quite successful in these markets.[81] For example, in collaboration with AKU (Algemene Kunstzijde Unie), a synthetic fiber factory in Arnhem, DSM began to produce caprolactam, a base material for nylon. The production of urea—a major base material for plastics that was related to fertilizer production from a chemical-technological angle—turned out to be a success.

In this way, processes of diversification gradually led to the establishment of large chemical complexes in Limburg (DSM) and near Rotterdam (Royal Dutch/Shell), whereby the large company labs played a pivotal role. These labs had to develop the new products and procedures that allowed their companies to compete, while the lab engineers and technicians pursued projects that determined the specific directions of diversification.

Our three excursions into technology history add to our insight into Dutch industrialization in several ways. First, it has become clear that the emergence of a strong group of industrial leaders (including Royal Dutch/Shell, DSM, Akzo Nobel) had a lot to do with the drive that emerged in the late nineteenth century to explore and exploit minerals and natural resources. The Dutch government began to view this as being in the national interest, an effort in which engineers played a crucial role.

Second, our three examples reveal more of the complexities involved in the story of industrial expansion and diversification. The expansion efforts initiated in the late nineteenth century ran up against technological boundaries, and the diversification strategies pursued by major Dutch companies (or their forerunners) in the first half of the twentieth century should not be viewed merely as a reaction to the sales crisis that followed the opting for expansion. In individual cases diversification preceding expansion can partly be explained by efforts to more fully exploit particular qualities of coal, salt, or oil. Developments in these large companies were also driven by the exploitation of networks of materials and technical competencies rather than the wish to produce on a larger scale. In addition, in particular before World War II, many medium-sized companies did not pursue maximum production capacity, but instead actively looked for the optimal scale for operations: increased flexibility realizable at smaller scales could be as important economically as volume of production. The net effect was a dual structure, with a few large companies and many medium-sized companies.

Finally, our technology history perspective makes it possible to shed new light on a turning point in the 1950s in Dutch industrial development, as occurred also in Dutch agriculture. In the 1950s for the first time there was an all-out effort to industrialize and expand, which was strongly reinforced by the discovery and rapid application of cheap natural gas. This effort created a powerful bulk chemicals industry requiring much capital and energy. In the aftermath of World War II, the already existing tendency toward industrial expansion, including an enhanced focus on the U.S. economy, was strengthened. The postwar "big-bang" pattern, then, is also observable from the angle of technology history and may even be more clearly delineated. At the same time, however, it is clear that industry's large-scale expansion did not automatically result from developments that started in the interwar period. Before World War II, scale increase was explicitly debated; it came with risks and the focus was always on realizing an optimal scale, whereby other factors affecting optimization, such as flexibility and load factor, were taken into account. In this sense the 1950s definitely constituted a break: in the first half of the twentieth century different concerns and approaches existed side by side in Dutch industrialization, but after 1950, scale increase or full-blown industrial expansion was embraced as the obvious approach of choice.

Active Consumers

Dutch consumers began to purchase and use durable goods on a massive scale in the 1960s, and this circumstance has motivated many socioeconomic histories of Dutch society to situate the rise of consumer society in this era, as a culmination of the process of economic growth and industrialization.[82] It is possible, however, to identify earlier cases of widespread diffusion of new consumer goods in the twentieth century. Prior to World War I, for instance, bicycles were sold widely, also to workers; in 1912 one of every ten Dutch citizens owned a bicycle.[83] In the interwar period, most working-class and middle-class households purchased a radio and a vacuum cleaner, and replaced their coal clothes iron with one that operated on gas or electricity. In 1930 10 percent of Dutch households had access to radio (a subscription to receive radio transmissions), a percentage that went up to 43 in 1935 and 65 in 1940. In 1940 citizens owned 1 million radio sets. An indicator for the popularity of radio among workers in the interwar period is that some workers paid their radio subscription before paying their rent.[84] Although the data on vacuum cleaners are less specific, estimates suggest that in 1930, 20 percent of lower-income and 60 percent of upper-class households owned a vacuum cleaner. Ten years later two thirds of lower-class households had a vacuum cleaner, and almost all other households owned one.[85] Middle-class households had a telephone, but only the wealthiest in the middle class owned an automobile. By 1940 the Netherlands had 300,000 telephone connections and 100,000 cars.

According to a 1938 survey commissioned by the German car industry, the Dutch market was saturated because all households that could afford a car had purchased one.[86] From the early 1950s, a rapidly growing number of households purchased a phone connection, a car, a television, a washing machine, a fridge, and a stove, resulting in a genuine democratization of ownership of durable consumer goods.[87] Kees Schuyt and Ed Taverne have correctly indicated that prosperity and production were linked to large-scale consumption fairly late in the Netherlands.[88]

The arrival of consumer society in the 1960s is corroborated by specific diffusion data, but the willingness to spend money on durable consumer goods was already in place.[89] Rapid growth was noticeable a decade or so earlier, and other data also suggest that consumer society was already in the making in the interwar period, when the upper middle class purchased fancy cars and workers bought expensive radios and vacuum cleaners. At the same time it is clear that in the interwar period households did have to make choices; they could not purchase everything they wanted, despite payments by installment and the flourishing second-hand market for many goods. How can we account for the choices made? Why in the early 1930s did a working-class family purchase a radio, a clothes iron, and a vacuum cleaner rather than a washing machine, and what made a middle-class family in that same period prepared to purchase an expensive car?

In the interwar period a new consumer society evolved in the Netherlands, featuring mass consumption of new products such as radio sets. This entertainment technology was often purchased sooner than many other household appliances because it could be used by the entire family while it contributed to what historians have labeled "domestication." Around 1928 Philips played to this trend with this staged photo to promote its new radio. The picture's underlying message is that women, too, will enjoy this new Philips product.

Radio's widespread diffusion among workers sometimes met with moral disapproval on the part of social scientists who studied the trend. In 1933, researchers studying the development of radio scolded,

> It is generally true that a too large percentage of people's income is spent on luxuries, even today. In 1929 and 1930 this was particularly reflected in the purchase of radio-sets. In many working-class families one will see such an appliance, which costs some 250 to 300 guilders, integrated into a fancy salon cabinet.[90]

A preference for purchasing a radio-set may well be tied to its potential to contribute to family life and its intimacy. Radio made it possible for entertainment, culture, information, and some edifying words to enter the living room via a single device and the entire household could enjoy listening. Women combined household chores with radio listening, children did homework and in the early evening the entire family would gather around the radio set.[91] Similarly, in the interwar period a car was used for business purposes (by civil servants, salesmen), but it was also seen as offering a possibility for middle-class families to spend their leisure time and tour the countryside together. After World War I, the Model T Ford became a huge success in part because it could be used for work *and* leisure. Precisely this double functionality caused cars to become so popular.[92] Both radio and cars brought new and modern worlds within reach; they offered the entire family entertainment or adventure in a safe way, behind glass.[93] The Dutch government called on families to offer a sense of security to young people and, within the context of the family's comfort and safety, to regulate the introduction and use of new consumer goods, so that youngsters would not be carried along helplessly in the maelstrom of the "roaring" 1920s.[94]

But what made Dutch households decide to purchase a vacuum cleaner, if price was not an impediment?[95] Two factors seem to have played a role. First, it made housecleaning less burdensome. Housewives experienced vacuum cleaning as a more effective, simpler, and less tiresome mode of cleaning than sweeping floors and shaking out rugs. In addition, according to the 242 women interviewed by H. Makkink, there was a prevailing desire to be a part of modernity, which could be expressed by, among other things, purchasing a vacuum cleaner.[96]

The promotional discourse of washing machine manufacturers advanced the same promises of convenience and modernity, but in this case Dutch housewives judged differently. Only 10 percent of households purchased a hand-driven or electric washer before World War II. Price was not a decisive

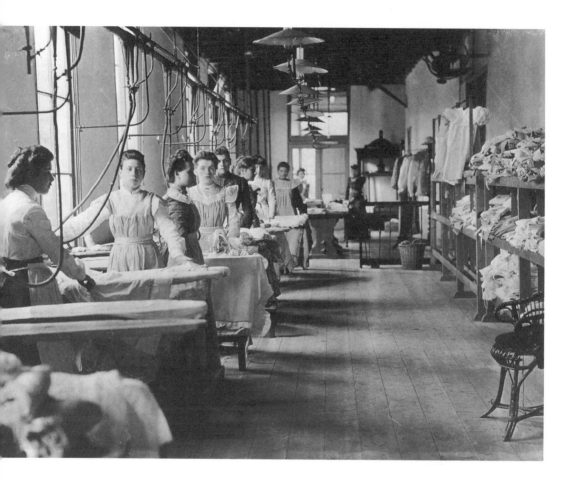

Young female employees of a company whose business was washing and ironing the laundry of wealthier customers, photographed in Dordrecht in 1905. Many consumer products and services that started to appear on the market in the late nineteenth century were geared to the wealthy, who had others to do household tasks such as laundry for them. Socialist and social-liberal women's organizations tried to set up cooperative laundries for the lower classes, but with little success. Doing laundry continued to be a burdensome household task performed almost exclusively by women.

factor, but several other aspects did play a role.[97] Although the first generation of washing machines, which consisted basically of a barrel of water with an agitator run by an electric motor, took over the heavy work of scrubbing by hand it still left women with lots of labor: getting water and heating it, filling the machine, soaking, rinsing and wringing, ironing, and drying. The washer, then, did not imply a major improvement on existing washing practices. In this period a washer was largely seen as a not very useful, luxurious device.[98]

This same reasoning explains the rapid diffusion of the new gas or electric iron. The new irons were successful because they tied in with and greatly enhanced women's existing ironing practice. Unlike the coal iron, the new iron did not come with the problem of soot, waste, or stench; housewives simply had to regulate the temperature and could go on ironing without interruption.[99]

These examples show that in the interwar period households made conscious decisions about consumption and that various new consumer goods were already being purchased on a large scale. Another striking aspect was the rejection of collective facilities for household tasks, such as laundries and central kitchens, despite successful experiments with these facilities.[100] At the start of the twentieth century collective solutions for household work seemed an attractive option. Businesses concentrated on suitable products for large-scale application in factories, while ignoring the consumer market. Their steam washers and vacuum cleaning installations were not suitable for use in private homes. Socialist and social-liberal women's clubs and a growing group of household professionals advocated collective facilities and experimented with them as a way to address the lack of servants and because thus working-class families would more quickly profit from this solution, speeding up their incorporation into civilized bourgeois society. The Netherlands Association of Housewives (Nederlandse Vereeniging van Huisvrouwen), founded in 1912, promoted cooperative kitchens, municipal laundry facilities, and user cooperatives. After World War I, individual solutions that made household chores less burdensome received more attention, and housewives' professionalization became a more important strategy to the aforementioned groups, but the option of collective performance of specific tasks was enthusiastically pursued as well. For a long time both options continued to exist side by side.

In the course of the 1930s, however, the shared use of all sorts of devices in kitchens, launderettes, and public baths by households became less and less attractive. Collective use became defined as a temporary option for households that could not yet afford their own facilities and appliances.

Many collective services were increasingly regarded as dated or even as backward. Thus, even before World War II the working and middle classes developed a desire to purchase their own new and "modern" products within the family context. The churches and most political opinion leaders defined marriage as the natural destination of each woman, and purchasing new consumer goods in a family context beckoned women to move toward this destiny. New household technologies became an active factor in a process of "domestication," whereby married women were expected to create a pleasant family atmosphere for their families.[101] Thus the new technologies were closely linked with the emergence of the housewife profession. Hominess took on moral but also technological content. After World War II the division of roles between men and women became fixed even further, while the nuclear family again was meant to compensate for the assumed emptiness and meaninglessness of modern mass culture.[102]

Although mass consumption was a reality in the interwar era and consumers contributed to shaping a rudimentary consumer society, the story about its further development is not just about decisions by manufacturers and consumers and how these might or might not agree with each other. Decisions about a product's development were frequently made prior to its market introduction, and choices by households were anticipated and influenced by the information and propaganda of a variety of businesses and social organizations, many of which had been founded by consumers themselves. The story of consumer society, then, is also one of intermediary actors and the development of a new "midfield" of storekeepers, wholesalers, advertising agents and a wide variety of consumer groups such as the General Dutch Bicyclists Union, the Netherlands Association of Housewives, the Dutch Women's Electricity Society, and the Dutch Household Council.[103] As intermediary actors these organizations found a niche between producers and consumers and worked hard to achieve better coordination between production and consumption. Because many products needed to meet legal standards, whose formulation was often promoted by intermediary actors, governments and parties such as insurance companies also became involved in this process of coordination. The two world wars stimulated government involvement because in these periods women's organizations were called upon to help secure the national food supply.

To understand the development of consumer society in the Netherlands one must understand the role of intermediary actors. As a group the various intermediary actors played three crucial coordinative roles. First, they formulated demands and posed questions regarding new products that came to the market or still had to be developed, thus partly shaping the market and

In the country, farmer's wives continued to contribute directly to the production process well into the fifties; the photo stems from 1958. In the cities things were different. During the interwar period in urban environments the tasks in and around the home increasingly became the exclusive domain of the housewife, where her productive contribution to the economy was seen to be taking care of her husband and family.

making it easier for entrepreneurs to estimate which products could be marketed successfully. This was important work because in the twentieth century consumers became increasingly more anonymous and thus invisible and harder for producers to identify in large-scale or newly developed markets, as in the case of gas and electricity companies. This process of setting standards consisted of formulating design criteria and communicating them to producers, but occasionally consumer organizations also worked on improvements on their own, so as to make new products more appropriate for use.

The establishment in 1926 of the Institute for Information on Household Work (Instituut voor Huishoudtechnisch Advies) by the Netherlands Association of Housewives is a case in point. Its aim was, among other things, to provide technical advice to manufacturers on how to improve products, because too little attention was paid to housewives' practical needs. The keyword was "efficiency." The institute also initiated research and set up a large network of contacts with other research institutions and businesses. It thus willingly relied on the expertise of the first generation of technically educated women, who performed tests and experiments at home or at domestic science schools, and thus became involved either directly or through surveys.

The housewives association and the institute developed various products. The best-known examples are perhaps the designs for model kitchens that would serve as basis for the very popular *Bruynzeelkeuken* designed later on.[104] The successful introduction of the "washing and ironing label" in 1952, an initiative of the Dutch Household Council, also was very important for the diffusion of washing machines. This label helped users, mostly women, in their choice of the right detergents and the proper use of washers. This illustrates the productive interaction between women's organizations and manufacturers.

Storekeepers constituted another group of intermediaries who were a source of detailed information about their customers that manufacturers could tap. For example, during the 1930s a Rotterdam branch manager of a washing machine manufacturer, Velo, provided extensive analyses of potential customers.[105] Philips collected information on consumers who bought a radio set in electrical appliance stores, and used it for new product development and for redesign. Unilever also pioneered market research by setting up its own marketing group in 1934, and also by hiring household experts and setting up a test kitchen and a test launderette. This underscores that manufacturers valued the input from intermediary actors because thus they could gain access to markets for which they had little feeling as of yet.

Second, intermediary actors taught housewives to operate new products such as irons, electric ovens, washers, and vacuum cleaners correctly, thus defining what constituted proper use. They conveyed knowledge about electricity and chemistry, and also indicated more efficient ways of doing things, which would leave mothers more time for raising children, family life, community activities, and personal development. This involved disciplining and the development of new skills, while it also implied the articulation of knowledge on food and hygiene that before was passed on directly from mother to daughter. Hundreds of household experts were occupied with answering questions from women and writing books, household guides and articles in popular magazines such as *Libelle*. They gave lectures and product demonstrations and taught at domestic science schools, where tomorrow's housewives received training. Finally, the Institute for Household Technical Advice issued a product quality seal, one that soon was sought after by manufacturers, who occasionally were willing to adapt their products in order to get it.[106]

The educational task of the intermediary is also evident in the work of the

General Dutch Bicyclists Union, which turned itself into a touring and auto-mobile club around 1900.[107] This organization played a crucial role in fashioning the relationship between car and user. The way you sit behind the wheel, crank up the engine, operate the clutch and accelerator—all had to be learned. This involved absorbing more than just the correct operation of the vehicle but also traffic rules, notably driver conduct vis-à-vis other traffic participants. The union insisted that drivers must be courteous and it developed many instructions for teaching proper conduct that were widely publicized.

Third, the intermediary actors tried to promote the creation of infrastructures necessary for making fruitful use of new products such as building one's own organization, creating training facilities, and codifying knowledge, and also contributing to the development of complementary technologies, regulation, and a material infrastructure. For example, the bicyclists union specifically lobbied for the development of government regulations, road construction, gasoline supply, road signs, a quality seal system of approved hotels, and so on.

To understand the development of Dutch consumer society it is, therefore, crucial to analyze the functioning of these intermediary actors. They organized decision processes and regulated consumption, occasionally voicing protests in public but much more often relying on silent diplomacy. Their activities may have been less visible in history and thus in history writing, they were nonetheless important. Economic historians have paid much attention to business, unions, advertising, government policies, and regulation of labor markets, but much less to actors that have guided businesses and consumers in their search for the best *fit* between product and market. Precisely this coordination, however, was of great importance for the emergence of mass consumption.

Jan Luiten van Zanden has introduced the notion of the "long twentieth century" to suggest that in terms of ideas and important cultural developments, this century actually "started" in the 1860s and is not yet quite finished—even though it is clear that after 1970 a transition to a new type of economy and society with other characteristics has been taking place. In essence this long twentieth century is marked by three institutional changes: the rise of multinational corporations, regulation of the labor market including the rise of labor unions, and the development of the welfare state. Van Zanden argues that these institutional changes have obviously been favorable to society in general in the case of the Netherlands. The country's economic growth between 1913 and 1995 was significantly stronger than in the liberal nineteenth century—about twice as high as that for the entire nineteenth century, in fact.

In his analysis Van Zanden entirely ignores a fourth major institutional development, namely the emergence of a "midfield," a group of various actors involved in coordinating mass consumption and mass production. This added factor does not lead to positing another periodization, because this institutional development also took off after 1870, picked up steam in the interwar period, and came to full flowering after World War II. It does suggest, however, a more complex picture of twentieth-century Dutch socio-economic history. Whereas Van Zanden argues that the economic miracle is accounted for by supply factors, capital formation, and technological change, our history-of-technology perspective makes clear that demand factors and institutional processes tied to market formation are also essential to explaining this miracle. Moreover, the development of new commodities and the rise of mass production did not precede the development of a new demand and mass production. Both processes occurred side by side, in a dynamic of mutual influencing and coevolution.

Do the 1950s also represent a structural break in the trajectory of the rise of consumer society? The answer is yes, because starting then the diffusion of consumer goods unmistakably accelerated. How this occurred can only be understood when we look at the framework put in place during the interwar period, when consumers first started gaining experience as such. The 1920s and 1930s had a character of their own, marked by a struggle about the nature of consumer society. Ultimately, however, collective services for households failed to gain a foothold as a valid alternative. Consumption of new durable consumer goods took place within individual households, which increasingly were viewed as havens from the negative effects accompanying the wave of modernization.

Final Discussion

Our analyses of changes in agriculture, industry, and consumption throw new light on the industrialization and wider modernization of the Netherlands. In the 1860s a long wave of industrialization set in, which surged after 1890, after World War I, and again after World War II. It reached a climax in the 1950s and 1960s, but this period also reflected a break as various major contradictions came to the fore. Their resolution implied an unequivocal choice in favor of large-scale production in both agriculture and industry. Although the growth of consumer expenditures had to be postponed for a while in the 1950s to allow for industrial progress, a society based on values associated with sustained consumption and sweeping industrialization represented the future, with the United States serving as the model to be imitated. A new, modern way of life was widely expected for the immediate future.[108]

Historians have long presented the 1950s as a period in which the prewar bourgeois and pillarized social relationships were restored. It was seen as an intermezzo between the violence of war and that of the cultural revolution of the 1960s, when the old pillarized social power relationships were discarded.[109] From the vantage point of a technology history, however, this assumed recovery largely functions as a façade hiding the country's true modernization from view. Both elites and the general public were oriented toward the modern future rather than the past. James Kennedy has shown that the Dutch elites believed their old-fashioned country to be desperately in need of modernization. He argues that although it was the young generation that caused such a stir in the 1960s, the existing elites constituted the dynamic force behind the era's cultural upheaval.[110]

Technological development and industrialization were soon seen as spearheading the attainment of modernization in the postwar period. Between 1949 and 1963 the Dutch government released eight industrialization memorandums that were written from a modernizing perspective. In retrospect, historians have argued that the main function of these influential policy papers, rather than articulating actual measures, was to provide fresh perspectives on a new future.[111] Schuyt and Taverne even speak of the emergence of a new national identity: that of a highly developed industrial nation.[112] Part of this view was that agrarian land and rural society had to be brought into line with the new industrialization paradigm, which ultimately resulted in the geographic distribution of industrial employment. Marshall Plan propaganda also stressed the importance of mass production as a precondition for mass consumption, or modernization following the American model.[113] The Netherlands was no longer supposed to base its economic life on its trade, agriculture, and colonies alone; it should focus more on technological progress and industrialization. Investments in R&D were substantial, and engineers and other experts received a broad mandate to rearrange the country, a mandate they have generously exploited in designing and planning polders, dikes, and storm surge barriers, housing, new roads, and new seaports and airports, and consolidating farmland.[114]

The aspiration to modernize had been around since the late nineteenth century. In the interwar period it took on the character of an "efficiency movement," which consisted of a rather diverse crew of engineers, accountants, management experts, budget inspectors, industrial consultants, women's organizations, psycho-technicians, architects, and artists who wanted to have

The nuclear family had long been a central target of the propaganda for new consumer goods and new food products. In the 1950s a uniform meal pattern developed in the Netherlands, but in the 1960s the emphasis on consumption in a family context was increasingly sidelined. Snacks became popular and consuming them became part of a new youth culture, also referred to as the *patatgeneratie*, the "French fries generation."

a leading part in adapting the Netherlands to the demands of the modern era. On hindsight, this proved to be a vanguard indeed because after World War II their aspirations were embraced by both the political and economic elites and the overall population. In the postwar era the efficiency movement's experiments with all sorts of planning, rational management, standardization, and use of new technologies, including new management techniques, were followed up on a large scale in companies, offices, and a wide array of social practices such as housing provision.

In retrospect the far-reaching social and cultural changes of the 1960s in the Netherlands can be viewed as a culmination of major developments in several domains that occurred after 1890 and that ended up providing a foundation for later rapid modernization. First, the country developed entirely new physical infrastructures for energy, transport, and communication, which was a basic condition for rapid and substantial diffusion of many technological products such as cars, refrigerators, television sets, and telephones in the 1950s and 1960s; various other consumer products—radios, bicycles, vacuum cleaners—already established attractiveness to consumers in the 1920s and 1930s.

Second, the country's rapid modernization was well prepared from an institutional angle as well. New organizations such as multinationals and new business infrastructures such as R&D systems were in place. A professionalized network of engineers and other experts had emerged that could be deployed. In addition, local and national governments increasingly began to feel responsible for the quality of economic and social life, and they were also more willing to intervene to effect desirable outcomes. Finally, the pillarized organizations had successfully taken up the intermediate space between state and citizens, thus significantly contributing to organizing education, health care, and leisure. In addition there existed a part of this "midfield" that was hardly pillarized, if at all, and which as a result has largely remained underexposed in political history writing. We have devoted considerable attention to these intermediaries between organizations and manufacturers of users of specific new technologies that engaged in socially embedding these technologies and regulating consumption. Having been very active in the interwar era, these intermediary actors were of crucial importance for effective coordination of production and consumption.

Third, swift modernization was culturally anticipated in the sense that the wish to modernize had been heard in all sorts of contexts since the late nineteenth century, which in time created a breeding ground for such views and aspirations. Although such aspirations were not shared by everyone, the confidence in technology and planning grew substantially, which was further strengthened by the experience of World War II: no one could ignore the interrelationship between the victory of the United States and that nation's technological superiority.

All of these elements provided the basis for unrelenting modernization after 1948, by which date the most urgent reconstruction of wartime damage was completed. As Schuyt and Taverne correctly note, policymakers, architects, and the public too were all obsessed by the idea of modernization, a fascination that cut across the various pillarized structures. This was in fact the most central aspect of the miracle of the 1950s: modernization was no longer contested.[115]

Of course, characterizing the period after 1890 merely as preparatory is anachronistic and reductionist—as if the outcome of the historical processes involved was known in advance. Moreover, our argument has underlined that modernization was contested until the 1950s. The interwar period is perhaps best described as "a searching time" (Piet de Rooy), one of major uncertainties—politically, but also economically and technologically.[116] Modernization was contested: in agriculture, where large-scale farming was consciously rejected in favor of different modernization strategies, and in industry, where electromotors made small and medium-sized business still viable. Whether the Netherlands might escape the modernization that capitalized on all-out industrial expansion, just as it had escaped World War I, was a burning issue. At the same time, some people were vigorously striving for a society built on industrial expansion and the development of a new culture of consumption.

What becomes visible here is that in history various paths of development exist alongside each other, and these path contained clear technological choices. This suggests the specific relevance of a contextual technology history perspective, since it stresses analysis of alternatives. In economic history, by contrast, ideals of economic growth and efficiency tend to come first, which leaves little room for a consideration of the viability and rationality of developmental paths are not so easily related to standards of economic efficiency and growth. For example, within a history of technology perspective lagging labor productivity in agriculture or big industry does not automatically suggest a disadvantage or lack of modernization resulting in scanty economic growth, because it can also be tied to decisions and choices aimed at retaining the family business or employment in agriculture, whereby variables other than economic ones may be relevant issues. In addition, when economic historians devote attention to choices, they mainly look at choices made by producers, companies, and, possibly, governments. Decisions by users and consumers thus remain underexposed. Paying attention to choices by these actors provides insight into the contested character of the overall modernization process.

After 1948 the Dutch elites and the general public indeed embraced modernization and industrialization, but there was still a felt need to control modernization, especially within the Netherlands. The large number of books that were published warning readers of the dangers of technology and modernization is but one indicator. For example, in 1958 the much-read sociologist P. J. Bouman wrote:

> Sports and club life, free interaction between the sexes, travel and camping out, radio in every household, newspapers and telephones, movie theaters and television, technology-oriented interests and the reduced number of hours spent within the family—one cannot judge all these things. But everyone knows a heavy toll was paid for much of this "progress." And we also know that we cannot turn back the clock. We can only try to prevent a few excesses from happening.[117]

The various Dutch "pillars"—Catholic, socialist, and Protestant—devoted much attention to finding ways of minimizing the harmful effects of modernization. They did so by expanding their constituencies and the Dutch nuclear-family culture and by imparting rules of conduct and self-control, or "discipline and asceticism" (J. C. H. Blom), to make the most out of modernization, but also contain consumerist impulses.[118] As De Rooy put it, "The living room had a lamp in the middle that threw a warm light on a table around which the entire family gathered contentedly."[119] The 1950s and early 1960s can be viewed as the heyday of the perfect family model, with the Dutch housewife in charge of relieving her husband's and children's stress, generated by their participation in the modernization process. This was her contribution to higher productivity.[120] The underlying ideal was rationally supported by government policies, such as the directive that women should quit working after getting married – which mainly confirmed a common practice that had emerged in the interwar years.

The ultimate goal, in the vocabulary of the elites, was "controlled modernization," or, as Irene Cieraad puts it (see chapter 10, this volume): the fragile balance between hedonism and control had to be maintained. A new generation of experts and sociologists set itself the task of guiding individual self-development, so as to closely monitor the development of the standards and values adopted by Dutch society.[121]

Quite soon, however, these efforts at controlling the process of modernization began to show cracks and fissures. As early as the 1950s a new youth culture emerged that hardly respected the norms and structures of pillarized society. In the 1960s it became clear that the attempt at controlled modernization had failed. At first criticisms were leveled at the system of pillarization that controlled Dutch life, yet increasingly modernization itself was denounced. This critique grew even louder when it appeared that the path toward sweeping industrialization, with its large-scale use of natural resources such as natural gas and petroleum, had also brought with it significant downsides. The energy-wasting economy resulted in economic risks and major environmental pollution. Euphoria about technology, modernization, and industrialization was transformed into fear and mistrust.

In the 1980s and 1990s fundamental criticism of modernization became more muted. There was a renewed focus on ongoing modernization, but with the caveat that it must be a modernization that is more sensitive to its effects. This meant incorporating issues such as pollution, democracy, global poverty, and energy conservation in new paths of growth. Modernization must be self-reflexive about its consequences.[122] Our analysis shows, however, that precisely in calling attention to these effects, continuities with the 1950s become apparent as well. Controlled modernization is again fashionable and leads to the articulation of goals that are shared more widely by experts, business, government, and grassroots groups such as the environmental movement.

Acknowledgments

We are grateful to Marja Berendsen, Liesbeth Bervoets, Mila Davids, Carianne van Dorst, Harry Lintsen, Ruth Oldenziel, Arie Rip, and Geert Verbong for their comments on and discussion of earlier versions of this chapter. Thanks to Arie Rip for various textual suggestions for the final draft of this chapter.

Notes

1 H. W. Lintsen, "Het verloren technische paradijs," in H. W. Lintsen et al., eds., *Geschiedenis van de Techniek in Nederland: De wording van een moderne samenleving, 1800–1890*, 6 vols. (Zutphen: Walburg Pers, 1992-1995), vol. 6, 33–50 . For a more detailed overview of Dutch industrial development, see Mila Davids, "De industrie in Nederland gedurende de twintigste eeuw," in Schot et al., *Techniek in Nederland*, vol. 6, 257–269.

2 Jan Luiten van Zanden and Arthur van Riel, *The Strictures of Inheritance: The Dutch Economy in the Nineteenth Century* (Princeton: Princeton University Press, 2005) 192 (table 6.2) and 263; it measures the share in terms of the three main sectors' added value as part of the gross domestic product. For an analysis of the late industrialization of the Netherlands, see also Lintsen et al., *Geschiedenis van de Techniek in Nederland*.

3 Measured in terms of value of production; see H. W. de Jong, *De Nederlandse Industrie*,

1913–1965: Een vergelijkende analyse op basis van productiestatistieken (Amsterdam: Aksant, 1999), 2–4.

4 By paying attention to the household sector, we transcend the traditional categories of agriculture, industry, and services.

5 The agricultural crisis of the 1880s was a result of the emergence of a new global economy in the second half of the nineteenth century, which affected all of Europe. It resulted from steeply falling prices for agricultural products and led to a restructuring of the agricultural sector and migration of agricultural laborers into towns and to other countries. The effect of the crisis on grain producers is the most instructive. European grain farmers had to face fierce foreign competition when prices fell drastically as a result of the opening up of vast grain-producing regions in North America, Argentina, Australia, and Russia, and transport costs decreased dramatically after the introduction of steam technology. See N. Stone, *Europe Transformed 1878–1919* (Oxford; Blackwell, 1999), 6–23.

6 Jan Bieleman, *Geschiedenis van de landbouw in Nederland, 1500–1950: Veranderingen en verscheidenheid* (Meppel: Boom, 1992), especially 210–212 and the map on 212.

7 For figures, see also W. H. Vermooten, *Stad en land in Nederland en het probleem der industrialisatie* (Amsterdam: H.J. Paris, 1949), 25. Vermooten shows that between 1910 and 1930 the number of businesses with a size of one to five, five to ten, and ten to twenty hectares increased, while the number of businesses of more than twenty hectares decreased.

8 Bieleman, *Geschiedenis van de landbouw in Nederland*, 210.

9 L. Douw, "Meer door minder: Ontwikkelingen in de agrarische structuur na 1950," in A. L. G. M. Bauwens, M. N. de Groot, and K. J. Poppe, eds., *Agrarisch bestaan: Beschouwingen bij vijftig jaar Landbouw-Economisch Instituut* (Assen: Van Gorcum, 1990), 35–53.

10 We rely in particular on the following sources: Van Zanden and Van Riel, *Structures of Inheritance*; Jan Luiten Van Zanden, *The Economic History of the Netherlands 1914–1995* (London: Routledge, 1997); Bieleman, *Geschiedenis van de landbouw in Nederland*; Merijn Knibbe, *Agriculture in the Netherlands 1851–1950: Production and Institutional Change* (Amsterdam: Nederlandsch Economisch-Historisch Archief, 1993); Merijn Knibbe, "Landbouwproductie en -productiviteit 1807–1997," in Ronald van der Bie and Pit Dehing, eds., *National goed: Feiten en cijfers over onze samenleving, 1800–1999*, Voorburg: Centraal Bureau Statistiek, 1999), 37–60; Bauwens, De Groot, and Poppe, *Agrarisch bestaan*; C. L. J. van der Meer, H. Rutten, and N. A. Dijkveld Stol, *Technologie in de landbouw: Voorstudies en achtergronden technologiebeleid* (The Hague: Dutch Scientific Council on Government Policies, 1991).

11 Knibbe, *Agriculture in the Netherlands*, 227–236.

12 J. Bieleman, "Landbouw en het sociaal-economische krachtenveld," in Schot et al., *Techniek in Nederland*, vol. 3, 19.

13 Ibid., 18.

14 P. R. Priester, "Honderdvijftig jaar mechanisatie—inleiding," in Schot et al., *Techniek in Nederland*, vol. 3, 65.

15 Ibid., 66.

16 Priester, "Het melkveehouderijbedrijf," 103.

17 Ibid., 101.

18 See Bieleman, "Landbouw en het sociaal-economisch krachtenveld."

19 Priester, "Het melkveehouderijbedrijf," 107.

20 P. R. Priester, "Het akkerbouwbedrijf," in Schot et al., *Techniek in Nederland*, vol. 3, 90–93.

21 Ibid., 90.

22 On the history of the agricultural equipment manufacturer Vicon in Nieuw Vennep, see R. Stokvis, *Ondernemers en industriële verhoudingen: Een onderneming in regionaal verband (1945–1985)* (Assen: Van Gorcum, 1989).

23 Priester, "Honderdvijftig jaar mechanisatie," 69.

24 J. Bieleman, "Bodemverbetering en waterbeheersing," in Schot et al., *Techniek in Neder-land*, vol. 3, 43; P. R. Priester, "Paarden en trekkers," in Schot et al., *Techniek in Nederland*, vol. 3, 73–81; and Priester, "Het melkveehouderijbedrijf," 117 passim.

25 Van Zanden, *The Economic History*, 140–141. See also M. N. de Groot and A. L. G. M. Bauwens, "Vijftig jaar landbouw?beleid in Nederland: Consensus en conflict," in Bauwens, De Groot, and Poppe, *Agrarisch bestaan*, 146–169, and A. van den Brink, *Structuur in beweg-ing: Het landbouwstructuurbeleid in Nederland, 1945–1985* (Wageningen: University of Wageningen, 1990), 22.

26 Van Zanden and van Riel, *Strictures of Inheritance*, 287; M. S. C. Bakker, "Boter," in Lintsen et al., *Geschiedenis van de Techniek in Nederland*, vol. 1, 103–134.

27 A. Maris, C. D. Scheer, and M. A. J. Visser, *Het kleine-boerenvraagstuk op de zandgronden: Een economisch-sociografisch onderzoek van het Landbouw-Economisch Instituut* (Assen: Van Gorcum, 1951), 12–13.

28 Bieleman, "Landbouw en het sociaal-economisch krachtenveld," 17–18.

29 Jan Douwe van der Ploeg, *De virtuele boer* (Assen: Van Gorcum, 1999), 263–265.

30 Priester, "Honderdvijftig jaar mechanisatie," 67, and Priester, "Melkveehouderijbedrijf," 104.

31 The history of this resistance is still to be written. For preliminary efforts, see A. T. J. Nooij, *De boerenpartij: Desoriëntatie en radicalisme onder de boeren* (Meppel: Boom 1969), and A. Houttuyn Pieper, "De acties der 'vrije boeren' in sociologisch perspectief," *Land-bouwkundig Tijdschrift* 74, no. 11 (1962): 449–460.

32 Van den Brink, *Structuur in Beweging*, 26 (table 2.5).

33 The added value is the market value minus the value of the raw and auxiliary materials. The food and foodstuffs industry, which is closely tied to agriculture, produced dairy, sugar, flour, margarine, cocoa, potato flour, beer, tobacco, canned vegetables, fruits, and fish, and so on. The metal industry can be divided into metallurgical companies (iron, steel, zinc, tin and aluminum), metal products (wire, sheet metal), machines, steel con-struction, transportation (shipbuilding, bicycles, bodyworks, and trucks, automobiles and airplanes).

34 To avoid confusion, we refer to names of companies as formally used in 1970, after merg-ers and name changes. Where relevant we also use earlier names of specific companies.

35 E. Bloemen, J. Kok, and J. L. van Zanden, *De top 100 van industriële bedrijven in Nederland, 1913–1990* (The Hague: Advisory Council on Science and Technology Policies, 1993), 17.

36 Ibid.; J. P. Smits, "Economische Ontwikkeling, 1800–1995," in Van der Bie and Dehing, *National goed*, 21

37 De Jong, *Nederlandse industrie*, 111, is therefore right to suggest that a separate study could be devoted to this issue. Although businesses were counted in 1930, 1950, 1963, and 1978, we have seen no extensive presentation of results yet. On the surveys themselves, see J. Atsma, "Structuur van het bedrijfsleven: Bedrijfstellingen 1930–1978," in B. Erwich and J. G. S. J. van Maarseveen, eds., *Een eeuw statistieken: Historisch-methodologische schetsen van de Nederlandse officiële statistieken in de twintigste eeuw* (Voorburg: Cen-traal Bureau Statistiek, 1999), 367–390. The results of the 1930 counting are analyzed in an article in *Economisch-Statistische Berichten* of December 1942; based on various sources, it maps developments after 1859. Other publications provide few numbers. In A. C. M. Jansen and M. de Smidt, *Industrie en ruimte: De industriële ontwikkeling van Ne-derland in een veranderend sociaal-ruimtelijk bestel* (Assen: Van Gorcum, 1974), reference is made to volumes 8 (1966), 9 (1967), 11 (1969), and 13 (1971) of *Maandstatistiek van de indus-trie*. De Vries, in his *Nederlandse economie tijdens de 20ste eeuw*, also provides a few num-bers, notably on 1963; in this year the percentage of small businesses is 16.6, medium-sized businesses represent 23.9 percent, and large companies 59.5 percent. His source is *Statistisch Zakboek 1968*, 79. Finally, we would like to refer readers to P. A. V. Janssen, *Groot*

en klein in de Nederlandse industrie, 1953–1968–1980: Een poging tot kwantificering van het concentratieverschijnsel en van de verschuivingen binnen het industriële patroon (The Hague: Toegepast Natuurwetenschappelijk Onderzoek and Stichting Maatschappij en Ondernemening, 1971).

38 See H. J. Scheffer, "Ontwikkeling van de ambachts- en fabrieksnijverheid in Nederland," *Economisch-Statistische Berichten* 27, no. 1402 (1942): 542–546.

39 These figures are not consistent with De Jong, *Nederlandse Industrie*, 111, who writes, "My impression is that in many industries it [average business size] decreased, in part as a result of the crisis measures."

40 Jansen and De Smidt, *Industrie en ruimte*, 49 (table 6).

41 See Dick van Lente, "The Crafts in Industrial Society: Ideals and Policy in the Netherlands, 1890–1930," *Economic and Social History in the Netherlands* 2 (1990): 99–119.

42 Keetie E. Sluyterman and Hélène J. M. Winkelman, "The Dutch Family Firm Confronted with Chandler's Dynamics of Industrial Capitalism, 1890–1940," *Business History* 35, no. 4 (1993): 152–183.

43 For this conclusion, see O. A. L. C. Atzema and E. Wever, *De Nederlandse industrie: Ontwikkeling, spreiding en uitdaging* (Assen: Van Gorcum, 1994), 57. For an excellent survey, see also E. Nijhof, "Industrialisatie en regionale identiteit," in Corrie van Eijl, Lex Heerma van Vos, and Piet de Rooy, eds., *Sociaal Nederland: Contouren van de twintigste eeuw* (Amsterdam: Aksant, 2001), 171–185.

44 Alfred D. Chandler, *Scale and Scope: The Dynamics of Industrial Capitalism* (Cambridge, Mass: Harvard University Press, 1990), 27. More than his followers, Chandler distinguishes among various sectors with different technologies and markets: "To repeat, different production technologies have different scale-scope economies. Costs decrease and increase more sharply in relation to volume in some production processes than in others. . . . So too the potential for exploiting the economies of scope varied widely from industry to industry." We are grateful to Frank Geels for pointing this out to us.

45 See Van Zanden, *Economic History*, 39, and chapter 8, by Jan Pieter Smits, in this volume.

46 Van Zanden, *Economic History*, 35, does explicitly mention the work by Sluyterman and Winkelman, but he does not use it in his analysis.

47 Sluyterman and Winkelman, "Dutch Family Firm," 154. See also Ben P. A. Gales and Keetie E. Sluyterman, "Outward Bound: The Rise of Dutch Multinationals," in Geoffrey Jones and Harm G. Schröter, eds., *The Rise of Multinationals in Continental Europe* (Aldershot: Edward Elgar, 1993), 65–98, where the authors devote attention to smaller firms as well.

48 Sluyterman and Winkelman, "Dutch Family Firm." Conclusions by Chandler on England have also been challenged: L. Hannah, "Scale and Scope: Towards a European Visible Hand," *Business History* 33, no. 4 (1991): 302.

49 Sluterman and Winkelman, "Dutch Family Firm," 176, indicate that family ties were also very important among the six leading companies.

50 See in particular De Jong, *Nederlandse Industrie*, chapter 7. See also Mila Davids, "Van stoom naar stroom: De veranderingen in aandrijfkracht in de industrie," in Schot et al., *Techniek in Nederland*, vol. 6, 271–283.

51 On the Marshall Plan, see Frank Inklaar, *Van Amerika geleerd: Marschall-hulp en kennisimport in Nederland* (The Hague: SDU Uitgevers, 1997), 47–50 (on mass production). See also Pien van der Hoeven, *Hoed af voor Marshall: De Marshall-hulp aan Nederland* (Amsterdam: Bakker, 1997), chapter 6 (on the specific importance of this aid for the Dutch economy). Finally we refer to chapter 4, in this volume, by Rienk Vermij, for more discussion of standardization and type limitation.

52 Karel Davids, "Familiebedrijven, familisme en individualisering: Nederland, ca. 1880–1990—Een bijdrage aan de theorievorming," *Amsterdams Sociologisch Tijdschrift* 24, no. 3–4 (1997): 536–537, 543–545.

53 See Hein A. M. Klemann, *Nederland, 1938–1948: Economie en samenleving in de jaren van oorlog en bezetting* (Amsterdam: Boom, 2002), especially chapter 7, on industry. Klemann demonstrates that industrialization efforts simply continued, notably in the first years of the occupation. Major investments were made to meet the German demand and as a substitute for imports. Companies also wanted to convert their money into permanent assets that would retain their value. Furthermore, machines and installations were barely replaced in the last years of war. During the war various businesses developed new products and production techniques and they invested R&D and human capital. By 1947, therefore, Dutch industrial output was already back at the prewar level.

54 Van Zanden, *Economic History*, 3. Van Zanden applies this pattern to the growth of multinationals and the welfare state.

55 B. P. Gales and J. P. Smits, "Een Nederlands scheppingsverhaal," in Schot et al., *Techniek in Nederland*, vol. 2, 29–43. On a similar nationalism on the part of chemical engineers, who wanted to create a large state chemical company, see Ernst Homburg, "De Eerste Wereldoorlog: Samenwerking en concentratie binnen de Nederlandse chemische industrie," in Schot et al., *Techniek in Nederland*, vol. 2, 317–331.

56 B. P. A. Gales and J. P. Smits, "Delfstofwinning in Nederland gedurende de twintigste eeuw," in Schot et al., *Techniek in Nederland*, vol. 2, 20.

57 Gales and Smits, "Nederlands scheppingsverhaal," 36–37.

58 Ibid., 40.

59 F. van der Most, J. W. Schot and B. P. A. Gales, "Zout," in Schot et al., *Techniek in Nederland*, vol. 2, 91–101.

60 E. Homburg, J. S. Small, and P. F. G. Vincken, "Van carbo- naar petrochemie, 1910–1940," in Schot et al., *Techniek in Nederland*, vol. 2, 355–356. See also E. Homburg, A. J. van Selm, and P. F. G. Vincken, "Industrialisatie en industriecomplexen: De chemische industrie tussen overheid, technologie en markt," in Schot et al., *Techniek in Nederland*, vol. 2, 389–390.

61 Van Zanden, *Economic History*, 37; F. Messing, "Het economisch leven in Nederland, 1945–1980," in D. P. Blok et al., eds., *Algemene geschiedenis der Nederlanden*, vol. 15 (1982), 178.

62 The following is based on G. P. A. Gales, ed. "Delfstoffen," in Schot et al., *Techniek in Nederland*, vol. 2, in particular chapter 4, "Olie en gas," 67–89, and chapter 6, "Het leven bruist in Nederlands bodem," 103–111; and G. P. J. Verbong, ed., "Energie," in Schot et al., *Techniek in Nederland*, vol. 2, especially chapter 7, "De revolutie van Slochteren," 203–219.

63 See also chapter 8, by Jan Pieter Smits, in this volume.

64 For these data, see G. P. J. Verbong, "Ter introductie," in Schot et al., *Techniek in Nederland*, vol. 2, 11–15.

65 For the United States, see Philip B. Scranton, *Endless Novelty: Specialty Production and American Industrialization, 1865–1925* (Princeton: Princeton University Press, 1997). See also Jonathan Zeitlin and Charles Sabel, "Historical Alternatives to Mass Production," *Past and Present*, no. 108 (1985): 133–176.

66 See chapter 4, by Rienk Vermij, in this volume.

67 M. Davids, "Het Philipscomplex," in Schot et al., *Techniek in Nederland*, vol. 6, 357–376.

68 See M. Davids, "Innovativiteit, complexvorming en arbeid," in Schot et al., *Techniek in Nederland*, vol. 6, 377–382.

69 R. Vermij, "Het gaat vanzelf—of niet? Industriële automatisering in Nederland," in Schot et al., *Techniek in Nederland*, vol. 6, 302–317.

70 See A. N. Hesselmans, G. P. J. Verbong, and H. Buiter, "Binnen provinciale grenzen: De electriciteitsvoorziening tot 1940," in Schot et al., *Techniek in Nederland*, vol. 2, 141–159.

71 See R. Oldenziel et al., "Het huishouden tussen droom en werkelijkheid: Oorlogseconomie in vredestijd, 1945–1963," in Schot et al., *Techniek in Nederland*, vol. 4, 126.

72 J. L. Schippers and G. P. J. Verbong, "De revolutie van Slochteren," in Schot et al., *Techniek in Nederland*, vol. 2, 214.

73 Homburg, van Selm, and Vincken, "Industrialisatie en industriecomplexen," 380.

74 Ibid., 377.

75 Ibid., 391, also graph 7.1.

76 Ibid., 390–396.

77 Van Zanden, *Economic History*, 27–28.

78 For a company such as Philips it was less important to exploit networks of materials than to capitalize on the technological competencies it had developed. For example, Philips began as a manufacturer of light bulbs, but during World War I the company faced problems related to the supply of glass bulbs. This is why it decided to set up it is own glass factory. Its experience with glass technology was of great importance for the firm's diversification into the fields of radio, X-ray and neon tubes. See Davids, "Philipscomplex."

79 The following discussion draws on Homburg, Small, and Vincken, "Van carbo- naar petrochemie," 341–44, and Homburg, Van Selm, and Vincken, "Industrialisatie en industriecomplexen," 381–385. See also H. Lintsen, ed., *Research tussen vetkool en zoetstof: Zestig jaar DSM Research* (Zutphen: DSM and Stichting Historie der Techniek, 2000), chapters 1 to 3.

80 Homburg, "Eerste Wereldoorlog," 324.

81 F. V. van der Most et al., "Nieuwe synthetische producten: Plastics en wasmiddelen na de Tweede Wereldoorlog," in Schot et al., *Techniek in Nederland*, vol. 2, 358–375.

82 "Active Consumers": We intentionally link "active" and "consumers" in our heading to counteract the common notion of consumption as passive. It is also possible to speak of "users" instead of consumers.

83 See A. Albert de la Bruhèze and F. C. A. Veraart, *Fietsverkeer in praktijk en beleid in de twintigste eeuw: Overeenkomsten en verschillen in fietsgebruik in Amsterdam, Eindhoven, Enschde, Zuidoost-Limburg, Antwerpen, Manchester, Kopenhagen, Hannover en Basel* (The Hague: Ministry of Traffic and Public Works, 1999), 44.

84 W. O. de Wit, "Radio tussen verzuiling en individualisering," in Schot et al., *Techniek in Nederland*, vol. 5, 220.

85 For the figures on radio, see W. O. de Wit, "Het communicatielandschap in de twintigste eeuw: De materiële basis," in Schot et al., *Techniek in Nederland*, vol. 5, 174. For the estimates on vacuum cleaners, we draw on H. Makkink, "Met een stofzuiger hoorde je erbij," in Ruth Oldenziel and Carolien Bouw, eds., *Schoon genoeg: Huisvrouwen en huishoudtechnologie in Nederland, 1898–1998* (Nijmegen: SUN, 1998), 105. The figures are based on 242 interviews; see also De Jong, *De Nederlandse Industrie, 1913–1965*, 2–4.

86 For data on phone connections, see De Wit, "Communicatielandschap in de twintigste eeuw," 168. For data on the diffusion of automobiles, see J. W. Schot et al., "Concurrentie en afstemming: Water, rails, weg en lucht," in Schot et al., *Techniek in Nederland*, vol. 5, 20–21, 31, 32. See also Peter-Eloy Staal, *Automobilisme in Nederland: Een geschiedenis van gebruik, misbruik en nut* (Zutphen: Walburg Pers, 2003), especially chapter 2.

87 Few accurate figures are available on the diffusion of devices and appliances in the interwar period. Available data can be found in H. Baudet, *Een vertrouwde wereld: 100 jaar innovatie in Nederland* (Amsterdam: Bakker, 1986). For the period after 1945 better data are available; see R. Oldenziel, "Epiloog," in Schot et al., *Techniek in Nederland*, vol. 4, 147–151, 148 (chart).

88 Kees Schuyt and Ed Taverne, *1950- Prosperity and Welfare* (Assen-Basingstoke: Van Gorcum-Palgrave/Macmillan, 2004), 116. For commentary, see Ruth Oldenziel, "Gender in zwart-wit anno 1950," *Tijdschrift voor Sociale Geschiedenis* 29, no. 1 (2003): 80–89.

89 For an earlier articulation of the argument that consumer society had already arrived in the Netherlands by the first half of the twentieth century, see Johan Schot, "Consumeren als het cement van de lange twintigste eeuw," *De Nieuwe Tijd* 9 (1997): 27–37. See also Adri Albert de la Bruhèze and Onno de Wit, "De productie van consumptie: De bemidde-

ling van productie en consumptie en de ontwikkeling van de consumptiesamenleving in Nederland in de twintigste eeuw," *Tijdschrift voor Sociale Geschiedenis* (special issue on the production of consumption) 28, no. 3 (2002): 257–272; and Adri Albert de la Bruhèze, and Ruth Oldenziel, *Manufacturing Technology, Manufacturing Consumers. The Making of Dutch Consumer Society* (Amsterdam: Aksant, 2009).

90 De Wit, "Radio tussen verzuiling en individualisering," 220.

91 Ibid., 221; R. Oldenziel and M. Berendsen, "De uitbouw van technische systemen en het huishouden: Een kwestie van onderhandelen, 1919–1940," in Schot et al., *Techniek in Nederland*, vol. 4, 47–48.

92 In this respect, automobile use followed bicycle use, which had the same dual character. For the development of automobile use, see G. P. A. Mom, J. W. Schot, and P. E. Staal, "Werken aan mobiliteit: De inburgering van de auto," in Schot et al., *Techniek in Nederland*, vol. 5 45–73, and Staal, *Automobilisme in Nederland*.

93 The notion of a "world within reach" is derived from chapter 10, by Irene Cieraad, in this volume.

94 The choice to use certain "family products" meant that other products that promised to save household work were less popular. A report by the Dutch Centraal Bureau Statistiek shows that at the start of the 1960s women had 25.4 hours of leisure each week, which was less than men, who had 32.4 hours each week. See chapter 8, by Jan Pieter Smits, in this volume.

95 Prices ranged from twenty to seventy guilders, significantly less than for a radio. See Makkink, "Met de stofzuiger hoorde je erbij."

96 Ibid., 111–112.

97 On the spread of washing machines in the Netherlands see Carianne van Dorst, *Tobben met de was: Een techniekgeschiedenis van het wassen in Nederland 1890–1968* (Eindhoven: Eindhoven University Press, 2007).

98 Oldenziel and Berendsen, "Uitbouw van technische systemen en het huishouden," 48. See also van Dorst, *Tobben met de was*. Joy Parr has also noted the slow spread of washing machines in Canada, because of their poor integration into women's existing laundry practices; see Joy Parr, *Domestic Goods: The Material, the Moral, and the Economic in the Postwar Years* (Toronto: University of Toronto Press, 1999).

99 R. Oldenziel, "Het ontstaan van het moderne huishouden: Toevalstreffers en valse starts, 1890–1918," in Schot et al., *Techniek in Nederland*, vol. 4, 26–27.

100 The following paragraphs are based on ibid., 11–151.

101 A. J. de Regt, *Arbeidersgezinnen en beschavingsarbeid: Ontwikkelingen in Nederland, 1870–1940—Een historisch-sociologische studie* (Meppel: Boom, 1995), 242.

102 See also Piet de Rooy, *Republiek van rivaliteiten: Nederland sinds 1813* (Amsterdam: Mets en Schilt Uitgevers, 2002), 218–220.

103 For a definition and characterization of the notion of the "midfield" (*middenveld*), see O. Munster, E. J. T. van den Berg, and A. van der Veen, *De toekomst van het middenveld* (The Hague: Uitgever DELWEL, 1996). Political history concentrates on actors that bridge the distance between state and citizens; in our argument, "midfield" actors try to close the gap between producers and consumers, with the state acting as a major third party. See also Johan Schot and Adri Albert de la Bruhèze, "The Mediated Design of Products, Consumption and Consumers in the Twentieth Century," in Nelly Oudshoorn and Trevor Pinch, eds., *How Users Matter: The Co-Construction of Users and Technology* (Cambridge Mass.: MIT Press, 2003), and De la Bruhèze and De Wit, "Productie van consumptie."

104 A. van Otterloo and M. Berendsen, "The 'Family Laboratory': The Contested Kitchen and the Making of the Modern Housewife," in De la Bruhèze and Oldenziel, *Manufacturing Technology, Manufacturing Consumers*, 115–138.

105 See R. Oldenziel and C. van Dorst, "De crisis: Kapitaal- versus arbeidsintensieve techniek,

1929–1940," in Schot et al., *Techniek in Nederland*, vol. 4, 77–78. See also Van Dorst, *Tobben met de was.*

106 Oldenziel, "Huishouden tussen droom en werkelijkheid," 111.

107 For more on this role of the bicyclists union, see Gijs Mom, Johan Schot, and Peter Staal, "Civilizing Motorized Adventure: Automotive Technology, User Culture, and the Dutch Touring Club as Mediator," in: de la Bruhèze and Oldenziel, *Manufacturing Technology, Manufacturing Consumers*, 139-158.

108 As a result, women could become familiar with the enticements of American consumer society while simultaneously being urged by the government to be economical. Industrialization was based on low wages and thus on an economy of mending and patching up. See Oldenziel, "Huishouden tussen droom en werkelijkheid."

109 On the image of the 1950s, see Hans Righart and Piet de Rooy, "In Holland staat een huis: Weerzin en vertedering over 'de jaren vijftig,'" in Paul Luykx and Pim Slot, eds., *Een stille revolutie: Cultuur en mentaliteit in de lange jaren vijftig* (Hilversum: Verloren, 1997), 11–18. See also J. C. H. Blom "'De Jaren Vijftig' en 'De Jaren Zestig'?," *Bijdragen en Mededelingen Betreffende de Geschiedenis der Nederlanden* 112, no. 4 (1997): 517–528.

110 James C. Kennedy, *Nieuw Babylon in aanbouw: Nederland in de jaren zestig* (Amsterdam: Boom, 1995), especially chapter 1.

111 See Van Zanden, *Economic History,* 144.

112 Schuyt and Taverne, *1950,* 19–20.

113 Pierre van der Eng, *De Marshall-hulp: Een perspectief voor Nederland* (Houten: De Haan, 1987), 245–246.

114 See Schuyt and Taverne, *1950,* in particular parts A and B. See also chapter 7, by Dick van Lente and Johan Schot, "Technology as Politics," in this volume. The experience of occupation during World War II, as well as the obsession with modernization, technology, and industrialization, was important for this breakthrough. This experience prompted a strong impetus toward social change and the ambition to put the country on the map. People felt a strong desire to put World War II and the preceding period of crisis and unemployment behind them as soon as possible, to make a new beginning and start looking ahead.

115 In this period modernization also became politically charged because it was seen a way to fend off Communism and totalitarianism.

116 On the term "a searching time," see Remieg Aerts et al., *Land van kleine gebaren: Een politieke geschiedenis van Nederland, 1780–1990* (Amsterdam: SUN, 1999), 182.

117 Kennedy, *Nieuw Babylon,* 42.

118 J. C. H. Blom, "Jaren van tucht en ascese: Enige beschouwingen over de stemming in Herrijzend Nederland, 1945–1950," in P. W. Klein and G. N. Plaat, eds., *Herrijzend Nederland: Opstellen over Nederland in de periode 1945–1950* (The Hague: Nijhoff, 1981), 125–158.

119 De Rooy, *Republiek van rivaliteiten,* 218.

120 Schuyt and Taverne, *1950,* 94.

121 Jan Willen Duyvendak, *De planning van ontplooiing: Wetenschap, politiek en de maakbare samenleving* (The Hague: SDU Uitgeverij, 1999), especially chapter 3.

122 For this analysis see Ulrich Beck, *Risk Society: Towards a New Modernity* (London: Sage, 1992).

Between Sensation and Restriction:
The Emergence of a Technological Consumer Culture

Irene Cieraad

A quotation from a book published in 1942 underscores the difference between the "masses" and the intellectual elite when it comes to the experience of technology. In that year the pioneering Hungarian psychoanalyst Franz Alexander, in *Our Age of Unreason: A Study of the Irrational Forces in Social Life*, wrote:

> While the masses unwittingly were enjoying with a naïve pride the blessings of technical advance, their radios, automobiles, Pullman cars, and airliners—scholars, teachers, philosophers, psychologists, social scientists, and writers were becoming increasingly concerned about the one-sided development ... which has led to our top-heavy technical civilization and has not contributed in the least to the improvement of human relationships.[1]

Although it is common for social classes or consecutive generations to attribute different meanings to technological devices and appliances, the enthusiasm of many about everyday life's new technological possibilities and the concern, if not revulsion, among a small vocal group about their negative influence on culture and civilization is indicative of a stark contrast indeed. One of these critics was Johan Huizinga, the influential Dutch historian who warned against culture's "galvanization," as he called it, caused by the actions of the "semicivilized" that were conquering the whole world. Movies and radio, Huizinga argued, have accustomed the semicivilized

A street in an old working-class neighborhood in 's-Hertogenbosch in 1965. In the early 1960s the income of the Dutch working population went up, allowing not just the middle classes but also the lower classes to buy luxury goods such as a car or television. The car may be third- or fourth-hand, like the ones in the photo, but the television set was usually brand new. Television sets were still black-and-white receivers of programs broadcasted mostly during evening hours by only two national public stations. A TV antenna on the roof was indispensable for good reception. In no time at all, parked cars and a forest of TV antennas on roof tops altered the appearance of streets, villages, and cities. Thus, cars and TVs (and their antennas) became the obtrusive icons of postwar consumer society.

to a questionable simplification of his mental perceptions. He does not see ... much more than the photographic caricature of a highly limited visual reality, or he only listens with half an ear to music that is wasted on him or to announcements that he might better have *read*. . . . The highly developed art of advertising has spread its power beyond the most lofty regions and made cravings for cultural offerings irresistible. In this cultural galvanization of an entire people, the country that was first completely conquered by technology, the United States of America, has provided the model, but all countries of Europe have followed it swiftly.[2]

Since the end of the nineteenth century, European intellectuals have interpreted American society as representing a frightening image of the future. In their writings and lectures, they warned in particular against the disastrous influence of technological developments that in line with the American model would be accompanied by unbridled materialism and moral decay.[3] Considering itself the keeper of the noble spiritual values of classic culture, Europe's intellectual leaders realized that new technological possibilities would alter society and culture. Throughout the twentieth century, this elite's dissenting voices have accompanied the rise of our technological consumer culture.

This is not to suggest that new technological possibilities were always accepted instantly. One of the assumptions of this technology history is that frequently a lengthy process of social embedding preceded the acceptance of a new technology. Not only did manufacturers have to find a market for their technological commodities, but also consumers had to be able to use them. After all, without power supply to power a new appliance or, as in the case of radio, without stations that broadcast programming, not much will happen. Conversely, new technological possibilities, in the shape of specific devices or appliances, somehow have to reach users or create a need among them in order to ensure their continued use, whether for entertainment, convenience, or utility. The concept of "cultural embedding" refers precisely to the emergence of this need and experience of user groups. Cultural embedding involves not only change in existing cultural ideas but also the articulation of new cultural notions such as modernity and speed. Specific notions of comfort, hygiene, beauty, adventure, self-development, identity, human relations, and, more generally, quality of life have changed through the use of devices, and they in turn have altered the ways in which technology is experienced.

Scholars and technology critics have argued that this process of cultural embedding has frequently evolved in problematic ways—and not only these groups. Similar views have also been expressed in newspapers, government

The Idema family from Assen, photographed in 1953 with their radio. "With this magical box called radio we have in fact brought a dangerous thing into our home," a 1951 household encyclopedia stated. A concerned cultural elite insisted on the hazards of spiritual decay, potentially caused by unselective use of the medium. These concerns were not shared by the proud owners. In 1953, before television became widespread, a fancy radio set was a symbol of prosperity in a resurrected country. For photographers, then, radio was the photogenic decor for portraying the modern family, as in the photo of the Idema family.

reports, and a host of other, quite divergent, publications. Consider, for instance, this call to selective radio use in a handbook for women from 1951:

> With this miracle box called radio we have in fact brought a dangerous object into our home…. The whole world is available to the listener through radio…. But all that enters family life through radio is not equally fit for all family members…. Another risk is that the outside world as it comes to us via radio is so full of divergent events that people run the risk of falling victim to moral decay…. By listening to the radio each and every day from early morning to late at night, thus bringing the outside world into your home, you will soon no longer discern anything about this world anymore; you will no longer be able to take in information…. Radio, however, is not meant for these people. They might as well keep the vacuum cleaner on all day …. Radio is meant to make us enjoy all there is to be enjoyed in the world; it is meant to increase our knowledge and sharpen our insight; it should

entertain us and we should let ourselves be educated and informed. But radio will only serve us well if we make a sensible selection from all it has to offer us. The possibilities *and* risks potentially tied to radio use are even larger in the case of television…. They can be both a danger and a civilizing factor. It all depends … on the usage we make of it, or, in other words, on our self-discipline.[4]

Such preaching may contain elements familiar to us, but more typically it makes us smile, particularly when it is about appliances that we take for granted today. Cars, phones, radios, television sets, and cameras—and others more recent—have become so inextricably bound up with our everyday life that we no longer notice their impact. This taken-for-granted-ness in our dealing with appliances has turned them into a kind of *tacit technology*: they have become an integral part of our behavioral and psychological patterns.[5] That is cultural embedding.

Cultural embedding—the establishment of the meaning, use, and behavioral pattern related to an appliance or technology—is part and parcel of the process of social embedding in which early adopters set a trend, but it has rarely involved automatic processes. When new technologies and appliances are introduced, the enthusiasm of the first curious users tends to play a pivotal role. For example, radio amateurs were already tinkering with radio before radios in living rooms could generate a single decent sound. Way into the early morning hours these amateurs would search the ether with their home-made devices mainly to generate a lot of cracking and beeping. Apparently, technologies and appliances, once introduced, managed to appeal to first users' drives and slumbering needs, which also accounts for their uncommon level of enthusiasm and persistence. For the general audience, however, the social embedding of radio in the Netherlands was what gave rise to the country's unique public broadcasting system based on a social segregation of listeners. From the start, in the 1920s and 1930s, Dutch radio listeners were expected to tune in exclusively to programming transmitted by the broadcast organization tied to their own wider religious or political community (be it Catholic, Protestant, socialist, etc.), but popular entertainment programs directed at other communities than one's own challenged this more or less voluntary restriction and increasingly turned a matter of conscience into needless torture. Although radical changes in Dutch public broadcasting during the second half of the twentieth century caused substantial changes in listening habits, radio as such was no longer a contentious issue. At most one occasionally heard complaints about the noise nuisance of portable radios in public spaces.

In the 1960s, protests by members of the Dutch intellectual elite against the influence of technology on society and culture fell on deaf ears. Their cultural struggle increasingly became a case of fighting a rearguard action. Its outcome perhaps accounts for the odd combination of contradictory attitudes in today's Dutch consumer society, marked as it is by the simultaneous longing for carefree enjoyment and an urge toward self-restraint and self-control. These elements of social embedding raise a number of questions: Where did the intellectual cultural critique of technology come from to begin with? What were its historical roots? What have its influence and effects been? Although "user" is a core concept in technology history, the actual role of users in the embedding of technology is either hard to track down or subject to multiple interpretations. This is why the research reported here largely has an exploratory character.

Our argument on both the role of Dutch consumers and the rise of new cultural notions relies in part on data derived from literary and other written sources that describe dealings with and views about specific technologies. For example, automobiles clearly stirred the imagination of Dutch authors in the opening decades of the twentieth century. Depictions of cars, cameras, radios, phones and television sets—as found in ads, photos, cartoons, and book illustrations—also constitute a rich source. After 1880 visual materials became increasingly abundant because of improved printing technologies. The popularity of the camera, together with the rise of amateur photography, generated another rich supply of images. Drawing on such materials, this chapter analyzes the relationship between specific appliances and their owners and users.[6] The emphasis is on the private sphere, in particular the home, family life, and leisure.

Consumers in the Making

Until the 1950s, cultural critics of technology especially targeted the "masses," whose members with their childish materialism and hedonism were held responsible for social ills such as moral decay and lack of community sense. This negative image dates back to the nineteenth century and should be considered against the backdrop of the era's urban development, whereby increasing industrialization caused large numbers of job seekers to move from the countryside to the cities. The elites and authorities experienced the ensuing urban overpopulation and visual massiveness as a threat to the established urban order, much as they interpreted the dire circumstances and poverty that drove the masses to the city as greed. Intellectuals and urban leaders thus

"Everyone takes pictures with cameras [*foto-toestellen*] from Mij. Guij de Coral & Co." This company boasted shops in Amsterdam and four other cities. More than for bicycles or sewing machines, manufacturers tried to create a mass market for cameras as early as 1900. The design of this advertising poster with its slogan "Everyone takes pictures," from 1901, may be called revolutionary. Until then manufacturers had geared their ads to a particular class or profession, but this ad captures the product's potential mass appeal in a single image: not only professional photographers but also servants, ladies, grandmas, and even children snap pictures. The old hierarchy of class and age is reduced here to a difference in price and quality of the cameras shown. Price, quality, and—by extension—taste became new, distinctive criteria in a society oriented more and more toward consumption.

created the threatening image of a criminal urban mass, an image that was counterbalanced by romantic idealizing of a rural population uncorrupted by industrialization and invested with an aura of morality and craftsmanship.[7]

It seemed only natural that factory workers, as the most visible accomplices of mechanization and industrialization, were the first to be stigmatized as the mechanical masses "driven by cogwheels."[8] By the late nineteenth century, however, this stereotyping began to shift from the sphere of production to that of consumption: shopping women gathering in the recently opened department stores were similarly depicted as a mindless crowd merely "driven by desire." If the social tensions involved in the burgeoning workers' movement accounted for this particular labeling of workers, the shift toward shopping women as targets of ire seemed to point to new tensions associated with the budding women's movement and its pursuit of equal rights.[9] In this sense, old social conflicts are at the root of contemporary views of consumer culture.

Moreover, the shift of focus of critical attention from factories and production to department stores and consumption occurred not only because critical intellectuals viewed these stores as a refuge for women (away from male company), but also, and in particular, because department stores through their spectacular display of goods appealed to the basic instinct of women, thus luring them into buying. Giving in to visual seduction and distraction, whether in department stores or movies or advertising, came to be seen as one of the masses' major vices, one that had long been marked as a typically female vice—by all males, not just by male intellectuals.

The widespread image of woman as driven by desire was contested in particular by Dutch feminists, but not in a consensual way. For example, the feminist who in 1920 took the initiative to set up an antifashion society out of sheer irritation about her gender's slavish consumer behavior caused quite a stir. In addition to making declarations of sympathy suggesting that only those in the masses have a desire for fashion "because they have no character and lack personality," other feminists criticized the fact that only women were scapegoated, while men, with their high tailors' bills, were exonerated. Moreover, feminists targeted the scrupulous seducers, the producers: "The stores are filled with the most misshapen objects, with silly soulless items invented by those who crave for profits and who make tin into gold and glass into pearls, because like children the masses grasp at all that shines and glows."[10]

Women's awareness of sin remained a sensitive issue in feminist circles. In the 1920s, expressions of guilt like the following triggered fierce reactions in the feminist press:

We, women of today, are spoilt and yet—or for this very reason— we long for more, for what is new. A modern home is stuffed with non-essential things, its kitchen having all sorts of mostly unusable electric appliances. Although these nice things do neither improve our life nor make it easier, we are still attached to them and ... we move heaven and earth to get hold of them.[11]

A new element slipped in as justification that was typical of the time: the guilty parties were not the women; industrialization as a system coerced them to buy and grasp for the new. A person was no longer an individual; as a cog in the wheel of industrialization he slavishly functioned as worker or con-

The interior of a Rotterdam branch of the De Gruyter grocery stores, around 1920. Consumer society is still associated more with food than with appliances or electronic devices, and the word "consumer" is associated with women rather than men. These associations are rooted in the past. Mass consumption first revealed itself in grocery stores, where most customers, and employees, were women. As early as the 1900s national chains emerged in this sector with branches that tried to outperform each other in terms of the fancy designs and features of these palaces of consumption. De Gruyter, of 's-Hertogenbosch, had an especially high reputation for the stores' exceptional décor (note the murals, decorative tilework, and crystal chandelier).

sumer. If he stops craving, hundreds of thousands have no more income. Do ninety-nine percent of the automobile drivers really have any purpose other than preserving the large factories of new and old world? ... The machine has become the demon that chases us, incessantly. Today's people have to be unfulfilled, erratic, changeable, craving, so that our gigantic factories will go on producing.[12]

With this era's rising unemployment the leaders of the Dutch working-class movement equally grew aware of this "demonic" relation between consumers and workers, between buying power and employment. In 1935 they proposed a buying-power incentive in their Labor Plan. It implied a heroic double role for workers: as workers *and* consumers they would make the economy grow again, to the benefit of everyone.[13]

After World War II the masses, conceived in various traditional ways, proved to be public enemy number one once more. A government report described the worrisome "social lawlessness of mass youth," and countless publications pointed to the danger of moral decay represented by the light entertainment of movies, radio, jazz music, and television.[14] Huizinga wrote:

> The industrial era has produced the semicivilized. The general edu-
> cation system, combined with a superficial leveling of classes and the
> prevailing looseness of spiritual and material interaction, has turned
> the semicivilized into our community's dominant type. The semiciv-
> ilized, however, is the archenemy of personality. By their sheer num-
> ber and homogeneity they smother the seed of the personal in the soil
> of culture.[15]

In Huizinga's view, the domains of culture and technology are irreconcilable and also represent an unbridgeable social contradiction between the semiciv-ilized and the fully civilized.[16] Like culture, personality was the result of active self-development and could only be attained by a select group with classic intellectual training. In the domain of technology, by contrast, the primacy of the future reigned supreme, as was true of the technological progress that would be at everyone's disposal and for which the masses did not have to do anything but pay. Furthermore, the egalitarian and material character of new technological developments inevitably led to spiritual laziness.[17] Huizinga used the term "semicivilized" to indicate that more education could perhaps lead to technological acuity but did not imply cultural civilization as well.

The social concerns in the 1950s largely took as their object youngsters and their idolizing of all things American. Sociologists and educational

VERBREEK DE NOODLOTS-KRING

Hoe minder men verdient, hoe minder men koopt.
Hoe minder men koopt, hoe groter de werkloosheid.
Hoe groter de werkloosheid, hoe minder men koopt, Enz., enz.

"Breaking the Wheel of Misfortune," a political poster of the social-democrats from the early thirties illustrating their view on the disastrous effects of the lowering of wages on the level of consumption and the economy as a whole. The lines within the image say: "The less one earns, the less one buys. The less one buys, the more unemployment. The more unemployment, the less one buys. Etc. etc." The economic hardships of the 1930s convinced not only leading economists but also the social democrats that consumers with spending power were of vital importance for employment in industry and that this interconnection formed the basis of a healthy economy. In 1935, in their Labor Plan, the social democrats proposed to stimulate spending power. Because of the war and the ensuing reconstruction of industry the plan was not implemented until the 1950s.

experts uncovered the weak mentality of the masses in their conduct and preferences.[18] They lacked personality: their appearance was determined by Hollywood movies, their leisure activities were marked by passivity, and their group behavior was inherently criminal. Sociologists wrote critical treatises on youngsters' spending of leisure, on the baneful influence of mass media on youth and the commercialization of everyday life.[19] This grimness constituted the harbinger of a new social conflict: the generation conflict, which in the 1960s erupted in full glory. At that time Dutch youths evolved into a new consumer group with outspoken preferences. Yet the young generation was not monolithic; there were major dividing lines. Those who were studying had time but no money, whereas those who worked in factories had money but no time. Patterns of consumption began to define specific group identities among Dutch youngsters, with those from the working class taking the lead.[20]

At about this same time social criticism became couched in terms of objectionable consumerism, detestable consumer society, and a materialist culture of consumption. One decade before, even a staunch pessimist such as Huizinga still had hopes that someday things would change: "The daily overload of movies and radio should eventually sadden the semicivilized masses as well. At one point, advertising, … this monster born from the

A 1960 photo of young "greasers" in Amsterdam showing off their new style, featuring smoking and taking up the street space on their motorbikes. In the late 1950s teenagers emerged as a new consumer group with explicit preferences. Although high school kids had more time than money, the reverse was true for those who had a factory job. Consumption patterns made it possible to define particular group identities. The greasers were young Dutch factory workers who defined themselves not only by their slicked-down hair but also by their "belly rubbers" (*buikschuivers*), or mopeds.

womb of the technological era, shall have to become ineffective because of the satiated public's disgust."[21] Although in the 1960s reference was made less and less to the passive masses in the plural and more often to the critical consumer in the singular, for the time being this consumer continued to be female.[22]

Consumer Culture: Embarrassment of Riches

That something had changed became clear in the 1970s. The term "masses" was seen to be denigrating and silently disappeared. Consider, for instance, the name of several academic disciplines: "mass communication" was renamed "communication studies" and "mass psychology" became "social psychology." Radio and television, however, were still labeled "modern mass media." In the then fashionable neo-Marxist literature, however, the original nineteenth-century image of the spineless, manipulated masses that had to be made aware of their role in the capitalist scheme of mass production and mass consumption continued to prevail. The critical writings of German leftist intellectuals who during Nazi rule had fled to the United States played a major role in neo-Marxism. In American society they again identified a

A cartoon from a mid-seventies brochure titled *Consumption Pressure* (*Konsumptie verplicht*), put out by the Dutch action group Strohalm (Straw). A woman approaches a supermarket to buy a kilo of sugar. The poster in the shopwindow shouts "BUY!" The woman exits the supermarket carrying a carton overflowing with all manner of products—but no sugar. She wonders, "What have I forgotten?" In 1972, *The Limits to Growth*, published by an international group of leading scientists, warned of the irreparable damage to mankind and the environment being inflicted by technological progress and industrial production. The report caused shockwaves around the world and gave legitimacy to small-scale initiatives such as Strohalm, which criticized consumerism and the power of advertising.

specter, now conceived as the preeminent perversity of late capitalism: full-blown commercialization.

The terms "consumer society" and "culture of consumption," which emerged in American scholarship of the 1960s, were also embraced by left-wing European intellectuals as critical characterizations of Western culture as a whole. In its material and industrial expansionism, Western consumer culture came to be seen as a threat to the spiritual non-Western cultures in which craftsmen and handicrafts still prevailed.[23] More than previously, intellectuals blamed this new culture on industrialists, and the publication of the Club of Rome report, *The Limits to Growth*, which outlined the critical situation of humanity and Western civilization in particular, was grist for their mill:

> Science and technology have brought us the threat of thermo-nuclear catastrophe but also health and prosperity; population growth and the drift to the city has led to new and humiliating forms of poverty and confinement in dirty, often culturally sterile, noisy and degenerated urbanized environments; electricity and propulsion have reduced the burden of physical labor, but have also drained people's job satisfaction; automobiles have brought us freedom of movement, but also gen-

erated poison in cities and machine fetishism. The unwanted consequences of technology are all too clear and cause a threat to our natural environment that could prove to be irreversible; some people are growing increasingly alienated from society and reject authority; drug addiction, crime and criminality are on the rise, belief is declining, not just in religion, which has supported people for centuries, but also in party-political processes and the effectiveness of social reform. These various difficulties appear to become larger with rising prosperity.[24]

Since the report's publication in 1972, "the environment" has become technology's opposite, if not its victim, which has to be protected from industrial pollution and the exhaustion of natural resources. In the context of ecological thinking, modernization, understood as production growth and technological progress, became a contested process. The doom scenario presented in this report soon gained ground among activist groups and also with government and industry, and its recommendations were taken serious. Technological and industrial innovation became leitmotifs in a critical public debate in which modernization and consumer culture became contested issues.

Social movements that today are widely accepted began to emerge from circles of grassroots activism of the 1970s, including movements associated with diverging concerns such as environmental and animal protection, small-scale agriculture, natural medicine, traditional trades, self-development, reuse of materials, exploration of sustainable sources of energy, and public campaigns for saving energy.[25] Many of these grassroots groups were initiated by the progressive leftist student movement and driven by an idealism that was personally motivated, animated by the belief that social change necessarily has to begin at the level of the individual.[26] The perceived need for curtailing industrial production was translated into a need to minimize individual consumption and stimulate recycling by purchasing secondhand goods. From the anticonsumption perspective all new items were suspect, which is why using secondhand items was promoted, and the business in antiques, classic design, curiosities, and old-timers has been flourishing ever since. Many of the era's alternative ventures have meanwhile evolved into established companies, while the anticonsumption ethic is still largely characteristic of an aging group of diehards only. The longing for what is old, as reflected—and corrupted—by the ongoing trend of retro design and making things look old, has persisted to this day.

Today consuming has clearly moved beyond its infancy. The most tangible influence of consumption criticism on current Dutch consumer behavior is the almost obligatory justification of consumer choices from the view

A secondhand store titled "Too Much" (Veel te Veel) in Amsterdam's Rozengracht, in 1981. Postwar consumer society had become a throw-away society, resulting in towering piles of waste. In response, recycling was touted as the environmentally preferable solution for curbing flows of waste. In the 1970s the young and critical generation actively pursued an image of soberness. By the early twenty-first century, the popularity of flea markets, secondhand stores, and recycling stores has become somewhat less environmentally motivated and driven more by a national fever to collect things.

that all manufacturers and consumers are responsible for protecting our natural environment. With respect to manufacturers this issue of environmental responsibility has become subject to legal regulation, which is perhaps the Club of Rome report's most positive effect. To consumers, however, only moral responsibility applies, occasionally stimulated by government through financial incentives. Although major Dutch consumer organizations show restraint in criticizing specific consumer conduct, they do provide information that allows individual consumers to make more or less responsible choices. The emphasis in this has been on "the best buy" as the optimal balance between price and quality. Consumer organizations and unions consider consuming a right to prosperity that cannot be denied to anyone, even if it means environmental damage. Buying power is more than ever the measure of the strength of the economy.

Consumption: From Class to Taste

The criticism of consumption that first emerged in the Netherlands in the late nineteenth century was an expression of changing social relationships, of a disintegrating class society in which not only income and education were tied to class and creed, but also housing, business, and the things one purchased. Gentlemen were recognized by their hats and workers by their caps, and women belonged to the honorable category of "mistresses" (*mevrouwen*)—those who had servants—or else were "misses" (*juffrouwen*), the category of married or unmarried women without servants, while workers' wives were just "women" (*vrouwen*).[27] A deep-rooted class awareness ensured that a worker would never buy a hat and that a worker's wife never pretended to be a mistress. All public places, from station waiting rooms and hospitals to trams and trains, were planned to accommodate a three-class system aimed at minimizing mutual social contact. Apart from the financial barriers linked to this system, it was largely the ingrained notion of class that determined people's conduct; choosing a lifestyle that suggested one belonged to a higher class was as bad as choosing one that suggested one had less status than one actually possessed.[28] This overall system of class coexistence was a major appeasing force in pillarized Dutch society, with church being in fact one of the few public spaces shared by people from different classes.

This particular classification of bourgeois, middle, and working classes was based on a male professional hierarchy, while the blue bloods—old landed nobility and ennobled patrician families—constituted a separate top layer.[29] The closed nature of upper-class society and the hardiness of this hierarchy rested on the interwovenness of descent, education, training, and profession. From around 1890, however, this structure's basic principles were increasingly undermined. Not descent but money earned in commerce or industry became a competitive status principle. This gave rise to a growing group of nouveaux riches, whose children could easily acquire good positions if they at least obtained the proper education. The only thing they still lacked was the proper cultural background—one marked by taste, refinement, and good manners.[30] In this newly emerging social hierarchy, consumption patterns began to function as group weapons.[31]

The publication of numerous etiquette books at the end of the nineteenth century betrayed the uncertainty of these nouveaux riches when it came to

Workers leaving the Philips factory in Eindhoven in 1910. The distinctive features of nineteenth-century class society, with gentlemen wearing hats and workers wearing caps, vanished only gradually in the twentieth century. An imposed sense of class kept workers from buying a hat, and judging from this photo, as of 1910 a cap was still de rigueur for workers.

participating in society.[32] Even if they tried to create an image of themselves as cultured in the classic tradition by purchasing art and antiquities, because of their commercial and industrial background they continued to be geared predominantly toward the future, technological progress, and modernization. They gave shape to a modern lifestyle without the burden of tradition, showing off their expensive automobiles, building spectacular mansions, and plunging into the fashionable life of Europe's coastal resorts.[33] This new economic elite added luster again to ostentatious wealth, but not without arousing aversion. The then popular singer-songwriter Koos Speenhoff said, car ownership was "the most shameless display of power of perfectly ordinary money."[34]

Similarly, the controversial study *The Theory of the Leisure Class* (1899) by Thorstein Veblen, an American of Norwegian descent, was prompted by aversion to the lifestyle of America's conspicuously wealthy, such as the legendary Vanderbilt family, and especially the role of women in this outward show.[35] Veblen's response, which started from a typical European bourgeois view of culture, was based on a comparison of the hedonist lifestyle of these wealthy Americans, the "leisure class" as he called them, with that of the earlier European aristocracy, the idle class of yore. Veblen amply documented the collapse of class society, and his zeroing in on the slavish and passive role of women in creating a new conspicuously consuming leisure class was also in line with the zeitgeist.

Although in Dutch society the optimism of American class mobility ("from newspaper boy to millionaire") would continue to be overshadowed by fatalism ("those who are born poor will remain poor all their lives"), it is true that in the early twentieth century the access to well-paying employment became increasingly determined by education instead of class. This new selection mechanism also met with criticism, however, as in this comment from 1921:

> Under the guise of striving for more equality, a dam [is being] put up between those who made it and those who did not make it. No century has seen so many *self-made men* as the nineteenth; in the near future this [will be rendered] impossible by the diplomas needed for each job. In no small measure the union leaders are responsible for this because most of them have a university background and as a result they cannot conceive of social climbing without book learning.[36]

This observation suggests that Huizinga had reason to look down on "general education, linked to a superficial leveling of classes."[37] It posed a threat to the cultural elite, which prided itself on slow yet noble cultivation of the mind, culminating in genuine civilization. It was impossible, after all, to achieve classic learning in just a few years of basic education.

In the 1930s the major Dutch poet Hendrik Marsman wrote about "tacky businessmen" who in their Buicks "drive expensive girls along the Lord's highways," while poets "sing among the bums of back alleys." This is fine, the poet comments, because "those who waste their fortune and seed / in the marble grave of palaces of lust" are bound to end up in the "complaints book of the muses" anyway.[38]

In the 1950s the Dutch cultural elite revolted once more. The familiar class structure showed more definitive signs of erosion, and now churches, too, appeared to be losing their stabilizing role.[39] Workers were no longer just workers: they turned into consumers with substantial buying power, and they enrolled in evening courses and correspondence courses as ways to increase their chances at social climbing.[40] The apparent acceptance of this American-style social mobility caused the fade-out of denigrating class references. Still, the cultural elite continued to feel a strong urge to distinguish itself and no concept was better suited to do so than the idea of the "masses," a term with an unambiguous self-exclusionary character. In 1951, the minister for education, arts, and sciences, Jo Cals, in a speech on the occasion of

An advertisement for Buick, from around 1928. In the 1920s car use became more common and most automobile owners also drove themselves instead of engaging a chauffeur. With this closer relation between car and owner brands became more important, as a particular brand name created an impression concerning those who owned it. For instance, the Buick came to be seen as typical of businessmen. The changing status hierarchy of preferences and taste patterns has become the dynamics of consumer society.

the first official Dutch television broadcast, expressed his concern regarding this mass communication technology:

> There is … ample reason to rejoice about this progress of technology, this new victory of the human mind over matter; precisely at a time when, it seems, we all have reason to reflect and ask ourselves whether this new attainment is indeed a victory of the mind. More than ever technology will rule our lives…. After mass labor, today it is mass recreation that besets human individuality and threatens to replace each activity and initiative and individual effort in the realm of the mind with passivity and fading strength of mind.[41]

"Mass culture" became the label for the doings of the man and woman in the street, who, having achieved some prosperity, expressed a consumption-friendly attitude by purchasing all sorts of commodities.[42] The masses lacked the taste whereby the cultural elite sought to distinguish itself. In the terms of the French sociologist Pierre Bourdieu, the cultural elite owned the grail of cultural capital and thus the self-assuredness of superior taste.[43] Status had become taste and economic capital only constituted the material precondition for the right taste. The elite looked down (as they still do) on the poor taste of the masses, with their passion for expensive-looking things and showing off. The cultural elite prolonged its elitist role and taste by taking recourse to relatively inaccessible areas, such as modern art or primitive art.[44]

Socially committed designers and interior designers, who after the war organized themselves in the Correct Living Association (Stichting Goed Wonen), made a sustained effort to teach the Dutch masses good taste, writing magazine articles, mounting exhibitions, producing info-documentaries, and creating model homes.[45] This effort amounted to a genuine propaganda offensive, but after 1968 it petered out. At this time, everyone was proclaiming his and her own taste and personal style; taste seemed democratized, if only for a moment. The prominent role of trendsetters and art authorities in the current taste hierarchy came into being in exactly this period; with the magic label "quality" they gave their preferences once again the illusion of superiority.[46] Even bad taste could be sold expensively when it got the quality label of the art authorities.[47] For example, their appreciation of pop art and the works of camp-style artists like Jeff Koons and the fortunes that modern art collectors and museums have paid for these art works have puzzled and offended many traditional lovers of good taste. That the tastes of the cultural elites in involving homes and automobiles are still markedly different from that of the economic elites is convincingly argued by Elleke de Wijs-Mulkens, who replicated Bourdieu's study for the Netherlands.[48]

Cultural Notions and Technology

The genesis of Dutch consumer society is marked not only by the transition from class society to taste hierarchy but also by the internationalization of the culture of consumption and the more general Western cultural notions that prompted it. Shared notions provided the basis for a new cultural infrastructure involving technological devices that could no longer be linked to a single country—notions that at one point also served as the driving force behind technology development, but that have become such a natural dimension of appliances and the technological structure that we do no longer realize it anymore. This naturalness has evolved in a long-term process of dealing with appliances and technologies; the introduction of a specific appliance or technology and its related design has, equally, been part of the socio-cultural infrastructure. In the next section we discuss the associations and meanings that various parties or groups of users have created around particular appliances. These interpretations were communicated in a variety of diverse forms, not only in design and images but also in spoken and written words. We will review typical Dutch meanings and associations in addition to more general cultural connotations of cars, phones, cameras, radios, and television.

The development of meanings has to be viewed against the historical backdrop of a disintegrating class society. Seen from this perspective, the cultural infrastructure has modernized in two respects: technologies and appliances have not only become part of this new cultural frame but also have shaped modern experiences. The significance of automobiles, telephones, cameras, radios, and televisions in the private sphere is mainly tied to the intermediary function of these technologies, which for individual consumers have brought the outside world within reach. As a result, automobiles and television have evolved into icons of twentieth-century consumer culture.

Speed and Movement

The pursuit of speed—rapid movement—as a new cultural notion gained importance only late in the nineteenth century. This pursuit was related to the rise of new means of transportation such as trains, bicycles, and automobiles. In time, automobiles, rather than trains and bicycles, became most associated with speed. In fact, however, the first car owners were not primarily interested in speed. The increase of car ownership was mainly due to the worsening of the urban climate through increased industrial activity and the large-scale influx of workers.[49] City residents of means escaped the city and took up residence in the urban periphery, preferably in or near the woods, to experience the healing power of nature. The first garages were built near these country homes.[50] At the start of the twentieth century, a wealthy commuter would have his own car with chauffeur to go to his office in the city. Yet the popularity or automobiles was tied less to commuting than to the pleasure of touring the countryside, to be in contact with nature.[51] Although in 1883 the General Netherlands Bicyclists Association (Algemene Neder-landsche Wielrijders Bond, ANWB), was founded to serve the interests of touring bicyclists, the organization soon evolved into a general touring club and promoter of automobile tourism—a status reinforced by the establishment in 1898 of a European league aimed at promoting international tourism.[52] In those early days, nature preservation, cultivating local history, and automobile tourism were not seen as conflicting interests.

That gasoline cars were notoriously unreliable and broke down frequently did not seem to be a major concern to the first users. In fact, this uncertainty increased the adventure of touring, even though some car owners would rather not take to the roads without a driver who doubled as mechanic. Besides, it took some muscle to get the car engine going with a crank. These features gave rise to a new category of male servant: the young chauffeur who was strong enough to crank up the engine, had enough technical skill to repair and maintain the engine, and whom the car owner also deemed civilized enough to have around. As this driver was usually the only one who could operate the vehicle, close physical proximity of a servant in the limited car space had to be taken for granted.

This particular context may also explain why in a disintegrating class society the relationship between driver and car owner, especially if she was female, provided such interesting tension from a literary angle. Having a car with chauffeur emphasized the owner's status of gentleman or master, but the drivers' technical skill gave them power over the well-being of passengers while their proximity led to intimacy and class conciliation.[53] Their being in control bestowed upon young drivers an aura of virility, which in combination with their leather clothing, uniform, strong male hands, and energetic posture provoked new erotic codes. Touring evolved into an activity that involved at least two people. Images of female drivers in the company of a chauffeur or female friend became common as well, notably in girls novels of the 1920s and 1930s, imagery that underlined the idea of free, independent women who also had technical skill.[54]

More adventurous young men who coupled financial means with technical competence turned racing into a sport. To them speed became the adventure of car driving: "The pleasure of being carried at high speed, driving one's car and being behind the wheel yourself—being in charge of this mechanically so exceedingly beautiful machine."[55] What started out as test drives aimed at checking the engine's reliability degenerated into speed races.[56] The risks were addressed in satirical ways in particular: "Festive inauguration of the newly set up graveyard especially for this race.... A deputation of gravediggers will sing an ode to the automobile sport."[57] Because these speed maniacs brought car driving into disrepute, the ANWB and other organizations tried to teach them manners with campaigns such as "The Ten Commandments for Car Drivers."[58] Safety for all road users became a major theme.[59]

Representing speed visually was a challenge, however. The camera's long shutter times only allowed for careful recording of what was static and immobile. Still, in those early days proud car owners liked to be photographed.

This proud car owner has himself photographed, around 1904, with his hands on the wheel to simulate action—a moving car, let alone speed, could not be recorded photographically—and perhaps also to suggest his mastery of the machine. In all likelihood, the younger man sitting next to him was the actual driver of this prestigious Darracq. Frequent problems with early gasoline autos created a new category of male servant: the young chauffeur who was strong enough to crank up the engine, had enough technical insight to repair and maintain it, and was also deemed civilized enough to have sitting next to one in the confined space of an automobile.

The act of actually driving the car was suggested by showing the driver's eyes fixed on the road and his hands on the wheel. Around 1900 a specific pictorial code evolved in drawings, whereby clouds of dust near the back wheels suggested movement, but a more elaborate pictorial code indicating speed had not yet been established. Plumes of steam as in depictions of speeding trains were hardly applicable, even if the first car owners referred to driving as "stoking" and to cars as "stoves."[60] The pictorial code of speed introduced in the early 1910s proved remarkably uniform. In illustrations produced for ANWB safety campaigns and automobile industry ads, speed was visualized by a low-angle perspective, as if the illustrator ran the risk of being run over. Giant front wheels defined the image and speed was suggested by stripes drawn from the wheels to far behind the car body.[61] Because the wheels did not quite

A poster from about 1920, advertising an auto made by Ansaldo—a brand produced in Italy from 1921 to 1931—was in many ways typical of its genre. The shutter speeds of early cameras were too slow to capture an image of an actually fast-moving vehicle—and if it had, the sense of speed might not have been conveyed. Around 1913, a pictorial code emerged to suggest a speeding car. Lines drawn from the front wheels to behind the rear wheals implied speed and direction, as did the extreme perspective created by drawing the the whole front area of the car and especially the wheels larger than the rear area. The car almost seems to fly rather than ride on the road surface. Commercial artists accentuated the adventure of car racing by depicting cars negotiating curvy mountain roads at full tilt.

touch the road surface the car seemed to glide. The image-defining character of the right front wheel, seen from the driver's angle, provided a suggestion of movement from left to right, at least from the perspective of the
beholder. In representing speed as a movement in time, it was more or less
a convention in European imagery that progress, or the future, was to the
right of the beholder and the past was to his left.[62] How and when the visual,
pictorial code for car speed exactly came into being is hard to trace, but
already in 1913 speed was visualized as we know it today in an ad for the
Spanish Hispano-Suiza luxury automobile.[63]

This appealing pictorial speed code became a source of inspiration not
only for car body designers but also for designers of other devices; it was
assumed that an image of speed could help raise sales.[64] The high rounded
shape of the car tire and the drawn speed stripes were translated in the 1920s
into a massively rising, streamlined design with rounded corners and flashy
chrome trims.[65] Although streamlining is often presented as the application
of the laws of aerodynamics, in this case it had little to do with it. A high
front conflicts with aerodynamics, which prescribes a nose-like shape (moving up from low to high) as in race cars and high-speed trains. This renders
it all the more likely that the pictorial speed code served as a source of inspiration for streamlining.[66]

Sense and Sensitizing

In this era's literature, speed was not described as screeching tires on a road
surface, but as euphoria or bliss for the driver, as a sensation that tickled all
senses. The "anti-automobilist" mainly stressed the unpleasant sensations of
noise, gasoline odor, air pressure shock, and dust in the eyes and mouth, and
felt harassed by the "shameless display of power of perfectly ordinary
money."[67] Although the noise of cars and other heavy machinery assaulted
the senses of nonusers, it turned out to have a sensitizing effect on the hearing of mechanics and experienced drivers in detecting mechanical imperfections. In a 1930 Lincoln ad this sensitivity was positively exploited by presenting the car as an extension of the body: "A perfectly 'healthy' car one
neither feels nor hears *when one is sitting in it*. And those who are sensitive
will notice each disruption immediately" (emphasis added).[68] In a strange
way has the frequent use of appliances whetted one of the senses, and made
it sharper. The use of cameras and, later on, film cameras sharpened our eyes,
much as phones and radiotelephony sharpened our hearing, yet the one-
sided improvement of either hearing or seeing also subordinated the other
senses. It explains the gradual predominance of seeing and hearing in our
dealings with appliances since the late nineteenth century.

DOOR DE TELEFOON.

The nature of telephone use contributed to the collapse of social class divisions. The greeting protocol that regulated social interaction across classes was invisible during telephone communication. In the early 1900s this led to much irritation about impoliteness on the phone. Hence this 1902 cartoon's legend, "The polite man: Please don't take it amiss, sir, that I keep my hat on, but it is so cold in here"

Early on, one of the inventors of telephony, Alexander Graham Bell, developed a fascination for the formation of sounds and for the speech development of deaf children and became a speech therapist. Both his mother and his wife were deaf, and he was also influenced by his father, who was a phonologist. Bell met his wife through his practice. His initial efforts were geared toward constructing a device that amplified resonance, which would allow deaf children to, for example, distinguish between the sounds "p" and "b," which are hard to distinguish via lip-reading.[69] By means of a primitive microphone the sounds articulated by the speech therapist were electrically amplified. The receiver with the amplified sound vibrations were placed against the forehead of the deaf child, for whom the sound difference became tangible. This one-way traffic of amplified sounds through a short copper wire also proved easily audible, and both Bell and his sponsor (and soon-to-be father-in-law) realized that the possibility of transferring the human voice over large distances could earn them a fortune.[70] After several years of intensive work on the development of the telephone, Bell subsequently concentrated on an audiometer, a device to measure sound volume, again at the service of the education of deaf children. This effort resulted in the creation of a unit for measuring of the intensity of sound, called the deci-Bell, later decibel.[71]

The practice of telephony, the transfer of voice over a long distance, thus made us all blind communicators, albeit unintentionally. Nevertheless, people were perfectly aware of this drawback, as suggested by this comment regarding the potential for misunderstandings in phone conversations:

> First one should realize that telephones operate without the aid of sight, which in any normal conversation, through gestures and facial expressions, supplements what is lacking in spoken words. Because in normal conversations misunderstandings are very common already, it is only natural that this is bound to happen much more often in phone conversations, so that offending each other is almost inevitable.[72]

The lack of gestures and facial expressions in phone communication had quite far-reaching social implications. For example, the everyday interaction between classes was based on an extensive greeting protocol: a lower-class person had to greet first, after which a gentleman would lift his hat with one hand and remove his cigar from his mouth with the other to subsequently make a light bow. Such a greeting ritual, including its implied social hierarchy, was rendered invisible in telephony.[73] Resentment of speakers regarding impolite treatment by operators, who mediated without regard to callers' status and showed no courtesy, was reciprocal. The songwriter J. Clinge Doorenbos expressed this in the lyrics of a 1915 song, "Miss Operator": "Folks who dare not say a thing at home here display their courage unpunished and invisible to Miss Operator."[74]

In cartoons as well as illustrated ads phone conversations were associated with anger and irritation, articulated by a somewhat stout gentleman.[75] This visual convention was retained in cartoons until well into the 1950s. In the early twentieth century the satirical weekly *De Ware Jacob* even had a separate rubric entitled "To the telephone" that poked fun at authorities in a so-called phone interview, a satirical genre that would be copied and that derives its humorous effect from the suggestive rendering of the "replies" from the other end of the line. Hearing those odd, one-sided phone conversations developed our competence to grasp a conversation's progression based on snippets.

Regarding women and phones, however, a different imagery developed: a woman with a receiver in her hand without the phone itself being shown. An early Dutch example is a 1908 untitled painting by J. P. ter Burg. It shows us a picture of a woman having, presumably, a confidential conversation with

A painting from 1908 by the artist J. P. ter Burg. Many representations exist of angry men speaking on the telephone, suggesting impatience between the speaker and, perhaps, the operator and thus antagonism between social classes. By contrast, the image of women using the telephone tended to communicate confidentiality between social peers, expressed by the act of listening. The more tilted the woman's head and the more relaxed her posture, the more informal the inferred relationship. After the late 1920s such images become more prevalent, but the earliest known example is this Ter Burg painting.

a relative or female friend, whereby the position of "the listening ear" expresses the intimacy between social equals. The more angled the head's position and the more casual the body's position, the more informal the relationship that can be inferred. In the later 1920s representations of confidential phone conversations became increasingly common. Advertisers also began to use this visual language in ads: the solution to painful problems was provided via telephone. After the large-scale automation of local telephone systems, operator interference in local calls became redundant and apparently the sense that another person could be listening in soon faded.[76]

The introduction of telephones, radio, and gramophones soon meant that voice and sound increasingly appealed to the imagination of listeners.[77] For example, occasionally operators received marriage proposals just because of their pleasant voices, which in the visible world of class distinction would have been unthinkable. Similarly, the panic, if not mass hysteria, that in 1938 erupted in the United States during the radio play *The War of the Worlds* is an example of the suggestive quality of sound. Listeners assumed that an invasion from space was occurring and fled their homes in large numbers. The first radio amateurs were excited by the sound snippets they caught with their receivers. Their scanning of the ether, in hopes of picking up signals from foreign stations, became a true sound adventure that could be enjoyed in simple attic rooms or even when out camping. Furthermore, the exhilaration of those who tuned in to a live broadcast of a radio concert for the first time arose less from the music they heard (already familiar from gramophone records) than from the bridging of distance and the illusion of "being present." This shared experience of a "live" broadcast is still a highly valued media phenomenon.

The voice—enunciation, articulation, and especially accent—became a new social indicator in auditory media. For this reason the speaking of Standard Dutch without local or regional accent was seen as increasingly important, especially in radio and television. In the 1950s radio listeners still reveled in stock characters with different accents. For example, the servant in the popular Dutch radio play *De familie Doorsnee* (*The Average Family*) spoke a broad Amsterdam dialect to underline her low social position. Even today we form an impression of our unknown interlocutors at the other end of the line on the basis of pitch, voice level, style of speech, and pronunciation; with those we know well we tend to focus on intonations that betray a specific state of mind.[78] We have developed this empathic power through the sensitizing of our hearing as a result of our repeated use of phones and other sound media.

Effects of Image and Sound

Photography derived its visual conventions from classical painting, and at first artists considered realistic photographic renderings and the simple ability to produce multiple images a great threat. Photography, they felt, supplied accurate yet superficial images—no more than reflections of light, in fact, without offering any penetration of character or landscape. In their view, photo studios were failed ateliers where extras posed in front of a painted decor, and the much-celebrated objective photographic renderings were merely manipulated visual arrangements. If the result did not live up to the classical ideal, the image was retouched. Such criticism, however, failed to

This 1904 cartoon, from the satirical magazine *De Ware Jacob* anticipated that the population's Protestant segment, known to be traditional, would have no objection to modern sound media if these helped spread the biblical message. It suggests a news real in the Church Newspaper in the distant future of 1948. The caption reads that the shortage of ministers has been effectively countered by the installation of so-called Odeon Disk-Speech machines on the pulpit. Moreover the purchase of the machine is stated to be much cheaper than hiring a minister. The cartoon was foresighted for in 1925 the Protestant-Christian Netherlands Christian Radio Association had been the first Dutch broadcaster. Moderation in everything worldly was essential. Still, in the early 1950s listeners were warned against moral decay resulting from too much radio listening—and of course listeners might tune in to programming from competing broadcast organizations with different ideologies. It was feared that this risk would increase with the new visual medium of television. After all, the Bible contained ample warnings against the temptation of images.

hinder the development of portrait photography, partly because photos were an affordable alternative to painted portraits.

This criticism was also hypocritical because in both coloring photos and retouching imperfections the photographer relied on painting. Painters had created a long tradition of ideal images, to which reality was subsequently adjusted—a very common interplay. Moving images, and later television, met with the same criticism. Huizinga criticized movies as "photographic caricature of a highly limited visual reality," and decades later the sociologist F. L. Polak voiced a similar critique regarding the effects of television when he said, "Television images literally move on the outer surface of the screen, leading to superficial people without substance."[79]

In the era when amateur photography began to flourish, people already possessed widespread visual experience in the painterly tradition of landscapes, group and individual portraits, still lifes, and panoramas. The popularity of cameras and taking pictures grew in the 1920s because of lower prices and the affordable photo-processing costs, stimulated by the many special promotional campaigns in which one could win or save money for a camera. The second photography boom, in the 1960s, coincided with newly affordable color photography.[80] Because the first simple box cameras were easy to handle, tourism and amateur photography rose in tandem.[81] From the very beginning the emphasis was on outdoor scenes that required no extra flash. Travel photos could not reproduce the movement of travel, but showed moments of pause and relaxation: the garden terrace, the picnic, the landscape, the mountain pass or ancient city gate; movement was suggested by the photographed car or bicycle. A significant aspect of sightseeing became looking at pictures at home.

In the 1950s, when cameras could be easily equipped with a separate flash, photographing domestic scenes became more popular and the genre of the baby photo album emerged, a tradition that evolved into elaborate family photo archives. Children were proudly photographed against the decor of newly achieved prosperity: the radio and television sets. For a long time the photographic realism of amateur images was accepted without skepticism, but from the perspective of the 1960s protest generation, family snapshots were seen to be manipulated, "falsely testifying of a happy childhood," as a leading Dutch protest singer, Boudewijn de Groot, put it at the time. Similarly, the feminist movement began to focus on stereotypical representation of women in ads, photography, and illustrations.[82] This visual critique failed to reduce the value of these childhood and family snapshots' emotional and personal significance—and this value has only increased since then. Frequently photo albums lead the lists of valuables that people strive to save in case of a fire. Photos have begun functioning as proof of a shared past, and having an undocumented childhood has become a modern trauma.[83]

As camera, film, and development technology improved, the shorter exposure times now possible influenced the perception of reality and made amateur photographers look differently at the world around them. It made them realize the uniqueness of moments that passed in a split second. These moments had to be captured in much the same manner as one shoots game: the photographer needed to be simultaneously aware of light, staging, and the disastrous effect of movement, ingredients that determine the magic of photos: the frozen here-and-now-moments that became history in the very second when the shutter cuts off the light, thus sealing time.[84]

As photos were shared among relatives, movies in movie theaters continued to be a sensation to be shared with friends. In the 1920s and 1930s, the movie theater targeted a group of young adult consumers looking for entertainment outside their home, in an ambiance of luxury and comfort with which theaters were trying to compete.[85] The adventures of screen heroes and heroines allowed the audience literally to "dream away," and flee the here-and-now. More than photos, movies managed to represent wishes, dreams, distant futures, or lost ideals.[86] This visible entertainment of growing theater audiences fanned the moral panic among confessionals, such that in 1926 there was support for a law aimed at combating "the moral and social risks of movie theaters."[87]

The preoccupation with manipulated photo and movie images was triggered by worries about the influence of visual stimuli. Invariably, the concern was with the objectionable role of images in luring and seducing the masses. Women, workers, and mass youth in particular lacked the self-control needed to resist the lure.[88] This fear was rooted in Protestantism with its iconoclastic past, in which the image stood for erotic seduction and the weak flesh.[89] In sermons sensuality and idolatry were denounced. The cultural elite especially flayed superficial, frivolous entertainment and the materialist, hedonist lifestyle displayed in ads, movies, and the illustrated press.

The written and spoken word, by contrast, had a nobler tradition and stood for active spiritual development, moral lessons, and mental discipline. This is why sound media, if they conveyed the right message, met with less resistance than visual media. In 1924 the Protestant Netherlands Christian Radio Association (Nederlandse Christelijke Radiovereniging, NCRV) was the first Dutch public broadcasting association, and it used radio for evangelizing: spreading the word.[90] In the imagery of the ad campaigns of the various radio broadcasters, the Netherlands Christian Radio Association symbolized the authoritative voice from the pulpit; Catholic Radio Broadcasting

A bedridden patient who has been given a television set thanks to a fund-raising drive of readers of the national newspaper *De Telegraaf* in 1957. More than radio the medium of television implied users' immobility and consequently was at first viewed as a therapeutic tool for those who were forced to stay at home or who were alone, such as the ill and elderly.

(Katholieke Radio-Omroep, KRO) likened radio listening to the concert situation; and the Association of Socialist Radio Amateurs (Vereniging van Arbeiders-Radioamateurs, VARA) promoted itself as the sound of the workers, the voice of protest from below, even if this protest hardly relied on the workers' vernacular. The first ad for radio sets, which showed the loudspeaker at ear height, expressed a sense of the friendly, informal "reliable friend of the family," whereby the listening woman might as well be mending stockings, which she would never do in a concert hall. Photos of radio listeners, largely women, in their living rooms reinforce this informal, homey relationship with radio. From the mid-1930s an even more intimate visual relationship came into being, in which a young woman bends over the radio as if it is her baby or lover. In the war, when owning a radio was prohibited by the Germans, listening to London-based Radio Orange (Radio Oranje) for information of the Dutch government in exile became an act of resistance, which added luster to radio as a heroic medium. After the war, photos of clandestine radio sets and people closely listening to Radio Orange evolved into icons of the Dutch resistance movement.

With concern for the moral decay of postwar Dutch youth running high, the new medium of television was greeted with fear and trembling. The only consolation seemed to be that television could keep youngsters at home because the programs could be enjoyed by the whole family.[91] When Jo Cals, the minister for education, arts, and sciences in 1951 inaugurated the first Dutch television broadcast, he sat in an armchair next to a coffee table, suggesting a domestic setting. Also to avoid associations with movie theaters, the TV stations retained the tradition of spoken program introductions, as was common in radio. If each radio station had a recognizable voice, now the related network would have a recognizable face: a seductive, nicely dressed and coiffed female presenter. However, this only confirmed the negative image of visual culture.

Initially Dutch television seemed to be the favorite medium for those who were tied to the home: the old and the sick. Their immobility justified the purchase of an expensive TV set and on Wednesday afternoons they could count on all the neighborhood children to come dropping in to watch the weekly special kids' programming. After the early 1960s almost everyone could afford a TV set, but the corny evening programming hardly appealed to youngsters, who preferred to go out with their friends. The networks subsequently tried to attract viewers by showing live sports broadcasts and popular music shows, soon turning soccer into the most favored TV sports. The new importance of the generation gap between parents and baby boomers meant that more and more teenagers gave priority to being around their friends and creating their own atmosphere away from family life.[92]

Dutch teenagers discovered alternative sounds in broadcasts by commercial foreign radio stations and so-called pirate stations located on ships in international waters off the Dutch coast. These commercial radio stations circumvented the strict Dutch broadcasting regulations and played the latest English-language pop and rock and roll. Dutch youths embraced English as their lingua franca, which to their less-educated parents was something like a secret code. Having your own stereo set was a major goal that had to be realized at all cost. Electronically amplified pop music became the deafening display of the male segment of this protest generation. Ever since the emergence of automobiles, strong auditory signals—whether involving roaring car engines, crackling mopeds, or whining electric guitars—had been a sign of male virility and emphasized ownership: young consumers' shameless display of power. Electronically amplified sounds were sheer music to the youths' ears, but deafening noise to their parents'. The generation gap was exemplified in this conflict on the interpretation of electronic sounds. From the youths' user perspective, pop music sensitized their hearing, while from the perspective of the ones who disliked electronic pop music it was the opposite. The youths' passion for pop music gave rise to an influential music business in which commercial interests would ultimately gain the upper hand.[93]

The Liberal Protestant Radio Broadcasting Organization (Vrijzinnig-Protestantsche Radio-Omroep, VPRO) was the first Dutch broadcaster to

In this photo, from 1960, the family gets ready to watch a major soccer game, their chairs arranged as if in a movie theater. But their daughter shows her lack of interest by turning her back to the TV set and ironing. In the early 1960s the broadcasting companies tried to get youngsters to watch TV by broadcasting soccer matches. On the whole they were not very successful, especially where teenage girls were concerned who in general preferred the cinema.

discard its confessional roots, in the mid-1960s, in favor of a leftist, critical attitude regarding society, first in radio and, later on, in television.[94] Progressive youths still dismissed watching TV as "boring" and "middle class." Their arguments were quite similar to the traditional criticism made by the cultural elite: watching TV numbed people, was meant to keep them quiet and fool them, so that the "hoi polloi" would not stand up to the bourgeois capitalist establishment. In this critical view, visual ads and TV commercials also catered to people's lower instincts.[95]

Television and its programming became a source of controversy, not only between the generations, but also with respect to the taste hierarchy among viewers. Not only the number of hours one watched but especially also one's tastes in programming and even the location of the TV set in the house—more or less prominent, high or low, central or hidden, adorned or concealed—came to be seen as an indicator of the role played by this medium in different households and classes.[96] In the 1970s, several commercial pirate radio stations managed to acquire legitimate broadcast status as a result of the support of their many young listeners. The first commercial network in the public system, the Television and Radio Broadcasting Foundation (Televisie-en Radio-Omroepstichting, TROS), broadcasted radio and TV programming that catered exclusively to popular taste. Among intellectuals and cultural

A 1902 cartoon, "In the Atelier," mocking the confrontation between culture and capital. "Painter: (to a rich parvenu who has asked him the cost of his painting by the square foot) "For you my painting is priceless." By the end of the nineteenth century, conspicuous wealth earned in trade or industry became a competitive status principle that provoked resistance among members of the established elites, who reestablished their social status by boasting of their cultural knowledge and aesthetic sensitivity. In the cartoon the prototypical capitalist, recognizable by his corpulence, garish signet ring, and cigar, is viewed with distaste by the artist—whose bony frame represents the straightened circumstances of a life spent in the service of art and culture.

critics it thus contributed to an even more negative stance vis-à-vis the deadening effects of mass media.[97]

Synchronization and Remote Control

By using cars, telephone, radios, televisions, and cameras, we have not only begun to experience distance in space differently, but also distance in time. Communication technology has accustomed us to the kick of the live effect, the synchronizing of experience.[98] The large-scale use of phones has only enhanced this effect. The booming use of cell phones, notably among young people, is indicative of the great need for synchronizing experience. This turns out to encourage face-to-face encounters and this in turn requires travel. Most cell phone interaction is between people who see each other every day, saw each other a little while ago, or will see each other shortly. This also explains why the content of cell phone conversations tends to be so unexciting to outsiders. The abolition of telecommunications might contribute to solving the traffic jam problem, if only commuter traffic did not spring from the same need for synchronizing experience, albeit in various locations: on the job with colleagues and at home with partner and children. In family contexts the cherished family snapshots stimulate the synchronizing of experience between parents and children.

This need, once triggered by telephones and later fostered by radio and television, will not be satisfied anymore by one-way communication, as in radio or television. The remote control of radio and TV may appear on the surface to be analogous to the modern phone, but it does not meet the ever-growing need, especially among young people, for interactivity and synchronization. Restless channel zapping possibly indicates that the interest in this one-way medium is evaporating. Furthermore, the importance of synchronization in social life is underscored by central radiographic time control from Paris, which makes its possible to synchronize European clocks, both outdoors and indoors.[99] The popularity of this type of clock suggests that the desire for synchronicity is widespread, and this explains why the audible time sign that used to mark the beginning of radio news broadcasts was recently abolished. Also, the awareness of time differences between Europe, North America, and Asia is more pervasive than ever, partly brought about by the need for synchronization of experience.

The use of the telephone in combination with radio and television has engendered new sensations and new needs—in other words, another experience of life. Although the need for synchronizing experience, the idea of wanting to be part of it, seems at odds with actively creating distance in tourism and the ever stronger need for privacy and the inclination to dissociate from others, these needs have come to entail each other. This is underlined by the reversibility and temporariness of tourism and privacy—otherwise tourism would become emigration, and privacy would become solitude. In essence, tourism is born from the need to be in another time, albeit temporarily: touring nature and the countryside as monument of a virgin past or urban trips as the world of the future or as model of ancient civilization. In this respect technology—in the form of cars, cameras, television, and radio—operates as a time machine. Today's seniors have grown accustomed to using these technologies their whole lives; they have become highly valued extensions of their ears, eyes, and legs that have brought the world within reach.

Technology Between Hedonism and Self-Control

A 1960 government report on regulating mass recreation typified the educational role postwar government took on in curbing youthful entertainment and individual pleasures in public space when it stated:

In this 1925 photo, taken at the "Polynesian beach"of a well-known swimming pool at the bank of the IJ in Amsterdam, the camera captured a group enjoying the new fun of listening to the radio. The new consumption-oriented lifestyle was publicized partly by photos in illustrated periodicals, giving rise to imagery inspired by the hedonist pleasures of the beau monde, the world of fashionable society. As the nouveaux riches sought entertainment especially in Europe's beach resorts, from Scheveningen to Nice, the beach, the boulevard, and the sea became the stereotypical setting for the newest technological playthings, such as fancy cars, airplanes, and radios.

> This is a major task: teaching the masses to spare each other when it comes to noise…. It is not just … that one should not sit on a train with a "portable radio" on; one ought to learn it is improper because it may bother other passengers. Bothering each other is uncivilized.[100]

The key feature of current Dutch consumer society has become precisely the delicate balance between hedonism and self-control.[101] In more recent government campaigns as well, individual pleasures are held in check to save lives by calling on self-control: "Enjoy your drink, but mind your alcohol consumption," "Smoking is harmful to your health" and "Mind overeating, mind eating too much fatty food." The creed "Enjoy the good things of life, but don't overdo it" appears to be the ultimate compromise after a century of rivalry between, on the one hand, a call on civilization and good taste by the cultural elite and, on the other hand, the need for entertainment, pleasure, and ownership, first by the nouveaux riches and later on by the masses. It has also resulted in the now accepted differences of taste and lifestyle between the cultural and the economic elite.

In the nineteenth-century bourgeois view, a hedonist lifestyle was still the dubious privilege of the aristocracy. The rise of both the nouveaux riches and the working class at the end of the nineteenth century posed a major threat to the social position of the bourgeois elite. The ensuing cultural battle between the opposing lifestyles had to be fought on multiple fronts, and this explains not only the fierceness but also the ambivalence of the arguments involved, whereby "progress" no longer automatically implied advancement and "modern" even became a curse.[102] In the hands of the uncivilized, technology as instrument and product of progress became synonymous with decline and deterioration. Thus, from a bourgeois point of view, hedonism and technology entered into a monstrous sensory alliance, one that was visible, smellible, and especially audible. The embittered reaction to car owners' urge to express themselves was only its first articulation.

The bourgeois class then discovered an effective weapon in its call to self-control. In urging others to control themselves—whether excessive phone use by women, objectionable use of the automobile as a race car, or nonselective radio listening—the cultural elite relied on media they were familiar with, the spoken and written word. By contrast, hedonism and its promotion largely took shape in visual media: magazine photos of the beau monde, pictures of lavish wealth, movie images of Hollywood's world of glamour, and so on.[103] Such images gave those of little means something to gaze at or offered Cinderellas the illusion of princes who would liberate them from their miserable existence. Such illusions flourished in a world in which family background was no longer the only path to fame and fortune.

Self-control and hedonism found expression not only in words and images but especially in the language of the body, in outward appearance, posture, and gesture. Partly on the basis of traditional ideal types found in classical paintings, in the late nineteenth century new bodily codifications were established. For example, portrait photography was deployed in theories of evolution to illustrate more or less developed human types and differences in appearance between the races.[104] Bodily features were understood to mirror character and disposition, both negatively and positively. Female beauty was represented by the classic ideal type of the modest Greek goddesses of ancient sculptures: well-rounded build, waving hair in a chignon, and a Grecian profile. The male ideal of beauty was classic as well: athletic, but also and especially modest, in the manner of Rodin's statue *The Thinker*.

In this view any beauty that wasn't classical beauty was dismissed as superficial. The classic, athletic ideal of beauty found its opposite in the caricature of the stout capitalist, the prototype of the nouveau riche: a corpulent middle-aged man with hat, cigar, and signet ring. For woman, two types emerged as counterpoints to the classical image of woman, one thin and one fat: the caricature representation of the feminist as a skinny older woman with a pointed face and glasses, and the caricature of the fat lady as a model of immoderation, the counterpart of the corpulent capitalist. Nineteenth- and early- twentieth-century portrait photography reflected classic ideals of modesty: rigid, earnest faces without a glimmer of a smile.

In the early twentieth century a radically different visual language emerged in photography that foregrounded hedonist elements rather than modesty. In essence this language merely continued an old popular visual culture tradition, one in which lust, pleasure, and corporality were represented unashamedly and often indecently and had always been set aside as vulgar and had been ignored by the bourgeois elite. In photography, beaches and other water scenes constituted the new setting of modern corporality. The building of hotels, boulevards, and piers in seaside villages scattered along Europe's coasts in the late nineteenth century was partly motivated by the healing effect attributed to sea bathing.[105] In this period the small fishermen's community of Scheveningen on the Dutch coast developed into a resort of international stature, where all was geared to entertaining the beau monde: from circus theaters, boulevard car races and, flight shows on the beach to the first exhibition of radios.[106] Around 1900 this stylish entertainment also began to attract curious day trippers who cautiously took a swim. Thus health, wealth, leisure, entertainment, and pleasure in combination with technology became the ingredients of a new physical beach culture that was no longer only for the rich.

In this trend the new feminine ideal type of the boyish, short-haired sporty woman developed alongside the persistent classic ideal image of the athletic man, and came to replace not only the image of the Greek goddess in technological contexts, but also the caricature of the skinny feminist. Initially, these well-rounded, emancipated young woman were represented on photos and advertising posters as drivers and as participating in the hedonist image of speed, sportiveness, and independence. But by the late 1920s they increasingly lapsed into serving as an elegant touch merely aimed at selling automobiles. This was probably modeled on the then popular *concours d'élégance*, automobile shows where chicly dressed car-owning women smoking cigarettes had themselves photographed with their pet hounds while casually leaning on their luxurious automobiles.[107] This passive visual relation between woman and car was also adopted in amateur photography of the 1920s and 1930s.

Controlled physical exercise fitted the classic-bourgeois tradition of a healthy mind belonging to a healthy body. Engaging in equipment-based

The cover of the 1938 New Year's issue of *Panorama*, which celebrated the first telephone connection between the Netherlands and the Dutch East Indies, also nicely evokes a consumption dynamic between young and old. The older couple is portrayed in a Dutch winter landscape using an old-fashioned phone model, while the young couple who clearly traveled to the other end of the globe is set in a tropical entourage handling the latest phone model. In the 1930s the image of consumer society was increasingly personified by active and mobile young people, who could afford all sorts of modern luxuries, from cars to radios and telephones. The image of the nineteenth-century parvenu, the corpulent middle-aged capitalist, was replaced by a younger and slimmer version. Older people were also portrayed increasingly as inactive, as passive spectators who no longer kept up with the pace of modern consumer society.

sports first became popular among young men. Their car or bicycle "races" were odes to speed, skill, virility, and bodily strength, but such races were also viewed as reckless.[108] In this respect, gymnastic exercise—a mandatory aspect of education in the 1910s and 1920s—was seen to reflect bodily control even more, a view that was particularly relevant to women and their "weak flesh." In photography, by contrast, the old visual convention of modesty, also concerning expressions of emotion, was discarded. Smiling or laughing as a sign of exuberant pleasure had once been seen as indicative of a lack of civilization and control. Now, the smile gained ground, a development that can be closely traced in the national illustrated press of the early twentieth century. For instance, in 1914 for the first time a toothpaste ad was run that depicted a broadly smiling child with the caption: "One of our customers." This portrait, entitled *The Smile*, had been the winning photo in an amateur photo contest the year before. The photo smile evolved into an accepted visual convention in both advertising and amateur photography to such an extent that we can scarcely understand why smiling or laughing for the camera should have been unacceptable. Over the years, however, movies and advertising have firmly established the new mimic code of the controlled smile. In contrast to the wide open mouth as a sign of exuberant pleasure, the controlled smile shows only the front teeth.

In the new visual conventions, the various forms of experience of time took literal shape. Youthfulness and the young slender body became paired with future, activity, eroticism, movement, and speed, while weakness, old age, and bodily decline were associated with decay and the past, passivity and slowness. Old age as connoting wisdom lost ground because being active became a vital part of the new code. Moreover, from the cultural elite's disciplining perspective, idleness—and, by extrapolation, comfort, ease, and luxury—was only legitimate if it involved sick, elderly, or disabled people. This explains not only the criticism of the passivity and striving for comfort and pleasure associated with mass youth, but also why "modern" developed into an ambivalent concept. On the one hand, this concept suggested an image of youthfulness as well as energy, sober and rigid forms, new materials, and future-mindedness; on the other hand, "modern" implied technological consumer products that were puffed up as "luxury and comfort," thus emphasizing a passiveness that signified the elderly.

New medical insights developed after the 1960s revealed that our physical condition suffered not only from excessive consumption—overeating, eating fatty foods, smoking, drinking alcohol—and lack of exercise but also from all too controlled behavior in which there was no role for pleasure, relaxation, and enjoyment. The medical plea for moderation, control, physical activity, and mental relaxation was propagated by information services as well as in product ads. The leftist, critical student movement of the late 1960s was already paving the path for the sober yet relaxed 1970s. In the movement's anti-consumption attitude, old age and wisdom were revalued, inactivity was associated with mind expansion, and thriftiness was put at the service of the environment.[109] Like the movement's masculine ideal was philosophical rather than athletic. Extra bodily activity was even avoided because bundles of muscles connoted stupidity. Above all, "modern" became an outdated word.

On the whole, students in the 1970s embraced an anti-materialist and anti-ambition, easygoing, lifestyle, in which car driving played a pivotal role. Many settled in remote villages and hardly comfortable farmhouses, preferably in groups and with their old car or van serving as a link to the outside world. The Frisian and French countrysides, seen as unspoiled and unaffected by mass tourism, were particularly popular. These renegades avoided

Europe's coast, which was for "beach cattle": the middle-class families with their cars stuffed with holiday luggage that rumbled off to the beaches of Spain. In this social critique, "middle class" stood for conformism and materialism, but especially for decent and hard-working people. This scorn for a notion that once represented honorableness and respectability was the final reckoning with the old class hierarchy and established the new hierarchy of taste in which the younger generation would set the tone. It shaped the different lifestyles of the new cultural and economic elite.

The onrushing generation of young urban professionals in the 1980s completely rejected the permissiveness and lack of ambition of the protest generation. Together with converts from the protest generation, they created the ambitious lifestyle of the new economic elite, with their strong sense of purpose and an Americanized work ethic, concentrated on careers whose success was defined by the size of the salaries and bonuses. For these young people their cosmopolitan lifestyle was based on urban settings rather than the countryside, and they did not shy away from showing off their expensive car or fancy loft apartment. Also in their sparse leisure these overambitious yuppies favored activity rather than idleness. This lifestyle gave rise to the athletic figure as beauty ideal for man and woman alike—an ideal that so far has lost little appeal. With the help of all sorts of new fitness equipment, from home trainers to rowing machines, this physical ideal can also be realized ever faster. For this generation, coast and beach have again become the favorite decor of sporting activities such as jogging and beach volleyball. They enjoy healing "moments of pampering" in, on, or near water: in bathrooms, private swimming pools, saunas, or on sailing yachts. Because external traces of stress and burnout have become taboo, pampering is viewed as a necessity instead of as a luxury. Increasingly, therefore, fast travel and frequent short vacations weigh more heavily than environmental concerns tied to increased air travel. After all, need has become necessity.

There is a constant urge to balance hedonism and self-control in Western economies. Although in the 1990s a Dutch cabinet minister even called on people to "slow down" as a preventive measure against a feared burnout threatening the working population as a whole, the present cabinet stresses the importance of long hours and later retirement to provide state pensions for the growing numbers of senior citizens in the coming decades. At the beginning of the twenty-first century, most Western governments stress consumers' self-control in the fight against global warming and the adverse effects of climate change. The melting of the ice caps and subsequent rise of the sea level will be especially dangerous to the Low Countries, as Al Gore so eloquently, points out in his documentary *An Inconvenient Truth*. There is a need not only for clean fuel but also for clean and fast cars. Energy saving is more readily accepted when it does not affect the comfort and pleasures we have become used to. Unlike in the 1970s, finding the right balance between hedonism and self-control is now perceived as being dependant on technological innovations and solutions.

Acknowledgements

I am greatly indebted to Johan Schot and Ruth Oldenziel for their confidence in my abilities. Furthermore, I would like to thank a number of colleagues who have been of great help and assistance. First of all, Harry Lintsen for his kind collaboration in the editing process, Onno de Wit and Peter Staal for bringing vital material to my attention, Giel van Hooff for his help in the search for illustrations, Jan Korsten and Roeslan Leontjevas for their managerial support, and last but not least my colleague and friend Liesbeth Bervoets for her moral support and stimulation.

Notes

1 Franz Alexander, *Our Age of Unreason: A Study of the Irrational Forces in Social Life* (Philadelphia: Lippincott, 1942), 21.

2 J. Huizinga, *Geschonden wereld: Een beschouwing over de kansen op herstel van onze beschaving* (Haarlem: H. D. Tjeenk Willink, 1945), 194–196.

3 Cf. J. Huizinga, *Man and the Masses in America: Four Essays in the History of Modern Civilization* (originally published 1918), in Johan Huizinga, *America: A Dutch Historian's Vision, from Afar and Near,* trans. Herbert R. Rowen (New York: Harper & Row, 1972).

4 J. B. Th. Spaan, "Luisteren naar de radio," in M. G. Schenk, ed., *Encyclopaedie voor de vrouw: Practisch handboek voor het dagelijks leven* (Amsterdam: De Bezige Bij, 1951), 378–381.

5 "Tacit technology" is a variation on the notion of "tacit knowledge," which in technology history is used to indicate knowledge that becomes visible in actions and technical skills but that usually remains unarticulated.

6 The distinction between ownership and usage is of crucial importance for a proper understanding of the user history of costly appliances and products in particular. For example, in the early days of the automobile, the owner and driver belonged to different social classes. See also Irene Cieraad, "De naaimachine in beeld: Over kleermakers, naaisters en modemaaksters," in Ruth Oldenziel and Carolien Bouw, eds., *Schoon genoeg: Huisvrouwen en huishoudtechnologie in Nederland, 1898–1998* (Nijmegen: SUN, 1998), 197–230, on the distinction between ownership and usage of the sewing machine.

7 See also by Irene Cieraad "De massa als vijandbeeld van cultuur," in A. J. J. van Breemen et al., *Denken over cultuur: Gebruik en misbruik van een concept* (Heerlen: Open University, 1993), 377–408; "De massa," *Evolutie* 28, no. 15 (1920): 113–116; and "De menigte als misdadigster," *Evolutie* 14, no. 25 (1907): 197–199.

8 See also Ernst Hijmans, *De psychose van machinisme en massa-productie* (Purmerend: Netherlands Institute for Efficiency, 1936).

9 William R. Leach, "Transformation in a Culture of Consumption: Women and Department Stores, 1890–1925," *Journal of American History* 71 (1984): 319–342; Cieraad, "Massa," 391–394, and Jan Hein Furnée, "Om te winkelen, zoo als het in de residentie heet: Consumptiecultuur en stedelijke ruimte in Den Haag, 1850–1890," in Marga Altena et al., eds., *Sekse en de city* (Amsterdam: Aksant, 2002), 28–55.

10 Mrs. Van Italie-Van Embden, quoted in "Beteekent mode alleen: Vrouwenkleedij?," *Evolutie* 27, no. 22 (October 4, 1920): 173–175.

11 "Binnen de grenzen," *Evolutie* 33, no. 10 (1925): 78, citing "Het vrouwenhoekje" in *De Maasbode*, 1925.

12 Ibid., 78.

13 H. van Hulst, A. Pleysier, and A. Scheffer, *Het Roode Vaandel volgen wij: Geschiedenis van de Sociaal Democratische Arbeiderspartij van 1880 tot 1940* (The Hague: Kruseman's Uitgeversmaatschappij 1969), 266–279.

14 M. J. Langeveld, *Maatschappelijke verwildering der jeugd: Rapport betreffende het onderzoek naar de geestesgesteldheid van de massajeugd: In opdracht van de Minister van Onderwijs, Kunsten en Wetenschappen samengesteld* (The Hague: Staatsdrukkerij, 1952).

15 Huizinga, *Geschonden wereld*, 194.

16 See also D. van Lente, *Techniek en ideologie: Opvattingen over de maatschappelijke betekenis van technische vernieuwingen in Nederland, 1850–1920* (Groningen: Wolters-Noordhoff/Forsten, 1988).

17 Irene Cieraad, *De elitaire verbeelding van volk en massa: Een studie over cultuur* (Tilburg: Tilburg University Press 1996), 99–112. Also Irene Cieraad, "Traditional Folk and Industrial Masses," in R. Corbey and J. Th. Leerssen, eds., *Alternity, Identity, Image: Selves and Others in Society and Scholarship* (Atlanta, Ga.: Rodopi, 1991), 17-36.

18 Cieraad, "Massa," 399–401.

19 The American concept of "mass society" came to stand for social evils such as rising crime rates resulting from moral decay in an urbanized society dominated by mass production, mass consumption, and mass communication. Mass culture became an unfavorable concept much like mass society. In 1960 the linking of "masses" to the sacred "culture" concept was still a difficult issue for the cultural historian P. J. Bouman; see *In de laagvlakten der cultuur: Rede uitgesproken bij de overdracht van het rectoraat der Rijksuniversiteit te Groningen op 19 september 1960* (Groningen: J. B. Wolters, 1960), 7. For more on the widespread nature of the concern about mass society in postwar Europe, see, for instance, Axel Schildt, "'Technik,' 'Masse' und 'Entfremdung'—kulturpessimistische Tendenzen und deren Kritik um 1950," in *Moderne Zeiten: Freizeit, Massenmedien und "Zeitgeist" in der Bundesrepublik der 50er Jahre* (Hamburg: Christians, 1995), 324–350.

20 Hans Righart, *De eindeloze jaren zestig: Geschiedenis van een generatieconflict* (Amsterdam: Uitgeverij De Arbeiderspers, 1995), 179–182.

21 Huizinga, *Geschonden wereld*, 219–220.

22 P. J. Bouman, *Cultuurgeschiedenis van de twintigste eeuw* (Utrecht: Het Spectrum, 1964), 166. For a critical commentary, see Ruth Oldenziel, "Man the Maker, Woman the Consumer: The Consumption Junction Revisited," in A. N. H. Creager et al., eds., *Feminism in Twentieth-Century Science, Technology, and Medicine* (Chicago: University of Chicago Press, 2001), 128–148.

23 In anthropological analyses of Western consumer culture from the late 1970s and early 1980s, the vision prevailed that all symbolic value had been lost and that superficial commercial values dominated everything. See Cieraad, *Elitaire verbeelding van volk en massa*, 114–123.

24 Dennis L. Meadows, *The Limits to Growth: A Report for the Club of Rome Project on the Predicament of Mankind* (New York: Universe Books, 1972), 12.

25 Cieraad, *Elitaire verbeelding van volk en massa*, 110–111, 129–131.

26 Feminist grassroots groups such as Dolle Mina also stimulated this development. See M. Veenis and R. Oldenziel, "Barsten in het bolwerk: De consumptie betwist, 1986–1980," in Schot et al., *Techniek in Nederland*, vol. 4, 133–145.

27 See Homme Wedman, "Hoeden en petten: Socialisme en burgerlijke cultuur in Nederland," in R. Aerts and H. te Velde, eds., *De stijl van de burger: Over Nederlandse burgerlijke*

cultuur vanaf de Middeleeuwen (Kampen: Kok Agora, 1998), 233, and Pieter R. D. Stokvis, *Terugblikken op het huiselijk leven in de twintigste eeuw: Een verzameling getuigenissen over veranderingen in levensstijl sinds 1920* (Voorburg: Die Haghe, 1999), 13–24.

28 Ileen Montijn, *Leven op stand, 1890–1940* (Amsterdam: Thomas Rap, 1998), 123–150.

29 In a context of disintegrating social classes and ongoing class positioning, the notions of "middle class," "modern," and "aristocratic" took on ambivalent meanings; see Henk te Velde, "Herenstijl en burgerzin: Nederlandse burgerlijke cultuur in de negentiende eeuw," Yme Kuiper, "Aristocraten contra burgers: Couperus's 'Boeken der kleine zielen' en het beschavingsdefensief rond 1900," in Aerts and Te Velde, *Stijl van de burger*, 173, 178, and 186–217.

30 See also Rolf Schuursma, *Jaren van opgang: Nederland 1900–1930* ([Amsterdam: Balans], 2000), 336–337.

31 Mary Douglas and Baron Isherwood, *The World of Goods* (New York: Basic Books, 1979), was one of the first studies emphasizing this.

32 Books that were predominantly written for and by women because it was a women's task to receive guests, pay visits, and conduct a respectable household. One of the oldest publications is by E.C.v.d.M. [Egbertina C. van der Mandele], *Het wetboek van mevrouw: Etiquette in 32 artikelen* (1893; reprint, Utrecht: H. Honig, 1910), and Johanna van Woude, *Vormen: Handboek voor dames* (Amsterdam: Van Holkema, [1898]). These books continued to be sold widely, except during World War I and the crisis of the 1930s. However, the title of another popular book by Amy Groskamp-ten Have, *Hoe hoort het eigenlijk?* [What is really proper?] (1937; reprint, Amsterdam: Becht, 1957), which went through twelve print runs, already conveyed a sense of uncertainty and suggests that social climbers wanted to conform to the old mores.

33 Louis Couperus, who wrote about the class and status concerns of upper-class families in The Hague in his cycle of novels *Boeken der kleine zielen* (Amsterdam: L. J. Veen, [1901–1903]), opted for a fashionable lifestyle as a dandy on the French Riviera, a lifestyle that his characters would have rejected as "decadent"; see Kuiper, "Aristocraten contra burgers," 198.

34 A. J. Q. Alkemade, "De auto in de Nederlandse literatuur en speelfilm," unpublished manuscript (Eindhoven: Nederlands Centrum voor Autohistorische Documentatie en Stichting Historie der Techniek, 2000), 51.

35 Thorstein Veblen, *The Theory of the Leisure Class: An Economic Study of Institutions* (New York: Viking Press, 1899).

36 See W. Drucker "Standen," editorial, *Evolutie* 29, no. 11 (1921): 81–84.

37 See Geert Mak, *De eeuw van mijn vader* (Amsterdam: Uitgeverij Atlas, 1999), 147. Mak wrote of this period that whenever possible children had to go on to college, if only to realize their parents' dreams of being educated and earning a respectable position.

38 Alkemade, "Auto in de Nederlandse literatuur en speelfilm," 55.

39 Righart, *Eindeloze jaren zestig*, 32–73.

40 Katona et al. argue that the "working class's adoption of a middle-class lifestyle" marked by "economic and social progressiveness, striving to improve one's position, better education for children, making plans for the future, or, in short, a more dynamic lifestyle" was hardly a general phenomenon in the 1960s. "The current behavior of the now prosperous workers and consumers in Western Europe . . . reminds one of a status or class society." See G. Katona, B. Strumpel, and E. Zahn, *Het psychologisch klimaat van de economie in de Verenigde Staten en West-Europa* (Utrecht: Het Spectrum, 1974), 38–39.

41 Speech quote printed in *Haagsche Courant*, October 3, 1951.

42 See Remieg Aerts, "Alles in verhouding: De burgerlijkheid van Nederland," in Aerts and Te Velde, *Stijl van de burger,* 296.

43 Pierre Bourdieu, *Distinction: A Social Critique of the Judgement of Taste* (London: Rout-

ledge, 1996), 56–57, presents the results of a large-scale quantitative research project of taste patterns in France in the late 1970s. Harry Ganzeboom, in *Leefstijlen in Nederland: Een verkennende studie* (Alphen/Rijn: Centraal Bureau Statistiek, 1988), has done preliminary work for a similar project in the Netherlands. Consumers' tastes were the object of research. Consumer preferences have of course also been the subject of studies by marketing companies; see Jaap van Ginneken, *De uitvinding van het publiek: De opkomst van het opinie- en marktonderzoek in Nederland* (Amsterdam: Otto Cramwinckel Uitgever, 1993), 9.

44 See Wessel Krul, "De burger als kunstenaar: Nederland in de jaren 1960," in Aerts and Te Velde, *Stijl van de burger*, 249.

45 See Irene Cieraad, "Milk Bottles and Model Homes: Strategies of the Dutch Association for Correct Living (1946–1968)." *Journal of Architecture* 9, no. 4 (Winter 2004): 431–443.

46 Barry Materman, "Hoe meer er verandert, hoe meer de kunstambtenaar op de voorgrond treedt!," in Theo Stokkink, ed., *De cultuur-elite van Nederland: Wie maken en breken de kunst* (Amsterdam: Uitgeverij Balans, 1989), 29–30.

47 Rudi Laermans, "Het mysterie van de goede smaak," in T. Quik, ed., *Smaak: Mensen, media, trends* (Zwolle: Waanders, 1999), 13–49.

48 Elleke de Wijs-Mulkens, *Wonen op stand: Lifestyles en landschappen van de culturele en economische elite* (Amsterdam: Het Spinhuis, 1999), 110–123.

49 Dick Schaap, *Een eeuw wijzer 1883–1983: 100 jaar Koninklijke Nederlandse Toeristenbond ANWB* (The Hague: Algemene Nederlandse Wielrijdersbond, 1983), 25.

50 See Jannes de Haan, *Villaparken in Nederland: Een onderzoek aan de hand van het villapark Duin en Daal te Bloemendaal 1897–1940* (Haarlem: Schuyt & Co., 1986), 135–136, and Irene Cieraad, "Droomhuizen en luchtkastelen: Visioenen van het wonen," in Jaap Huisman et al., eds., *Honderd jaar wonen in Nederland, 1900–2000* (Rotterdam: 010-Publishers, 2000), 216–218.

51 See G. P. A. Mom, J. W. Schot, and P. E. Staal, "Werken aan mobiliteit: De inburgering van de auto," in Schot et al., *Techniek in Nederland*, vol. 5, 45–74; W. J. Lugard, *Veertig jaar: Uit het archief der Kampioen-redactie* (The Hague: Algemene Nederlandse Wielrijdersbond, 1923), 242–243; Gijs Mom, "De auto van avonturenmachine naar gebruiksvoorwerp," in M. S. C. Bakker et al., eds., *Techniek als cultuurverschijnsel* (Heerlen: Open University, 1996), 104.

52 Lugard, *Veertig jaar*, 239.

53 This tension is nicely evoked in Jan Feith's, *Billy: De gedenkschriften van m'n chauffeur* (Amsterdam: Scheltens & Giltay, 1912), in which the chauffeur and a rich young lady, the fiancé of his employer, fall in love. See also the many examples in Alkemade, "Auto in de Nederlandse literatuur en speelfilm," 16, 19–20, 24, 42–43.

54 See, for example, girls' novels such as those by Cissy van Marxveldt, *Joop van Dil–ter Heul* (Amersfoort: Valkhoff, 1923), and *Een zomerzotheid* (Amersfoort: Valkhoff, 1927), and by Nellie Wesseling, *Een tante die chauffeert: Een amusant verhaal voor jonge meisjes* ([Helmond: Boek & Handelsdrukkerij, 1934]). See Alkemade, "Auto in de Nederlandse literatuur en speelfilm," 40–48.

55 Tom Schilperoort, "Automobilisme en autotoerisme," *Panorama* 6, June 4, 1919, 4–5, and Robert Peereboom, *Dat snelle ding: Een klaxonate* (Amsterdam: Andries Blitz, 1937), 41.

56 See Mom, "Auto van avonturenmachine naar gebruiksvoorwerp," 105.

57 "Automobilisme: Gordon-Bennet Cup," *De Ware Jacob* 3, no. 18 (1904): 147.

58 See "De tien geboden voor den automobilist," *De Auto* 19 (April 13, 1922): 52–54.

59 See also Schaap, *Eeuw wijzer*, 173.

60 Louis Couperus, *Boeken der kleine zielen*, vol. 2: *Het late leven* (Amsterdam: L. J. Veen, 1902), 352–353; see also Alkemade, "Auto in de Nederlandse literatuur en speelfilm," 14.

61 Frank M. van der Heul, *Auto-advertenties uit de jaren 1890–1940* (Rijswijk: Elmar, 1991); Angela Zatsch, "Reich, schnell, mobil: Automobilwerbung zu Beginn des 20. Jahrhunderts,"

in Peter Borscheid and Clemens Wischermann, eds., *Bilderwelt des Alltags: Werbung in der Konsumgesellschaft des 19. und 20. Jahrhunderts* (Stuttgart: Franz Steiner Verlag, 1995), 284, shows a German poster for a 1904 automobile exhibition in which this visual code of the right front wheel was used as well, albeit without speed stripes.

62 See Roland Marchand, *Advertising the American Dream: Making Way for Modernity, 1920–1940* (Berkeley: University of California Press, 1985), 254–255, which makes clear that American visual clichés differ from European ones.

63 Remarkably, in this same year the Futurist Giacomo Balla painted his abstract work *Futur*, which depicts rotating movements and speed. The similarity with large front wheels in car ads is striking and Balla also used speed stripes produced by the spinning "wheel." In the same year Balla painted a canvas entitled *The Speed of Car + Light + Noise*. In art history the representation of speed and movement is commonly associated with the introduction of moving images, rather than with the invention of the automobile; see also Stephen Kern, *The Culture of Time and Space 1880–1918* (Cambridge, Mass.: Harvard University Press, 1983), 117–120.

64 Jeffrey L. Meikle, *Twentieth Century Limited: Industrial Design in America, 1925–1939* (Philadelphia: Temple University Press, 1979), 7–18.

65 Ibid., 187.

66 See also James J. Flink, *The Automobile Age* (Cambridge, Mass.: MIT Press, 1988). I am solely responsible for drawing this relationship between this visual code of speed in illustrations and streamlined design.

67 Fons Alkemade, *Het beeld van de auto 1896–1921: Verslag van een speurtocht door Nederlandse collecties* (Deventer: Kosmos Uitgevers, 1996), 169, shows these countervoices from the satirical weekly in the 1908 volume of *De Ware Jacob*.

68 Frank M. van der Heul, "Lincoln," in *Auto advertenties uit de jaren 1890–1940* (Rijswijk: Elmar, 1991). The German car manufacturer BMW has a special sound lab where technicians and psychologists collaborate in the creation of psychologically satisfactory sounds attuned to the preferences of target consumers of its different car types. See Bard van de Weijer, "De sound van BMW: Kalm tuffen of bronstig cruisen," *De Volkskrant*, February 3, 2007, p. 9.

69 See also Steven Lubar, *Infoculture: The Smithsonian Book of Information Age Inventions* (Boston: Houghton Mifflin, 1993), 121.

70 Ithiel de Sola Pool et al., "Foresight and Hindsight: The Case of the Telephone," in Ithiel de Sola Pool, ed., *The Social Impact of the Telephone* (Cambridge, Mass.: MIT Press, 1977), 127–157.

71 See Robert V. Bruce, *Bell: Alexander Graham Bell and the Conquest of Solitude* (Ithaca: Cornell University Press, 1973), 394. This subtitle suggests a slightly different interpretation than the one in Onno de Wit, *Telefonie in Nederland 1877–1940: Opkomst en ontwikkeling van een grootschalig technisch systeem* (Amsterdam: Otto Cramwinckel Uitgever, 1998), 31–50.

72 "Toelichting op de brochure," in Veritas, *Vlugschrift ter bereiking van eenen beteren toestand voor geabonneerden en telefonisten te Amsterdam door Veritas*, Amsterdam, 1901, in the collection of the Museum of Communication, The Hague, box 13A, folders 6–9.

73 The historian Piet de Rooy, in Sander van Walsum, "Het ontbreekt ons aan historische ijkpunten," *Het Volkskrant Magazine* 13 (April 2002): 10, recalls how his aunt would take off her apron before answering the phone.

74 From G. Hogesteeger and R. A. Korving, *De juffrouw van de telefoon* (Zwolle: Waanders, 1993).

75 See "Klachten over de telefoon," *Panorama*, February 11, 1920), 4–5. In 1936 the PTT tested telephone devices for their durability in the case of angry outbursts; see "De gekwelde telefoon," *PTT Nieuws* 5 (1936): 84–86. For a numerical indication of the type of phone

conversations and the phone conversation in Dutch literature, see De Wit, *Telefonie in Nederland*, 187, 206–230.

76 De Wit, *Telefonie in Nederland*, 164–187, 230–257.

77 See also Henry M. Boettinger, "Our Sixth-and-a-Halfth sense," in De Sola Pool, *Social Impact of the Telephone*, 200–207.

78 Ph. H. Kleingeld, *Telefoneren: Practisch handboek voor de telefoniste/receptioniste* (Baarn: Bosch en Keunig, 1987), 17–19.

79 See also Schuursma, *Jaren van opgang*, 397–406; F. L. Polak, quoted in F. van Vree, "Massacultuur en media," in H. Wijfjes, ed., *Omroep in Nederland: Vijfenzeventig jaar medium en maatschappij, 1919–1994* (Zwolle: Waanders, 1994), 31.

80 See H. Baudet, *Een vertrouwde wereld: 100 jaar innovatie in Nederland* (Amsterdam: Uitgeverij Bert Bakker, 1986), 132–134.

81 A survey undertaken in May 1965 among readers of the ANWB's journal, *Kampioen* (*Champion*), confirmed this picture. The readers owned significantly more cameras and went on vacation more often than the average Dutch person. See *Kampioen: Lezerskringonderzoek in opdracht van de Koninklijke Toeristenbond ANWB* (The Hague: ANWB, 1965). See also David E. Nye, "The Emergence of Photographic Discourse: Images and Consumption," in D. E. Nye and C. Pederson, eds., *Consumption and American Culture* (Amsterdam: Free University Press, 1991), 32-48.

82 Erving Goffman, *Gender Advertisements* (London: Macmillan, 1979).

83 See Jaap Boerdam and Warna Oosterbaan Martinus, "Het fotogenieke van het samenleven," *Amsterdams Sociologisch Tijdschrift* 5, no. 1 (1978): 3–36.

84 See Scott McQuire, *Visions of Modernity: Representation, Memory, Time and Space in the Age of the Camera* (London: Sage, 1998), 4.

85 In this period movie theaters were mushrooming in large cities; see Schuursma, *Jaren van opgang*, 397–406.

86 The experience of movie watching is also traced back to the train or car passenger's outward gaze. See McQuire, *Visions of Modernity*, 210.

87 Van Vree, "Massacultuur en media," 26.

88 "The worker who suddenly realized prosperity has a much more vulnerable position in life. From the very first moment of his emancipation as consumer, he experiences the pressures of subtle consumption stimulation. Advertising, installment plans—there is so much that encourages anticipating what one will actually earn as income." See Bouman, *Cultuurgeschiedenis van de twintigste eeuw*, 162.

89 Although visual images had always had a religious-didactic function in Catholic tradition and many Dutch Catholics subscribed to *Katholieke Illustratie* (an illustrated magazine that first came out in 1867), until 1926 the bishops rejected film's moving images out of fear of moral decay. See L. A. C. Jentjens, *Van strijdorgaan tot familieblad: De tijdschriftjournalistiek van de 'Katholieke Illustratie' 1867–1968* (Amsterdam: Otto Cramwinckel Uitgever, 1995), 139–140.

90 Hendrik Algra, Cornelis Rijnsdorp, and Ben van Kaam, *Vrij en gebonden: 50 jaar NCRV* (Baarn: Bosch en Keunig, 1974), 10, 63, 138–139.

91 An illusion triggered by the example of the United States, where in the 1950s television proved to have a cohesive effect on families. See Lynn Spigel, "Making Room for TV," in D. Crowley and P. Meyer, eds., *Communication in History: Technology, Culture, Society* (New York: Sage, 1995), 272–280.

92 See Righart, *Eindeloze jaren zestig*, 100–113, as well as Mark van den Heuvel and Hans Mommaas, "Oorden van vrijheid en vermaak: De herstructurering van de vrije-tijd," in Ger Tillekens, ed., *Nuchterheid en nozems: De opkomst van de jeugdcultuur in de jaren vijftig* (Muiderberg: Coutinho, 1990), 148–164, and Stokvis, *Terugblikken op het huiselijk leven in de twintigste eeuw*, 51.

93 See Willem Alfred Dolfsma, *Valuing Pop Music: Institutions, Values, and Economics* (Delft: Eburon, 1999). Since 2005 the number of publications on the senses has exploded. See, for instance, the book series *Sensory Formations* (Berg) and the journal *The Senses and Society.*

94 Jan Bank, "Televisie in de jaren zestig," *Bijdragen en mededelingen betreffende de geschiedenis der Nederlanden* 101 (1986): 52–75.

95 The newspaper *De Waarheid*, which was linked to the Dutch Communist Party and was read by many progressive youths at that time, carried no ads.

96 Jean Baudrillard, *For a Critique of the Political Economy of the Sign* (St. Louis, Mo: Telos Press, 1981), 54–57.

97 See also Huub Wijfjes, *Hallo hier Hilversum! Driekwart eeuw radio en televisie* (Weesp: Van Dishoeck, 1985).

98 Pauline Terreehorst, *Langzame stad, snelle mensen: Leven in een informatietijdperk* (Amsterdam: Van Gennep, 1997).

99 The Netherlands still had three systems of timekeeping in 1892: Greenwich time, used by Western-European railway and telegraph services; Amsterdam time, used in Amsterdam and other Dutch cities; and other local time systems used by local tramway services. To make travel easier, in 1909 a law was passed requiring national synchronization of the time. In 1940 the German occupiers imposed Central European time, which we still have today. See Hans Knippenberg and Ben de Pater, *De eenwording van Nederland: Schaalvergroting en integratie sinds 1800* (Nijmegen: SUN, 1988), 81–82.

100 A 1960 government report aimed at regulating recreational activities, cited in Van den Heuvel and Mommaas, "Oorden van vrijheid en vermaak," 151.

101 Colin Campbell, *The Romantic Ethic and the Spirit of Modern Consumerism* (Oxford: Basil Blackwell, 1989), 69, characterizes the combination of self-control and hedonism as "modern hedonism."

102 Arnold Labrie, *Zuiverheid en decadentie: Over de grenzen van de burgerlijke cultuur in West-Europa, 1870–1914* (Amsterdam: Het Spinhuis, 2001), 100, 118–123.

103 The illustrated photo journal *Het Leven* documented the emergence of Dutch consumer society more completely than any other source. See F. Bool, H. Overduin, and G. J. de Rook, *Het Leven, 1906–1941: Een weekblad in beeld* (The Hague: Haags Gemeentemuseum, 1981).

104 See Mary Cowling, *The Artist as Anthropologist: The Representation of Type and Character in Victorian Art* (Cambridge: Cambridge University Press, 1989).

105 See Alain Corbin, *The Lure of the Sea* (Berkeley: University of California Press, 1994), and Thomas A. P. van Leeuwen, *The Springboard in the Pond: An Intimate History of the Swimming Pool* (Cambridge, Mass.: MIT Press, 1998).

106 In a short experimental novel by F. Bordewijk, *Knorrende beesten* (Utrecht: De Gemeenschap, 1933), the relationship between woman, automobile, and seaside resort takes center stage; see Alkemade, "Auto in de Nederlandse literatuur en speelfilm," 25.

107 In the 1920 and 1930s, smoking cigarettes in casual postures was limited to young men and women who thus also emphasized they owned the phone, car, or radio. Such a posture was highly improper for non-owners; it could even be grounds for firing servants, chauffeurs, or secretaries on the spot.

108 In sports the relationship between human and machine, between motor system and mechanism, has always been expressed in terms of speed, and in both industry and the household context this relationship was expressed by the then fashionable notion of efficiency.

109 See Diederik van Loggem, *Jeugd: Probleem van ouderen?* (Baarn: Ambo, 1989), 210.

List of Abbreviations

AKU	Algemene Kunstzijde Unie — General Rayon Union
ANSI	Algemeen Syndicaat van Suikerfabriekanten in Nederlands Indië — Sugar Producers in the Dutch East Indies
ANWB	Algemene Nederlandse Wielrijdersbond — General Netherlands Bicyclists Association (now a general Dutch touring association)
BASF	Badische Anilin- und Soda-Fabrik (German chemical company)
BENISO	Bond van Eigenaren van Nederlans-Indische Suikerondernemers — Union of Owners of Netherlands Indies Sugar Mills
DSM	De Staatsmijnen — State Mines (company)
ETIL	Economisch Technologisch Instituut Limburg — Economic-Technological Institute Limburg
EU	European Union
GE	General Electric
GNP	gross national product
GSM	Groupe Spécial Mobile — Global System for Mobile Communication
HCNN	Hoofdcommissie voor de Normalisatie in Nederland — High Commission for Standardization in the Netherlands
HVA	Handelsvereeniging Amsterdam — Trade Association of Amsterdam
ICI	Imperial Chemical Industries
ISEW	Index of Sustainable Economic Welfare
KLM	Koninklijke Luchtvaart Maatschappij — Royal Dutch Airlines
KNAW	Koninklijke Nederlandse Akademie van Wetenschappen — Royal Netherlands Academy of Sciences
KNZ	Koninklijke Nederlandse Zoutindustrie — Royal Dutch Salt Company
NACO	Netherlands Airport Consulting Office; later, Netherlands Airport Consultants
NatLab	Natuurkundig Laboratorium (Philips) — Natural Sciences Laboratory
NEN	Nederlandse Norm — Netherlands Standards
NGSF	Nederlandsche Gist- en Spiritusfabriek — Netherlands Yeast and Spirits Works
NHM	Nederlandsche Handel-Maatschappij — Netherlands Trading Company
NIMBY	"Not in my back yard"
NS	Nederlandse Spoorwegen — Netherlands Railroad Company
NVVH	Nederlandse Vereniging van Huisvrouwen — Netherlands Association of Housewives
OECD	Organization for Economic Cooperation and Development

OK&W	Ministerie van Onderwijs, Kunsten en Wetenschappen — Education, Arts, and Sciences Ministry
OVO	*onderzoek, voorlichting, onderwijs* (research, information, education)
PTT	Posterijen Telegrafie Telefonie — Postal, Telegraphy and Telephony Authority
SBB	Stikstofbindingsbedrijf — Nitrogen Fixing Works
TFP	total factor productivity
TNO	Toegepast Natuurwetenschappelijk Onderzoek — Netherlands Organization for Applied Scientific Research
VOC	Verenigde Oostindische Compagnie — United East India Company
ZWO	Nederlandse Organisatie voor Zuiver Wetenschappelijk Onderzoek — Netherlands Organization for Pure Scientific Research

Further Reading

This book is a translation of the seventh volume of Techniek in Nederland in de Twintigste Eeuw, a series published between 1998 and 2003 by the Stichting Historie der Techniek (Foundation for the History of Technology) and Walburg Pers (Walburg Press). Unfortunately, the six earlier volumes have not been translated into English. To provide access to some of this rich work, this list of further readings contains a wide range of articles and books published in English and other non-Dutch languages.

Albert de la Bruhèze, Adri A. "Manufacturing Snacks and Snackers: Unilever and Dutch Snack Consumption." In Adri A. Albert de la Bruhèze and Ruth Oldenziel, eds., *Manufacturing Technology, Manufacturing Consumers: The Making of Dutch Consumer Society*. Amsterdam: Aksant, 2009.

Albert de la Bruhèze, Adri A., and Ruth Oldenziel, eds. *Manufacturing Technology, Manufacturing Consumers: The Making of Dutch Consumer Society*. Amsterdam: Aksant, 2009.

Albert de la Bruhèze, Adri A., and Anneke H. van Otterloo. "The Milky Way: Infrastructures and the Shaping of Milk Chains." *History and Technology* 20, no. 3 (September 2004) (special issue: *Networked Nation: Transforming the Netherlands in the 20th Century*): 249–270.

———. "Snacks and Snack Culture in the Rise of Eating Out in the Netherlands in the Twentieth Century." In Marc Jacobs and Peter Scholliers, *Eating Out in Europe: Picnics, Gourmet Dining and Snacks Since the Late Eighteenth Century*. Oxford: Berg Publishers, 2003.

Baggen, Peter. "Technischer Vollzeitunterricht in den Niederlanden am Beispiel der Berufsfachschulen." *Technikgeschichte*, Bd. 68 (2001) Heft 2.

Belt, H. van den. "Networking Nature, or Serengeti Behind the dikes." *History and Technology* 20, no. 3 (September 2004) (special issue: *Networked Nation: Transforming the Netherlands in the 20th Century*): 311–330.

Bervoets, Liesbeth. "A Utopia in Stone: Mediating in the Name of Working-Class Collectivism." In Adri A. Albert de la Bruhèze and Ruth Oldenziel, eds., *Manufacturing Technology, Manufacturing Consumers: The Making of Dutch Consumer Society*. Amsterdam: Aksant, 2009.

Bervoets, Liesbeth, and Ruth Oldenziel. "Speaking for Consumers, Standing Up as Citizens: The Politics of Dutch Women's Organizations and the Shaping of Technology, 1880–1980." In Adri A. Albert de la Bruhèze and Ruth Oldenziel, eds., *Manufacturing Technology, Manufacturing Consumers: The Making of Dutch Consumer Society*. Amsterdam: Aksant, 2009.

Bieleman, Jan. "Technologica Innovation in Dutch Cattle Breeding and Dairy Farming, 1850-2000." Agricultural Review 53, no. 2 (2005): 229-250.

Bogaard, Adrienne van den, and Cornelis Disco. "Die Stadt als Innovationsknotenpunkt." Technikgeschichte, Zeitschrift der Verein Deutscher Ingenieure 68, no. 2 (2001): 107-132.

Buiter, Hans. "Constructing Dutch Streets: A Melting Pot of European Technologies." In Michael Hard and Tom Mis eds., Urban Machiner: Inside Modern European Cities. Cambridge, Mass.: MIT Press, 2008.

———. The History of Power Cable in the Netherlands. (The Hague and Eindhoven: Foundation for the History of Technology, 1997.

Davids, Mila. "The Fabric of Production: The Philips Industrial Network." History and Technology 20, no. 3 (September 2004) (special issue: Networked Nation: Transforming the Netherlands in the 20th Century): 271–290.

Davids, Mila, and Hans Schippers. "Innovations in Dutch Shipbuilding in the First Half of the Twentieth Century." Business History, 50 (March 2008): 205–225.

Disco, Cornelis. "Remaking 'Nature': The Ecological Turn in Dutch Water Management." Science, Technology and Human Values 27, no. 2 (2002): 206–235.

Disco, Cornelis, and Jan. C. M. van den Ende. "Strong, Invincible Arguments? Tidal Models as Management Instruments in Twentieth Century Dutch Coastal Engineering." Technology and Culture 44, no. 3 (2003: 502-535.

Disco, Cornelis, and Erik van der Vleuten. "The Politics of Wet System Building: Balancing Interests in Dutch Water Management from the Middle Ages to the Present." Knowledge, Technology & Policy 14, no. 4 (2002): 21–40.

Driel, Hugo van. "The First Mechanisation Wave in Coal and Ore Handling as an Example of Patterns of Technological Innovation in the Port of Rotterdam." In Reginald Loyen, Erik Buyst, and Greta Devos, eds., Struggling for Leadership: Antwerp-Rotterdam Port Competition Between 1870 and 2000. Heidelberg: Physica-Verlag, 2003.

Driel, Hugo van, and Ferry de Goey. Rotterdam Cargo Handling Technology 1870–2000. Zutphen: Walburg Pers, 2000.

Driel, Hugo van, and Johan Schot. "Regime-transformatie in de Rotterdamse graanoverslag." NEHA-jaarboek voor economische, bedrijfs- en techniekgeschiedenis 64 (2001): 286–318.

———. "Radical Innovation as a Multilevel Process: Introducing Floating Grain Elevators in the Port of Rotterdam." Technology and Culture 46, no. 1 (January 2005): 51–76.

Empelen, Louis van, Geert Verbong, and Ton Hesselmans. "Die Entwicklung des Holländischen Stromnetzes von 1939 bis 1950 und der Verbund mit dem RWE." In H. Maier, ed., Elektrizitätswirtschaft zwischen Umwelt, Technik und Politik: Aspekte aus 100 Jahren RWE-Geschichte 1898–1998. Forschungsheft Series D 204, Geschichte. Freiberg, Germany: Technical University Freiberg Bergakademie, 1999.

Gales, Ben, Paolo Malanima, Astrid Kander, and Mar Rubio. "North Versus South: Energy Transition and Energy Intensity in Europe over 200 Years." European Review of Economic History 11 (2007): 219–253.

Hermans, Janneke. ICT in Information Services: Use and deployment of ICT in the Dutch Securities Trade 1860–1970. Rotterdam: Erasmus Institute of Management, 2004.

Hermans, Janneke, Nachoem Wijnberg, Jan van den Ende, and Onno de Wit. "Vertical Integration as a Remedy to Imbalances in the 'Porterian' Value Chain: The Dutch Securities Industry at the Beginning of the 20th Century." *Scandinavian Journal of Management* 20, no. 4 (December 2004): 357–374.

Hermans, Janneke, and Onno de Wit. "Bourses and Brokers: Stock Exchanges as ICT Junctions." *History and Technology* 20, no. 3 (September 2004) (special issue: *Networked Nation: Transforming the Netherlands in the 20th Century*): 227–248.

Homburg, Ernst. "Explosives from Oil: The Transformation of Royal Dutch/ Shell During World War I from Oil to Petrochemical Company." In Brenda J. Buchanan, ed., *Gunpowder, Explosives and the State: A Technological History* Aldershot, U.K.: Ashgate, 2006.

———. "Operating on Several Fronts: The Trans-National Activities of Royal Dutch/ Shell, 1914–1918." In R. M. MacLeod and J. A. Johnson, eds., *Frontline and Factory: Comparative Perspectives on the Chemical Industry at War, 1914–1924*. Dordrecht: Springer, 2007.

———. "Boundaries and Audiences of National Histories of Science: Insights from the History of Science and Technology of the Netherlands." *Nuncius: Annali di storia della scienza* 22, no. 2 (2008): 309–345.

Homburg, Ernst, and Arjan van Rooij. "Die Vor- und Nachteile enger Nachbarschaft: Der Transfer Deutscher chemischer Technologie in die Niederlande bis 1952." In R. .Petri, R. ed., *Technologietransfer aus der Deutschen Chemieindustrie (1925–1960)*. Berlin: Duncker & Humblot, 2003.

Knecht-van Eekelen, Anne-Marie, and Anneke H. van Otterloo. "The Introduction and Diffusion of Glass Preserved Food in the Dutch Food Pattern 1945–1995." In M. R. Schärer and A. Fenton, eds., *Food and Material Culture: Proceedings of the Fourth Symposium of the International Commission for Research into European Food History*. East Linton, Scotland: Tuckwell Press, 1998.

———. "What the Body needs: Developments in Medical Advice, Nutritional Science, and Industrial Production in the Twentieth century." In A. Fenton, ed., *Order and Disorder: The Health Implications of Eating and Drinking in the Nineteenth and Twentieth Centuries*. East Linton, Scotland: Tuckwell Press, 2000.

Lagaaij, Alexander, and Geert Verbong. "Different Vision of Power: The Introduction of Nuclear Power in the Netherlands 1955–1970." *Cantaurus* 41, nos. 1–2 (1999): 37–63.

Lente, Dick van. "The Crafts in Industrial Society: Ideals and Policy in the Netherlands 1890–1930." *Economic and Social History in the Netherlands* 2 (1990): 99–119.

———. "Ideology and Technology: Reactions to Modern Technology in the Netherlands 1850–1920." *European History Quarterly* 22 (1992): 383–414.

———. "Machines and the Order of the Harbour: The Debate About the Introduction of Grain Unloaders in Rotterdam, 1905–1907." *International Review of Social History* 43 (1998): 79–109.

Lente, Dick van, and Ernst Homburg. "Ambient Technology and Social Progress: A Critical View." In E. Aarts and S. Marzano, eds., *The New Everyday: Views on Ambient Intelligence*. Rotterdam: 010 Publishers, 2003: 30–33.

Lintsen, Harry W. "Two Centuries of Central Water Management in the Netherlands" *Technology and Culture* 43, no. 3 (2002): 549–568.

Mokyr, Joel. "Techniek in Nederland in de Twintigste Eeuw." (English review) Tijdschrift voor Sociale en Economische Geschiedenis 1, no. 1 (2004): 154-156.

Mom, Gijs. "Civilized Adventure as a Remedy for Nervous Times: Early Automobilism and Fin-de-Siècle Culture." *History of Technology* 23 (2001): 157–190.

———. "Frozen History: Limitations and Possibilities of Quantitative Diffusion Studies." In A. A. Albert de la Bruhèze and Ruth Oldenziel, eds., *Manufacturing Technology, Manufacturing Consumers: The Making of Dutch Consumer Society*. Amsterdam: Aksant, 2009.

Mom, Gijs, Johan Schot, and Peter Staal. "Civilizing Motorized Adventure: Automotive Technology, User Culture, and the Dutch Touring Club as Mediator." In A. A. Albert de la Bruhèze and Ruth Oldenziel, eds., *Manufacturing Technology, Manufacturing Consumers: The Making of Dutch Consumer Society*. Amsterdam: Aksant, 2009.

Oldenziel, Ruth. "Man the Maker, Woman the Consumer: The Consumption Junction Revisited." In A. N. H. Creager, E. Lunbeck, L. Schiebinger, and C. R. Stimpson, eds., *Feminism in Twentieth-Century Science, Technology, and Medicine*. Women in Culture and Society Series. Chicago: University of Chicago Press, 2002.

Oldenziel, Ruth, and Adri A. Albert de la Bruhèze. "Theorizing the Mediation Junction for Technology and Consumption." In Adri A. Albert de la Bruhèze and Ruth Oldenziel, eds., *Manufacturing Technology, Manufacturing Consumers: The Making of Dutch Consumer Society*. Amsterdam: Aksant, 2009.

Oost, Ellen van. "Aligning Gender and New Technology: The Case of Early Administrative Automation." In Cornelis Disco and Barend van der Meulen, eds., *Getting New Technologies Together: Studies in Making Sociotechnical Order*. Berlin: Walter de Gruyter, 1998.

Otterloo, Anneke van. "The Low Countries." In K. F. Kiple and K.C. Ormelas, eds., *The Cambridge World History of Food*, vol. 2 (Cambridge: Cambridge University Press 2000).

———. "The Other End of the Food-Chain in Long-Term Perspective." In C. A. A. Butijn, J. P. Groot-Marcus, M. van der Linden, eds., *Changes at the Other End of the Chain: Everyday Consumption in Multidisciplinary Perspective*. Wageningen: Wageningen University, Department of Agrotechnology and Food Sciences, 2002).

Otterloo, Anneke, and Marja Berendsen. "The 'Family Laboratory': The Contested Kitchen and the Making of the Modern Housewife." In Adri A. Albert de la Bruhèze and Ruth Oldenziel, eds., *Manufacturing Technology, Manufacturing Consumers: The Making of Dutch Consumer Society*. Amsterdam: Aksant, 2009.

Raven, Rob, and Geert Verbong. "Dung, Sludge and Landfill: Biogas Technology in the Netherlands 1970–2000." *Technology and Culture* 45, no. 3 (July 2004): 519–539.

Ravenstein, Wim, L. Hermans, and Erik van der Vleuten. "Participation and Globalization in Water System Building." *Knowledge, Technology & Policy* 14 (2002): 4–12.

Rooij, Arjan van, and Ernst Homburg. *Building the Plant: A History of Engineering Contracting in the Netherlands* Zutphen: Walburg Pers, 2002.

Schot, Johan. "The Usefulness of Evolutionary Models for Explaining Innovation: The Case of the Netherlands in the Nineteenth Century." *History and Technology* 14, no. 3 (1998): 173–200.

———. "Einleitung Kontextualistische Technikgeschichte in den Niederlanden." *Technikgeschichte* 68, no. 2 (2001): 93–106.

———. "Themenheft Kontextualistische Technikgeschichte in den Niederlanden." *Tech-*

nikgeschichte 68, no. 2 (2001): 93-179.

Schot, Johan, and Adri A. Albert de la Bruhèze. "The Mediated Design of Products, Consumption and Consumers in the Twentieth Century." In N. Oudshoorn and T. Pinch, eds., *How Users Matter: The Co-Construction of Users and Technology.* Cambridge, Mass: MIT Press, 2005.

Verbong, Geert. "Wind Power in the Netherlands 1970–1995." *Cantaurus* 41, no. 1–2 (1999): 137–160.

———. "Dutch Power Relations: From German Occupation to The French Connection." In Erik van der Vleuten and Arne Kaijser, eds., *Networking Europe: Transnational Infrastructures and the Shaping of Europe, 1850–2000.* Sagamore Beach, Mass.: Science History Publications, 2006.

Verbong, Geert, and Erik van der Vleuten. "Under Construction: Material Integration of the Netherlands 1800–2000." *History and Technology* 20, no. 3 (September 2004) (special issue: *Networked Nation: Transforming the Netherlands in the 20th Century*): 205–226.

Vleuten, Erik van der. "Étude des conséquences sociétales des macro-systèmes techniques: Une approche pluraliste." *Flux Cahiers scientifiques internationeaux réseaux et territoires* 43 (2001): 42–57.

———. "In search of the Networked Nation: Transforming Technology, Society and Nature in the Netherlands in the 20th Century." *European Review of History* 10, no. 1 (2003): 59–78.

———. "Infrastructure and Societal Change: A view from the Large Technical Systems Field." *Technology Analysis & Strategic Management* 16, no. 3 (2004): 395–414.

———. "Networking Technology, Networking Society, Networking Nature." *History and Technology* 20, number 3 (September 2004) (special issue: *Networked Nation: Transforming the Netherlands in the 20th Century*): 195–204.

Vleuten, Erik van der. "Understanding Network Societies: Two Decades of Large Technical Systems Studies." In Erik van der Vleuten and Arne Kaijser, eds., *Networking Europe: Transnational Infrastructures and the Shaping of Europe, 1850–2000.* Sagamore Beach, Mass.: Science History Publications, 2006.

Vleuten, Erik van der, and Geert Verbong, eds. *Networked Nation. Transforming the Netherlands in the 20th Century.* Special issue of *History and Technology.* September 2004.

Vleuten, Erik van der, and Cornelis Disco. "Water Wizards: Reshaping Wet Nature and Society." *History and Technology* 20, no. 3 (September 2004) (special issue: *Networked Nation: Transforming the Netherlands in the 20th Century*)": 291–310.

Wachelder, J. "TIN-20: Techniek in Nederland in de twintigste eeuw." (review) *Technology and Culture*, 46, no 1 (January 2005): 187-191.

Wijnberg, Nachoem, Jan van den Ende, and Onno de Wit. "The Relation Between Systematic Management and Scientific Management: Two Cases from the Financial Services." *Group & Organization Management: An International Journal* 27, no. 3 (2002): 408-429.

Wit, Onno de. "Corporate Mediation Junctions: Philips and Media in the Netherlands." In A. A. Albert de la Bruhèze and Ruth Oldenziel, eds., *Manufacturing Technology, Man-*

ufacturing Consumers: The Contested Making of Dutch Consumer Society in the Twentieth Century. Amsterdam: Aksant, 2009.

Wit, Onno de, Adri A. Albert de la Bruhèze, and Marja Berendsen. "Ausgehandelter Konsum: Die Verbreitung der modernen Kueche, des Kofferradios und des Snack Food in den Niederlanden." *Technikgeschichte* 68, no. 2 (2001): 133–156.

Wit, Onno de, and Jan van den Ende. "The Emergence of a New Regime. Business Management and Office Mechanisation in the Dutch Financial Sector in the 1920s." *Business History* 42, nr. 2 (2000): 50–72.

Wit, W. O. de, J. C. M. van den Ende, J. W. Schot, and E. van Oost. "Innovation Junctions: Office Technologies in the Netherlands, 1880–1980." *Technology & Culture* 43, no. 1 (2002): 50–72.

Sources of Tables, Graphs, and Maps

All tables, graphs, and maps were created for use in this book. They are based on a wide variety of sources, listed here. The graphs were made by Camiel Lintsen from Kade05 Graphic Design.

Map 2-1. Development of natural gas and electricity networks.

Verslag der Staatscommissie ingesteld bij Koninklijk Besluit van 14 juli 1911, no 60, om van advies te dienen omtrent de vraag, welke maatregelen genomen kunnen worden om te bevorderen, dat de behoefte aan elektrische kracht, welke in verschillende streken des lands en met name ten plattelande bestaat, op zoo doeltreffend en zoo economisch mogelijke wijze worde voorzien [Report of the national commission on the royal decree of July 14, 1911, concerning steps to be taken to extend the distribution of electric power] (Leiden: IJdo, 1914), 28, 81.

J. C. van Staveren, "De electriciteitsvoorziening," *Tijdschrift van het Koninklijk Nederlandsch Aardrijkskundig Genootschap* 50 [special issue on twentieth-century traffic in the Netherlands] (1933): 592–607.

F. C. Wirtz Cz., "Gasvoorziening in Nederland," *Het Gas* 51, no. 19 (1931): 355–356.

J. J. Suyver, "Hoe en waarom kwam men tot het nieuwe koppelnet," *Electrotechniek* 48 (1970): 3.

H. Meijer and Th. Woudstra, *Kortfattet oversight over Nederlands geografi* (Utrecht and The Hague: Informatie- en Documentatiecentrum voor de Geografie van Nederland and Udenrigsministeriet, 1974), 29;

C. Klaassen, "Gasunie-transportnet: Enkele aspecten van het transportnet van de N.V. Nederlandse Gasunie," *Het Gas* 90 (August 1970): 282.

Map 2-2. Development of transportation networks.

A. Bosch and W. van der Ham, *Twee eeuwen Rijkswaterstaat, 1798–1998* (Zaltbommel: Europese Bibliotheek, 1998) 62, 75.

J. J. Stieltjes, "Het verkeer te land: De spoor- en tramwegen," *Tijdschrift van het Koninklijk Nederlandsch Aardrijkskundig Genootschap* 50 (1933): 477.

F. L. Schlingemann, "II Het verkeer te water," *Tijdschrift van het Koninklijk Nederlandsch Aardrijkskundig Genootschap* 50 (1933): 334–419.

R. Loman, "Het verkeer te land: De wegen voor gewoon verkeer en het gebruik daarvan," *Tijdschrift van het Koninklijk Nederlandsch Aardrijkskundig Genootschap* 50 (1933): 478–545.

J. M. Kan, "Het luchtverkeer," *Tijdschrift van het Koninklijk Nederlandsch Aardrijkskundig*

Genootschap 50 (1933): 566–591.

J. M. Fuchs, "Verkeer en vervoer," in *Eerste Nederlandse systematisch ingerichte encyclopaedie*, vol. 7 (Amsterdam: 1950), 250–325.

H. Meijer and Th. Woudstra, *Kortfattet oversight over Nederlands geografi* (Utrecht and The Hague: Informatie- en Documentatiecentrum voor de Geografie van Nederland and Udenrigsministeriet, 1974), 35.

Map 2-3. Development of communication networks.

W. Ringnalda, "De Rijkstelegraaf in Nederland: Hare opkomst en ontwikkeling (1852 – 1 December 1902)," *Geïllustreerd gedenkboek ter herinnering aan haar vijftig-jarig bestaan* (Amsterdam: Scheltema & Holkema, 1902).

Hans Knippenberg and Ben de Pater, *De eenwording van Nederland: Schaalvergroting en integratie sinds 1800* (Nijmegen: SUN, 1988), 64–65.

J. H. Warning, "De automatisering van het Nederlandse Telefoonnet," *De Ingenieur* 42 (1934): E143.

H. Mol, "Telegrafie en telefonie," in *Eerste Nederlandse systematisch ingerichte encyclopaedie*, vol. 8 (Amsterdam: 1950), 354–364.

J. Th. R. Schreuder, "Telecommunicatie via satellieten," *De Ingenieur* 82, no. 2 (1970): E2.

G. A. van der Knaap, *A Spatial Analysis of the Evolution of the Urban System: The Case of the Netherlands* (Rotterdam: EUR, 1978), 169.

P. Vijzelaar, *70 jaar radio-omroepzenders in Nederland: Ontwikkeling van het Nederlandse zenderpark op lange- en middengolf in de periode 1919–1989* (Deventer: Kluwer, 1991), 32, 78.

Map 2-4. Military infrastructure.

J. P. C. M. van Hoof, "Met een vijand als bondgenoot: De rol van het water bij de verdediging van het Nederlandse grondgebied tegen een aanval over land," *Bijdragen en Mededelingen betreffende de Geschiedenis der Nederlanden* 103, no. 4 (1988): 636, 643.

J. R. Beekmans, "De gevolgen van de inundaties," in J. R. Beekman and C. Schildt, eds., *Drijvende stuwen voor de landsverdediging: Een geschiedenis van de IJssellinie* (Zutphen: Walburg Pers, 1997), 171.

E. J. J. van Vliet-van Schie, M. M. Bolk, and B. H. A. T. Straatman, eds., *40 jaar DPO, Defensie Pijpleiding Organisatie* (n.p., 1997), 14, 19.

Table 5-1. Number of students in technical daytime education.

Centraal Bureau Statistiek (Statistics Netherlands), *De ontwikkeling van het onderwijs in Nederland, Editie 1966* [Development of education in the Netherlands, 1966 edition], vol. 1 (The Hague: 1966), tables 150 and 221.

Centraal Bureau Statistiek (Statistics Netherlands), *Bevolking van Nederland naar geslacht, leeftijd en burgerlijke staat 1830–1969* [The population of the Netherlands by sex, age, and civil status, 1830–1969] (The Hague: 1970), tables 1 and 2.

Graph 5-1. Combined investments in education and R&D as percentage of GDP.

Bart van Ark and Herman de Jong, "Accounting for Economic Growth in the Netherlands Since 1813," *Economic and Social History in the Netherlands* vol. 7 (1996): 220.

Graph 5-2. Number of scientists in Dutch industry.
J. J. Hutter, *Toepassingsgericht onderzoek in de industrie: De ontwikkeling van kwikdam-plampen bij Philips, 1900–1940* (Eindhoven: Eindhoven Technical University, 1988), 21

Table 6-1. Suppliers of vacuum installations to the sugar industry on Java in 1929, by number of installations supplied.
G. J. Schott, "Ketelinstallaties op de Java-suikerfabrieken in de jaren 1925 en 1929," *Archief voor de Suikerindustrie in Nederlandsch-Indië: Mededeelingen van het proefstation voor de Java-suikerindustrie* 18 (1929): 949–955

Graph 6-1. Years of first deployment of vacuum installations in Javanese sugar mills, 1884–1929.
G. J. Schott, "Ketelinstallaties op de Java-suikerfabrieken in de jaren 1925 en 1929," Archief voor de Suikerindustrie in Nederlandsch-Indië: Mededeelingen van het proefstation voor de Java-suikerindustrie 18 (1929): 949–955.

Graph 6-2. Value of machines and machine parts imported to the Dutch East Indies, 1891–1940.
W. A. I. M. Segers, "Manufacturing Industry 1870–1942," in P. Boomgaard, ed., *Changing Economy in Indonesia: A Selection of Statistical Source Material from the Early 19th Century up to 1940*, vol. 8 (Amsterdam: Royal Tropical Institute, 1987), 50–54.

Graph 6-3. Ethnicity of evaporators in the Javanese sugar mills, 1922–1932.
Philip Levert, *Inheemsche arbeid in de Java-suikerindustrie* (Wageningen: 1934), 118.

Graph 6-4. Ethnicity of boilers in the Javanese sugar mills, 1922–1932.
Philip Levert, *Inheemsche arbeid in de Java-suikerindustrie* (Wageningen: 1934), 122.

Table 8-1. Dutch percentage of the total number of patents applied for by foreigners in the American market, 1880 to 1993, by decade.
Jan Luiten van Zanden, *Een klein land in de 20e eeuw: Economische geschiedenis van Nederland, 1914–1995* (Utrecht: Het Spectrum, 1997), 65 (data derived from the annual reports of the U.S. Commissioner of Patents).

Table 8-2. Average annual percentage growth of labor productivity in the Netherlands and northwestern Europe, 1870–1994.
Jan Pieter Smits, Edwin Horlings, and Jan Luiten van Zanden, *Dutch GNP and Its components, 1800–1913* (Groningen: Growth and Development Centre, 2000).
Bart van Ark and Herman de Jong, "Accounting for Economic Growth in the Netherlands Since 1813," *Economic and Social History in the Netherlands* 7 (1996): 201.
Angus Maddison, *Dynamic Forces in Capitalist Development: A Long-Run Comparative View* (Oxford: Oxford University Press, 1991), appendixes A and C.

Table 8-3. Competitiveness of the Dutch economy as percentage of total world trade accounted for by Dutch products, by product, in 1986.

D. Jacobs, P. Boekholt, and W. Zegveld, *De economische kracht van Nederland: Een toepassing van Porters benadering van de concurrentiekracht van landen* (The Hague: Stichting Maatschappij en Onderneming, 1990), 29.

Table 8-4. Relative advantage of Dutch industry, 1906–1950.

For 1906 see U.S. Department of Commerce and Labor, Bureau of Statistics, *Statistical Abstract of Foreign Countries*, Parts I–III *Statistics of Foreign Commerce, October 1909* (Washington, D.C.: 1909).

NOTE: Data for the period 1928 to 1950 are derived from Ingvar Svennilson, *Growth and Stagnation in the European Economy* (Geneva and The Hague: United Nations, Economic Commission for Europe, 1954), 293.

Table 8.5: Relative advantage of Dutch Industry, 1970, 1980, and 1998.

Organization for Economic Cooperation and Development, *Foreign Trade by Commodities*, various years (Paris: OECD, 1970, 1980, and 1998

Table 8-6. Average annual percentage productivity growth in several industrial sectors from 1865 to 1913 and the use of steam power, expressed as number of steam engines in use as percentage of the total number of machines, in 1890.

H. W. Lintsen et al., eds., *Geschiedenis van de techniek: De wording van de moderne samenleving*, vol. 6 (Zutphen: Walburg Pers, 1995), 269–279.

Jan Pieter Smits, Edwin Horlings, and Jan Luiten van Zanden, *Dutch GNP and Its Components, 1800–1913* (Groningen: Growth and Development Center, 2000).

Table 8.7: Average annual percentage growth of real output, labor, capital and Total Factor Productivity (TFP) of Dutch industry, 1921–1960.

H. J. de Jong, *De Nederlandse industrie, 1913–1965: Een vergelijkende analyse op basis van de productiestatistieken* (Amsterdam: Nederlandsch Economisch-Historisch Archief, 1999), 194.

Table 8-8. Investments in education and R&D, 1921–1992, as percentage of their contribution to GDP.

Bart van Ark and Herman de Jong, "Accounting for Growth in the Netherlands Since 1813," *Economic and Social History in the* Netherlands 7 (1996): 220.

Table 8-9. Average annual percentage growth of GDP, by component (labor, capital, human capital), investments in R&D, and Total Factor Productivity [TGP], 1947–1994.

Bart van Ark and Herman de Jong, "Accounting for Growth in the Netherlands Since 1813," *Economic and Social History in the* Netherlands 7 (1996): 211.

Table 8-10. Comparions of percentage growth, by technological and economic indicator, between 1971–1975 and 1986–1990 periods (OECD = Organization for Economic Cooperation and Development).

M. Pianta, "Technology and Growth in OECD Countries, 1970–1990," in Daniele Archibugi and Jonathan Michie, eds., *Trade, Growth and Technical Change* (Cambridge: Cambridge University Press, 1998) 93.

Table 8-11. Percentage of men's and women's time spent in formal and informal labor, 1955–1997.
Centraal Bureau Statistiek, *Vrije-tijdsbesteding in Nederland 1962–1963* [Leisure time in the Netherlands 1962–1963], vol. 8 (Zeist: De Haan, 1966).
Centraal Bureau Statistiek, *De tijdsbesteding van de Nederlandse bevolking: kerncijfers 1988* [How the Dutch spend their time: basic 1988 data] (Voorburg: 1991).
W. P. Knulst and P. van Beek, *Tijd komt met de jaren, Onderzoek naar tegenstellingen en veranderingen in dagelijkse bezigheden van Nederlanders op basis van tijdbudgetonderzoek* (Rijswijk: Centraal Bureau voor de Statistiek, 1990).

Table 8-12. Individual sectors' environmental damage as percentage of total damage, 1950–1995.
J. P. Smits, "Omvang en oorsprong van de milieuschade," in Ronald van der Bie and Pit Dehing, eds., *Nationaal goed: Feiten en cijfers over onze samenleving (ca.) 1800–1999* (Voorburg: Centraal Bureau voor de Statistiek, 1999), 235–256, table on 246.

Table 8-13. Gap between the GNP and the Index of Sustainable Economic Welfare (ISEW) as a percentage
E. Horlings and J. P. Smits, "De welzijnseffecten van economische groei in Nederland, 1800–2000," *Tijdschrift voor Sociale Geschiedenis* 27, no. 3 (2001): 266–280, table on 277.

Graph 8-1. Dutch percentage of world trade.
J. P. Smits, "Economische ontwikkeling, 1800–1995," in Ronald van der Bie and Pit Dehing, eds., *Nationaal goed: Feiten en cijfers over onze samenleving (ca.) 1800–1999* (Voorburg: Centraal Bureau voor de Statistiek, 1999) 15–33, 17.
Brian R. Mitchell, *European Historical Statistics, 1750–1975*, 2nd ed. (London: Macmillan, 1981); Ingvar Svennilson, *Growth and Stagnation in the European Economy* (Geneva and The Hague: United Nations, Economic Commission for Europe, 1954); International Monetary Fund, *Direction of Trade Statistics Quarterly*, various editions from 1950 to 2009)(Washington D.C.: IMF, 1950–2003).

Graph 8-2. Appointments to boards of directors and advisory committees accepted by chemistry professors at Delft Technical University and other universities.
Jasper Faber, *Kennisverwerving in de Nederlandse industrie 1870–1970* (Amsterdam: Aksant, 2001), 37.

Graph 8-3. Volume of environmental damage in the Netherlands, 1910–1995, indexed to a sustainability standard of 100.
J. P. Smits, "Een macro-economische analyse van de omvang en oorsprong van de milieuschade in Nederland 1945–1995," in R. van der Bie and P. Dehing, eds., *Nationaal goed: Feiten en cijfers over onze samenleving (ca.) 1800–1999* (Voorburg: Centraal

Bureau voor de Statistiek, 1999), 235–256.

Graph 8-4. Comparison of per capita GDP with Index of Sustainable Economic Welfare (ISEW), 1800–2000, using a semi-logorithmic scale (in constant 1990 guilders).

E. Horlings and J. P. Smits, "De welzijnseffecten van economische groei in Nederland," *Tijdschrift voor Sociale Geschiedenis* 27, no. 3 (2001): 276.

Illustration Credits

Permission to reproduce the illustrations and photographs in this book has generously been granted by the institutions and collections named here, and is gratefully acknowledged by the editors and contributors, the MIT Press and Walburg Pers.

12   Photo: R. Collette.

14   T. de Rijk, *Het elektrische huis* (Rotterdam: Uitgeverij 010, 1998), 90.

16   The Hague Municipal Archives.

20   Rotterdam Municipal Archives (photo: J. H. F. Roovers).

24   Center for Historical Studies Region Eindhoven.

26   DSM Central Archive, Heerlen.

29   Foundation for the History of Technology, Eindhoven.

31   Weert Municipal Archives.

34   Print Room, Leiden University Library.

35   The Hague Municipal Archives.

37   Reproduced by permission from the photographer, Bas Klimbie.

40   Keuringsdienst voor Elektrotechnische Materialen, Arnhem.

41   Rijkswaterstaat, Directorate Utrecht.

46   Ministry of Housing, Spatial Planning and the Environment (VROM), Directorate-General for Spatial Development, The Hague.

48   Corporate Service Rijkswaterstaat, Lemcke Collection, Utrecht.

57   Service for Infrastructure Rijkswaterstaat, Utrecht.

66   Zeeland Archives, Middelburg, Zeeuws Genootschap ZI III 1079.

70   *Natuurbeleidsplan* (The Hague: 1990).

76   *Rapport inzake de electriciteitsvoorziening van Nederland* (The Hague: 1919).

80   *Philips Technisch Tijdschrift* 15 (1953): 165.

83   NV Nederlandse Gasunie, Groningen.

93   Shell Pernis, Rotterdam.

98   Top: J. P. Wiersma, *Erf en Wereld* (Leeuwarden: 1959) no page number; bottom: Ministry of Agriculture, Nature, and Food Quality, The Hague.

102   Museum for Communication, The Hague.

110   NV Nederlandse Gasunie, Groningen.

124   Foundation for the History of Technology, Eindhoven.

126   Reproduced by permission of the photographer, Nil Disco.

127   Amsterdam Municipal Archives.

129   Amsterdam Municipal Archives.

136   Noud de Vreeze, ed., *6,5 miljoen woningen* (Rotterdam: Uitgeverij 010, 2001), 44.

139   Maria Austria Institute, Amsterdam (photo: Ad Windig).

141   Erfgoedcentrum DiEP (Museum, Monuments and Archives Center), Dordrecht.

143   Tilburg Regional History Center.

145   KLM Royal Dutch Airlines Archives, Amstelveen.

147   KLM Royal Dutch Airlines Archives, Amstelveen.

150   Rotterdam Municipal Archives, Poster Collection.

153   Schiphol Airport Archive, Amsterdam.

155   Nederlands Fotomuseum (Netherlands Photo Museum), Rotterdam, item no. 2849/2 (photo: Frits J. Rotgans).

157   Foundation for the History of Technology, Eindhoven.

160   The Utrecht Archives.

162   The Hague Municipal Archives.

167   Foundation for the History of Technology, Eindhoven.

170   Amsterdam Municipal Archives.

178   NORIT Nederland BV.

181   *N.V. Nederlandsche Gist- en Spiritusfabriek Delft: de ontwikkeling der onderneming in zestig jaren 1870-1930* (Delft, 1930).

183   Stichting Historische Verzamelingen van het huis Oranje-Nassau (Foundation for the Historical Collections of the House of Oranje-Nassau), Royal Archive, The Hague.

185   Corus, IJmuiden.

190   The Utrecht Archives.

193   Koos van Weringh, *Albert Hahn* (Amsterdam: 1969), 140.

197   Spaarnestad Photo, Haarlem.

203   Fryske Academy Collection, Leeuwarden.

207   Vapro-OVP, Leidschendam.

212   Rotterdam Municipal Archives, Rotterdam.

215   Food and Consumer Product Safety Authority, The Hague.

221   Foundation for the History of Technology, Eindhoven.

226   *Noord-Brabantsch Nijverheid in beeld* (Haarlem: n.d.[1918]).

228   Econosto Nederland BV, Rotterdam.

229   *Katholieke Illustratie* 64 (1929–1930): 166.

234   P.A. Koolmees, *Vleeskeuring en openbare slachthuizen in Nederland 1875-1985* (Utrecht: 1991), 40.

238   Sint Antonius Binnenweg Nursing Home, Rotterdam.

240   Foundation for the History of Technology, Eindhoven.

242   Museum for Communication, The Hague.

252   Inholland, University of Applied Sciences, Haarlem.

256   Erfgoedcentrum DiEP (Museum, Monuments, and Archives Center), Dordrecht.

260   Tebodin, The Hague.

264   Foundation for the History of Technology, Eindhoven.

267   *Klei* 5 (1913): 199.

272   The Hague Municipal Archives.

276   *Gedenkschrift ter gelegenheid van het 50-jarig bestaan der Middelbare Technische School "Amsterdam"* (Amsterdam: 1928), 36.

279   Vapro-OVP, Leidschendam.

283   DSM Central Archive, Heerlen.

285   *De zeepindustrie in beeld* (Haarlem: 1914), no page number.

289   Philips Company Archives, Eindhoven.

294   DSM Central Archive, Heerlen.

297   Akzo Nobel, Arnhem.

299   Phytopathological Agency, Wageningen.

305   Foundation for Fundamental Research on Matter, Archive, Utrecht.

308   The Hague Municipal Archives.

323   Royal Tropical Institute, Amsterdam, Photo Bureau.

327   Royal Tropical Institute, Amsterdam, Photo Bureau.

332   Royal Tropical Institute, Amsterdam, Photo Bureau.

335   Royal Tropical Institute, Amsterdam, Photo Bureau.

337   Wageningen UR Library, Special Collections.

340   Royal Tropical Institute, Amsterdam, Photo Bureau.

344   Royal Tropical Institute, Amsterdam, Photo Bureau.

346   Library of the University of Wageningen, Special Collections.

349   Historical Center Overijssel, Zwolle, Stork Photo Collection, 1868-1968, album A52, no. 350.

352   Historical Center Overijssel, Zwolle, Stork Photo Collection, 1868-1968, album A144, no. 77.

356   Royal Netherlands Institute for Southeast Asian and Caribbean Studies, Leiden.

359   Rotterdam Municipal Archives, Poster Collection.

364   Van den Broek en Bakema BV, Architects, Rotterdam.

368   Delft Technical University.

371   Foundation for the History of Technology, Eindhoven.

373   Erfgoedcentrum DiEP (Museum, Monuments, and Archives Center), Dordrecht.

376   The Hague Municipal Archives.

380   Marga Coesèl, *Natuurlijk Verkade* (Warnsveld/Zaandam; 1999), 62.

383   Private collection; reproduced with permission.

387   Nieuwland Museum, Archive, and Study Center, item no. RYP 9001694 (photo: Maaskant).

392   Amsterdam Municipal Archives.

395   Private collection; reproduced with permission.

398   Ministry of Transport and Water Management, Beeldbank V&W, The Hague.

402   Left: Nieuwland Museum, Archive, and Study Center, item no. RYP 9001498; right: *Van korre tot koren* (The Hague: 1968), 66.

407   *Bouw* 3 (1948), 401.

411   Amsterdam Municipal Archives.

413   Ministry of Transport and Water Management, Beeldbank V&W, The Hague.

417   The Utrecht Archives.

423   National Archives, The Hague, Delta Commission Archive, 2.16.45, item no. 36.

432   Rijksmuseum, Amsterdam.

437   Provincial Archive of North Holland, Haarlem.
443   Staatliche Museen zu Berlin–Preußischer Kulturbesitz, Kunstbibliothek, Berlin.
446   Swinkels Family Archive, Stiphout/Helmond.
450   Zeeuws Documentation Center, Middelburg.
453   Enschede Municipal Archives.
458   Gastec Holding NV, Apeldoorn.
460   Historical Center Overijssel, Zwolle (photo: Dolf Henneke).
463   The Hague Municipal Archives.
465   *Elektriciteit voor Nederland* (Arnhem: 1977), 92.
471   Unilever Archives, Rotterdam/Iglo Nederland BV, Hendrik-Ido-Ambacht.
473   The Utrecht Archives.
484   DSM Central Archive, Heerlen.
486   Foundation for the History of Technology, Eindhoven.
488   Foundation for the History of Technology, Eindhoven.
490   Foundation for the History of Technology, Eindhoven.
492   International Institute for Social History, the AdvertisingArsenal, Amsterdam.
496   Brabant Historical Information Centre, 's-Hertogenbosch.
497   Dienst Landelijk Gebied (Land and Water Management Service), Arnhem.
500   Spaarnestad Photo, Haarlem.
501   Foundation for the History of Technology, Eindhoven.
504   Enschede Municipal Archive.
508   Akzo Nobel, Hengelo.
511   DSM Central Archive, Heerlen, item no. 13107.
515   *W.P. voor het bedrijfsleven*, vol. 2 (Amsterdam/Brussels: 1957), 276.
520   Philips Company Archives, Eindhoven.
522   Erfgoedcentrum DiEP (Museum, Monuments, and Archives Center), Dordrecht.
525   Keuringsdienst voor Elektrotechnische Materialen, Arnhem.
530   Nationaal Archief, The Hague/the Netherlands, image bank 914-5231.

542   's-Hertogenbosch Municipal Archives.
545   Van Gorcum, Assen.
548   Collection of Peter van Dam.
550   's-Hertogenbosch Municipal Archives.
552   Eric de Lange, *Geen officiële, maar levende schonheid* (The Hague: 1983), 41.
553   Amsterdam Municipal Archives.
554   *Konsumptie verplicht* (book, np, 1976), 36
556   Private collection, reproduced by permission.
558   Philips Company Archives, Eindhoven.
560   Nederlands Centrum voor Autohistorische Documentatie (Netherlands Center
        for Documentation of the History of the Automobile), Helmond.
564   Private collection; reproduced by permission.
565   Frank M. van der Heul, *Auto-advertenties uit de jaren 1890-1940* (Rijswijk: 1991),
        no page number.
567   *De Ware Jacob* 2 (1902–1903), 98.
569   Museum for Communication, The Hague.

571   *De Ware Jacob* 3 (1903–1904), 240.

574   International Institute for Social History, Amsterdam (photo: Ben van
         Meerendonk/AHF).

576   Netherlands Institute for Sound and Vision, Hilversum.

577   *De Ware Jacob* 1 (1901–1902), no. 35.

579   Leonard de Vries, *Het Leven: Een fascinerende selectie uit de jaargangen 1906–*
         1940 (Laren: 1972), 70.

582   *Panorama*, 31 December 1937.

Contributors

Peter Baggen studied philosophy at the University of Amsterdam. He earned his doctorate in social science at the University of Nijmegen, writing his a dissertation on the development of Dutch university education between 1815 and 1960. He has been editor of the Dutch journals *Comenius* and *Pedagogiek*. He is Senior Policy Adviser for the Association of Dutch Research Universities and serves as a board member of the Netherlands Graduate Research School of Science, Technology, and Modern Culture.

Adrienne van den Bogaard earned her doctorate on the history of economic modelling in the Netherlands at the University of Amsterdam. She led a research project on the History of Information Technology in the Netherlands that resulted in the 2008 publication of *De eeuw van de computer* (*The Century of the Computer*). She is a mathematics teacher at a Dutch high school and a researcher with the project Software for Europe.

Irene Cieraad is a cultural anthropologist and a Senior Researcher in the Department of Architecture at the Technical University Delft, where she holds the chair for Interiors. She has published widely on the history of Dutch vernacular interiors and related issues of household technology, material culture, and consumption; recent publications include "Gender at Play: Décor Differences Between Boys' and Girls' Bedrooms," in Emma Casey and Lydia Martens, eds., *Gender and Consumption: Domestic Cultures and the Commercialisation of Everyday Life* (2007); "The Milkman Always Rings Twice: The Effects of Changed Provisioning on Dutch Domestic Architecture," in David Hussey and Margaret Ponsonby, eds., *Buying for the Home: Shopping for the Domestic from the Seventeenth Century to the Present* (2008); and "The Radiant American Kitchen: Domesticating Dutch Nuclear Energy," in Ruth Oldenziel and Karin Zachmann, eds., *The Cold War Kitchen: Americanization, Technology, and European Users* (2009); and (as volume editor) *At Home: An Anthropology of Domestic Space* (1999).

Jasper Faber is Consultant on Climate Policy for Aviation and Maritime Transport at the Technical University Delft and has also consulted with Nyfer. His dissertation, at the Free Universiteit Amsterdam, was titled "Kennisverwerving in de Nederlandse industrie in de twintigste eeuw" ("Knowledge Acquisition by Dutch Industry in the Twentieth Century").

Ernst Homburg is Professor of History of Science and Technology at the University of

Maastricht. His dissertation, at the University of Nijmegen, was on the rise of the German chemical profession between 1790 and 1850. He co-edited two book series on the history of technology in the Netherlands and was one of the chemistry editors of the eight-volume *New Dictionary of Scientific Biography* (2007). Other recent publications include *Groeien door kunstmest: DSM Agro 1929-2004* (*Growth Through Fertilizer: DSM Agro, 1929–2004*), and (as co-editor) *De geschiedenis van de scheikunde in Nederland 3: De ontwikkeling van de chemie van 1945 tot het begin van de jaren tachtig* (*The History of Chemistry in the Netherlands: From 1945 to the Eighties*) (both 2004). He is book editor of *Ambix* and serves as Chairman of the Working Party on History of Chemistry of the European Association for Chemical and Molecular Sciences. He is currently working on a book on the 150-year history of the Solvay company.

Harry Lintsen studied physics at the Eindhoven University of Technology, earning his doctorate in 1980 with a dissertation on the history of the engineering profession in the Netherlands. In 1990 he was appointed Full Professor in the History of Technology at Eindhoven University of Technology and the Technical University Delft and since 2003 he has held a full professorship at the School of Innovation Sciences at the TU Eindhoven. Having led the research and editorial teams that resulted in the six-volume *Geschiedenis van de Techniek in Nederland* (*History of Technology in the Netherlands 1800-1890*), Lintsen is considered the father of the history-of-technology discipline in the Netherlands. Lintsen was one of the main editors and chairmen of the board of . the *Techniek in Nederland in de Twintigste Eeuw (History of Technology in the Netherlands in the Twentieth Century)* book series and also was lead editor of *Twee Eeuwen Rijkswaterstaat (Two Centuries of Rijkswaterstaat)* (1998). He is the author of *Made in Holland: Een techniekgeschiedenis van Nederland (Made in Holland: A Technological History of the Netherlands)*. He is a member of the Royal Netherlands Academy of Arts and Sciences

Harro Maat is Lecturer on Technology and Agrarian Development in the Social Science Department of Wageningen University. He studied agricultural science and his doctorate, from Wageningen University, was on the history of agricultural science in the Netherlands and its former colonies. He has published on the history of wheat breeding in the Netherlands, field experimentation in agriculture, and agricultural extension in colonial Indonesia. His current research focuses on the long-term dynamics of international rice research, in particular rice genetics. In 2007 he edited a special issue of *IDS Bulletin*, on science, technology, and participation, and is currently editing a special issue on rice science for the journal *East Asian Science, Technology and Society*. He is a member of the Tensions of Europe and Commodities of Empire collaborative research networks, and in 2009 and 2010 will chair the International Outreach Committee of the Society for the History of Technology.

Arie Rip studied chemistry and philosophy at the University of Leiden. After serving as Professor of Science Dynamics at the University of Amsterdam from 1984 to 1987, he is now Professor of Philosophy of Science and Technology at the University of Twente and also holds a visiting professorship at the University of Stellenbosch, South Africa. He was President of the Society for Social Studies of Science from 1988 to 1989, and Scientific Director of the Dutch Graduate Research School on Science, Technology and Modern Culture from 2000 to 2004. Recently he has served as chairman of the panel assisting the team evaluating the Research Council of Norway (2001); as Président du Comité d'Évaluation, Expérience pilote sur les vignes transgéniques, INRA, Paris), 2002–2003); and as a member of the EU High-Level Expert Group on the New Technology Wave (2004) and of the Expert Group on Science and Governance (2005–2006). He was one of the main editors of *Techniek in Nederland in de Twintigste Eeuw (History of Technology in The Netherlands in the Twentieth Century)*. Currently he leads a research program on Technology Assessment of Nanotechnology (as part of the Dutch R&D program known as NanoNed). His research focuses on science and technology dynamics and science and technology policy analysis, including issues of ongoing changes in modes of knowledge production, controversies and expertise, and constructive technology assessment. He has published widely in the sociology of science and technology, in science, technology and society studies, and in science, technology, and innovation policy studies.

Johan Schot is Professor of Social History of Technology at the Eindhoven University of Technology. He is research director of the Foundation for the History of Technology, and of the Foundation for System Innovation and Transitions Towards Sustainable Development. He is a fellow of the N. W. Posthumus Institute for Social and Economic History. He is cofounder and co-chair (with Ruth Oldenziel) of the Tensions of Europe collaborative network and research program. He was the program leader and main editor of *Techniek in Nederland in de Twintigste Eeuw (History of Technology in the Twentieth Century)* In 2002 he was awarded a VICI grant under the Innovational Research Incentives Scheme for talented scholars (highest category) by the Netherlands Organization for Scientific Research for his proposal Transnational Infrastructures and the Rise of Contemporary Europe. He is a member of the Royal Netherlands Academy of Arts and Sciences. His interests range from the history of technology to Dutch history to European history to sustainability studies.

Jan-Pieter Smits is Assistant Professor in the Economics Department of the University of Groningen and works as Program Manager and senior statistical researcher at the department of macro economic statistics of Netherlands statistical bureau CBS. He studied history at the Free University in Amsterdam and wrote his doctorate, "Economic Growth and Structural Change in the Dutch Service Sector," in 1995. His research focuses on the interrelationship among technological development, institutional change, and economic growth. In the Groningen Growth and Development Center he coordinates a website on comparative historical national accounts.

Dick van Lente is Lecturer in Cultural History at the Department of History and Arts, Erasmus University, Rotterdam. He studied modern history at the University of Amsterdam and Erasmus University. After finishing his dissertation, on the ideological responses to technological change in the Netherlands from 1850 to 1914, he joined the editorial team that produced a six-volume *History of Technology in the Netherlands in the Nineteenth Century.* He is the coauthor of a book on Western societies since 1750, and contributed to the encyclopedia of Dutch book history, *Bibliopolis.* Currently he is collaborating on a project to study popular responses to the atom during the fifties and sixties in eight countries and is writing a book on Dutch responses to several new technologies in the same period.

Rienk Vermij is Assistant Professor in the History of Science Department at the University of Oklahoma. He studied history at Utrecht and earned his doctorate at the Institute for the History of Science at Utrecht University. Among his books are *The Calvinist Copernicans: The Reception of the New Astronomy in the Dutch Republic (1575–1750)* (2002); *Kleine geschiedenis van de wetenschap (A Short History of Science* (2005); *David de Wied: Toponderzoeker in polderland* (2008), a biography of David de Wied, a pioneer of neuroendocrinology.

Erik van der Vleuten is a Research Associate at the School of Innovation Sciences, Eindhoven University of Technology. His main research interest is the mutual shaping of infrastructural technologies and societal change in the nineteenth and twentieth centuries. He has focused on specific infrastructures—Danish electrification, Dutch wet infrastructure, and European electrical and food transport infrastructure—and on the development of broader syntheses and concepts, such as the shaping of the networked nation, the networking of Europe, and the possibility of a transnational history of technology. He was the coeditor of "Globalization and Participation in Water System Building" (2002); "Networked Nation: Transforming Technology, Society and Nature in the Netherlands in the Twentieth Century" (2004); and *Networking Europe: Transnational Infrastructures and the Shaping of Europe, 1850–2000* (2006). He is currently working on two books, on infrastructure and its implications for contemporary European history.

Index:

names of persons, geographical names and subjects

Note: page numbers in italic refer to captions.

Technology and the Making of the Netherlands

The research presented in this book was supported by the Netherlands Organisation for Scientific Research (NWO), the Foundation for the History of Technology (SHT) and Eindhoven University of Technology. NWO has also supported the translation of the Dutch edition into English. This book is a translation of the seventh volume of the book series *Techniek in Nederland in de Twintigste Eeuw*, edited by Johan Schot, Harry Lintsen, Arie Rip, Adri Albert de la Bruhèze, Liesbeth Bervoets, Jan Bieleman, Aad Bogers, Mila Davids, Nil Disco, Ben Gales, Ernst Homburg, Eddy Houwaart, Jan Korsten, Ruth Oldenziel, Anneke van Otterloo, Geert Verbong, and Onno de Wit. Giel van Hooff was responsible for the picture editing. This book series was the main product of a research program supported by 7 universities, many industrial sponsors and governmental agencies, and NWO.

The production of the program and book series was initiated and coordinated by the Foundation for the History of Technology (SHT). SHT aims to develop and communicate knowledge that increases our understanding of the critical role of technology in the history of the Western world. Since 1988 the foundation has been supporting scholarly research in the history of technology. This has included large-scale national and international research programs and numerous individual projects, many in collaboration with Eindhoven University of Technology. SHT also coordinates the international research network Tensions of Europe: Technology and the Making of Europe. For more information see: www.histech.nl or www.tensionsofeurope.eu.